Umesh N. Gandhi
Sebastian Goris
Tim A. Osswald
Yu-Yang Song

Discontinuous Fiber-Reinforced Composites

Fundamentals and Applications

HANSER
Hanser Publishers, Munich
Hanser Publications, Cincinnati

The Authors:

Umesh N. Gandhi, Ann Arbor, Michigan, USA

Sebastian Goris, St. Paul, Minnesota, USA

Tim A. Osswald, Madison, Wisconsin, USA

Yu-Yang Song, Ann Arbor, Michigan, USA

MIX
Papier aus verantwortungsvollen Quellen
FSC
www.fsc.org
FSC® C016439

Distributed in the Americas by:
Hanser Publications
414 Walnut Street, Cincinnati, OH 45202 USA
Phone: (800) 950-8977
www.hanserpublications.com

Distributed in all other countries by:
Carl Hanser Verlag
Postfach 86 04 20, 81631 Munich, Germany
Fax: +49 (89) 98 48 09
www.hanser-fachbuch.de

Library of Congress Control Number: 2020930382

Editor: Dr. Julia Diaz Luque
Production Management: Jörg Strohbach
Coverconcept: Marc Müller-Bremer, www.rebranding.de, Munich
Coverdesign: Max Kostopoulos
Typesetting: Kösel Media GmbH, Krugzell
Printed and bound by Druckerei Hubert & Co GmbH und Co KG BuchPartner, Göttingen
Printed in Germany

ISBN: 978-1-56990-694-1
E-Book ISBN: 978-1-56990-695-8

Preface

This book is written at an intermediate to advanced level to provide a background on discontinuous fiber-reinforced composites to practicing engineers as well as graduate students. The technical information and practical examples enable the reader to understand the underlying physics and the complex behavior of this unique fiber composite material. Discontinuous fiber-reinforced polymer composites are a growing class of composite materials that are appealing to the aerospace and automotive industries because they are easy to process into components and structures of complex shapes in an automated fashion, using traditional molding techniques or extrusion processes. Compared to the processing of continuous fibers, the automated processes such as injection or compression molding used for discontinuous fibers are quite low cost and suitable for high production rates, which makes them attractive for automobiles and other consumer goods.

Designing with such discontinuous fiber composite material can be quite challenging because the molding process strongly affects the final state of the fibers in the manufactured part. The fiber orientation, fiber length, and fiber concentration can show a large degree of heterogeneity throughout the molded part, especially for complex shaped structures. These variations of the fiber microstructure have a profound impact on the performance of the finished product, evident in the heterogeneous and anisotropic structural properties. This is a challenging situation when compared to isotropic metal parts or continuous fiber-reinforced composites, where the fiber orientation and fiber length are known and are uniform throughout the part. The goal in this book is to provide a theoretical and practical background to address these challenges and provide know-how that will help design parts that are made with discontinuous fiber-reinforced composites, in order to fully exploit the potential of this class of composites.

In the first part of this book the various aspects that lead to the anisotropic properties of the finished parts are covered. Fundamentals of polymeric materials and fibers are discussed, followed by a survey of manufacturing processes used in the industry. The microstructure of fiber-reinforced polymers, with attention to discontinuous fibers, is presented, and fundamental relations between the microstructure and mechanics of fiber-reinforced polymers are developed. The second part of the book explains how the mechanics of fiber-reinforced composites at the micro level can be translated to structural analysis programs, usually at millimeter scale, using a multiscale modeling approach. Furthermore, the fundamentals of mold-filling simulation for injection and compression molding for short and long fibers are included. The third part of the book introduces the reader to practical examples covering compression molding, injection molding, hybrid structures, and joining. The case studies of practical examples take the readers through all the steps necessary to arrive at a full structural model of a component and compare the predictive models to actual measurements.

Some of the work presented here started in 2012 when the Toyota Research Institute North America (TRINA) began to develop know-how to design lightweight components for automotive applications using discontinuous fiber-reinforced composites. Realizing many fundamental technical challenges, TRINA approached Professor Tim Osswald from the Polymer Engineering Center (PEC) at the University of Wisconsin-Madison for help. What followed was an outstanding collaboration between industry and academia, resulting in significant improvements in the know-how to design using discontinuous fiber-reinforced composites, which is the essence of this book.

The authors would like to acknowledge the invaluable help of many during the preparation of this manuscript. We would like to thank Dr. Huan-Chang Tseng, Dr. Jim Hsu, and Dr. Anthony Yang of CoreTech System for contributing the chapter on process simulation (Chapter 7) and other members of the CoreTech team for their continuous support in many other areas. We are grateful to Tobias Mattner for his outstanding job in not only drawing the figures, but also making excellent suggestions on how to present the information more clearly. We would like to offer special thanks to Prof. Noboru Kikuchi, president of Toyota Central R & D Lab, for the unceasing encouragement to write this book. Takeshi Sekito, Hidetoshi Okada, Masaya Miura, and Yoshinori Suga of Toyota Motor Company for serving as sounding boards and for their technical input. Prof. Uday Vaidya and his group at the University of Tennessee are thanked for their help with the processing technology. We are grateful to Dr. Vlastimil Kunk, from Oakridge National Laboratory, for the help in the characterization of compression molded parts. Dr. Suresh Shah of the Society of Plastics Engineers is thanked for sharing many practical insights and suggestions, and Dr. Danil Prokhorov from TRINA for his continuous cheering and support. Dr. Roger Assaker and his team at e-Xstream Engineering are thanked for their valuable assistance with multiscale modeling. Thanks are due to Dr. Mark Smith and Dr. Julia Diaz Luque of Carl Hanser Verlag for their valuable expertise in editing this book, as well as Jörg Strohbach for his support throughout this project. Above all, the authors would like to thank their families for their continued support of their work and their input throughout the writing of this book.

Umesh N. Gandhi and *Yu-Yang Song* January 2020
Ann Arbor, Michigan, USA

Tim A. Osswald
Madison, Wisconsin, USA

Sebastian Goris
St. Paul, Minnesota, USA

About the Authors

Dr. Umesh N. Gandhi

Umesh Gandhi is Executive Scientist at the Toyota Research Institute North America (TRINA) in Ann Arbor, MI, and was formerly Staff Engineer at General Motors. He holds a Ph. D. from the University of Michigan. His research interests are lightweight and programmable mechanical systems. He has authored more than 50 publications and holds over 40 issued patents.

Prof. Prof. hon. Dr. Tim A. Osswald

Tim Osswald teaches at the University of Wisconsin-Madison and is co-Director of the Polymer Engineering Center. He also holds honorary professorships at the Friedrich Alexander Universität in Erlangen, Germany and the Universidad Nacional de Colombia. He teaches and does research in polymer engineering, a field where he has published 11 books and over 300 papers.

Dr. Sebastian Goris

Sebastian Goris is Sr. Research Engineer at the Corporate Research Process Laboratory at 3M in St. Paul, MN, where he focuses on new technology development in the area of automotive electrification. He received his Ph. D. in Mechanical Engineering at the University of Wisconsin-Madison under Prof. Tim Osswald.

Dr. Yu-Yang Song

Yu-Yang Song is Senior Scientist at the Toyota Research Institute North America (TRINA) in Ann Arbor, MI. His research focuses on lightweight composite materials and textile fabric materials for future mobility applications. He received his Ph. D. in the area of defects detection of aerospace composite at Wayne State University in Detroit, MI.

Contents

1 Introduction

Composites are engineered materials composed of two or more components with substantially different properties to synergistically produce a product that is better than the individual components. Discontinuous fiber-reinforced composites are a special subcategory of composite materials that are used due to the ability to process them into parts and structures of complex shape in an automated fashion. This chapter will present a historic perspective of composites and discontinuous fiber-reinforced composites along with a general classification of composite materials and trends in the composites field.

1.1 Historical Background

Composite materials have been around for thousands of years, such as in the construction materials shown in Figure 1.1 in the form of casted *adobe*[1] brick used as construction material for a house in Panajachel, Guatemala. This form of discontinuous fiber composite building material is one of the oldest forms of construction still being used in some places around the world. Adobe bricks or casted walls are made of a mixture of clay, sand or stones, and natural fibers such as straw, small wooden sticks, or *caña brava*, a form of wild cane. The fibrous materials throughout the brick or wall help the clay stay together, increase its strength, and reduce the incidence of cracks and crumbling.

Plywood, another form of building material in the form of a laminated composite, was first used in ancient Egypt over 5000 years ago by gluing together several thin layers of wood. The early Egyptian technology bonded the layers of thin wood with glue made from boiling animal hide and bones, and, to press the layers together, they used the weight of hot sandbags [1]. In the 1860s, Immanuel Nobel, who with his son Alfred Nobel invented dynamite, is credited for inventing modern plywood.

[1] The word adobe dates back at least 4000 years and is pronounced relatively the same in most languages.

In his process, he glued crossed veneers under steam and pressure, resulting in a laminated wooden plate that was stronger than the original wood, creating strength in the direction of the fiber orientation of each layer [1, 2].

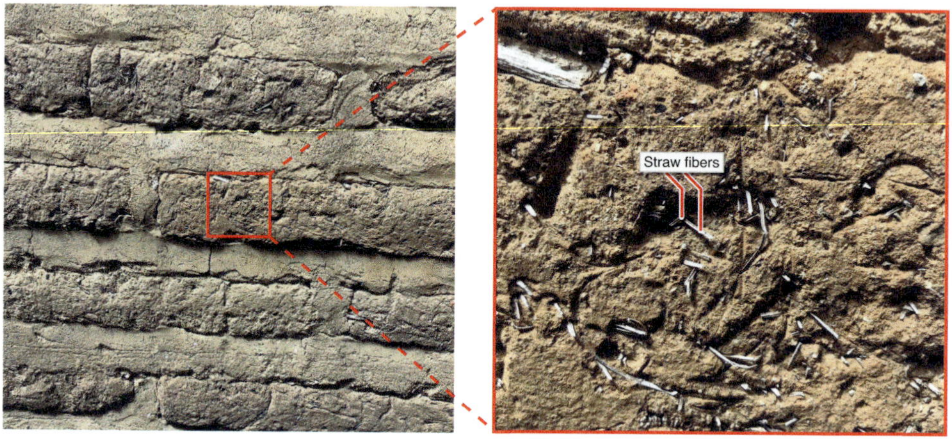

Figure 1.1 Straw fiber-reinforced mud composite casted bricks used for construction in Panajachel, Guatemala. Photograph by Tim Osswald © 2018

In the manufacture of plywood and other composite materials, hot sandbags were replaced with the screw press, invented by the Greeks over 2000 years ago. For the next two millennia, these devices were the standard form of generating large pressures between two platens. These press platens were eventually heated by running steam through channels. In the second half of the 19th century, steam-heated screw presses were used to compression mold early plastics and plastic composite items. During that same time, steam presses were used to mold decorative items made of shellac, an amber-like polymer derived from Asian beetles, reinforced with wood fiber [3]. Finally, the steam-heated screw press was replaced by a hydraulic press (Figure 1.2) with steam-heated platens, patented by F. B. Northrup in 1915 [4]. This invention was based on a four-column hydraulic press invented in the late 1700s by the British inventor Joseph Bramah. Today's compression molding machines are clear descendants of Northrup's hydraulic press.

One of the first plastic composites dates back to 1856, when François Charles Lepage patented in France and England a plastics material composed of wood fiber and albumen, a material he called *Bois durci*[2]. Bois durci is a mixture of sawdust, usually from a hardwood such as ebony or rosewood, and albumen from blood or egg. The wood fibers were mixed with vegetable oils, mineral or metallic fillers, and the albumen with a gelatinous substance diluted in water. The dry and wet components were mixed and compressed into finished parts in a steel mold held in

[2] Bois: wood; durci: hardened.

a steam-heated screw press. Figure 1.3 presents two decorative medallions of this early plastic composite material compression molded using a screw press.

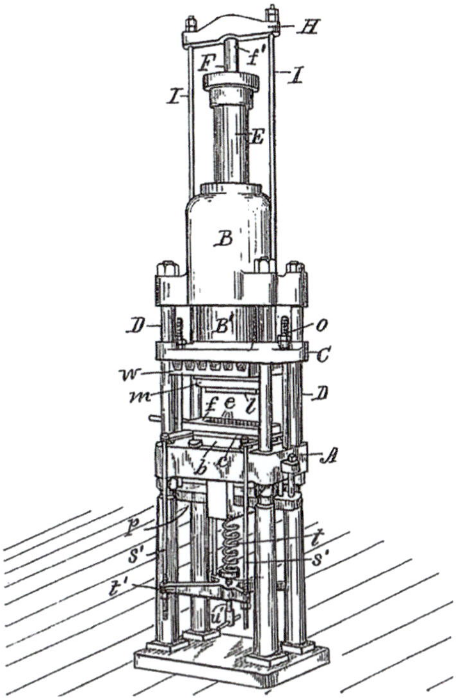

Figure 1.2 Northrup's hydraulic press

Figure 1.3 Commemorative Bois durci medallions of Queen Victoria and Prince Albert from the 1860s (The Kölsch Collection). Photograph by Tim Osswald © 2010

A big breakthrough in the field of modern polymer composites occurred at the beginning of the 20th century when Leo Baekeland developed phenol-formaldehyde, the first synthetic plastic, and a polymer that is still in use today, as a reinforced and unreinforced plastic. In his *Heat and Pressure Patent*, Leo Baekeland already suggested the use of wood and asbestos fibers to fill and reinforce phenol-formaldehyde plastics [5].

Bakelite not only helped shape the streamlined pre-WWII years, but it also presented a material that made it possible to mass-produce items that made life easier in the home, office, street, field, and factory. The ability to mass-produce plastic products helped create the myth that plastics are a cheap replacement for materials of higher quality. The reality was that phenolics, especially those that were fiber-reinforced, were materials of superior mechanical and electrical properties with higher chemical resistance than those of the materials they replaced. For example, tough, lightweight, and stiff phenolic-cloth composite propellers replaced wooden aircraft propellers that easily cracked, causing catastrophic failures in the early years of aviation. This was the first implementation of composites in aircraft applications. In his 1933 textbook, L. F. Rahm [6] presented properties for the filled phenolic resins of the time. He lauded the increase in stiffness and strength of reinforced phenolics, and the significantly lower shrinkage after processing asbestos fiber filled plastics. Wood fiber filled Bakelite replaced wood as the material of choice for radio housings (Figure 1.4), not necessarily because it was a cheaper material, but because it made it possible to mass produce finished products that were tougher than the handmade lower production alternatives. Furthermore, these new materials gave the designer the freedom to produce parts of more complex shape with a wide palette of available colors.

Figure 1.4 Bakelite Fada radio from 1945 (The Kölsch Collection). Photograph by Tim Osswald © 2010

In the 1930s, in order to use up the discarded wood residues from sawmills and lumberyards, which accounted for 60% of felled trees, the German inventor Max Himmelheber developed particle board. This highly filled composite material is made up by mixing milled wood residue with phenolic resin and pressing it into plates. This composite material distinguished itself from other composites in that very small amounts of phenolic were needed, since the phenolic's only task was to bind the wooden chips together. Today, particle boards are extensively used in construction as, for example, backing for vinyl flooring applications.

That same decade, in 1938, glass fiber was successfully spun into fiberglass wool and implemented, along with unsaturated polyester resin, into flat panels which were used for military aircraft during the Second World War. By the end of the war over 7 million pounds of fiberglass materials were being used by the military industry. In the 1940s, to take advantage of the low cost of lightweight opportunity, various attempts were made to introduce the fiberglass into the automotive industry. These included automotive body panels by Ford Motor Company (Figure 1.5). As so many subsequent plastic composites projects, the Ford body panels were not introduced commercially because of warpage[3]. Most noteworthy among the early composite vehicles designs is Project Y, also known as 1946 Stout Scarab. This was the result of collaborative work between William B. Stout, an aircraft engineer, and Owens Corning to demonstrate the use of fiberglass material to build an automobile body. This vehicle was built based on the existing design of Scarab and is considered the first vehicle where the entire body panels and structure were made from fiber-reinforced composites.

Figure 1.5 Henry Ford hitting the fiberglass trunk lid of a Ford with an axe to demonstrate its toughness (1940s)

[3] The photograph depicted in Figure 1.5 clearly shows how the seam between the trunk lid and the rest of the vehicle changes around the circumference.

A big breakthrough occurred in 1953, when General Motors Corporation introduced the Corvette, fitted with fiberglass body panels on a steel frame (Figure 1.6). In 1964, GM suppliers started manufacturing the Corvette's body panels using compression molding of sheet molding compound (SMC), still a widely used discontinuous fiber-reinforced composite material. It was not until the 1970s that the automotive industry became the number one industry in the composites market, passing the marine industry, which had occupied that place throughout the 1960s.

Figure 1.6 The 1953 Corvette, first fiber composite production car

However, from the beginning, the aircraft manufacturing industry has occupied a prominent role in the composites industry. While the initial composites applications in aircraft during the Second World War were to overcome the shortage of aluminum, the actual reason why composite materials are so desirable for aircraft manufacture is their light weight. In fact, in German, the field that studies composite materials and their implementation into their final applications is called *"Leichtbau"*, which can be translated into *light construction*. Hence, the desire to make airplanes lighter has resulted in an increased role composites play in their manufacture. Figure 1.7 shows how the use of composite structural components in airplanes has increased in the past four decades. The graph shows how the industry has grown from less than 5% of composites by weight in the original Boeing 747 and Airbus 300 in 1970, to the Boeing 787 with 50% composites by weight and the Airbus 350 with 52%.

A major advancement came in the mid-2000s with the development of the Boeing 787 Dreamliner (Figure 1.8), whose complete fuselage is manufactured with carbon-epoxy composites, using an automated tape laying technique, a widely accepted continuous fiber-reinforced composites area.

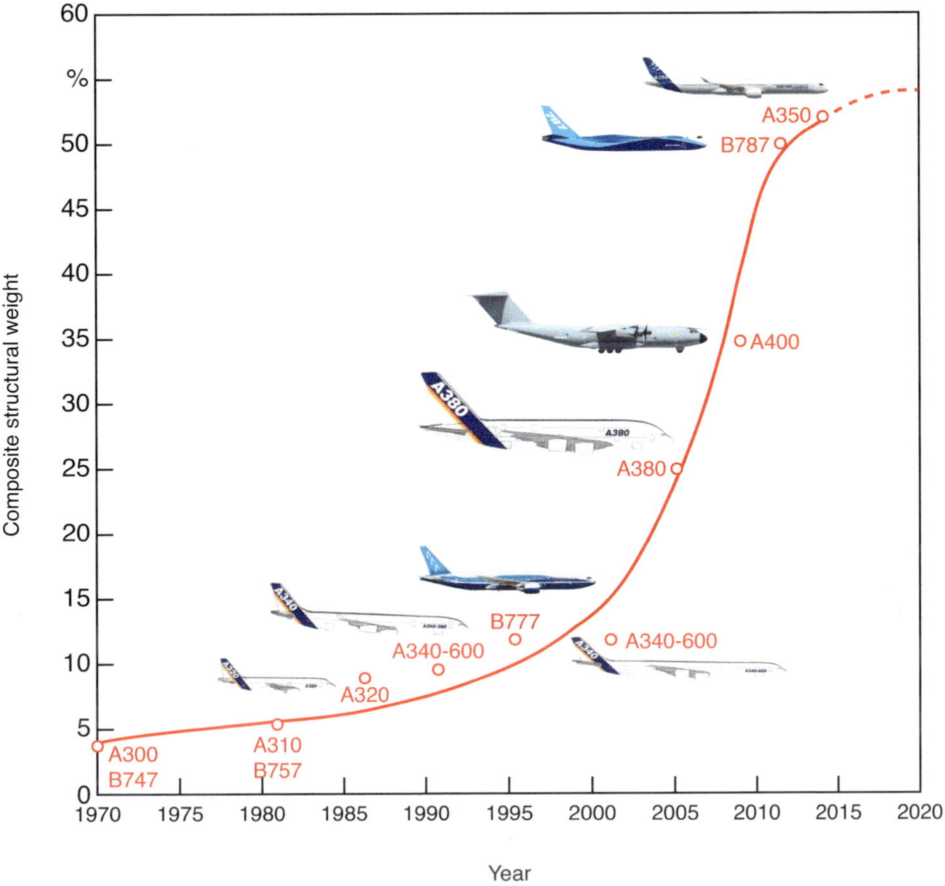

Figure 1.7 Growth of composite structural components in aircraft design (source: Boeing and Airbus)

1.2 Fiber-Reinforced Composites

Fiber-reinforced composites (FRP) are a distinctive class of materials defined by the combination of at least two constituents: the matrix material and the reinforcing fibers. By merging the separate phases, the composite has enhanced properties that exceed the performance and capabilities of the individual constituents. Fiber-reinforced composites are classified into *continuous* and *discontinuous* fiber-reinforced composites.

Continuous fibers as a reinforcement can be used in many forms such as woven, knitted as well as laminated composites. Laminated composite consists of multiple layers of unidirectional fibers each oriented differently to optimize performance at laminate level. Laminated composite is quite commonly used for high-performance applications due to its superior performance properties. The Boeing 787 depicted in Figure 1.8 is one such example. The fuselage is made with wrapped carbon fiber-reinforced epoxy tape. The wing and vertical fin structures are also manufactured using an automatic carbon-epoxy tape lay-up technique. There are also numerous continuous non-woven fiber mat structures as well as woven textiles made of glass and carbon fibers used in the Boeing 787.

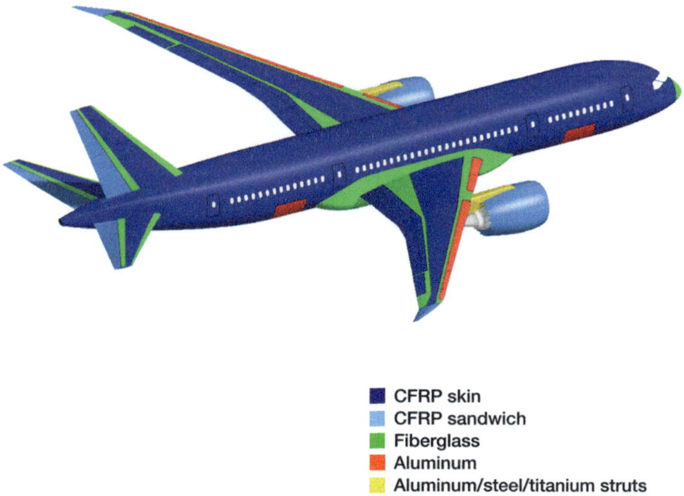

Figure 1.8 Implementation of composites in the Boeing 787 Dreamliner. Copyright © Boeing

In contrast to the laminates, the woven composite consists of woven fabric material infused with resin. Resin transfer molding (RTM) is one such process. In this process, the thermosetting resin is infused throughout the fiber layout or woven prepreg secured within a closed cavity. The photos shown in Figure 1.9 present a sequence of steps in the manufacture of a pillar structure or column of a Super R8 sports car for Audi in Germany. In the first step the fiber mats are cut to a preform. The preform is shaped in a press and then transferred to a mold, at which point the resin is forced through the fibers using high pressure at the gate or gates, and vacuum around the edges. Once the resin cures, the part is removed from the cavity and water jets are used to cut the holes and shape final features. With the rapidly developing technology in resin cure control, this process has a cycle time of 3 minutes. This makes it competitive with other high-volume processing techniques.

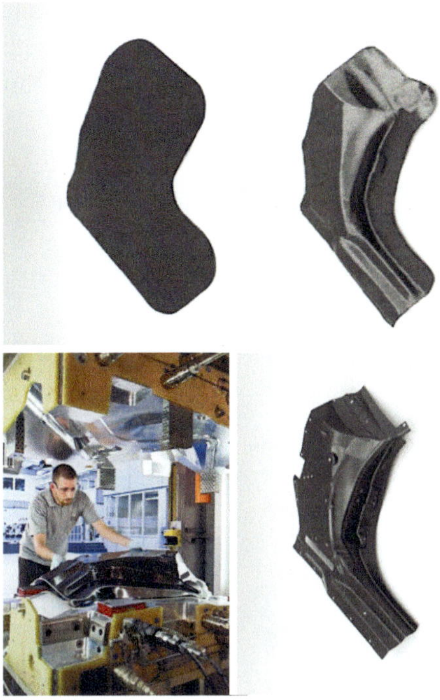

Figure 1.9 Resin Transfer Molding (RTM) of the B-pillar of the R8 Audi Super sports car (Courtesy Audi AG)

Alternatively to the RTM process are vacuum assisted resin infusion (VARI) techniques. Here, the fibers are laid on a composite mold cavity, covered with a vacuum bag, and infused using several vacuum ports, such as schematically depicted in Figure 1.10 for the resin infusion of a boat hull. In general, the RTM process is slower and preferred for a smaller volume production.

Discontinuous fibers are another form in which fibers can be used to improve the resin properties. Injection molding and compression molding are the most common methods to manufacture components from such materials. The key advantages of discontinuous fibers are their outstanding material properties, lower manufacturing costs, and potential of high-volume production. Such fiber-reinforced materials are more commonly applied as a substitute for metal to achieve a reduction of the overall weight of parts, especially in the high-volume applications. Discontinuous fibers are characterized by the fact that the fibers are chopped and change their configuration as a direct response to the deformation experienced by the material during processing. The final state of the fibers determines the local and global properties of the molded part. The configuration of the fibers significantly changes during mold filling of discontinuous fiber-reinforced materials, reflected in the mechanisms referred to as fiber attrition, fiber alignment, and fiber matrix separation [21]. The process-induced change of the fiber microstructure introduces a substantial heterogeneity to the final part. The concentration of fibers, the length of the fibers, as well as the alignment of fibers can vary within the part.

Figure 1.10 Vacuum Assisted Resin Infusion (VARI) of a boat hull. Left: laying the fiber structure into the mold. Right: Inlet tubes (dark with resin) and vacuum tubes during the infusion process. Photograph by Tim Osswald © 2014

The development of anisotropy is one of the major factors that determine the mechanical properties and the dimensional stability of a molded part. The anisotropic properties of the molded part are a consequence of the fact that fibers show a preferential alignment based on the flow conditions during mold filling [23]. Another important aspect is the reduction of the length of the fibers during processing, referred to as fiber attrition or fiber breakage. In injection molding, the material is injected and forced through a gate changing the material's configuration. As the material is forced to flow, the fibers are subjected to extensive stresses that cause them to deform, buckle, bend, and break, resulting in fiber breakage that significantly reduces their reinforcing effect [1]. Lastly, the concentration of the fibers can vary within a part due to migration and separation effects during processing. The mechanical properties of a part, however, are directly related to the fiber concentration. Processing can cause a local variation of the fiber concentration, which may lead to a change along the flow path of the material as well as through the thickness of the molded part.

Discontinuous fiber-reinforced composites can be further classified as short fiber-reinforced thermoplastics (SFT) and long fiber-reinforced thermoplastics (LFT). In general, if the fibers' length is such that they are straight and have little interaction with each other, they are called short fibers, whereas if fibers are curved and show higher interaction with each other, they are considered long fibers. For the glass fiber and thermoplastic material case, typically when the average aspect ratio is of less than 80–100, the material behaves like short fibers. Whereas when aspect ratio is greater than 100, the material behaves like long fibers. It should be pointed that such definition of short and long fibers is subjective and can vary with the fiber volume as well as the material system. The performance and cost of LFT materials usually places them between continuous fiber-reinforced composites used for high-performance applications and SFT compounds, because LFT can be processed economically using compression or injection molding while providing superior mechanical properties to SFT. Both SFT and LFT are available as pellets as shown in Figure 1.11. Another popular method called in-line compounding is also used to produce the long fiber thermoplastic material for both injection and compression molding.

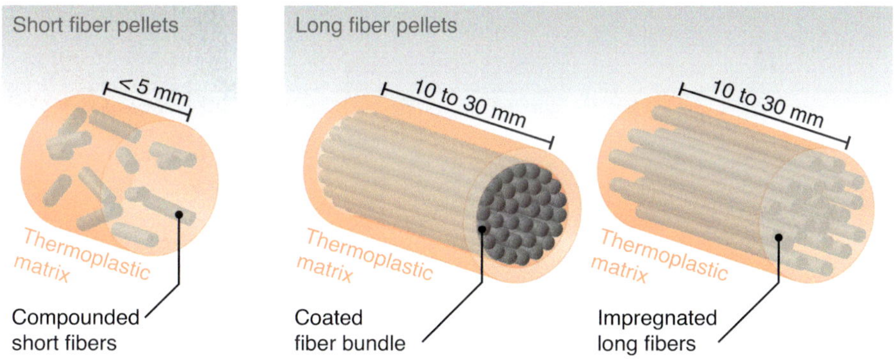

Figure 1.11 Illustration of short fiber pellets (left) and long fiber pellets: coated fiber bundles (center) and pultruded fibers (right)

Finally, the third approach in using the fibers as reinforcement with polymers is hybrid materials. In this, the major volume of the component is manufactured using low cost discontinuous fibers composite and then the areas experiencing higher load are augmented using continuous fibers, or in some cases steel or other metal inserts as reinforcements. Such approach is quite effective with complex parts for automotive applications. The most common manufacturing approach is to start with a lightweight steel sheet insert, or a fiber or fabric reinforced thermoplastic preform and over-mold them with the discontinuous fiber-reinforced material, resulting in a structural part that is much lighter, tougher, and stronger than the traditional single material steel components it replaces.

Hybrid parts have been particularly important in the automotive industry. One of the first hybrid structures was the front end of the Audi A6 in 1998, which lowered the cost by 10% and the weight by 15% [25]. In 1999, Ford introduced the hybrid front end for their Focus, making it 40% lighter and 20% cheaper than the steel part it replaced. This front end integrated steel inserts over-molded with 30% glass fiber filled polyamide 6 features. The front end is depicted in Figure 1.12.

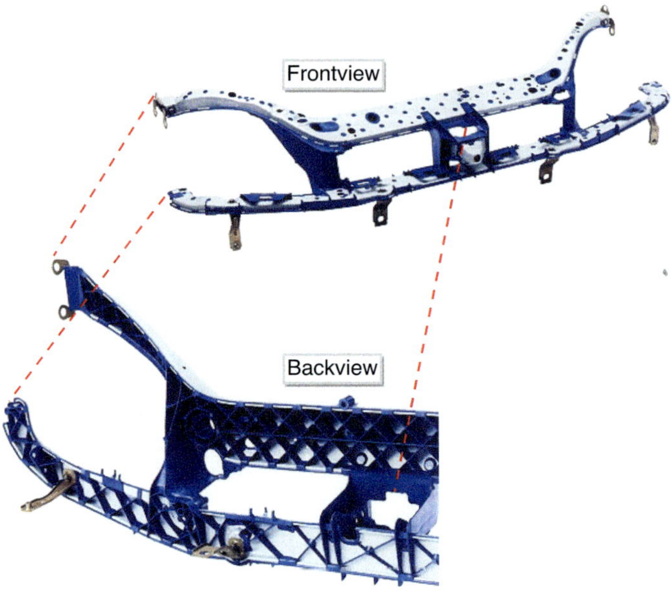

Figure 1.12 Ford-Focus hybrid front end structure made of steel inserts over-molded with 30% by weight glass reinforced polyamide 6 (Courtesy Dynamit Nobel Kunststoff)

■ 1.3 Lightweighting in Automotive Applications

In recent years, global temperatures have been rapidly rising, pushing noticeable changes in the environmental patterns and landscape, including sea levels. At this moment, the earth environment is still relatively stable. However, many scientists fear that, if the current trends continue, the environment may reach a tipping point where significant irreversible shift in the earth environment may be triggered, causing large changes in the landscape and human disaster of unknown proportions. There are many theories on what is causing the global warming. The majority of the scientist believes that the increased emission of carbon dioxide from the fossil fuels such as petroleum, natural gas, and coal are major contributors. In the past 100 years, since the advent of industrial revolution, the use of such fossil fuels has grown tremendously (9 kTWh in 1910 to 123 kTWh in 2010) due to the increase in population and changes in the lifestyle.

Automobiles in the U.S. are considered among the largest emitters of carbon dioxide. As shown in Figure 1.13, almost 28% of greenhouse gases are coming from transportation. This directly correlates with the energy consumption in the transportation sector at 28% of the total energy consumed in the U.S., as shown in Figure 1.14. Therefore, it is imperative that we reduce the fuel consumption in the transportation industries to lower the greenhouse gases overall.

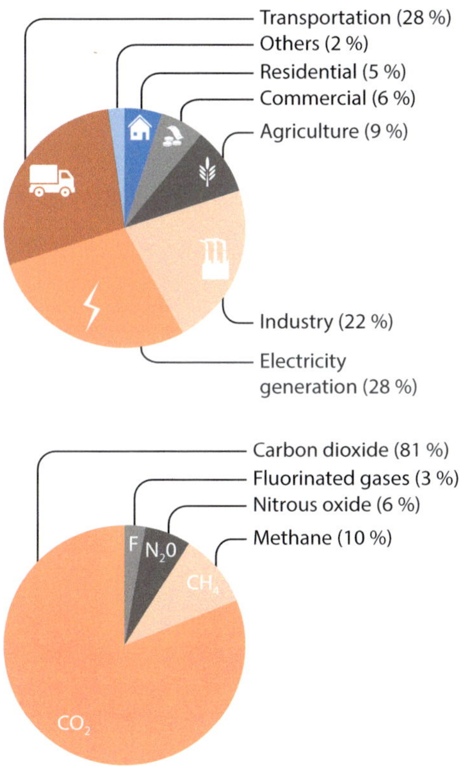

Figure 1.13 Source and breakdown of content of greenhouse gases in the U.S.; note that 28% of greenhouse gases are coming from transportation [7]

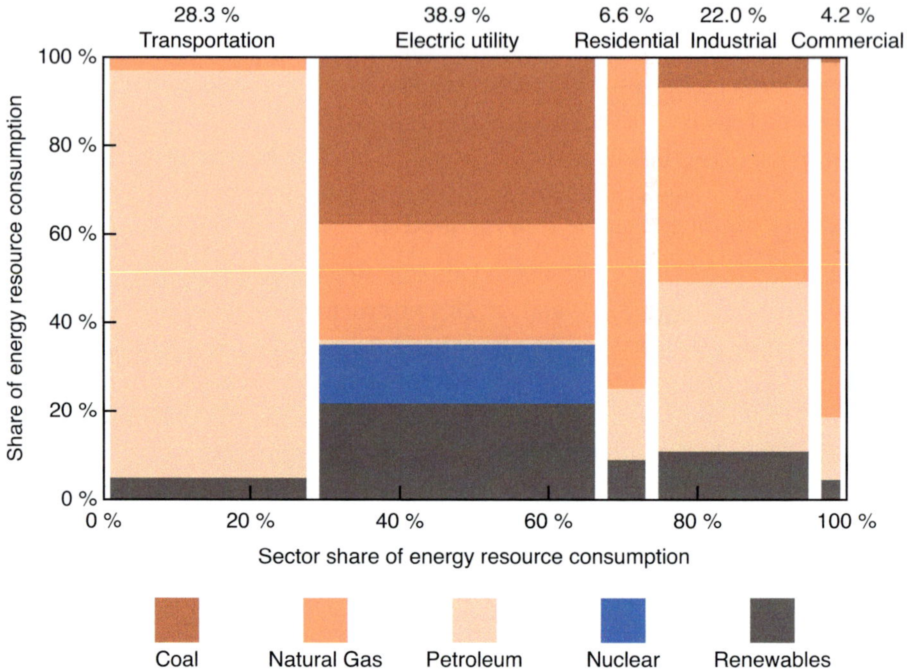

Figure 1.14 U.S. energy consumption by sector; 28% of the energy is consumed by transportation sector; most of it comes from petroleum [8]

The fuel consumption can be reduced by improving the powertrain efficiency, reducing the aerodynamic drag, and using lighter vehicles. Many studies conducted over time in the U.S. [9–11] indicate that for every 10 percent reduction in vehicle weight, fuel consumption can be reduced by approximately 4–8 percent. Actual reduction can vary based on the vehicle specifics, but the trend holds true in general. The main reason for such a direct relationship is engine size. As the vehicle weight is reduced, the power to accelerate the vehicle is also reduced; hence, smaller engines are needed. The smaller engine size becomes the key contributor in reducing the fuel consumption. When the vehicle body is made of lighter materials to reduce weight, it has cascading effects; the size and weight of auxiliary components such as brakes, wheel axles, transmission, and engine are reduced as smaller amounts of weight are supported. As a result, for an engine operating under similar conditions, it can be smaller and use less fuel. The flow of energy from fuel is presented in Figure 1.15: almost 2/3 of energy is used in the engine, while the remaining energy is used in overcoming the friction and other losses; only about 10–12 percent of the energy goes into moving the vehicles. The weight reduction is even more important for electric vehicles. This is because as the vehicle's weight is reduced the energy required to travel is also reduced, hence realizing significant improvements in the range [12].

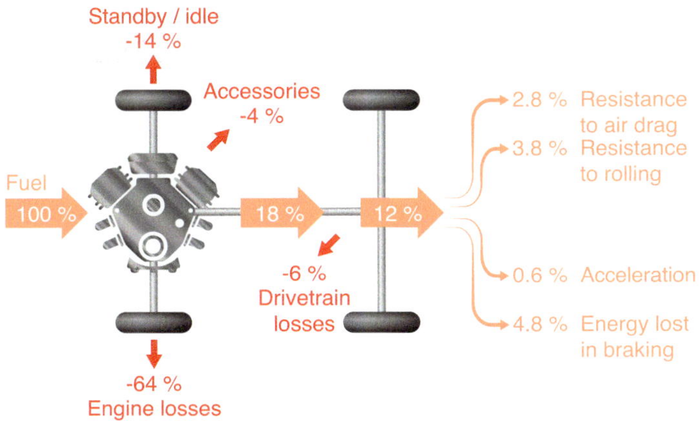

Figure 1.15 Source of fuel usage in a typical vehicle and how weight reduction can help

Therefore, weight reduction of the vehicle body structure has become one of the key focus areas. The history of a typical passenger vehicle's weight since 1975 is presented Figure 1.16. It is interesting to observe that after the fuel crisis in the late 1970s the vehicles became significantly lighter in the early 1980s and then the weight started to increase. Trends in the past 10 years show that the weight of vehicles have stabilized. If we consider the Toyota Corolla, a typical Corolla that weighed about 950 kg in 1983 increased its weight to 1350 kg in 2010. The increase in the vehicle weight was due to many safety regulations rolled out in the 1990s and 2000s, which resulted in improvements in the vehicle structure to absorb crash energy, sensors to detect the impact, and airbags to protect the occupants. These all improved safety but also added weight to the car. In addition, the demand for more comfort and entertainment has increased significantly in the past few years, resulting in larger vehicles with many comfort features and gadgets. During the same period the powertrain technology has also significantly improved. Surprisingly, the result is that the fuel consumption per mile has only increased modestly. In recent years (since 2005) we observe that the weight increase has stabilized while technology to improve fuel efficiency continues to improve; as a result, fuel economy has continued to improve [13].

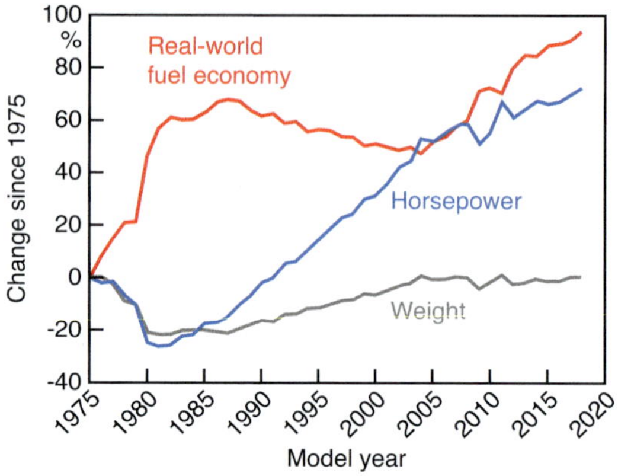

Figure 1.16 History of typical passenger vehicle weight fuel economy and horsepower [13]

Before looking for ways to reduce a vehicle's weight, it is important to understand where the weight is distributed in a typical vehicle. A rule of thumb used in the automotive industry is that about 1/3 of the weight is in the vehicle body, 1/3 in the powertrain, and the remaining 1/3 in the interior and all the other items. Of course, the proportion will change depending on the type of car, such as luxury or economy, and intended usage. An approximate weight breakdown of the key components for a wide range of smaller vehicles is presented in Figure 1.17. For hybrid and electric vehicles, battery and related components can be a source of additional weight.

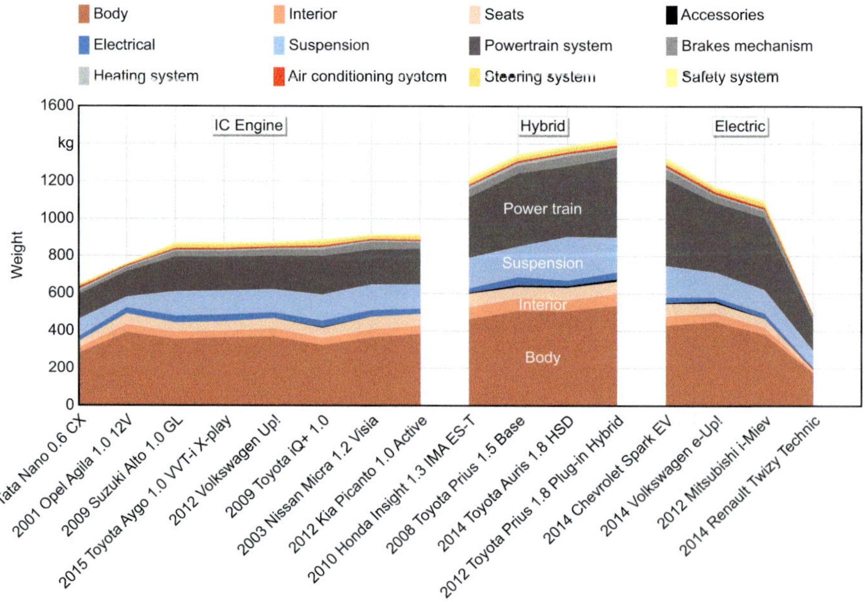

Figure 1.17 Weight distribution comparison of lightweight vehicles in different categories

Efforts to reduce a vehicle's weight are being considered for many years. The most common approach is material replacement, which is to replace the heavier materials with the lighter material components. For example, steel with higher strength steel, aluminum with magnesium, steel with fiber-reinforced polymers, etc. The potential lightweight materials that can be used for automobiles are presented in Figure 1.18. The relative cost, potential for weight reduction, and impact of CO_2 are also compared. Each material has its own strength and weakness; typically, as lighter weight is required, the cost starts to increase exponentially. Since the high-volume automotive market is very sensitive to cost, developing lightweight solutions is a tough balancing act between the targeted performance, cost, and weight reduction. We observe that steel is a deeply entrenched material. Mainly due to lower cost, availability in quantity, established processing methods, and know-how of how to design with steel. For other materials, it is challenging to compete with steel.

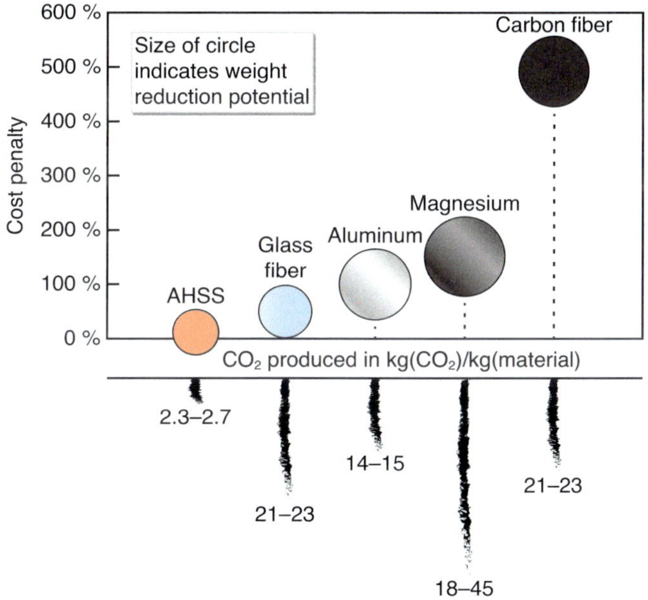

Figure 1.18 Vehicle weight reduction materials and their relative comparison; AHSS stands for advanced high strength steel

However, in recent years there is an increased demand for lower weight and, therefore, studies to understand vehicle weight reduction have been conducted. These studies were often funded by U.S. government agencies and conducted in partnership with the industry to explore potential weight reduction possibilities. These studies included some considerations for manufacturing constraints and performance, but not to a level where one would invest in production of such vehicle technology. It should also be pointed out that the Toyota Venza and Honda Accord studies were conducted by independent groups. The results in the reports are not recognized by automakers and, therefore, considered as the opinion of the researchers who conducted these studies. Our interpretation of the studies is presented in Table 1.1. It is claimed that about a 30% to 50% weight reduction is possible if a lighter material can be used and the vehicle is significantly redesigned. However, there are many practical challenges in using such lightweight materials, such as cost of the material and fabrication of new designs. Both are expected to add significant cost. The relative cost of the different materials is presented in Figure 1.18. Also, key performance factors such as safety, comfort, ride and handling, etc. for such lightweight vehicle design need to be proven.

Table 1.1 Summary of Selected Studies to Identify Weight Reduction Potential of Passenger Vehicles in the U. S. [14–18]; DOE: Department of Energy, NHTSA: National Highway Transportation Safety Administration

Vehicle investigated	Ford Fusion			Toyota Venza			Honda Accord		
Model year	2013			2009			2011		
Funding source	DOE			Energy Foundation			NHTSA		
Project timing	2011–2015			2009–2012			2012–2014		
Source	MMLV[1]			Lotus[2]			EDAG[3]		
Weight	Total wt kg	Low reduction	High reduction	Total wt kg	Low reduction	High reduction	Total wt kg	Low reduction	High reduction
Body structure/interior	869	23%	50%	911	20%	37%	803	16%	33%
Chassis	350	28%	53%	379	27%	43%	308	27%	29%
Powertrain	340	22%	53%	410	13%	13%	368	19%	19%
Potential weight reduction		23%	51%		20%	33%		19%	29%
Potential vehicle weight (kg)	1569	1208	769	1700	1364	1140	1479	1195	1058

1: MMLV: Multi-material lightweight vehicle project [14, 15]
2: Lightweight study by Lotus [16]
3: EDAG study [18]

The other challenge with lightweight materials can be availability in the large volume that may be required. For example, magnesium production today is enough for only a small volume of vehicles. Finally, the manufacturing methods used in the automotive industry can pose serious limitations. Since Henry Ford, the automotive assembly process has been highly optimized. The process has its own science and any change can be quite complex. Using lighter materials can pose many challenges. First, the manufacturing of these components can be slower, and cost can be higher. Furthermore, the assembly process that involves lighter material components can be different than those used for traditional steel parts. This may require significant changes in the assembly process and layout, e.g., certain resin parts have temperature limitations which could limit their usage in hot areas within the assembly line.

The knowledge of how to design with the lightweight material is also very important. The industry has been using steel for many years and over time the know-how of how to design components from steel is well developed and understood. The most common approach is to use CAE (computer aided engineering) methods to predict the performance of the components. In the past 30 years, finite element method software has evolved, and hardware has improved significantly. The computer follows Moore's law and increases computing power by a factor of two every two years. Also, the algorithms used in the computational mechanics as applied to the structural studies have also improved significantly. As a combined effect, it is possible to compute the structural performance accounting for detailed geometry and complex loading conditions. The approach has been used for steel parts and, over the years, know-how is developed to get a good correlation with the actual tests. This has improved confidence in using the CAE methods to analyze, predict performance, and optimize the design with steel parts. A few important points need to be made here for the steel. The steel material properties used in the CAE are assumed to be simpler; for example, the elastic modulus is constant, contrary to recent studies showing that the modulus varies [19]. Also, the material is assumed to be isotropic and follow simple stress-strain relations. This makes CAE faster. Also, there are imperfections and variation in the material that are not always possible to include in the model. Traditionally, these limitations are addressed through small correction parameters used in the CAE. Such correction parameters for steel as applied to automobiles have been developed through the years of "know-how" and experience. Consequently, the confidence in using CAE with steel in the automobile industry is quite high.

When we consider using lightweight materials such as aluminum, magnesium, and fiber-reinforced polymers, etc., we recognize that designing with such lightweight materials is a major challenge, along with the challenges of higher cost and a manufacturing process for large volume productions. Each material has unique properties and, therefore, unique challenges must be addressed; for example, lightweight material such as aluminum has a very different post yield behavior that requires material models that are suited for aluminum. Discontinuous fiber-reinforced composites are heterogeneous materials, where material properties primarily depend on the fiber orientation, fiber length, and volume within the polymers. Therefore, when CAE is used for components made of fiber-reinforced polymers it is important to account for the fiber condition. This is a major challenge. In this book, we are presenting the “state of the art” technology to address these challenges with a focus application in large volume low cost products such as automobiles.

■ 1.4 Background and Challenges in Designing with Discontinuous Fibers

Discontinuous fiber-reinforced composites used for automotive applications commonly consist of a thermoset or thermoplastic matrix material. Sheet molding compound (SMC) is the most prominent type of discontinuous fiber-reinforced thermoset and is processed in compression molding. A variety of thermoplastic matrices and processes are available for discontinuous fiber-reinforced composites [20]. Engineering thermoplastics, such as polyamide (PA) or polypropylene (PP), represent many matrix materials used for fiber-reinforced composites due to their superior properties compared to commodity plastics [21]. Glass fibers are frequently used as reinforcements due to their availability, low cost, and high strength. Carbon fibers can offer improved performance; although their usage has been limited in the past due to higher cost, in recent years the cost of carbon fiber is coming down, and interest in using them is gaining momentum.

Since our motivation to add fiber in the thermoset or thermoplastic material is to improve the mechanical properties, it is very important to develop a clear understanding of the relation between the fibers and the material properties. The first fiber condition is its geometry, i. e., length and diameter. Typically, glass fibers are around 14–17 microns in diameter and carbon fibers are 5–7 microns in diameter. The length of the fiber can vary from 100 microns to a few millimeters. In 1 cubic millimeter there can be thousands of fibers. Also, the length of fiber can vary, which means there will be a length distribution. Studies have been done in the past to relate the fiber length with the mechanical performance. The key trend of fiber length on strength, stiffness, and energy absorption during impact is presented in Figure 1.19(A). We observe that for a given set of fibers and polymeric resin material, if the fibers are well dispersed in the matrix, the mechanical properties increase and then reach a maximum value. To realize most benefit from the fibers it is important to use the optimum length of the fibers; usually, longer fibers are preferred, as they result in higher mechanical properties. However, as fibers become longer, they tend to interact with each other and in many cases start to form bundles; the mechanical properties show a drop, as shown in Figure 1.19(D). This is mainly because fibers are not well dispersed and, hence, are not as effective in improving strength and stiffness; they may continue to improve the impact or energy absorption. Also, a longer fiber tends to make the melt viscous and increase resistance to flow, which limits the shape of molded parts.

The second fiber condition that plays an important role is fiber orientation. Typically, fiber orientation is expressed as a fiber orientation tensor, which is a mathematical expression representing a numerical value for all the fibers in a selected volume. As the parts are manufactured using injection or compression molding, the mixture of fiber and resin flows and fills the mold cavity. During the mold filling process, depending on the flow vectors, the fibers are aligned; subsequently, in the solidified part the fibers show a large variation in the orientation through the volume of the part. In general, as shown in Figure 1.19(C), the mechanical properties strongly depend on the fiber orientation. The material shows higher mechanical properties in the direction more fibers are aligned, while it has reduced properties in the direction where fibers are not well aligned.

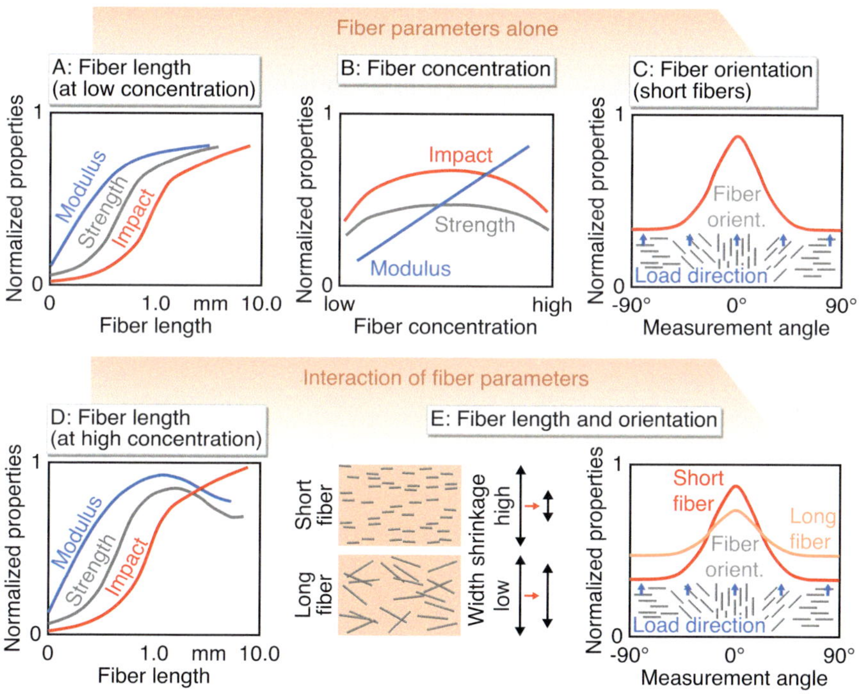

Figure 1.19 Effect of fiber condition such as length, concentration, and orientation on the mechanical properties

The third key fiber condition is fiber volume fraction or concentration. In general, as the concentration of fibers is higher, the mechanical properties increase up to a level beyond which dispersion of fibers may not be possible and improvement with increasing fibers may diminish, as shown in Figure 1.19(B). In addition to the above key factors, the fiber curvature, fiber to resin interface, etc. are also important microstructure conditions that play an important role in defining the performance of fiber-reinforced composite materials. Also, the fiber length, fiber orientation, and concentration have cross influences on each other; for example, longer fibers tend to interact with each other, resulting on less alignment and lesser anisotropy, as shown in Figure 1.19(E).

Since processing discontinuous fiber-reinforced composites has a profound impact on the arrangement of the fibers–i.e., length, orientation, and concentration distribution in the finished part–it is highly desired to develop an in-depth understanding of effect of processing on the fiber condition in the finished parts and develop know-how of the process-microstructure-property relationship. Such capabilities can be used in optimizing the process parameters to achieve excellent mechanical properties and produce lightweight components using discontinuous fiber-reinforced composites. To attain this goal, it is important to consider all the aspects of the process-microstructure-property relationship while estimating the structural performance of such material. Although fiber-reinforced composites have been used for decades, ignorance persists on the underlying physics and phenomena present during processing, including fiber attrition, fiber matrix separation, and fiber alignment. Consequently, the method to account for such process-induced effect of fiber in calculations is still under development. Therefore, in practice it is common to use larger safety factors, resulting in higher weight and hence limited applications of such fiber-reinforced materials.

For instance, reducing fiber attrition is necessary to exploit the superior reinforcement of the longer fibers. For example, Inoue et al. [22] found that the screw design in the compression zone has a substantial impact on fiber breakages. They found that an optimized screw design (low shear screw with a variable pitch) reduced the fiber breakage compared to a standard screw. Nevertheless, little research has been published discussing optimized machine design. Mostly, the LFT injection molding is done on conventional injection molding machines and mold designs, offering lots of potential for optimization and improvements in the process and tooling.

It is extremely important to establish a fundamental understanding of the process-microstructure relationship and extend it to structure-property assessment in the structural analysis. This can be realized by establishing the interdependencies between length, orientation, and concentration of the fibers during processing and quantifying their impact on the structural properties. In order to achieve this goal, experimental capabilities to characterize the material, as well as modeling and simulation of such complex material, must be achieved. Today both areas for discontinuous fibers are under development. Methods to measure fiber length and orientation efficiently have many limitations. The interrelationship between the three microstructural variables have not been thoroughly investigated. In particular, there is a lack of experimental data providing a full three-dimensional (3D) characterization of the entire fiber configuration.

At the same time, methods to model the material properties for the fiber-reinforced material are also a challenge. The main reason is that fibers are at micron scale while most of the components are at millimeter scale. To account for the effect of fiber, it is important to include fiber conditions such as orientation, length, as well as concentration. This means that we need to consider multiscale material models for such parts where the fiber condition is modeled at microscale and part geometry can be modeled at millimeter scale.

To fully exploit the potential of discontinuous fiber-reinforced composites it is crucial to develop capability to generate models that can predict the fiber microstructure in the finished part and also provide clues about how such microstructure can be optimized to improve performance. Also, it is important to be able to measure the microstructure to verify the simulation. It is important as well to relate the fiber microstructure to the finished part's mechanical properties, which can be achieved through a multiscale simulation approach.

1.5 Structure and Objectives of this Book

Injection molding and compression are the most commonly used processes to achieve the high volume and lower piece costs for the fiber-reinforced polymer components in the industry. Both processes are well established and extensively used to produce polymeric and fiber-reinforced polymeric parts for various applications; however, their usage for the critical structural applications has been limited compared to continuous fibers. This is because the fibers used in injection and compression molding are discontinuous and their contribution to the structural strength and stiffness is not well established as compared to continuous fibers. In order to gain significant weight reduction in automobiles at reasonable cost, it is important to increase the use of fiber-reinforced polymers. In the past few years, efforts are being made to address this challenge by improving the manufacturing processes as well as properties of fibers and polymeric materials to improve the mechanical properties while reducing the cost.

While the method to manufacture discontinuous fiber-reinforced composites is improving, it is also very important to understand the mechanics of discontinuous fibers and develop predictive tools to estimate their influence on the final components. Modeling discontinuous fibers can be very complex compared to continuous fibers, because after injection molding or compression molding, the discontinuous fibers in the final part can vary in terms of fiber length, fiber orientation, and fiber concentration, from one location to another through the geometry of the part. This means that the discontinuous fiber-reinforced part will be heterogeneous and anisotropic in structural properties. This is a quite complex situation compared to continuous fibers, where the fiber orientation is generally known and is consistent through the part.

Continuous fiber materials are well established and have been in use for structural performance for many years in aerospace and weight critical structural applications. As a result, a lot of literature on mechanics of continuous fibers along with the methods to analyze structural performance and design is already available. For the discontinuous fiber-reinforced polymers, the story is a little different. Due to the complexities in defining the condition of the fiber microstructure, i.e., variations in length, orientation, and concentration through the part, the existing literature on analytical methods and design is very limited. Also, traditionally, such material has been used mostly for non-load critical parts; therefore, the need to estimate the load carrying capacity has been limited in the past.

We believe that the following key capabilities are needed to address the above challenges:

1. Fast, efficient, and accurate methods to measure the microstructure, which is the fiber length, fiber orientation, and fiber concentration. Measurement of fiber curvature, voids, etc. are also desired.
2. Modeling and simulation to duplicate the processing and predict the microstructure in the finished part.
3. A method to relate the microstructure to the physical properties of the material.
4. A multiscale modeling approach, which can account for the microstructure details at micron scale and also integrate the effect of the microstructure to the material properties of the component of interest, often at millimeter scale.

The overall objective of this book is to share our combined know-how on addressing these challenges in modeling and characterization of the discontinuous fiber-reinforced composite materials.

The book is organized in three sections:

- Section 1, comprising Chapters 2, 3, 4, and 5, provides fundamental knowledge and the foundation required for a fiber-reinforced composites study.
  - Chapter 2 presents information on the latest materials and their fundamental properties used for the discontinuous fiber-reinforced polymers.
  - Chapter 3 is on different processing methods, including the strength and weakness of each process. The goal of these two chapters is to provide background knowledge about the manufacturing technology materials for the discontinuous fiber-reinforced polymers that would help in analytical studies.

- Chapter 4 is on measurement methods for the microstructure. We have observed that, understanding microstructure such as fiber orientation, fiber concentration, fiber length distribution, etc. is very important in understanding the effect of processing as well as relating the effect of processing on the structural properties of the final part. Microstructure measurement technology is still evolving. In this chapter, we have included know-how accumulated by the authors over years of using the existing methods. Also, we have added improvements developed to address some of the current limitations.
- Chapter 5 includes the fundamental mechanics of fiber-reinforced materials with a comprehensive coverage of discontinuous fibers. This chapter provides the foundation behind the heterogeneous material model to represent fiber and resin as two different phases. The fundamental laws of mechanics are very important in addressing the complex structures discussed in the next chapters.

- Section 2, comprising Chapters 6 and 7, focuses on using the fundamental principles and presents a framework for modeling and simulation.
  - Chapter 6 starts with the fundamentals of the finite element method (FEM), which is presented in a very intuitive fashion, and the approach is extended to heterogeneous systems such as fiber-reinforced material, using multiscale modeling. The concept of homogenization is discussed in detail. The simple but comprehensive coverage of the topic is intended to provide sufficient know-how to understand and use commercial finite element codes available for multiscale modeling correctly.
  - Chapter 7 mainly focuses on mold fill simulation methods. Both injection molding and compression molding simulation are discussed. Compression molding is an evolving area and key challenges in compression molding of bulk as well as sheet material are discussed. The current mold fill simulation methods in practice are mainly based on the Hele-Shaw model for the flow and the Folger-Tucker approach for the fiber orientation. This is a practical approach, which provides good results in a timely manner; however, it is also recognized that the fiber representation in such models has many limitations, because fibers are represented as an aggregate with some specific properties in a given volume. The influence of each fiber is not considered. We have also discussed the direct fiber model, where each fiber is individually modeled with a number of degrees of freedom, as an alternate evolving method. We recognize that such approach is not practical at this time, but it may offer specific insights and help address complexities such as jamming of the fibers.
- Section 3 consists of Chapters 8, 9, and 10. The main focus of these chapters is to demonstrate how the multiscale modeling approach, discussed earlier, can be applied to real life complex, three-dimensional parts. This is very important because a fiber-reinforced part with a complex shape offers many unique challenges.

The case studies presented in this chapter provide very good reference information on how such challenges can be addressed. A unique simulation strategy to address specific challenges is developed, the results are compared with actual measurements to verify the approach, and insights on how to improve the quality of prediction is shared.

We recognize the famous quote by G. Box, *"Essentially, all models are wrong but some are useful"*, which is quite relevant for the modeling of discontinuous fiber-reinforced polymers. The approach presented in this book is practical, i.e., it uses the theory and experience-based know-how to develop capabilities to analyze the discontinuous fiber-reinforced polymer structures. We are presenting the learning from years of collaborative work between the industry and academia in an easy-to-use, comprehensive manner.

References

[1] Dung, N. and Pfeiffer, E., *Bent Ply: The Art of Plywood Furniture*, Princeton Architectural Press, New York (2003)

[2] Day, L. and McNeil, I., *Biographical Dictionary of the History of Technology*, Routledge, London (1996)

[3] Davis, B.A., Gramann, P.J., Osswald, T.A., and Rios, A.C., *Compression Molding*, Hanser, Munich (2003)

[4] Northrup, F.B., U.S. Patent 1,158,830 (1915)

[5] Baekeland, L.H., United States Patent number 942699, "Method of Making Insoluble Products of Phenol and Formaldehyde" (1907)

[6] Rahm, L.F., *Plastic Molding*, McGraw Hill Company, New York (1933)

[7] U.S. EPA, Inventory of US greenhouse gases and sinks (2018)

[8] U.S. Energy Information Administration (Nov, 2016)

[9] Wohlecker, R., Johannaber, M., and Espig, M., Determination of weight elasticity of fuel economy for ICE, hybrid and fuel cell vehicles (No. 2007-01-0343). SAE Technical Paper (2007)

[10] Cheah, L.W., Cars on a diet: the material and energy impacts of passenger vehicle weight reduction in the US, Doctoral dissertation, Massachusetts Institute of Technology (2010)

[11] Anderson, M.L. and Auffhammer, M., Pounds that kill: The external costs of vehicle weight, *Review of Economic Studies*, 81(2), pp. 535–571 (2013)

[12] Mruzek, M., Gajdáč, I., Kučera, Ľ., and Barta, D., Analysis of parameters influencing electric vehicle range, *Procedia Engineering*, 134, pp. 165–174 (2016)

[13] U.S. EPA Automotive Trends Report (2018)

[14] Skszek, T.W., Zaluzec, M., Conklin, J., and Wagner, D., MMLV: project overview (No. 2015-01-0407), SAE Technical Paper (2015)

[15] Bushi, L., Skszek, T., and Wagner, D., MMLV: Life Cycle Assessment (No. 2015-01-1616), SAE Technical Paper (2015)

[16] Peterson, G. and Peterson, A., Cost-effectiveness of a lightweight BIW design for 2020–2025: An assessment of a midsize crossover utility vehicle body structure (No. 2013-01-0667), SAE Technical Paper (2013)

[17] Lutsey, N.P., Review of technical literature and trends related to automobile mass-reduction technology (2010)

[18] Singh, H., Economic light-weighting options for high volume production vehicle structures for year 2020. In Presentation in Great Designs in Steel Seminar (2012, May)

[19] Chen, Z., Gandhi, U., Lee, J., and Wagoner, R.H., Variation and consistency of Young's modulus in steel, *Journal of Materials Processing Technology*, 227, pp. 227–243 (2016)

[20] Osswald, T. and Hernández-Ortiz, J.P., *Polymer Processing: Modeling and Simulation,* Hanser, Munich, pp. 1–651 (2006)

[21] Osswald, T.A. and Menges, G., *Materials Science of Polymers for Engineers*, Hanser, Munich (2012)

[22] Inoue, A., Morita, K., Tanaka, T., Arao, Y., and Sawada, Y. Effect of screw design on fiber breakage and dispersion in injection-molded long glass-fiber-reinforced polypropylene, *Journal of Composite Materials*, 49(1), pp. 75–84 (2015)

[23] Goris, S. and Osswald, T.A., Progress on the Characterization of the Process-Induced Fiber Microstructure of Long Glass Fiber-Reinforced Thermoplastics, *Plastics Engineering*, 73(1), pp. 46–47 (2017)

[24] Phelps, J.H., El-Rahman, A.I. A., Kunc, V., and Tucker III, C.L., A model for fiber length attrition in injection-molded long-fiber composites, *Composites Part A: Applied Science and Manufacturing*, 51, pp. 11–21 (2013)

[25] Al-Sheyyab, A., PhD Thesis, LKT, University of Erlangen-Nuremberg (2007)

2 Materials

The major components of a composite material are the reinforcing phase and the surrounding matrix. The reinforcing phase can be particles, agglomerates, platelets, or fibers. When considering fibers, they are either continuous or discontinuous while the continuous fibers can also be woven or braided. In general, the reinforcing phase is joined to the matrix with sizing materials, which depend on the type of matrix used, which can be either thermoplastic or thermosetting. This chapter will present an overview of the reinforcing phase and the matrix.

2.1 Reinforcing Phase

Composite materials are a distinctive set of materials characterized by the combination of individual components to achieve an enhanced material, which exceeds the capabilities of each individual phase. Due to the unique properties of composites, these materials are found in many industries and unique processes have been developed that correspond to the various types of reinforcement. In their most general form, composites are the result of embedding particles of one material in a surrounding matrix of another material. The combination of the particles with the matrix produces a favorable mix of properties that cannot be achieved with either of the constituents acting alone. There are many types of composite materials and Figure 2.1 presents a general classification of fiber-reinforced composites. The shape and dimensions of the reinforcing phase highly affect the mechanical properties of the composite and, in general, composite materials are classified based on the geometry of the reinforcement.

Figure 2.1 is only one way of classifying fiber-reinforced composite materials and a more general and comprehensive classification can include non-fibrous particles or hybrid compositions. Within the larger classification of fibers we can subdivide those as continuous and discontinuous fibers.

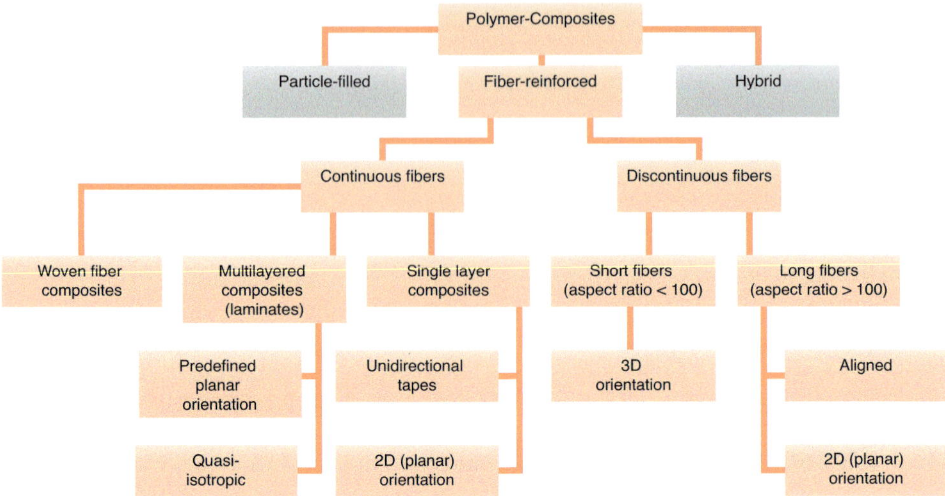

Figure 2.1 Classification of composite materials based on the type of reinforcement

2.1.1 Particle-Reinforced Composites

Particles are those reinforcing materials that do not have a long dimension and can have a spherical, ellipsoidal, obloidal, polyhedral, or even irregular shape. Particle-reinforced composites are some of the oldest types of composite materials. Particulate fillers can increase the stiffness of composite parts, but the level of reinforcement is much smaller than that of fiber-reinforced composites, as the small particles lead to stress concentrations that make the combined materials brittle. For example, a polymer that is tough in its unfilled state becomes brittle when volume fractions of the particles exceed 10% [1]. More often, particles are used to enhance properties such as thermal and electrical conductivity, surface hardness, and wear resistance. The most popular particle used in particle-filled polymers is carbon black. This inorganic filler has been used in the rubber industry since the advent of the rubber tire in the form of rubber/carbon-black composites. Carbon-black, which when dispersed is composed of 10 nm to 100 nm particles, serves not only to stiffen and reinforce rubber, but also to increase its thermal and electrical conductivity. Besides enhancing the properties of the composite, particles can be considerably cheaper than the matrix material. Thus, particles can also be simply used to reduce cost. In the field of inorganic particulate fillers, an active research area in the composites industry has emerged, which concentrates on making polymers more thermally conductive, in some cases by a factor of 100 [2]. This allows the use of inexpensive manufacturing techniques to mold particle-filled plastic parts that conduct heat out of critical automotive components, computers, and heat exchangers [3] as well as in mechatronics applications.

A new class of particle-reinforced composites, nanocomposites, has its roots in the rubber industry. By definition, nano-sized particles have at least one dimension in the nanometer scale. The nano-fillers have an exceptionally high surface to volume ratio and, compared to their bulk size equivalents, nanoparticles have significantly different properties. Nanoparticles greatly improve the properties of the composite and sometimes only small amounts of nanoparticles can lead to improved performance.

Although particle-reinforced composites are a major class of composites, the discussion in this chapter focuses on fiber-reinforced polymer composites. For a detailed discussion on particle-filled composites, the authors suggest [4].

2.1.2 Continuous Fibers

Continuous fiber-reinforced composites are used in structural applications where increased strength and stiffness are required. For example, by stacking single plies of continuous fibers, composites are often made into laminates. The fibers in each ply have a defined orientation and the arrangement of the plies is tailored to enhance the strength of the composite in the primary load directions. For continuous fiber-reinforced composites, the orientation of the fibers is pre-defined by assembling preforms and typically does not change during processing. Several identical or different layers may be bonded together, which is also referred to as *laminate*. The orientation of the fibers in each layer is designed to match the load of the final part. Such types of continuous fiber composites have been traditionally used in aerospace applications, such as in the Boeing 787 discussed in Chapter 1, as well as in lower volume production automobiles such as some panels and structural elements like the B-pillar of the R8 Audi Super sports car, also presented in Chapter 1.

Textile composite structures are an advanced subcategory of continuous fiber composites, which can even further improve the performance of a composite part. The three main types of fabric used in textile composites are: woven, braided, and knitted. Knitted fabric results from knitting of fibers. Generally, such fabrics are stretchable with low tensile and compressive properties, which makes them ideal for gloves, shoes, etc. Their usage in composites has been limited so we will focus on the other two types in the next section.

2.1.2.1 Woven Fabrics

Resin impregnated woven fabrics are widely used in hybrid structures, where they are over-molded with discontinuous fiber-filled ribbed structures. The simplest form of a woven fabric is the one that has a plain weave and is schematically depicted in Figure 2.2 (left). In a weave, the longitudinal threads are called the warp, and the lateral threads are called the weft or the filling. A woven fabric is typically manufactured using a loom that interlaces the warp and weft yarns. The longitudinal threads, or warp threads, are held stationary and in tension, and the weft or fill is inserted over and under the warp threads.

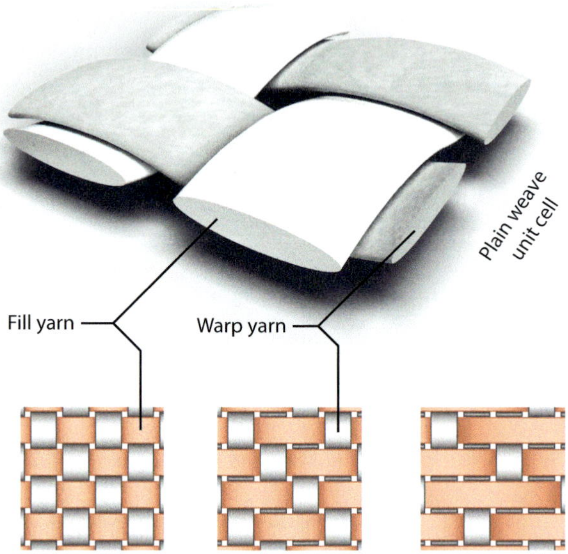

Figure 2.2 Most common types of fabrics: plain (left), twill (center), and satin (right)

Three major types of weaves are used in the composites industry [5]. These are plain, twill, and satin weave. The plain weave is the type of weave with the shortest float where warp and weft are interlaced at every intersection. In a twill fabric, the weft skips or floats over one or more warp threads. A satin is the type of fabric where the weft skips over more than four warp threads. With increasing float, the resistance to deformation decreases, which can have a negative impact when handling the fabric but can improve its drapability. This makes twills and satins more likely to take on shapes that have double curvatures. If the weave of two neighboring weft threads is shifted by one thread, a characteristic diagonal pattern is created such as shown with the twill and satin weaves in Figure 2.2 (center and right) [6].

In the manufacture of hybrid structures, woven fabrics must often be conformed to complex three-dimensional surfaces such as shown in Figure 2.3. The figure depicts an X-ray micro-computed tomography (μCT) scan as an example of a thermoplastic impregnated fabric, draped over a complex tool. As can be seen, such a draping process poses many problems such as those resulting from wrinkling and shearing of the fabric during shaping. Hence, the drapability of these fabrics becomes one of the major issues during the manufacture of complex parts [5]. Twills and satins are best suited for such applications.

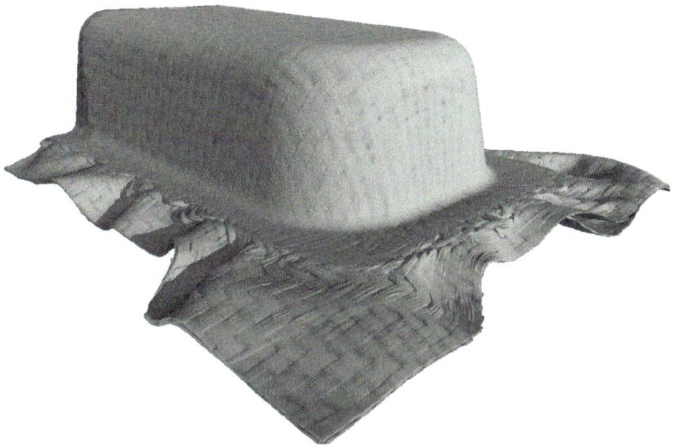

Figure 2.3 μCT scan of a part made of thermoplastic pre-impregnated fabric material [5]

2.1.2.2 Braiding

Braiding technology was already used to manufacture candlewicks in the 19th century and became a process to make preforms for fiber-reinforced composite structures in the middle of the 20th century. The unique ability of a braided tubular structure to conform to any geometry, just like the "Chinese finger trap", makes braiding particularly well-suited to make the reinforcing structure of complex hollow shapes in the composites industry.

In braiding techniques, two or three sets of rovings are interlaced to form a tubular structure. In the most common braiding techniques, two sets of roving carriers rotate in opposite directions from each other in a "maypole dance" pattern, or biaxial braid architecture, such as schematically depicted in Figure 2.4. Here, each roving creates a helical pattern along the tubular structure. Depending on the size of the tube, braiding machines such as shown in Figure 2.4 can have up to 600 carriers. The systems that run with three sets of rovings generate a triaxial braiding architecture, where the third set of rovings run in the axial direction of the tube. Due to the added reinforcement in the lengthwise direction of the tube, a triaxial braiding architecture generates properties similar to those achieved with typical hand layup preform. Figure 2.5 shows a picture of a braiding system at the Institute of Lightweight Engineering and Polymer Technology (ILK), Dresden, Germany.

Other braided structures involve spatial interlacing between multiple rovings that result in bulk 3D textiles. Some 3D weaving processes have been used to generate profiles such as an I-beam.

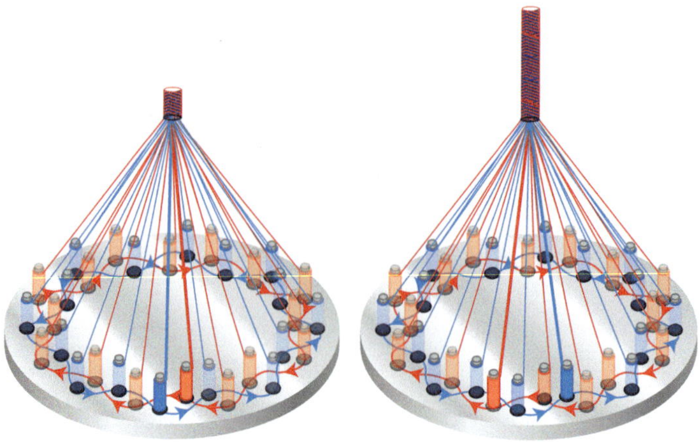

Figure 2.4 Schematic of a biaxial braiding process

Figure 2.5 Braiding machine. Photograph by Tim Osswald © 2015

2.1.3 Discontinuous Fibers

Discontinuous fiber-reinforced composites commonly consist of a thermoset or thermoplastic matrix material. The sheet molding compound (SMC) is the most prominent type of discontinuous fiber-reinforced thermoset and is processed in compression molding [7]. A variety of thermoplastic matrices and processes are available for discontinuous fiber-reinforced composites [5]. Engineering thermoplastics, such as polyamide (PA) or polypropylene (PP), represent the majority of matrix materials used for FRP, due to their superior properties compared to commodity plastics [6]. Glass fibers are frequently used for reinforcement due to their availability, low cost, and high strength. Although carbon fibers can offer improved performance, the cost increase often does not justify their use as substitute in FRP for all applications [7]. In addition, Aramid, Kevlar, and many other natural fibers are also considered; however, their usage in automotive applications is quite limited. In this book we will be focusing only on glass and carbon fibers.

Discontinuous fiber-reinforced composites can be further classified as short fiber-reinforced thermoplastics (SFT) and long fiber-reinforced thermoplastics (LFT). The distinction between LFT and SFT is made by the average fiber aspect ratio (length to diameter). A fiber-filled material with an average aspect ratio of less than 100 is defined as short fiber-reinforced, while long fiber-reinforced composites have an average aspect ratio of more than 100 [1]. Generally speaking, the performance and cost of LFT materials places them between continuous fiber-reinforced composites used for high performance applications and SFT compounds, because LFT can be processed economically using injection molding while providing superior mechanical properties to SFT.

The raw material for SFT or LFT processes is supplied in pellet form, as illustrated in Figure 2.6. LFT pellets are either produced by pultrusion or a coating process. Coated LFT pellets consist of fiber bundles that are surrounded by the thermoplastic material and the fibers are not fully impregnated with the thermoplastic matrix (Figure 2.6, center). Pultruded pellets are manufactured by a continuous pultrusion process and the matrix completely impregnates the fibers within the pellet (Figure 2.6, right). The fibers in LFT pellets are aligned along the major axis in both types of pellet and have a uniform length ranging from 10 mm to 30 mm [19, 20]. SFT pellets are produced by a compounding process, which results in significantly shorter fibers and a non-uniform length (Figure 2.6, left). Current trends in the field of discontinuous fiber-reinforced composites aim to use composite materials with longer fibers in order to increase the reinforcing characteristic of the fibers [1, 8, 9].

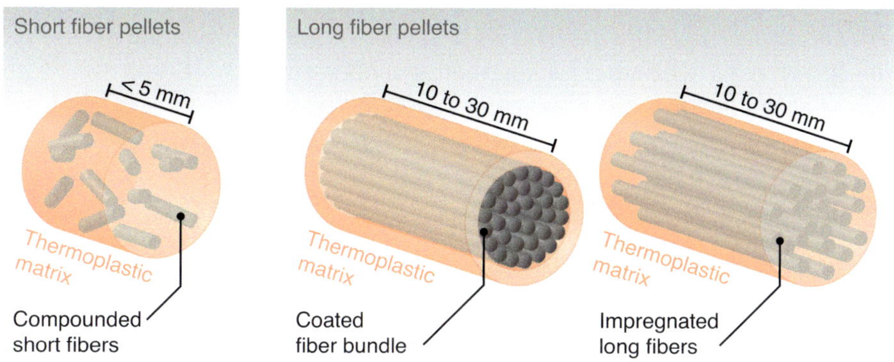

Figure 2.6 Classification of discontinuous fiber-reinforced pellets: short fiber pellets (left) and long fiber pellets (right)

LFT materials can be processed with conventional processing methods like injection molding or compression without major modifications [19]. However, reducing fiber attrition is necessary to exploit the superior reinforcement of the longer fibers. For example, Inoue et al. [10] found that the screw design in the compression zone has a substantial impact on fiber breakages. They found that an optimized screw design reduced the fiber breakage compared to a standard screw. Nevertheless, little research has been published discussing optimized machine design and most LFT injection molding is done on conventional injection molding machines and mold designs [11, 12], offering lots of potential for optimization and new developments.

2.1.4 Sizing

The surface treatment of fibers, often referred to as sizing, is crucial in the performance of composite materials. Sizing is a chemical coating on the fiber's surface, which is known to aid in the adhesion process between the fiber and the matrix. When using glass fibers, common sizing agents, also referred to as coupling agents, are organosilanes. Organosilanes contain both silica and organic functional groups. The silica is compatible with the glass fiber and the organic functional groups are compatible with the resin. For example, organosilanes with epoxy or amino groups have been known to work well with epoxy matrices. However, the actual chemistry of sizing materials is often unknown and considered a trade secret by the manufacturer of SFT or LFT pellets.

The sizing material has a profound effect on the interfacial shear strength between the fiber and the matrix. As reported by Zhao and Takeda [13], water sized or untreated fibers had an interfacial shear strength of 53 MPa, while fibers treated with an epoxy-silane had an interfacial strength of 72 MPa. Also reported by Zhao and Takeda was that the sizing increased the tensile strength of the fiber. Water sized fibers had an average strength of 2.24 GPa and a standard deviation of 0.97 GPa, while fibers sized with epoxy-silane had a higher strength of 3.57 GPa with a standard deviation of 1.29 GPa. Other coupling agents, such as a mixture of γ-aminoxypropyltrimethoxysilane, urethane, and a lubricant, increased the fiber strength to 4.59 GPa with a standard variation of 0.95 GPa. However, this lubricated sizing agent decreased the interfacial shear strength to only 20 MPa, lower than with an untreated fiber surface.

As expected, sizing influences the strength of the composite, which is a crucial performance indicator for fiber-reinforced composites and justifies the additional cost of the surface treatment of the fibers [13]. Unidirectional fiber composites with a 54.2% fiber volume fraction had an ultimate strength of 671 MPa with a maximum strain of 1.67%. On the other hand, an epoxy-silane treaded fiber composite had an ultimate strength of 873 MPa with a maximum strain of 2.10%. While increasing the volume fraction of the fibers results in an increase in strength, the maximum strain reduces with an increase in fiber volume fraction. These aspects will be discussed in more detail in Chapter 5.

2.2 Matrix

The purpose of the matrix for fiber-reinforced composites is to keep the fibers in place and to effectively transfer the load to the fibers. This section will present the most common matrix materials used in today's fiber-filled composites, along with the important properties, allowing the reader to acquire a feel for them, and giving enough information required for calculations in the subsequent chapters. Because of their molecular structure, polymers might be the most complex fluids engineers will encounter. Therefore, what seems trivial to engineers dealing with Newtonian fluids is a complex experimental and mathematical exercise for a rheologist dealing with polymeric melts and becomes even more difficult by adding fibers to the polymer. An important aspect of any rheologist's work is to find relations between deformation and stresses for various well-defined conditions, such as transient shear flows, step strain, creep, and oscillatory shear flow, to name a few. These relations, also called material functions, are determined using different types of rheometric techniques. This chapter will introduce the reader to the causes of the various phenomena only observed with plastics and the relation of these effects to the molecular architecture of polymer melts.

2.2.1 Thermoplastics

Thermoplastic polymers are macromolecular structures that are typically generated synthetically but sometimes also through natural processes. These macromolecules start as monomers, such as ethylene, schematically represented in Figure 2.7. Once these monomers are polymerized, they take the form of long polymer chains with thousands of repeating units. The ethylene of Figure 2.7 becomes polyethylene, as schematically depicted in Figure 2.8. Polymer molecules may also be represented symbolically or by using a cartoon as depicted in Figure 2.9.

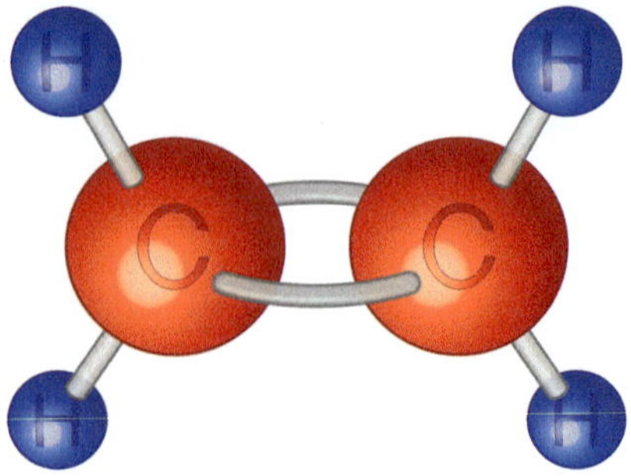

Figure 2.7 Schematic representation of an ethylene monomer

Several forces hold a polymeric material together on the molecular level. The most basic acting forces are covalent bonds that hold together the backbone of a polymer molecule, such as the C-C bond. The energy holding together two carbon atoms amounts to approximately 350 kJ/mol, which is equivalent to a polymer component strength between 14 and 19 GPa and a stiffness of up to 300 GPa. However, because the long polymer chains are typically not aligned, the strength of polymers ranges only between 10 and 100 MPa and their stiffness is approximately 1 GPa.

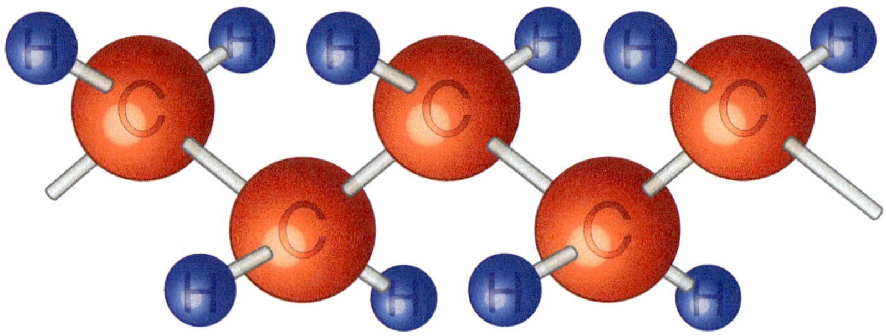

Figure 2.8 Schematic representation of a polyethylene molecule

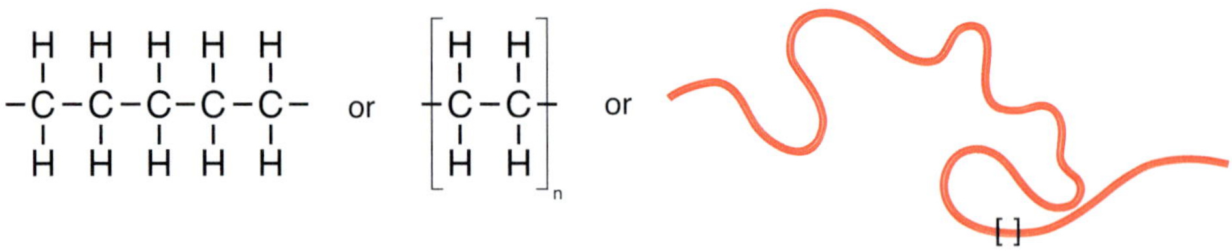

Figure 2.9 Symbolic representation and cartoon of a polyethylene molecule

The comparatively low strength found in polymer components indicates that the forces holding a polymer component together do not originate from the C-C bonds but are intermolecular forces, or the so-called *van der Waals forces*. The energy that generates the intermolecular attraction between two polymeric molecules increases with decreasing distance of the molecules as described by Eq. (2.1)

$$\mathcal{E} \sim \frac{1}{r^6} \tag{2.1}$$

where $\mathcal{E} < 10\,\text{kJ}\,/\,\text{mol}$ and r is the distance between the molecules. Thus, it becomes clear that as a polymer sample is heated, and the distance between the molecules increases with increasing vibration amplitude of the molecules, the intramolecular forces diminish. This increase in free volume allows the molecules to move more freely, enabling the material to flow at the macroscopic level. An increase in free volume, and consequently more ease of movement, can also be achieved by introducing solvents or *plasticizers* between the molecules. As these low molecular weight plasticizers wedge themselves between the molecules, they generate free volume, just as an increase in temperature would.

The properties of polymeric materials are strongly linked to the number of repeat units, n, in the molecular chain, or the molecular weight of the polymer, as shown schematically in Figure 2.10. A polymer such as polystyrene is stiff and brittle at room temperature with a degree of polymerization of 1000, or $n = 1000$. At a degree of polymerization of 10, polystyrene is sticky and soft at room temperature. While the stiffness properties reach an asymptotic maximum, the viscosity increases steadily with molecular weight.

Figure 2.11 shows the relationships between molecular weight, temperature, and properties of a typical polymeric material. The glass transition temperature region increases with molecular weight. Therefore, the temperature range at which the polymer can flow (processing temperature), which for most polymers is 45 to 100 K above the glass transition, also increases with molecular weight. On the other hand, the degradation temperature steadily decreases with increasing molecular weight. Hence, it is necessary to find the molecular weight that renders ideal material properties for the finished polymer product, while providing flow properties that make it easy to shape the material during the manufacturing process. It is important to mention that the temperature scale in Figure 2.11 corresponds to a specific time scale, e. g., time required for a polymer molecule to flow through an extrusion die. If the time scale is reduced (e. g., by increasing the extruder throughput), the molecules do not have enough time to relax, or get used to the shape of the die. Now a somewhat higher temperature is required to assure flow. In fact, at a specific temperature, a polymer melt may behave like a solid if the time scale is reduced sufficiently. Hence, for this new time scale the stiffness properties and flow temperature curves must be shifted upward on the temperature scale. A limiting factor is the fact that the thermal degradation curve remains fixed, forcing processing conditions to remain above certain time scales.

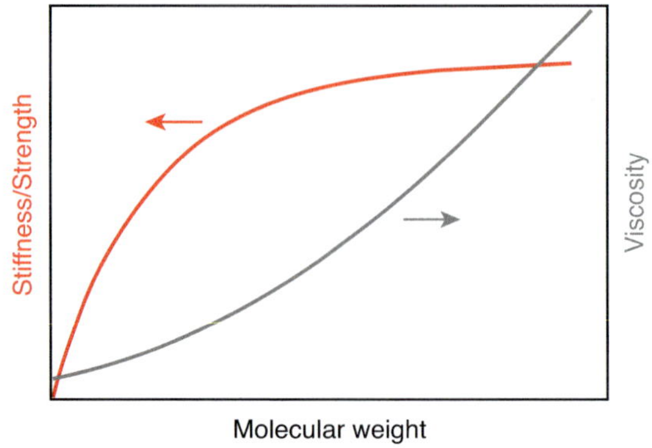

Figure 2.10 Influence of molecular weight on rheological and mechanical properties

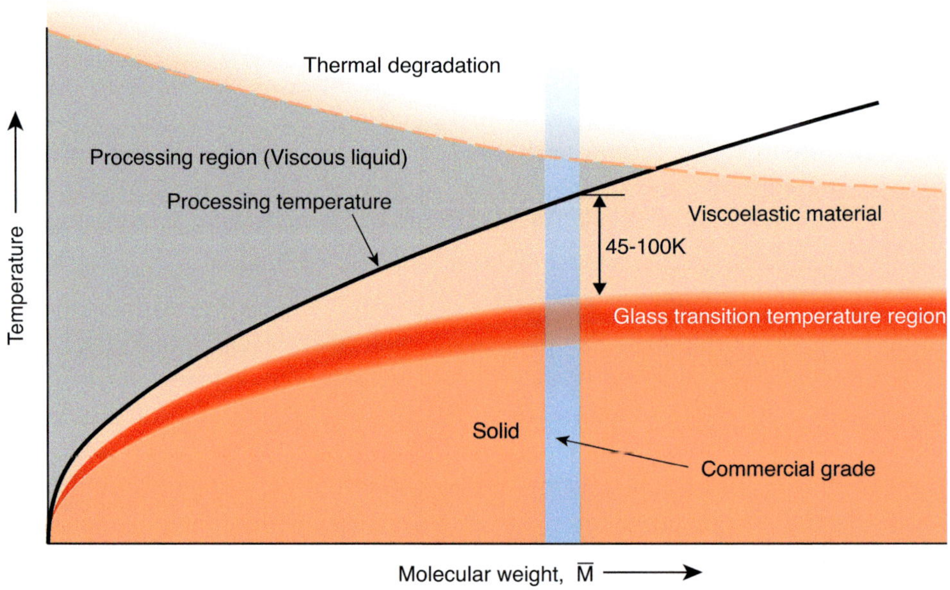

Figure 2.11 Diagram representing the relation between molecular weight, temperature, and properties of a typical thermoplastic

To illustrate the shear thinning behavior, picture the randomly oriented entangled macromolecules depicted in Figure 2.12. At low rates of deformation, the entangled molecules have a hard time sliding past each other, resulting in a relatively high viscosity. Polymer melts exhibit this behavior over a range of small rates of deformation. This constant high viscosity is referred to as a Newtonian plateau, depicted in Figure 2.13. However, as the rate of deformation increases, the corresponding increased shear stress is sufficiently high to disentangle and unravel the molecules. The disentangled molecules can slide past their neighbors with more ease, resulting in an overall lower melt viscosity. This phenomenon is referred to as *shear thinning behavior, pseudoplasticity,* or *structural viscosity*. Eventually, the molecules can no longer stretch out any further. This point is seen in a second Newtonian plateau at higher rates of deformation, reflecting the lowest possible viscosity.

The same effect depicted for shear in Figure 2.13 occurs in elongation flows as well as in combined deformation flows of various shear and elongations. Therefore, the viscosity of a polymer is a function of the magnitude of the rate of deformation tensor

$$\tau_{xy} = \eta(\dot{\gamma}, T)\dot{\gamma}_{xy} \tag{2.2}$$

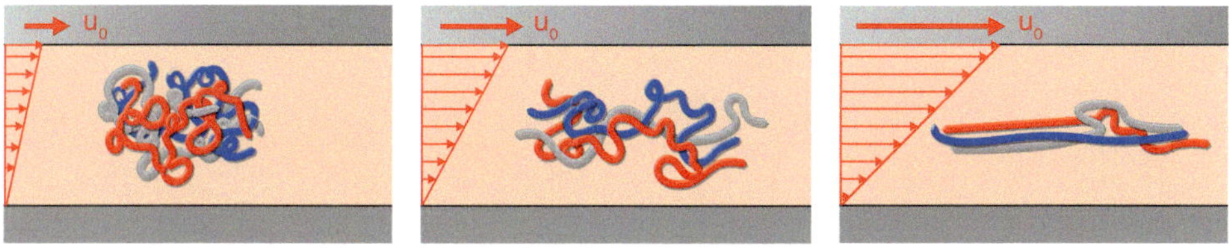

Figure 2.12 Molecular structure with increasing shear

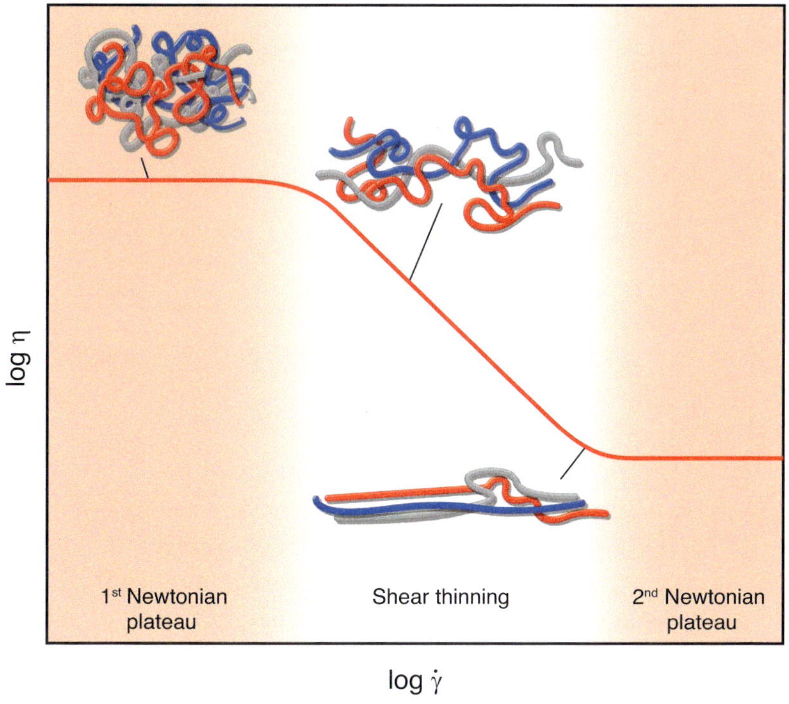

Figure 2.13 Shear thinning behavior or pseudoplasticity

Because the free volume between the molecules increases with temperature, the intra-molecular friction decreases, resulting in an overall lower viscosity as shown in Figure 2.14 for various polymers. Figure 2.14 also presents the range of rate of deformation experienced by polymer melts when using various processing techniques.

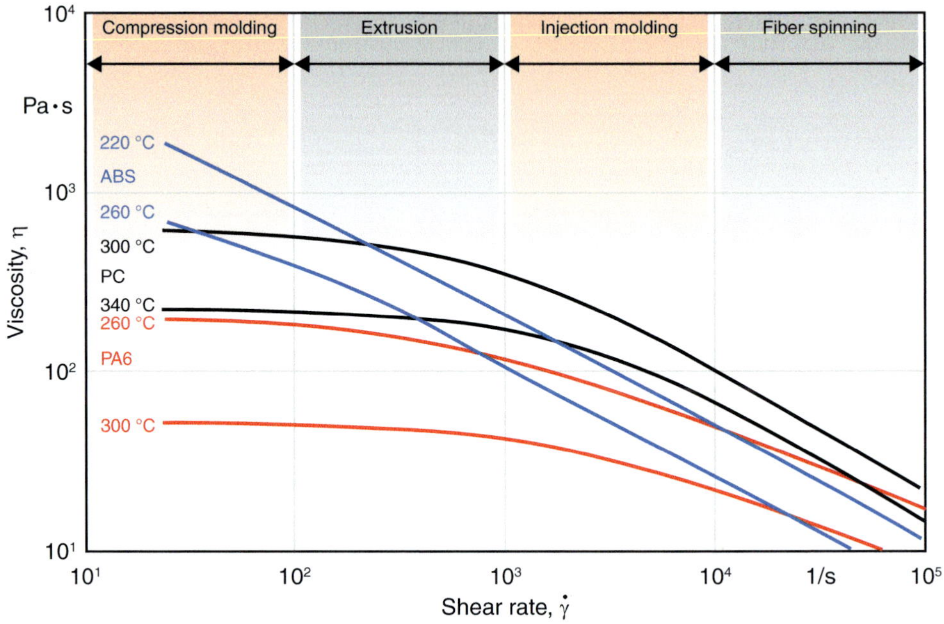

Figure 2.14 Viscosity curves for various polymers at different temperatures

Figure 2.15 shows the general effect of molecular weight variations on the shear thinning behavior of thermoplastics. The shear rate independent plateau is shifted to higher viscosities with increasing molecular weight as the friction between chains increases. Side branching leads to a lower viscosity (shear thinning) caused by the lubrication effect of the side group chains. With increasing width of the molecular weight distribution, shear thinning sets in at higher shear rates, which can be attributed to the smaller shear thinning effect of short chains.

As thermoplastic polymers cool, they take on two different types of structure: amorphous or semi-crystalline. Amorphous polymers are those where the molecules solidify in a random arrangement, whereas some of the molecules in semi-crystalline polymers align with their neighbors, forming regions with a three-dimensional order. In semi-crystalline polymers, the molecules that do not align into crystals remain amorphous structures.

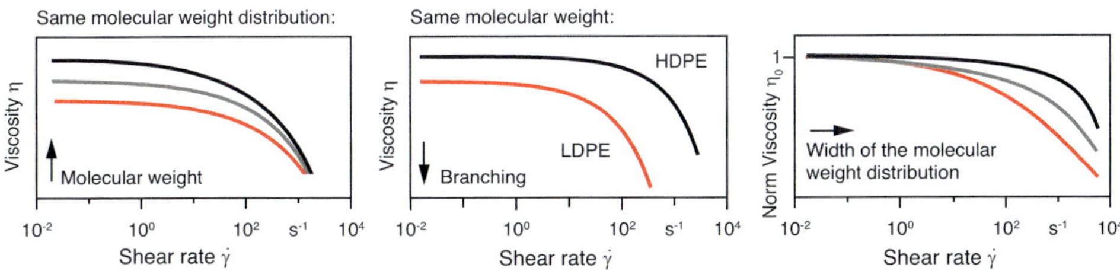

Figure 2.15 Schematic representation of the influence of molecular weight, branching, and molecular weight distribution on the shear rate dependent viscosity of polyolefins [6]

2.2.1.1 Amorphous Thermoplastics

Amorphous thermoplastics, with their randomly arranged molecular structure, are analogous to a bowl of spaghetti. Due to their random structure, the characteristic size of the largest ordered region is on the order of a carbon-carbon bond. This dimension is much smaller than the wavelength of visible light and so generally amorphous thermoplastics are transparent.

Figure 2.16 [4] shows the shear modulus, G', versus temperature for polystyrene, one of the most common amorphous thermoplastics. The figure shows two general regions: one where the modulus appears fairly constant, and one where the modulus drops significantly with increasing temperature. With decreasing temperatures, the material enters the glassy region where the slope of the modulus approaches zero. At high temperatures the modulus is negligible and the material is soft enough to flow. Although there is not a clear transition between "solid" and "liquid", the temperature that divides the two states in an amorphous thermoplastic is referred to as the glass transition temperature, T_g. For the polystyrene in Figure 2.16 the glass transition temperature is approximately 110 °C. At temperatures below the glass transition temperature the material behaves like a viscoelastic solid; the area in the curve is denoted as energy elastic region. Once the material is above the glass transition temperature, it enters the so-called entropy elastic region. At this point the material can be more easily deformed, such as is done during the thermoforming process. However, in order for the material to flow, its temperature must be above a softening temperature, T_S, at which point it behaves like a viscoelastic fluid. The softening temperature defines the point where the viscous forces during deformation (loss) exceed the elastic forces during deformation (storage), and for amorphous thermoplastics it typically ranges about 50 K above the glass transition temperature.

It should be mentioned here that the curve shown in Figure 2.16 was measured at a constant frequency. If the frequency of the test is increased–reducing the time scale–the curve is shifted to the right, because higher temperatures are required to achieve movement of the molecules at the new frequency. Figure 2.17 demonstrates this concept by displaying the elastic modulus as a function of temperature for polyvinyl chloride at various test frequencies. The relation between time scale and temperature, also known as the time-temperature superposition principle, is treated in detail in the literature [1, 14]. Increasing the molecular weight of the material has a similar effect on temperature to that observed with the time scale and temperature relation. The longer molecules have more difficulty sliding past each other, thus requiring higher temperatures to achieve "flow".

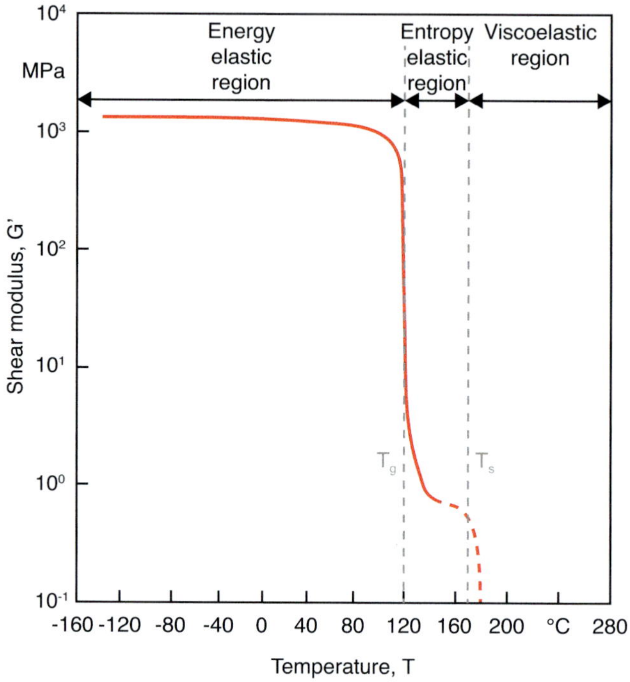

Figure 2.16 Shear modulus of polystyrene as a function of temperature

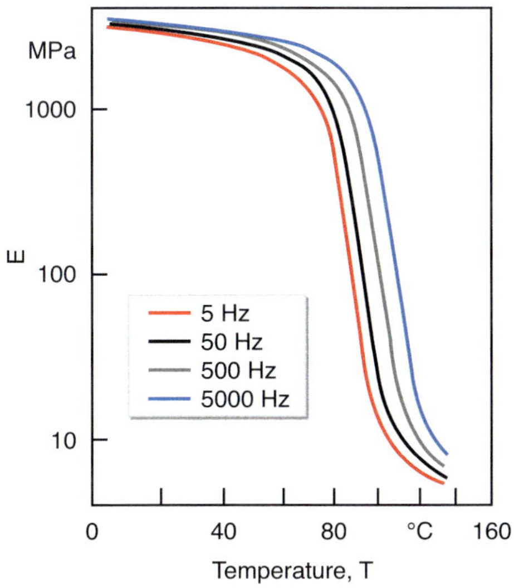

Figure 2.17 Modulus of polyvinyl chloride as a function of temperature at various test frequencies

2.2.1.2 Semi-crystalline Thermoplastics

Semi-crystalline thermoplastic polymers show more order than amorphous thermoplastics. The molecules are arranged in an ordered crystalline form as shown for polypropylene in Figure 2.18. The crystalline structure shown in the photograph is composed of spherulites, which are formed by lamellar crystals. The schematic in Figure 2.19 shows the general structure and hierarchical arrangement in semi-crystalline materials, using polyethylene as an example. The spherulitic structure is the largest domain with a specific order and has a characteristic size of 50 to 500 µm. Some of the polypropylene spherulites in the micrograph presented in Figure 2.19 are about 150 µm in diameter. The size of spherulites is much larger than the wavelength of visible light, making semi-crystalline materials translucent and not transparent.

However, the crystalline regions are very small with molecular chains comprised of both crystalline and amorphous regions. The degree of crystallinity in a typical thermoplastic will vary from grade to grade. For example, in polyethylene the degree of crystallinity depends on the branching and the cooling rate. A low density polyethylene (LDPE) with its long branches can only crystallize to approximate 40–50%, whereas a high density polyethylene (HDPE) crystallizes to up to 80%.

Figure 2.20 [1] shows the dynamic shear modulus versus temperature for a high density polyethylene, the most common semi-crystalline thermoplastic. Again, this curve presents data measured at one test frequency. The figure clearly shows two distinct transitions: one at about −110 °C, the glass transition temperature, and another near 140 °C, the melting temperature, T_m. Above the melting temperature, the shear modulus is negligible and the material will flow. A crystalline arrangement begins to develop as the temperature decreases below the melting point. Between the melting and glass transition temperatures, the material behaves like a leathery solid. As the temperature decreases below the glass transition temperature, the amorphous regions within the semi-crystalline structure solidify, forming a glassy, stiff, and in some cases brittle polymer.

Figure 2.18 Micrograph of the spherulitic crystalline structure in a polypropylene polymer (courtesy of the Institute of Plastics Technology, LKT, University of Erlangen-Nuremberg)

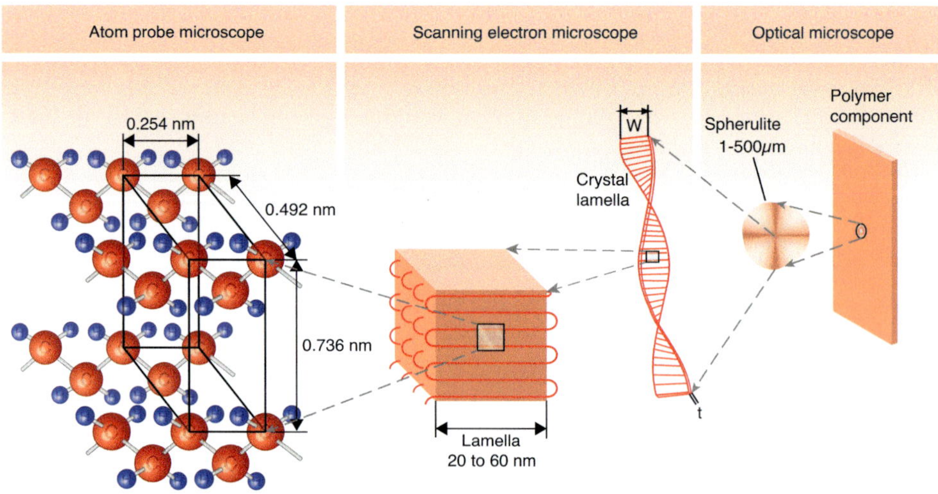

Figure 2.19 Schematic representation of the general molecular structure and arrangement of typical semi-crystalline materials

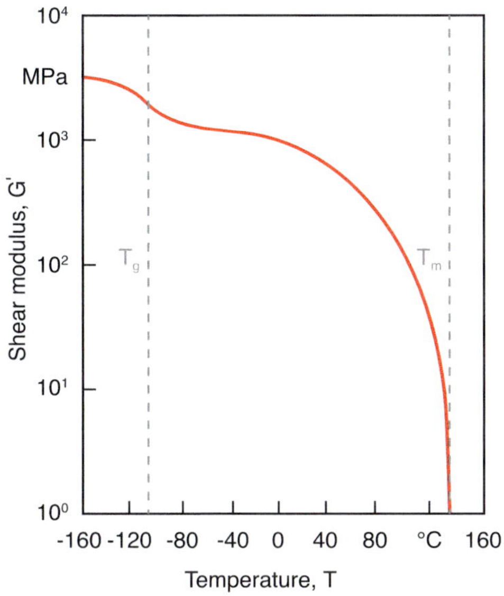

Figure 2.20 Shear modulus of a high density polyethylene as a function of temperature

2.2.2 Thermosets

In a cured thermoset, the molecules are cross-linked and rigid, formed by short groups that are connected by randomly distributed links. The fully reacted or solidified thermosetting polymer does not react to heat as observed with thermoplastic polymers. A thermoset may soften somewhat upon heating, but then it will degrade at high temperatures. Due to the high cross-linking density, a thermoset component behaves as an elastic material over a large range of temperatures. However, it is brittle with breaking strains of usually 1 to 5%.

Thermosets can be identified in three categories: thermosets that cure via condensation polymerization, those that undergo addition polymerization, and those that cure via free radical polymerization.

Condensation polymerization is defined as the growth process that results from combining two or more monomers with reactive end-groups together with by-products such as alcohol, water, or acid, such as in the phenol formaldehyde reaction shown in Figure 2.21. The by-product of the phenolic curing reaction is water.

An example of a cross-linking reaction of a thermoset by free radical reaction is the co-polymerization of unsaturated polyester with styrene molecules, shown in Figure 2.22. The molecules contain several carbon-carbon double bonds, which act as cross-linking sites during curing. Another important free radical reaction is encountered when curing vinyl ester resins such as depicted in Figure 2.23. Both unsaturated polyester and vinyl ester resins use a styrene monomer as a cross-linking reactant. While unsaturated polyester is the older, less expensive, and traditional matrix material found throughout the auto industry, vinyl esters are gaining ground because of their superior physical properties such as toughness and corrosion resistance. This is in great part due to the fact that the reactive bonds in vinyl esters are at the ends of the chains, while the reactive double bonds in unsaturated polyesters are in every repeat unit throughout the chain.

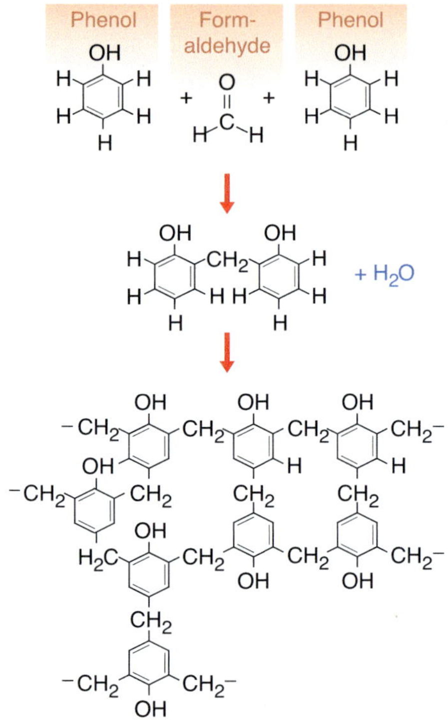

Figure 2.21 Symbolic representation of the condensation polymerization of phenol-formaldehyde resins

A curing reaction is an irreversible process that results in a structure of molecules that are more or less cross-linked. These exothermic reactions are accelerated by heat; however, the cross-linking can sometimes also be activated by adding a catalyst or initiator, or by the absorption of moisture or radiation, such as ultraviolet, electron beam, and laser energy sources [21].

From a processing point of view, thermosets are divided into three distinct categories, namely those that undergo a *heat activated cure*, those that are dominated by a *mixing activated cure*, and those that are activated by the *absorption of humidity or radiation*. Examples of heat activated thermosets are phenolics and of mixing activated cure are polyurethanes (Figure 2.24). However, depending on the additives, such as catalysts or accelerators, a thermoset can be either heat activated or mixing activated. For example, epoxies that cure at room temperature, right after mixing, are mixing activated, while high temperature cure epoxies are typically heat activated. A typical epoxy cross-linking reaction is shown in Figure 2.25.

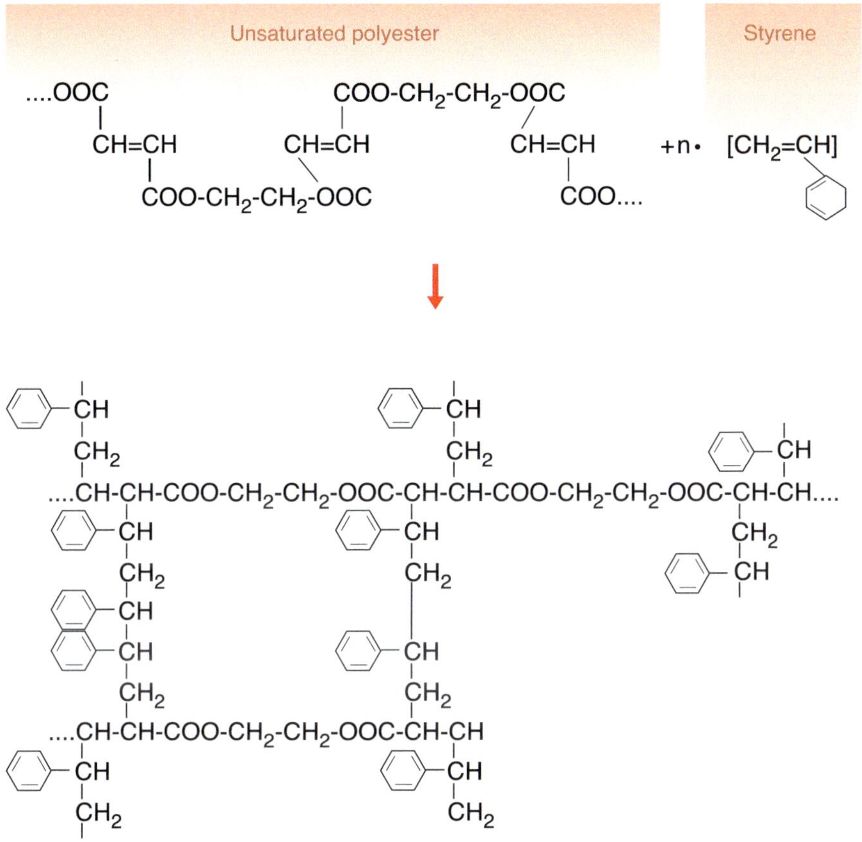

Figure 2.22 Symbolic and schematic representations of curing process of unsaturated polyester

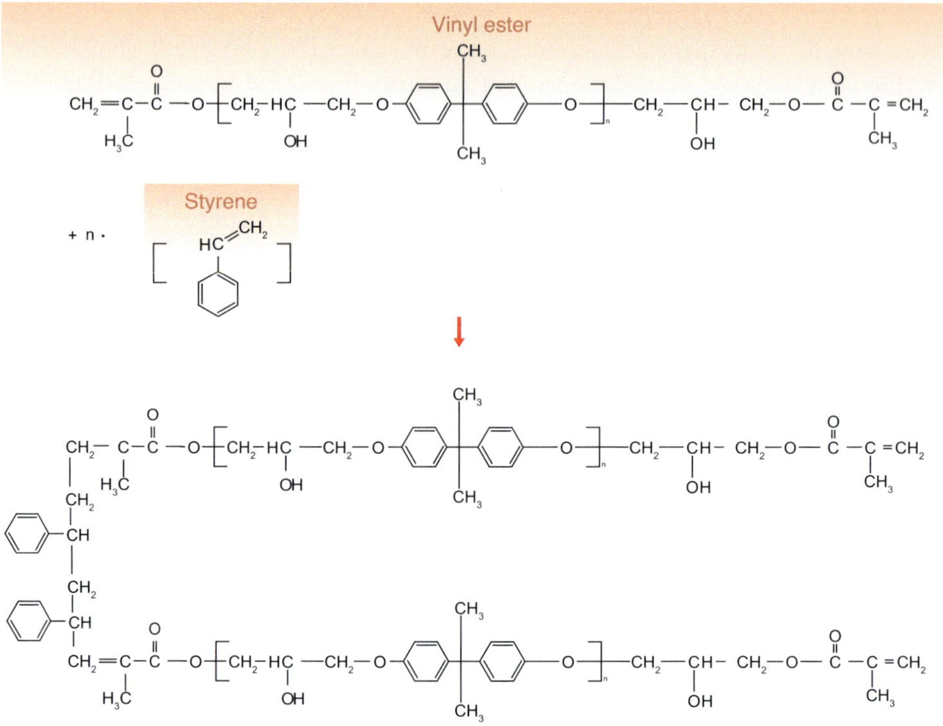

Figure 2.23 Symbolic and schematic representations of curing process of vinyl esters

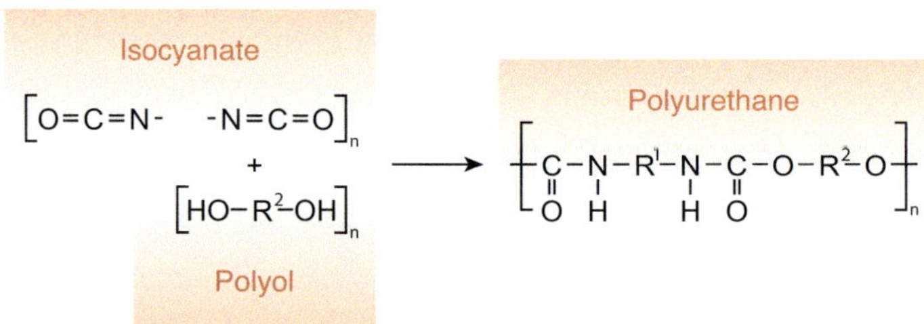

Figure 2.24 Symbolic and schematic representations of curing of a urethane

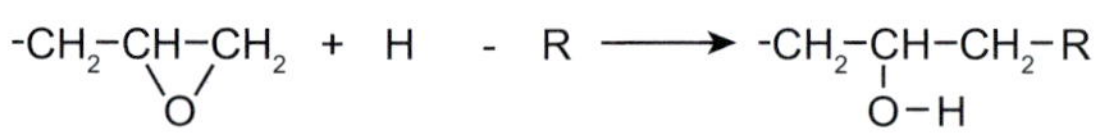

Figure 2.25 Symbolic and schematic representations of curing of epoxy

2.2.2.1 Curing Reaction

No matter which category a thermoset belongs to, heat or mixing activated cure, its curing reaction can be described by the reaction between two chemical groups denoted by A and B that link two segments of a polymer chain. The reaction can be followed by tracing the concentrations C_A or C_B of unreacted As or Bs. If the initial concentrations of As and Bs are defined as C_{A_0} and C_{B_0}, respectively, the degree of cure can be described by

$$c = \frac{C_{A_0} - C_A}{C_{A_0}} \tag{2.3}$$

The degree of cure or conversion, c, equals zero when there has been no reaction and equals one when all As have reacted and the reaction is complete. However, it is impossible to monitor reacted and unreacted As and Bs during the curing reaction of a thermoset polymer. It is known, though, that the heat released during curing can be used to monitor the conversion, c. When several small samples of unreacted thermoset polymer are placed in a differential scanning calorimeter (DSC), each at a different heating rate, every sample will release the same amount of heat, Q_T. This happens because every cross-linking that occurs during a reaction releases a small quantum of energy in the form of heat. Figure 2.26 [1] shows the rate of heat released during the isothermal cure of a vinyl ester at various temperatures.

The degree of cure can be defined by the following relation

$$c = \frac{Q}{Q_T} \tag{2.4}$$

where Q is the heat released up to an arbitrary time t, and is defined by

$$Q = \int_0^t \dot{Q}dt \tag{2.5}$$

DSC data is commonly fitted to empirical models that accurately describe the curing reaction. Hence, the rate of cure can be described by the exotherm, $\dot{Q}$, and the total heat released during the curing reaction, Q_T, as

$$\frac{dc}{dt} = \frac{\dot{Q}}{Q_T} \tag{2.6}$$

With the use of Eq. (2.6), it is now easy to take the DSC data and find the models describing the curing reaction.

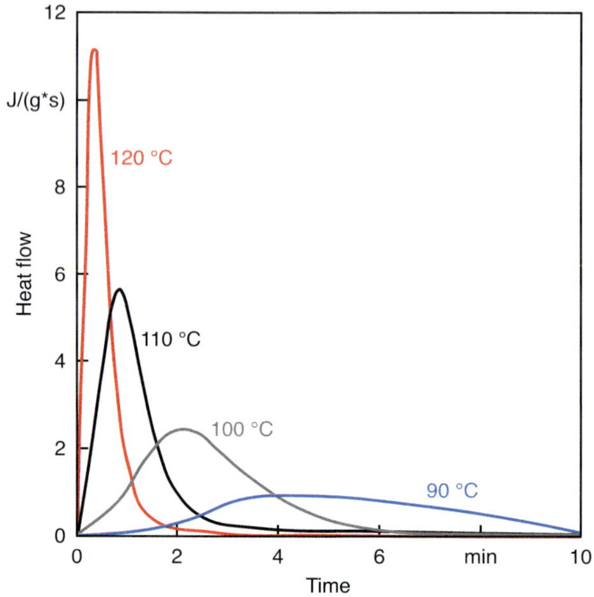

Figure 2.26 DSC scans of the isothermal curing reaction of vinyl ester at various temperatures

During cure, thermoset resins exhibit three distinct phases: viscous liquid, gel, and solid. Each of these three stages is marked by dramatic changes in the thermomechanical properties of the resin. The transformation of a reactive thermosetting liquid to a glassy solid generally involves two distinct macroscopic transitions: molecular gelation and vitrification. Molecular gelation is defined as the time or temperature at which covalent bonds connect across the resin to form a three-dimensional network that gives rise to long range elastic behavior in the macroscopic fluid. This point is also referred to as the gel point, where $c = c_g$. As a thermosetting resin cures, the cross-linking begins to hinder molecular movement, leading to a rise in the glass transition temperature. Eventually, when T_g nears the processing temperature, the rate of curing reduces significantly, eventually dominated by diffusion. At this point the resin has reached its vitrification point. Figure 2.27, which presents the degree of cure as a function of time, illustrates how an epoxy resin reaches a maximum degree of cure at various processing temperatures. The resin processed at 200 °C reaches 100% cure because the glass transition temperature of fully cured epoxy is 190 °C, less than the processing temperature. On the other hand, the sample processed at 180 °C reaches 97% cure and the one processed at 160 °C only reaches 87% cure.

Figure 2.26 and Figure 2.27 also illustrate how the curing reaction is accelerated as the processing temperature is increased. The curing reaction of thermally cured thermoset resins is not immediate, thus the blend can be stored in a refrigerator for a short period of time without initiating any significant curing reaction.

The behavior of curing thermosetting resins can be represented with the generalized time-temperature-transformation (TTT) cure diagram developed by Enns and Gillham [15]; it can be used to relate the material properties of thermosets as a function of time and the processing temperature as shown in Figure 2.28.

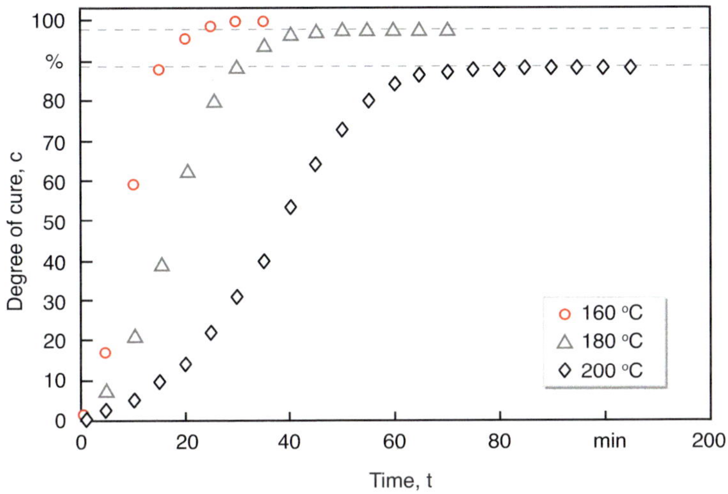

Figure 2.27 Degree of cure as a function of time for an epoxy resin measured using isothermal DSC

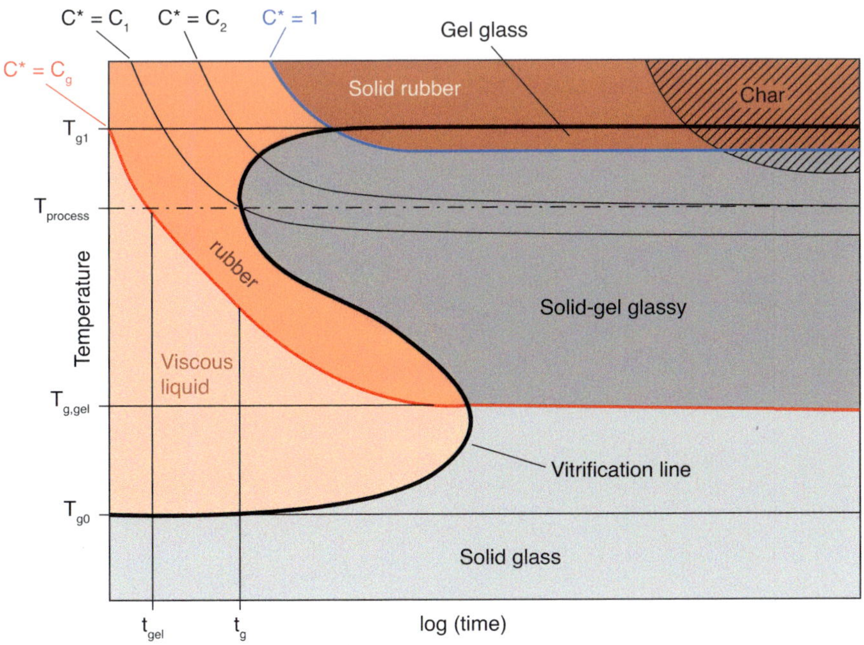

Figure 2.28 Time-temperature-transformation (TTT) diagram for a typical thermoset

The various lines in the diagram represent constant degrees of cure. The curve labeled $c = c_g$ represents the gel point and $c = 1$ the fully cured resin. Both curves have their corresponding glass transition temperatures, T_{g1} and $T_{g,gel}$, for the fully cured resin and for the resin at its gel point, respectively. The glass transition temperature of the uncured resin, T_{g0}, and an S-shaped curve labeled "vitrification line" are also depicted. The *vitrification line* represents the boundary at which the glass transition temperature becomes the processing temperature. Hence, to the left of the vitrification curve the curing process is controlled by a very slow diffusion process. The TTT-diagram shows an arbitrary process temperature. The material being processed reaches the gel point at $t = t_{gel}$ and the vitrification line at $t = t_g$. At this point the material has reached a degree of cure of c_1 and the glass transition temperature of the resin is equal to the processing temperature. The material continues to cure very slowly (diffusion controlled) until it reaches a degree of cure just below c_2. There are also various regions labeled in the diagram. The one labeled "viscous liquid" represents the resin from the beginning of processing until the gel point has been reached. The flow and deformation that occurs during processing or shaping must occur within this region. The region labeled "char" must be avoided during processing, because at high processing temperatures the polymer will eventually undergo thermal degradation.

A general cure kinetic equation that describes the reaction can have the form

$$\frac{dc}{dt} = ae^{-\frac{E}{RT}} f(c) \tag{2.7}$$

where the function $f(c)$ describes the nature and order of the reaction. The various reaction orders are described by

$$\text{Zero-order reaction: } f(c) = 1 \tag{2.8}$$

$$\text{First-order reaction: } f(c) = 1 - c \tag{2.9}$$

$$\text{Second-order reaction: } f(c) = (1 - c)^2 \tag{2.10}$$

$$n\text{th-order reaction: } f(c) = (1 - c)^n \tag{2.11}$$

$$\text{First-order autocatalytic reaction: } f(c) = (B + c)(1 - c) \tag{2.12}$$

$$\text{Second-order autocatalytic reaction: } f(c) = (B + c)(1 - c)^2 \tag{2.13}$$

$$n\text{th-order autocatalytic reaction: } f(c) = (B + c^m)(1 - c)^n \tag{2.14}$$

where B is a fitting constant and represents the effect of the catalyst. The curing reactions of mixing activated polyurethane resins are best modeled using a second-order reaction, sometimes referred to as the Castro-Macosko model [16, 17]. The reaction of epoxies and vinyl esters that cure at room temperature can be represented with an nth-order autocatalytic reaction. Kamal and Sourour [18] developed an autocatalytic model that has been used in the past decades to simulate the curing reactions of various unsaturated polyester resins, such as those encountered in compression molding of sheet molding compound (SMC), injection-compression molding of bulk molding compound (BMC), and in pultrusion processes. The model is a modified version of the nth-order autocatalytic reaction given in Eq. (2.14) and is given by

$$\frac{dc}{dt} = (k_1 + k_2 c^m)(1 - c)^n \tag{2.15}$$

where k_1 and k_2 are Arrhenius-type temperature functions similar to the one given in Eq. (2.15), given by

$$k_1 = a_1 e^{-E_1/RT} \tag{2.16}$$

and

$$k_2 = a_2 e^{-E_2/RT} \tag{2.17}$$

The constants a_1 and a_2 are fitting parameters, E_1 and E_2 are activation energies, and R is the ideal gas constant. However, many heat activated polymers such as some epoxy resins, as the one shown in Figure 2.27 and schematically reflected in the TTT-diagram in Figure 2.28, require a different treatment that accounts for the diffusion controlled reaction once the glass transition temperature reaches the process temperature during the reaction. To account for this, the original functions k_1 and k_2 of the Kamal-Sourour autocatalytic cure kinetic model are replaced by $k_1^{'}$ and k_2^{*} and are given by

$$\frac{1}{k_i^*} = \frac{1}{k_i} + \frac{1}{k_D} \tag{2.18}$$

where k_D is the diffusion rate constant defined as

$$k_D = a_D e^{-E_D/RT} e^{-b/f} \tag{2.19}$$

The constants a_D and b are adjustable parameters, E_D is the activation energy of the diffusion process, and f is the equilibrium fractional free volume given by

$$f = 0.00048\left(T - T_g\right) + 0.025 \tag{2.20}$$

where T_g is the instantaneous glass transition temperature during cure. Equation (2.18) shows that the overall rate constant is governed at one extreme by the Arrhenius rate constant (when $k_D \gg k_i$), which is the case prior to vitrification, and at the other extreme by the diffusion rate constant (when $k_D \ll k_i$), which is the case well after vitrification. For a system exhibiting a unique one-to-one relationship between the glass transition temperature and conversion, DiBenedetto's equation [19] is one of the easiest approaches for stoichiometric ratios to express this relationship using only a single parameter as

$$T_g = T_{g0} + \frac{\left(T_{g1} - T_{g0}\right)\lambda c}{1 - \left(1 - \lambda\right)c} \tag{2.21}$$

where T_{g0} is the glass transition temperature of the uncured resin, T_{g1} is the glass transition temperature of the fully reacted network, and λ is a structure-dependent parameter theoretically equated to

$$\lambda = \frac{\Delta C_{p0}}{\Delta C_{p1}} \tag{2.22}$$

The values of ΔC_{p0} and ΔC_{p1} are the differences in the heat capacity between the glassy and rubbery state for the uncured resin and the fully cured network, respectively. However, the parameter λ can also be used as a fitting parameter. A typical development of glass transition temperature during cure for an unsaturated polyester resin is depicted in Figure 2.29 [20]. The glass transition temperature was measured using a differential scanning calorimeter. The graph shows how the glass transition rises as the sample is post-cured at different temperatures.

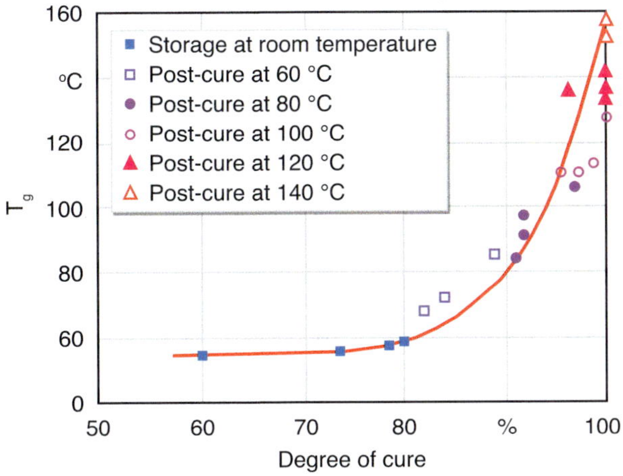

Figure 2.29 Glass transition temperature as a function of degree of cure for an unsaturated polyester resin [20]

References

[1] Osswald, T. A. and Menges, G., *Materials Science of Polymers for Engineers*, 3rd ed., Hanser, Munich (2012)

[2] Heinle, C., Brocka, J., Hülder, G., Ehrenstein, G. W., and Osswald, T. A. Thermal Conductivity of Polymers Filled with non-Isometric Fillers: A Process Dependent, Anisotropic Property, in Proceedings of the Annual Technical Conference of the Society of Plastic Engineers (2011)

[3] Mulholland, T., Goris, S., Boxleitner, J., Osswald, T. A., and Rudolph, N., Fiber Orientation Effects in Fused Filament Fabrication of Air-Cooled Heat Exchangers, *JOM*, vol. 70, no. 3, pp. 298–302 (2018)

[4] Rothon, R. N., *Particulate-Filled Polymer Composites*, 2nd ed., Smithers Rapra Press, Shrewsbury, UK (2003)

[5] Goris, S., John, P., and Osswald, T. A., Chapter 33: Polymer Composites Manufacturing Processes, in *Manufacturing Engineering Handbook*, 2nd ed., McGraw-Hill Education (2015)

[6] I. V. K. AVK, *Handbuch Faserverbundkunststoffe: Grundlagen, Verarbeitung, Anwendungen*, Vieweg+Teubner, Wiesbaden (2010)

[7] Vaidya, V., *Composites for Automotive, Truck and Mass Transit: Materials, Design, Manufacturing*, 1st ed., DEStech Publications, Inc., Pennsylvania, US (2011)

[8] Phelps, J. H., Abd El-Rahman, A. I., Kunc, V., and Tucker, C. L., A model for fiber length attrition in injection-molded long-fiber composites, *Compos. Part Appl. Sci. Manuf.*, vol. 51, pp. 11–21 (2013)

[9] Rohde, M., Ebel, A., Wolff-Fabris, F., and Altstädt, V., Influence of Processing Parameters on the Fiber Length and Impact Properties of Injection Molded Long Glass Fiber Reinforced Polypropylene, *Int. Polym. Process.*, vol. 26, no. 3, pp. 292–303 (2011)

[10] Inoue, A., Kazuya, M., Tanaka, T., Arao, Y., and Sawada, Y., Effect of screw design on fiber breakage and dispersion in injection-molded long glass-fiber-reinforced polypropylene, *J. Compos. Mater.*, vol. 49, no. 1 (2015)

[11] Kumar, K.S., Patel, V., Tyagi, A., Bhatnagar, N., and Ghosh, A.K., Injection Molding of Long Fiber Reinforced Thermoplastic Composites, *Int. Polym. Process.*, vol. 24, no. 1, pp. 17–22 (2009)

[12] Metten, M. and Cremer, M., Langfaserverstaerkte Thermoplaste spritzgiessen – Verfahrensparameter beeinflussen die Faserlaenge, *Kunststoffe*, vol. 90, no. 1 (2000)

[13] Zhao, F.M. and Takeda, N., Effect of interfacial adhesion and statistical fiber strength on tensile strength of unidirectional glass fiber/epoxy composites. Part I: experiment results, *Compos. Part Appl. Sci. Manuf.*, vol. 31, no. 11, pp. 1203–1214 (2000)

[14] Osswald, T. and Rudolph, N., Polymer Rheology – From Molecular Structure to Polymer Process, in *Polymer Rheology*, Hanser, Munich, pp. I–XI (2015)

[15] Enns, J.B. and Gillham, J.K., Time-temperature-transformation (TTT) cure diagram: Modeling the cure behavior of thermosets, *J. Appl. Polym. Sci.*, vol. 28, no. 8, pp. 2567–2591 (1983)

[16] Castro, J.M. and Macosko, C.W., Studies of mold filling and curing in the reaction injection molding process, *AIChE J.*, vol. 28, no. 2, pp. 250–260 (1982)

[17] Castro, J.M. and Macosko, C.W., Kinetics and Rheology of Typical Polyurethane Reaction Injection Molding Systems, in Proceedings of the Annual Technical Conference SPE, New York, NY, pp. 434–438 (1980)

[18] Kamal, M.R. and Sourour, S., Kinetics and thermal characterization of thermoset cure, *Polym. Eng. Sci.*, vol. 13, no. 1, pp. 59–64 (1973)

[19] DiBenedetto, A.T., Prediction of the glass transition temperature of polymers: A model based on the principle of corresponding states, *J. Polym. Sci. Part B Polym. Phys.*, vol. 25, no. 9, pp. 1949–1969 (1987)

[20] Ehrenstein, G.W., *Polymeric Materials: Structure, Properties, Applications*, 1st ed., Hanser, Munich (2001)

[21] Randell, D.R., *Radiation Curing Polymers*, Burlington House, London (1987)

3 Manufacturing Processes

This chapter presents processes used to manufacture discontinuous fiber-reinforced composite parts. The major part of the chapter is divided into the two types of resin used: thermoset and thermoplastic.

3.1 Fiber-Reinforced Thermoset Molding Processes

Thermosets, such as phenolics, unsaturated polyesters, epoxy resins, and polyurethanes, are polymeric materials that irreversibly solidify due to cross-linking between molecules. The solidification process of thermosets is dominated by an exothermic chemical reaction called the *curing reaction*. During curing, the polymer solidifies due to molecular cross-linking as a result of the presence of carbon-carbon double bonds as well as epoxy rings, allowing the molecules to cross-link with their neighbors. The curing reaction can be activated by heat, mixing, or both. As curing progresses and the molecules begin to cross-link, the viscosity of a thermoset increases due to the emerging three-dimensional molecular network and less mobility of the molecules. The competing mechanisms of flow and curing reaction must be considered during processing of thermosetting resins. Due to the exothermic reaction, the thermosetting resin's temperature is raised and its viscosity decreases, which facilitates fiber impregnation. Simultaneously, high temperatures result in an accelerated cross-linking reaction, which causes the viscosity to increase faster. During this transition, the viscosity increases until the resin ceases to flow, a stage known as the *gel point*. At the gel point, the molecular weight goes to "infinity". Hence, it can be said that at this point all the molecules are interconnected.

A major advantage of thermosets in composite manufacturing is the low initial viscosity of the matrix, which leads to an easy and fast impregnation of the fibers. In general, thermosets are harder, more rigid and brittle, and their mechanical properties are less temperature dependent. However, since the solidification process is a chemical curing reaction, thermosets typically have longer processing times compared to thermoplastics.

3.1.1 Compression Molding of Sheet Molding Compound

Sheet molding compounds (SMC) are thin sheets of semi-finished composites consisting of an uncured thermosetting matrix and discontinuous fibers. The most common SMC material is unsaturated polyester reinforced with long glass fibers (~25 mm). The fibers are randomly oriented in the sheets and the fiber concentration can reach 50% by weight. Different processes exist to manufacture the sheets and Figure 3.1 illustrates a typical SMC production line. The SMC compounding process is a continuous operation and starts with mixing the thermosetting resin, fillers, and other additives into a paste. The paste is spread onto a conveyor with an upper and lower polymeric carrier film. The chopped fibers are distributed onto the lower paste layer forming a dry bed of fibers, also called a *mat*. The upper and lower layers are subsequently merged into a sandwich consisting of a layer of chopped fibers between two outer layers of paste. In the compaction belt, the sandwich is calendered into the semi-finished SMC sheets.

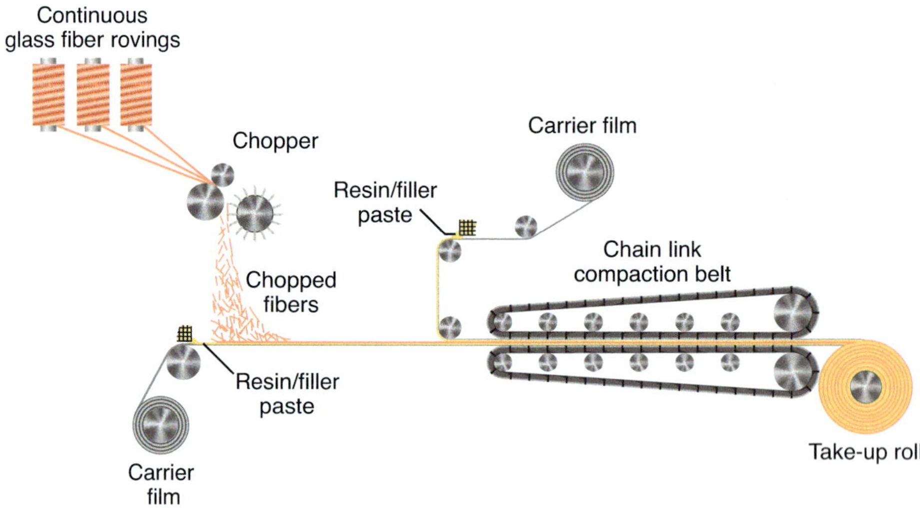

Figure 3.1 SMC sheet production line

Usually, SMC is processed in compression molding and Figure 3.2 illustrates this process. The semi-finished sheets are cut into pieces, stacked to a pre-defined charge, and placed into the mold. Due to the heterogeneous nature of the semi-finished SMC, it is difficult to gauge the weight of the charge from its area. Hence, small pieces are often withdrawn from the charge, or are added on top of the charge to achieve the desired weight of the final part. The charge usually covers 30–70% of the mold surface. The mold is first closed rapidly until both mold halves touch the SMC charge. Then, it slows down, causing the material to flow and fill the mold cavity. Once the mold is completely filled, a constant force is maintained on the material as the curing of the thermosetting resin progresses. When the reaction is almost completed, the part is removed and allowed to cool. Next, the cycle is repeated when a new charge is placed on the heated mold surface. At this stage, some one-sided heating may take place. Preheating the charge speeds up the curing process and is therefore desirable in some applications.

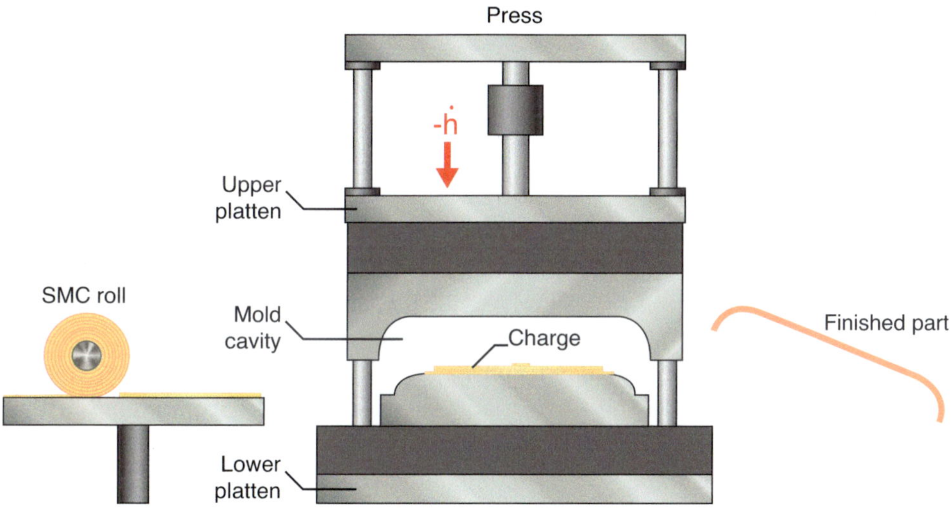

Figure 3.2 SMC compression molding process

The SMC compression molding cycle is the sequence of events during the compression molding of an SMC sheet, as shown in Figure 3.3. The cycle begins when the prepared charge is placed on the mold cavity, and it ends when the finished part is removed from the mold. As can be seen, the dominant event during the cycle is the curing process. Depending on the thickness, type of matrix, and processing conditions, the curing time ranges from 20 s to over 1 min. Charge placement and part removal each take 10 s. Mold closing, compression, and mold opening each take approximately 5 s.

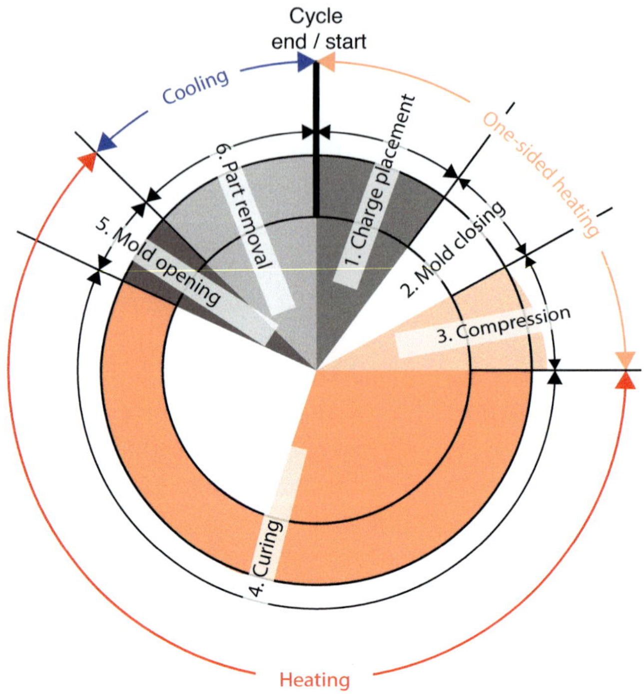

Figure 3.3 SMC compression molding cycle

As in any polymer process that includes discontinuous fibers, the deformation of the material forces the fibers to orient in a particular manner. The final orientation of the fibers determines the mechanical properties of the part and any warpage is due to the difference in thermal expansion coefficients between the fibers and the matrix. Figure 3.4 depicts the fiber orientation distribution within a plate where the initial charge coverage was 33%. The measurements show that a majority of the fibers are aligned in the flow direction, indicating a high degree of fiber orientation.

Distinctive alignment of fibers in discontinuous fiber composites is common in compression molding or injection molding and leads to high degrees of anisotropy throughout the entire part (see Chapter 4 for more information on fiber alignment). Usually, it is desirable to introduce a lower degree of fiber orientation into the finished part. In SMC compression molding, this can be attained when there is little to no flow inside the mold cavity. However, if too little flow occurs during mold filling, the part may have a poor surface finish. This is detrimental when molding Class-A automotive panels. To illustrate the effect of fiber orientation on material properties of the final part, Figure 3.5 shows how the deformation and resulting orientation from 33%, 50%, 66%, and 100% mold coverage affects the stiffness of the plate.

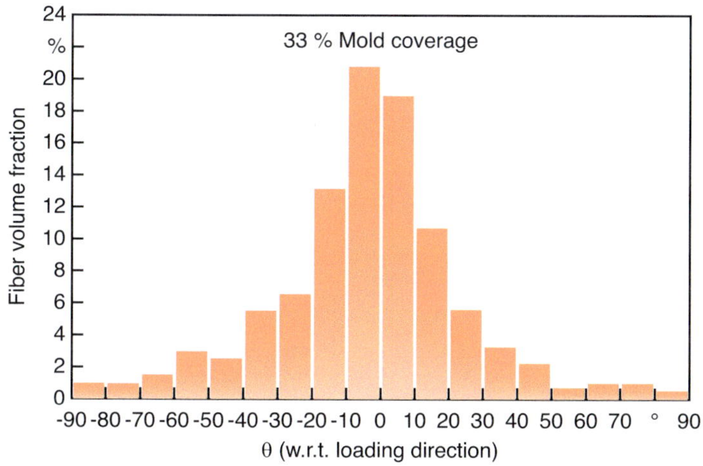

Figure 3.4 Measured fiber orientation distribution histogram in a plate with 33% initial mold coverage and extensional flow during mold filling

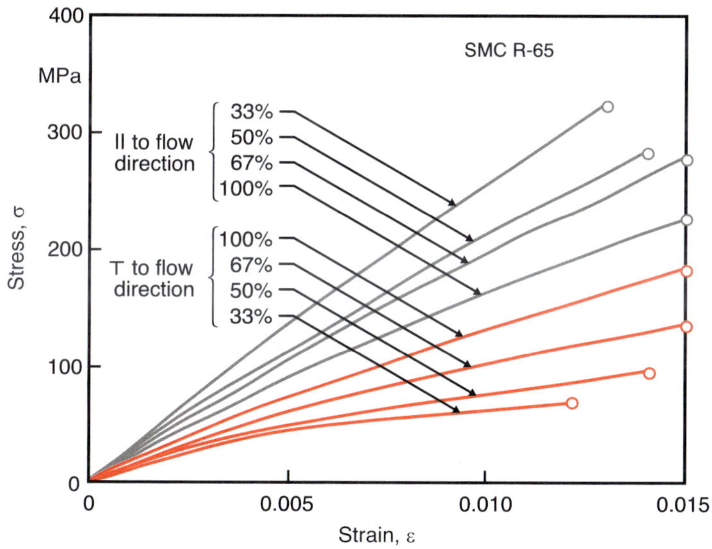

Figure 3.5 Stress-strain curves of 65% glass by weight SMC for various degrees of deformation

In conjunction with the anisotropy caused by the fiber orientation, residual stresses are developed during the cooling stage and the curing reaction due to phase change and chemical shrinkage of the thermosetting resin. These processing events can cause the part to shrink and warp. Additionally, if two or more resin fronts of material meet during mold filling, knit lines or weld lines form. These knit lines have little to no fiber content, becoming the weakest areas in the part. Flow and deformation also force air voids out of the material, eliminating some porosity inside the sheet. This will have a direct influence on the quality of the surface in the finished product.

SMC is used in several industries due to the highly flexible production process and adjustable material properties. Especially in the automotive industry, SMC compression molding is used to produce large, thin, and lightweight parts with favorable mechanical properties. Moreover, it is used for exterior and interior body panels due to the great surface finish that can be achieved in SMC compression molding *(Class-A surface finish)*. It is also used to produce household goods and in the electrical industry due to its design freedom and highly automated production lines.

3.1.2 Injection-Compression Molding of Bulk Molding Compound

Injection-compression molding is another alternative to processing charges of thermoset matrices reinforced with discontinuous fibers. The material used in the injection-compression molding process is called bulk molding compound (BMC). BMC is composed of an unsaturated polyester resin reinforced with a 20–25% volume fraction of randomly oriented 10 mm long fibers.

The BMC material is compounded in an internal batch mixer and subsequently extruded into the shape of a large strand, also known as a *BMC charge*. This charge is then fed into a cylinder above the BMC press. The cylinder injects the charge into the mold of the press. Closing the press causes the material to flow and fill the mold cavity. Figure 3.6 illustrates the process steps of BMC molding.

The main benefit of injection-compression molding over compression molding is automation, which makes it a suitable process for large batch productions. Additionally, the combination of injection and compression molding can lead to a lower degree of fiber alignment and fiber attrition compared to injection molding. BMC is used in the production of smaller items, such as housings, valve covers, and other engineering applications that require strength.

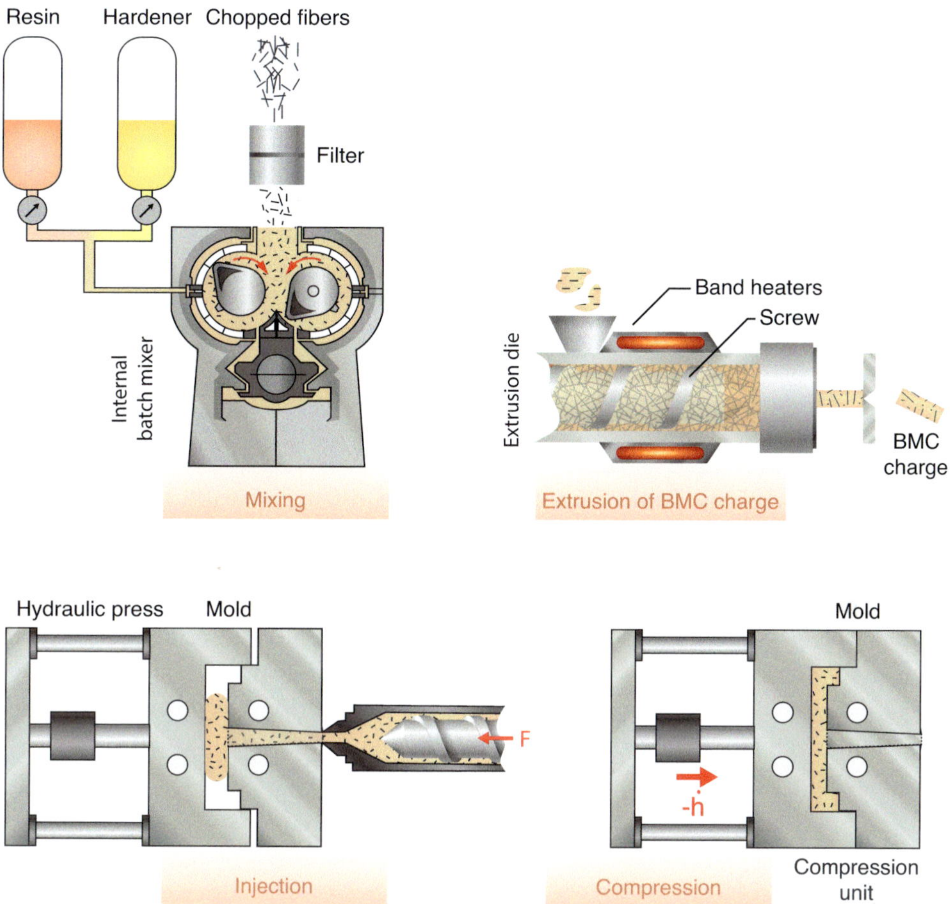

Figure 3.6 BMC injection-compression molding process

■ 3.2 Fiber-Reinforced Thermoplastics

In a thermoplastic polymer, the molecules are not cross-linked with each other. Instead, the individual molecules are held in place by weaker intermolecular forces such as van der Waals forces and hydrogen bonds. In polymer processes, thermoplastics solidify as they are cooled from the molten state and the long molecules are no longer allowed to move freely. When reheated, these materials regain the ability to "flow" and the molecules are able to slide past each other with ease. Thermoplastic polymers are divided into two classes: amorphous and semi-crystalline polymers. Amorphous thermoplastics have molecules that remain in disorder as they cool, leading to materials with a fairly random molecular structure. An amorphous polymer hardens, or vitrifies, as it is cooled below its glass transition temperature. Semi-crystalline thermoplastics, on the other hand, solidify while forming a crystalline structure. As the material is cooled, the molecules begin to arrange in a regular order once the temperature drops below the melting temperature. However, in semi-crystalline polymers small amorphous regions remain where the molecules are not ordered. These amorphous regions within the semi-crystalline domain lose their "flowability" below their glass transition temperature. Because most semi-crystalline polymers have a glass transition temperature at subzero temperatures, they behave like rubbery or leathery materials at room temperature.

Discontinuous fiber-reinforced thermoplastics have been used in injection or compression molding for a long time. Compared to SMC or BMC, fiber-reinforced thermoplastics show advantages such as higher impact strength, lower cycle times, and recyclability. However, the high viscosity of a thermoplastic resin makes it more difficult to impregnate the fibers and the surface finish is poor compared to SMC. Different processes have been developed and several material combinations exist. Depending on the application, the matrices may be of polypropylene, polyamide, or any other engineering thermoplastic. Currently, glass fibers have the biggest market share, although the applications with carbon fibers are increasing. The following sections summarize the three main processes in discontinuous fiber-reinforced thermoplastics: injection molding, compression molding, and extrusion-compression molding.

3.2.1 Injection Molding

Injection molding is the most important process used to manufacture plastic products. Today, more than one-third of all thermoplastic materials are injection molded and more than half of all polymer processing equipment is for injection molding. The injection molding process is ideally suited to manufacture mass-produced parts of complex shapes requiring precise dimensions. The process goes back to 1872, when the Hyatt brothers patented their stuffing machine to inject cellulose into molds. However, today's injection molding machines are mainly related to the reciprocating screw injection molding machine patented in 1956. A modern injection molding machine with its most important elements is shown in Figure 3.7. The components of the injection molding machine are the plasticating unit, clamping unit, and the mold.

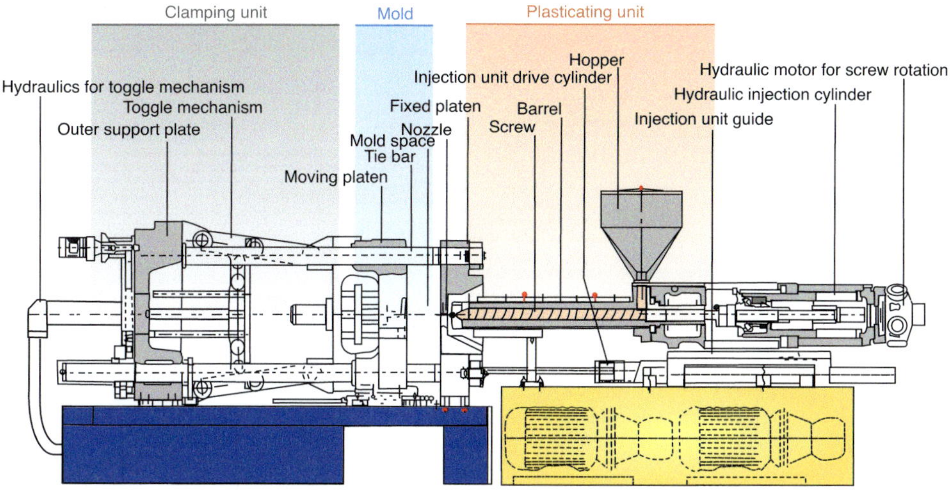

Figure 3.7 Schematic of an injection molding machine

Today, injection molding machines are classified by the following international convention[1]

$$\text{Manufacturer } T\,/\,P \tag{3.1}$$

where T is the clamping force in metric tons and P is defined as

$$P = \frac{V_{max}\,p_{max}}{1000} \tag{3.2}$$

where V_{max} is the maximum shot size in cm^3 and p_{max} is the maximum injection pressure in bar. The clamping force T can be as low as 1 metric ton for small machines, and as high as 11,000 tons.

The sequence of events during the injection molding of a plastic part, as shown in Figure 3.8, is called the injection molding cycle. The cycle begins when the mold closes, followed by the injection of the polymer into the mold cavity. Once the cavity is filled, a holding pressure is maintained to compensate for material shrinkage. In the next step, the screw turns, feeding the next shot to the front of the screw. This causes the screw to retract as the next shot is prepared. Once the part is sufficiently cool, the mold opens and the part is ejected.

[1] The old U. S. convention uses Manufacturer *T-v*, where *T* is the clamping force in British tons and *v* the shot size in ounces of polystyrene.

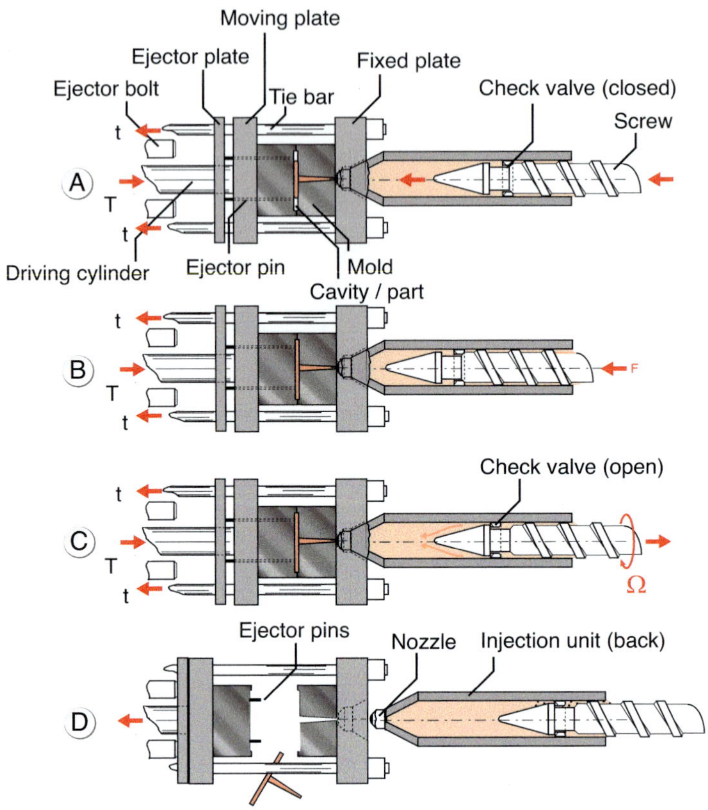

Figure 3.8 Sequence of events during an injection molding cycle

Figure 3.9 schematically illustrates the sequence of events during the injection molding cycle. The figure shows that the cycle time is dominated by the cooling of the part inside the mold cavity. The total cycle time can be calculated using

$$t_{\text{cycle}} = t_{\text{closing}} + t_{\text{cooling}} + t_{\text{ejection}} \tag{3.3}$$

where the closing and ejection times, t_{closing} and t_{ejection}, can last from a fraction of a second to a few seconds, depending on the size of the mold and machine. The cooling times, which dominate the process, depend on the maximum thickness of the part. The cooling time for a plate-like part of thickness h can be estimated using

$$t_{\text{cooling}} = \frac{h^2}{\pi\alpha} \ln\left(\frac{8}{\pi^2} \frac{T_m - T_w}{T_D - T_w} \right)$$

and for a cylindrical geometry of diameter D using

$$t_{\text{cooling}} = \frac{D^2}{23.14\alpha} \ln\left(0.692 \frac{T_m - T_W}{T_D - T_W}\right) \tag{3.4}$$

where T_m represents the temperature of the injected melt, T_W the temperature of the mold wall, T_D the average temperature at ejection, and α the thermal diffusivity.

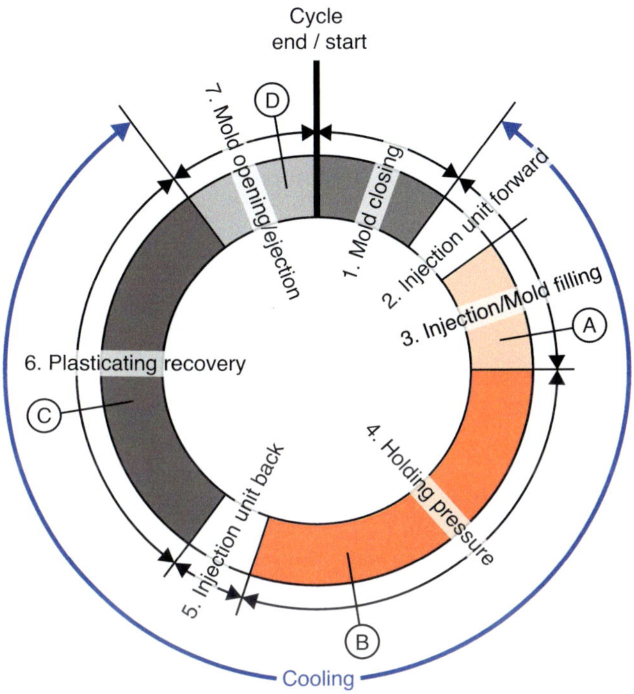

Figure 3.9 Injection molding cycle

Using the average part temperature history and the cavity pressure history, the process can be followed and assessed using the *PvT* diagram as depicted in Figure 3.10 [1]. To follow the process on the *PvT* diagram, we must transfer both the temperature and the pressure at matching times. The diagram reveals four basic processes: an isothermal injection (0–1) with pressure rising to the holding pressure (1–2), an isobaric cooling process during the holding cycle (2–3), an isochoric cooling after the gate freezes with a pressure drop to atmospheric (3–4), and then isobaric cooling to room temperature (4–5).

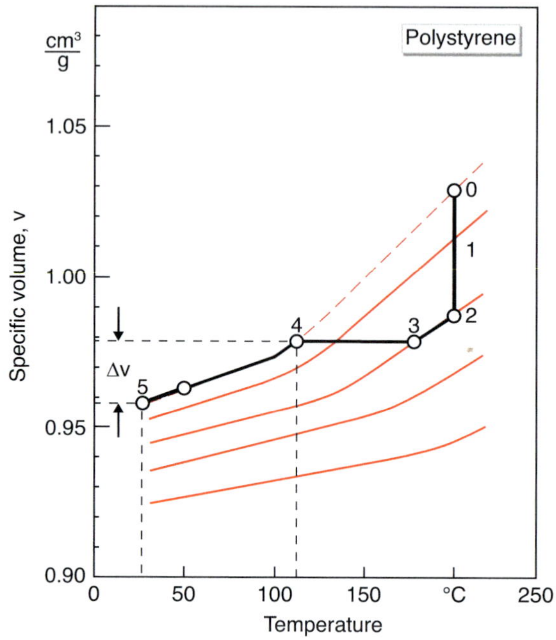

Figure 3.10 Trace of an injection molding cycle in a *PvT* diagram

The point on the *PvT* diagram where the final isobaric cooling begins (4) controls the total part shrinkage, Δv. This point is influenced by the two main processing conditions–the melt temperature, T_m, and the holding pressure, P_H, as depicted in Figure 3.11. Here the process in Figure 3.10 is compared to one with a higher holding pressure. Of course, there is an infinite combination of conditions that render acceptable parts, bound by minimum and maximum temperatures and pressures. Figure 3.12 presents the molding diagram with all limiting conditions. The melt temperature is bound by a low temperature that results in a short shot or unfilled cavity and a high temperature that leads to material degradation. The hold pressure is bound by a low pressure that leads to excessive shrinkage or low part weight, and a high pressure that causes flash.

Flash results when the cavity pressure force exceeds the machine clamping force, leading to melt flow across the mold parting line. The holding pressure determines the corresponding clamping force required to size the injection molding machine. An experienced polymer processing engineer can usually determine which injection molding machine is appropriate for a specific application. For the untrained polymer processing engineer, finding this appropriate holding pressure and its corresponding mold clamping force can be difficult.

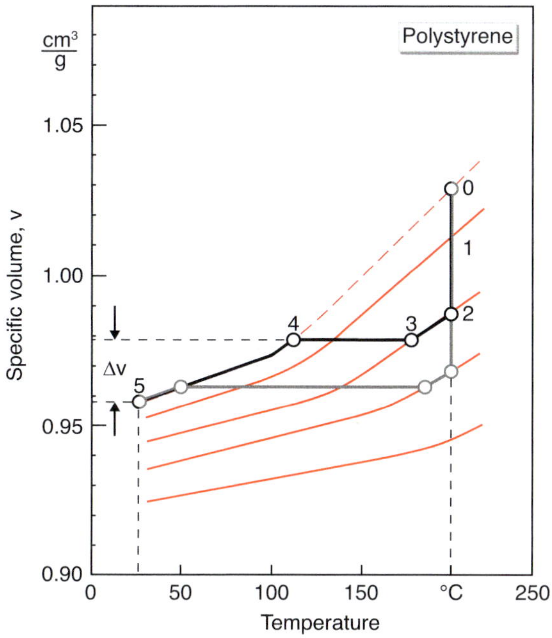

Figure 3.11 Trace of two different injection molding cycles in a *PvT* diagram

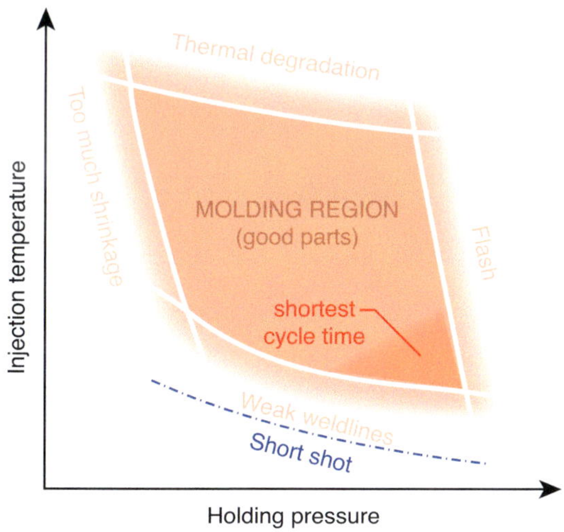

Figure 3.12 The molding diagram

With difficulty one can control and predict the component's shape and residual stresses at room temperature. For example, sink marks in the final product are caused by material shrinkage during cooling, and residual stresses can lead to environmental stress cracking under certain conditions [1].

Warpage in the final product is often caused by processing conditions that lead to asymmetric residual stress distributions through the part thickness. The formation of residual stresses in injection molded parts is attributed to two major coupled factors: cooling and flow stresses. The first and most important is the residual stress formed as a result of rapid cooling, which leads to large temperature variations.

The central point in an injection molding machine is the mold. The mold distributes polymer melt into and throughout the cavities, shapes the part, cools the melt, and ejects the finished product. As depicted in Figure 3.13, the mold is custom-made and consists of the following elements:

- Sprue and runner system
- Gate
- Mold cavity
- Cooling system (thermoplastics)
- Ejector system

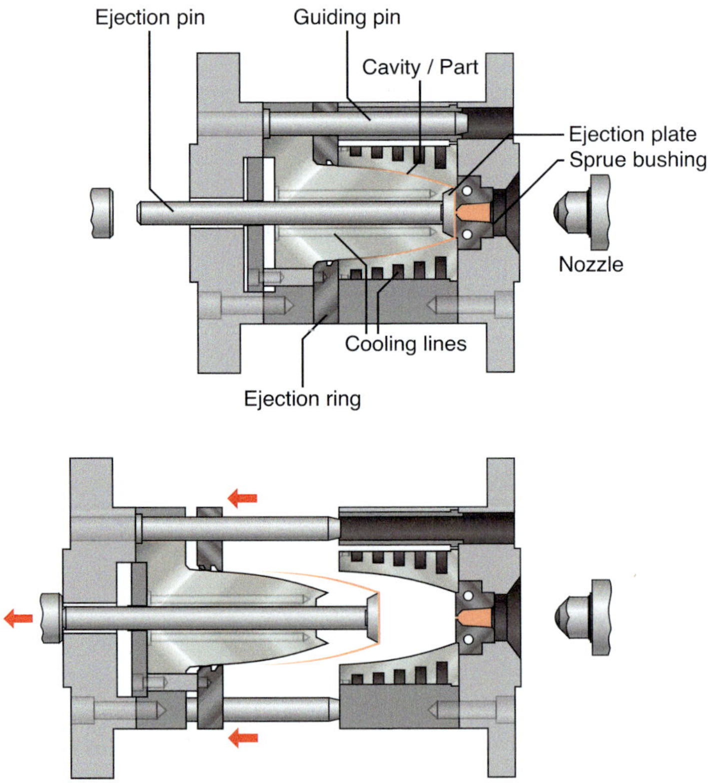

Figure 3.13 Schematic of an injection mold, highlighting the major components of the mold (top) and the mold opening process (bottom) to eject the part

During mold filling, the melt flows through the sprue and is distributed into the cavities by the runners, as seen in Figure 3.14.

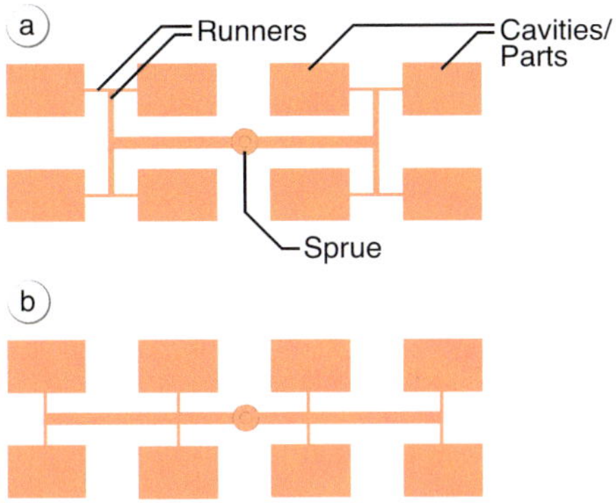

Figure 3.14 Schematic of different runner system arrangements

The runner system in Figure 3.14(a) is symmetric, where all cavities fill at the same time, causing the polymer to fill all cavities in the same way. The disadvantage of this balanced runner system is that the flow paths are long, leading to high material and pressure consumption. On the other hand, the asymmetric runner system shown in Figure 3.14(b) leads to parts of different quality. Equal filling of the mold cavities can also be achieved by varying runner diameters. There are two types of runner systems–cold and hot. Cold runners are ejected with the part, and are trimmed after mold removal. The advantage of the cold runner is lower mold cost. The hot runner keeps the polymer at its melt temperature. The material stays in the runner system after ejection, and is injected into the cavity in the following cycle. There are two types of hot runner system: externally and internally heated. The externally heated runners have a heating element surrounding the runner that keeps the polymer isothermal. The internally heated runners have a heating element running along the center of the runner, maintaining a polymer melt that is warmer at its center and possibly solidified along the outer runner surface. Although a hot runner system considerably increases mold cost, its advantages include elimination of trim and lower pressures for injection.

When large items are injection molded, the sprue sometimes serves as the gate, as shown in Figure 3.15. The sprue must be subsequently trimmed, often requiring further surface finishing. On the other hand, a pin-type gate (Figure 3.15) is a small orifice that connects the sprue or the runners to the mold cavity. The part is easily broken off from such a gate, leaving only a small mark that usually does not require finishing. Other types of gates, also shown in Figure 3.15, are film gates, used to eliminate orientation, and disk or diaphragm gates for symmetric parts such as compact discs.

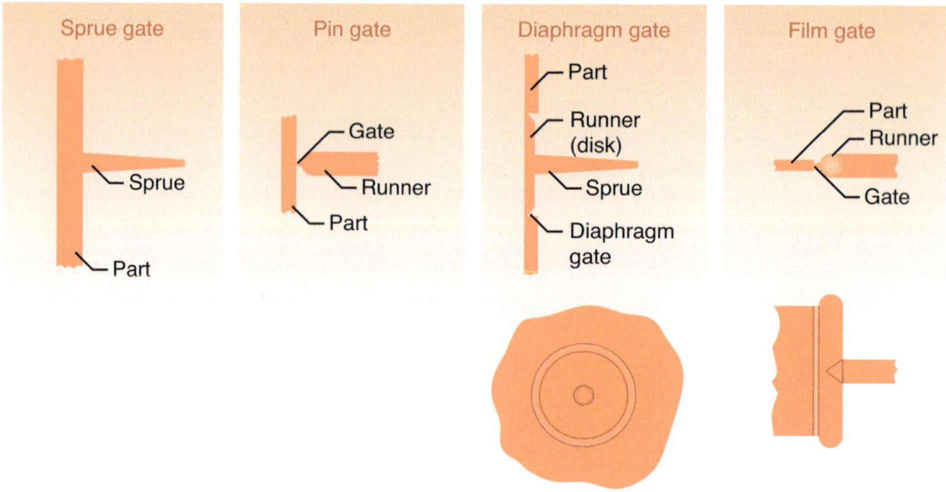

Figure 3.15 Schematic of different gating systems

One of the main challenges with injection molding fiber-filled compounds is the high degree of fiber alignment throughout the part resulting from the high shear rates during mold filling and the long flow paths. As will be discussed in more detail in Chapter 4 of this book, in injection molded parts a distinctive fiber orientation emerges, which can be divided into seven layers, schematically represented in Figure 3.16 [1]. The seven layers may be described as follows:

- Two thin outer layers with a biaxial, random orientation in the plane of the disk
- Two thick layers next to the outer layers, with a main orientation in the flow direction
- Two thin randomly oriented transition layers next to the center core
- One thick center layer, with a main orientation in the circumferential direction

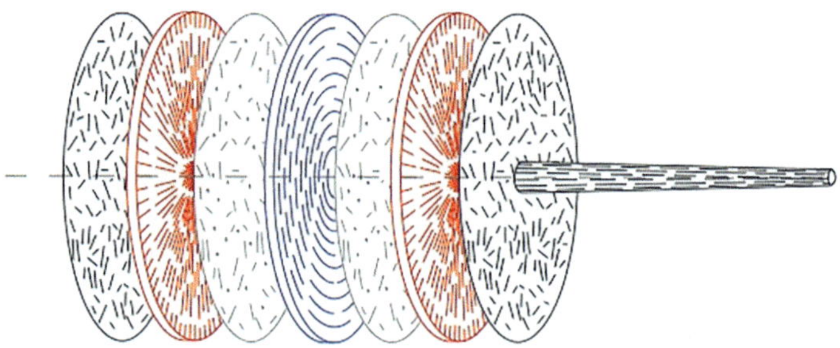

Figure 3.16 Fiber orientation in seven layers of a centrally injected disk

The *fountain flow effect* and the *radial flow* are the main reasons there are high degrees of orientation in injection molded parts. The *fountain flow effect* [2, 3] is caused by the no-slip condition on the mold walls, which forces material from the center of the part to flow outward to the mold surfaces as shown in Figure 3.17 [4]. As the figure schematically represents, the melt flowing inside the cavity freezes upon contact with the cooler mold walls. The melt subsequently entering the cavity flows between the frozen layers, forcing the melt skin at the front to stretch and unroll onto the cool wall, where it freezes instantly. The molecules and the fibers that move past the free flow front are oriented in the flow direction and meet the cooled mold surface, which freezes them into place.

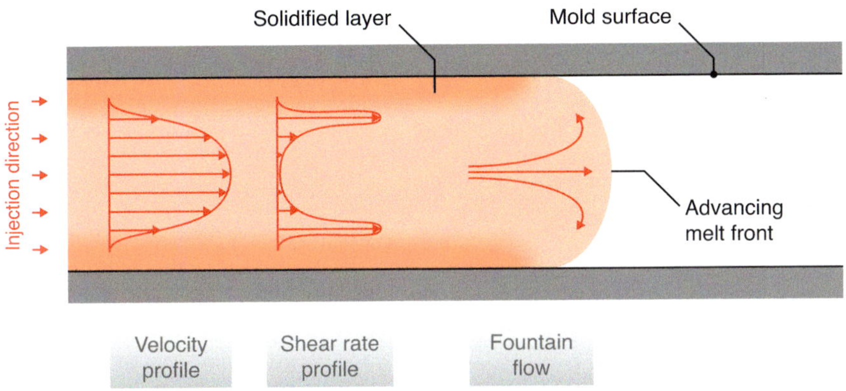

Figure 3.17 Flow and solidification mechanisms through the part thickness in injection molding

Radial flow is the second mechanism that often leads to orientations perpendicular to the flow direction in the central layer of an injection molded part. This mechanism is illustrated in Figure 3.18. As the figure suggests, the material entering through the gate is transversely stretched while it radially expands as it flows away from the gate. This flow is well represented in today's commonly used commercial injection mold filling software programs (see Chapter 7 for more detail).

Besides high degrees of fiber alignment, fiber attrition is another challenge in injection molding. Fiber attrition is a result of high shear stresses. As the polymer is melted, pumped inside the plasticating unit, and forced through the narrow gate, most fibers significantly shorten in length (see Chapter 4 for more detail). This results in reduced mechanical performance in the final part (e.g., stiffness and strength). The fiber damage that occurs in the plasticating region of the injection molding machine and during mold filling can reduce initial fiber length from over 10 mm to a sub-millimeter range [4–6].

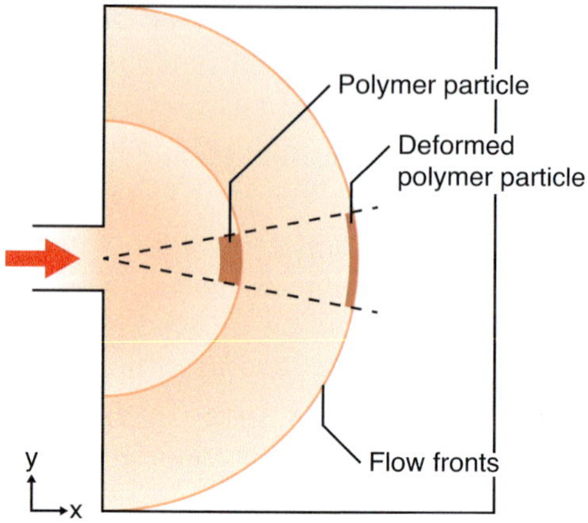

Figure 3.18 Deformation of the polymer during injection molding

3.2.2 Extrusion Compression Molding

Compression molding is widely used in the automotive industry to produce parts that are large, thin, lightweight, strong, and stiff. It is also used in the household goods and electrical industries. In general, the extrusion compression molding process can be divided into two stages. In the first stage, the fiber-filled compound is plasticized in an extruder and the charge is placed in a mold cavity. Subsequently, the mold closes and the material is compression molded into the final shape. Today, there are various extrusion compression molding systems available, ranging from in-line direct processing of long fiber-reinforced thermoplastic (D-LFT) to the extrusion of prefabricated long fiber-reinforced thermoplastic granulates (LFT-G). The main distinction between D-LFT and LFT-G is the type of material initially used. Next, both processes are explained and schematics of the two variations of the LFT molding process are provided.

In the LFT-G process, the feed material is pellets that consist of a thermoplastic matrix and incorporated fibers. As depicted in Figure 3.19, the LFT-G process starts by feeding the semi-finished pellets into an extruder to plasticize and homogenize the compound. If necessary, additives are added during the plasticating phase. In the second stage of the process, the compound is extruded through a die and cut into a charge for the molding stage.

In the D-LFT process, the semi-finished product stage of the LFT-G process is skipped. The base materials for this process are raw glass fibers and thermoplastic pellets. The materials are fed into the extruder in subsequent steps as depicted in Figure 3.20. In a typical D-LFT process, the resin is fed into a single or twin screw extruder, where it is melted and mixed together with additives, pigments, and fillers to obtain a homogeneous material. The glass fibers are fed from continuous rovings and added to the extruder. The fibers are pulled in by the rotation of the screw and cut downstream in the barrel. After the mixing phase, the compound is extruded, cut into a strand–called the LFT charge–and placed into a mold where it is shaped into a final part.

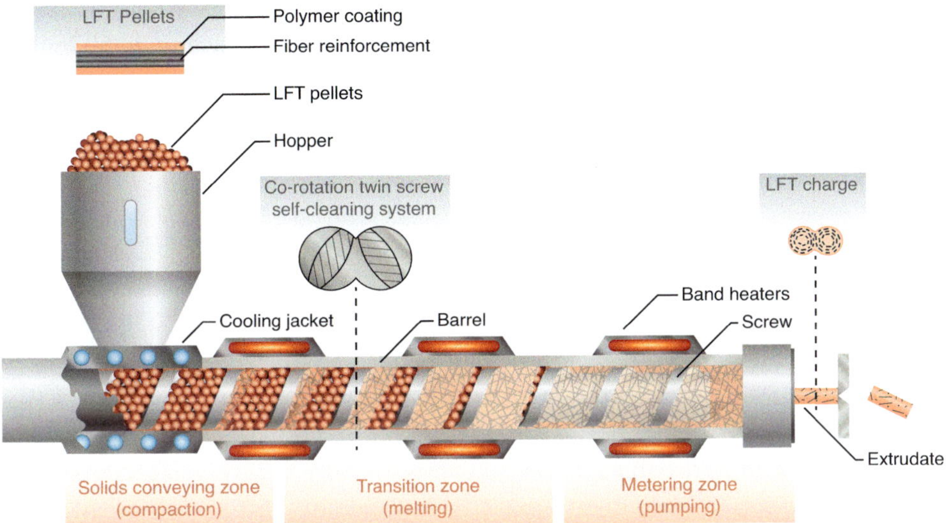

Figure 3.19 Plasticating stage of processing long fiber-reinforced thermoplastic granulates (LFT-G), depicting the process from raw pellet feed to cutting the extrudate (LFT charge) for the subsequent compression molding step in a co-rotating twin screw extruder

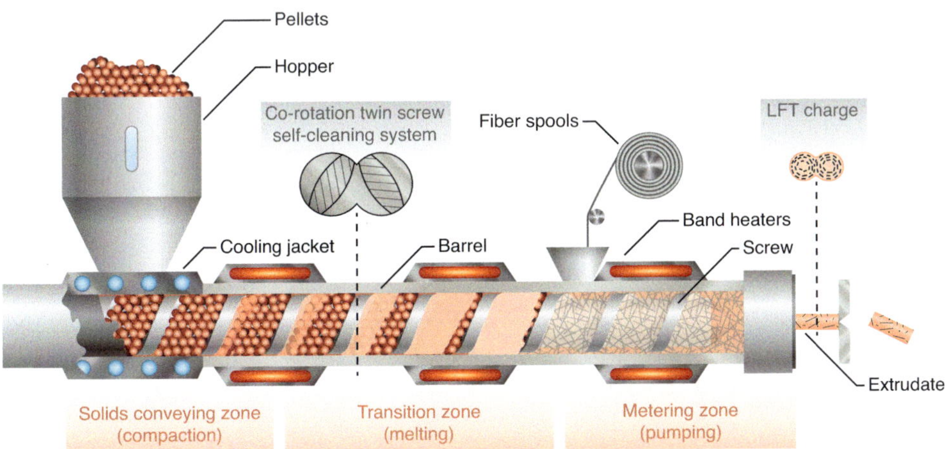

Figure 3.20 Direct processing of long fiber-reinforced thermoplastic (D-LFT). The thermoplastic pellets are fed separately from the reinforcing fibers before forming the LFT charge

The D-LFT process can provide significant advantages over conventional molding processes of pre-compounded materials for several reasons: i) commodity grade materials can be used, ii) longer fiber length distributions can be achieved compared to LFT-G, and iii) the process can be highly automated. Compared to injection molding of fiber-reinforced thermoplastics, compression molding is a gentle process regarding fiber attrition. The fiber length distribution in the finished part is significantly higher and, thus, the mechanical properties in LFT compression molding are better. Particularly in D-LFT, the fibers can reach up to 60 mm in length due to feeding of continuous fibers from rovings.

3.2.3 GMT Compression Molding

Glass mat-reinforced thermoplastics (GMT) are semi-finished sheets produced by combining a thermoplastic resin with continuous, woven, or chopped fiber mats. For most applications, a combination of polypropylene resin and glass fibers is used. The manufacturing process of GMT is divided into two distinctive stages. In a first stage, the semi-finished GMT sheets are produced and, if necessary, packed and stored. In a second stage, the GMT sheets are re-heated and molded into the final form in a compression molding process. Usually, a supplier produces the GMT sheets and sells them to a customer company that processes the sheets and manufactures the final part.

The semi-finished GMT sheets can be produced by melt impregnation of non-woven glass mats, or by mixing sliced fibers with polymer powder in a fluid medium, followed by straining, drying, and consolidation. For the impregnation of the reinforcing-fibers mats with the thermoplastic resins, high temperatures and pressures are needed due to the high viscosity of the resin. Figure 3.21 shows a schematic of the GMT sheet production via melt impregnation.

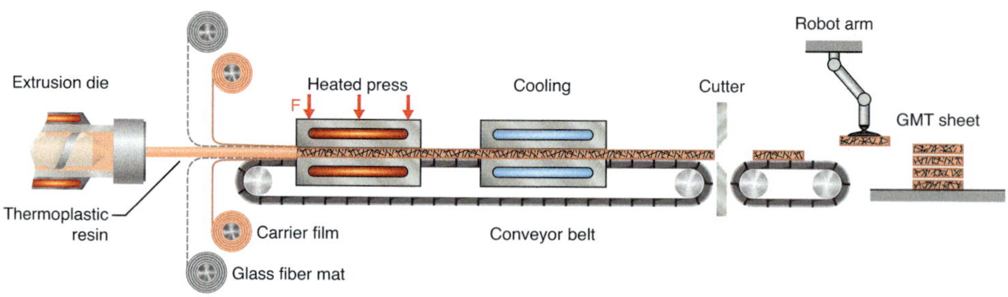

Figure 3.21 GMT sheet production

The second stage is the actual molding of the material, which begins by cutting the GMT sheets and stacking them up to a charge. The charge is then guided on a conveyor belt through a heating unit, which heats the material to the molten state. The heating step is critical in a GMT process since a homogeneous temperature distribution through the charge is required for reprocessing. However, the heating step has to be completed rapidly to maintain short cycle times for a cost-efficient process. The heating unit may consist of radiative heating, contact heating, convective heating, or a combination of all three. After heating, the charge is lifted from the conveyor belt and placed on the cooler mold surface. The mold is closed rapidly until both mold halves touch the charge. Then, the mold slowly compresses the charge, filling the mold cavity as the part cools down. When the mold cavity is completely filled, the part is cooled further under pressure, allowing it to attain its structural integrity. Once the part reaches dimensional stability, the mold opens and the finished part is removed from the mold cavity. The compression molding process of GMT is schematically depicted in Figure 3.22.

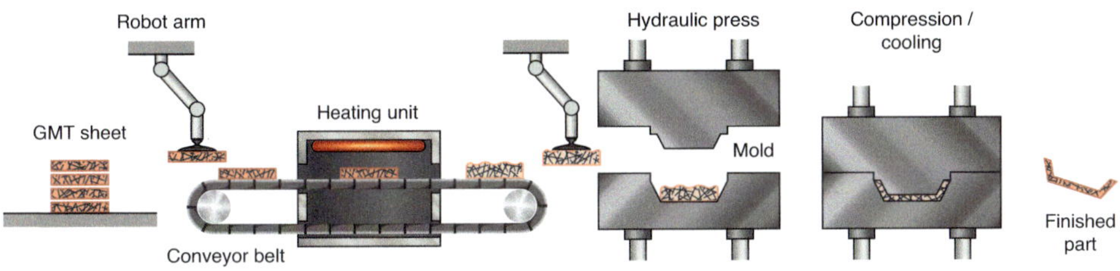

Figure 3.22 GMT compression molding process

GMT compression molding process consists of the steps illustrated in the process cycle in Figure 3.23. The figure clearly shows that the dominant event is the heating phase of the GMT charge. Usually, the heating cycle takes between 60 s and 90 s, but can be even longer in some cases.

GMT compression molding is a widely used process to manufacture semi-structural parts due to the mechanical properties that can be achieved. Due to the low cycle time and the high automation of the process, GMT compression molding offers economic advantages in large batch operations. In the automotive industry, typical applications are front ends, bumper beams, seat structures, under-body panels, and battery trays. However, parts made of GMT do not offer a class-A surface finish, unlike parts manufactured of SMC. Thus, GMT is limited to applications where the surface finish is not important.

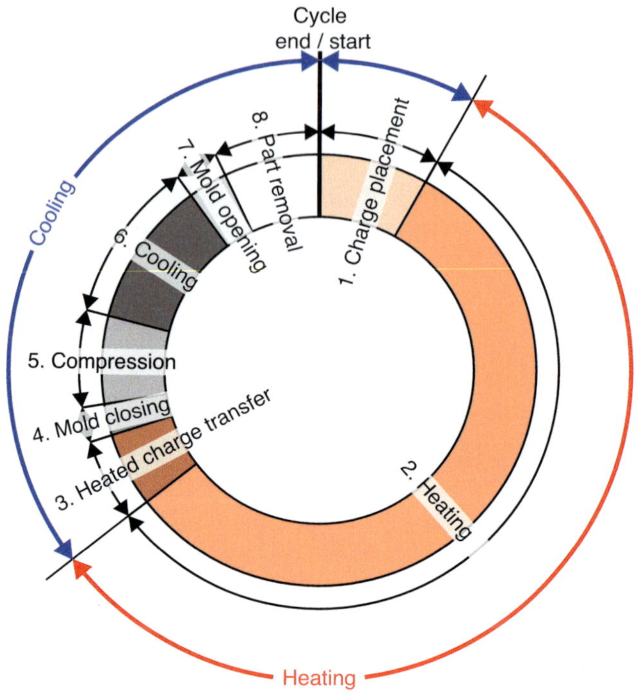

Figure 3.23 GMT compression molding cycle

■ 3.3 Additive Manufacturing

Additive manufacturing (AM) or 3D printing describes a class of manufacturing technologies that produces near net-shape parts directly from CAD data [7]. The original idea of AM is more than 100 years old with a patent granted to J. E. Blanther in 1892 [8] for a method of producing topographical contour maps by cutting individual wax sheets followed by stacking the wax sheets layer by layer to generate a three-dimensional (3D) topography. Currently, a variety of AM technologies exist and more are emerging, which share the fundamental principle of producing 3D parts layer by layer from a digitally designed CAD model.

The general additive manufacturing process is schematically presented in Figure 3.24. The process begins by transforming a CAD model into a surface-based file format such as an STL format, which is still a standard today. The STL file comprises the complex CAD shape as a surface approximated with triangles and the 3D shape is divided into two-dimensional (2D) slices, which are then physically reproduced using the AM process. The thickness of the slices depends on the AM technology and determines the resolution of the printed part, which can be a few nanometers to some millimeters.

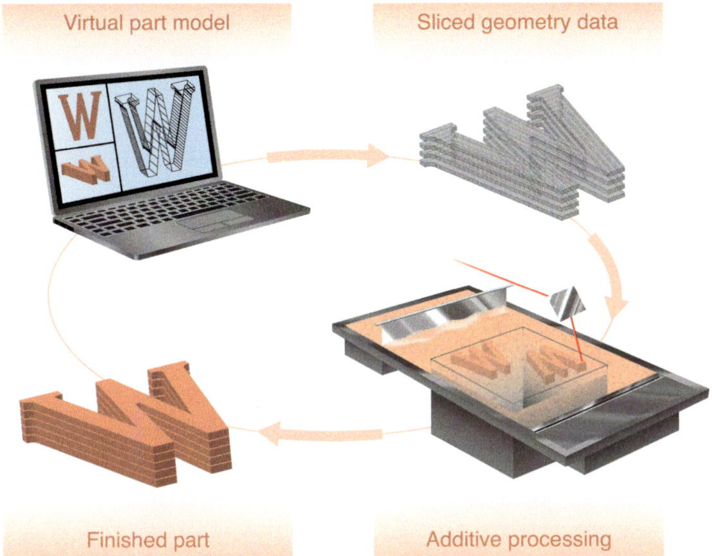

Figure 3.24 Basic workflow of typical additive manufacturing techniques

The material extrusion or fused filament fabrication (FFF) technique, often known by the trade name abbreviation for fused deposition modeling, FDM™, is the most widely used additive manufacturing technique for polymer composite materials. In the FFF process, a polymer filament is molten and extruded as a bead through a heated nozzle, as schematically depicted in Figure 3.25. The applied filament fuses to the preliminary layer and subsequently cools down and solidifies. The heart of the material extrusion process is the extruder, also known as the liquefier, which melts the polymer and pushes it through the capillary to form the bead. Today, do-it-yourself FFF printer kits can be purchased for less than US$200, which leads to a completely new way of usage and knowledge of this technology. The application areas of plastic FFF can be seen in many businesses, from the production of small implants to big components in vehicles.

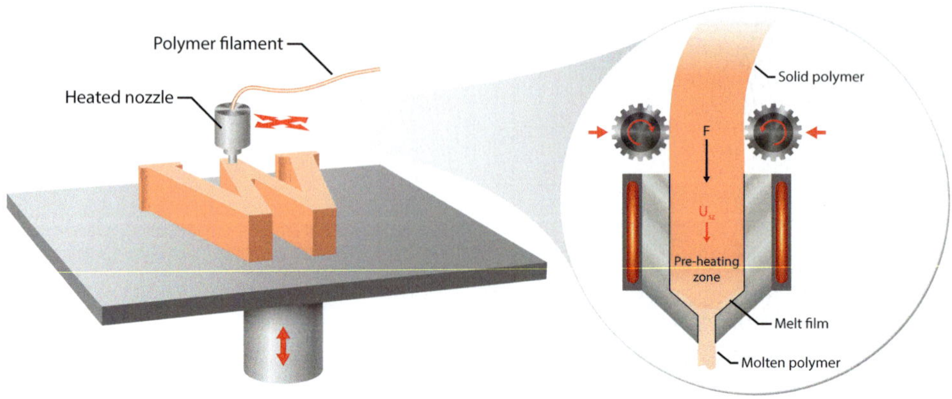

Figure 3.25 Schematic diagram of the fused filament fabrication process (left) and a detailed view of the printing head

The feed material for FFF is commonly supplied as polymer filament and a wide variety of polymer types are available, including acrylonitrile butadiene styrene (ABS), polylactic acid (PLA), high-impact polystyrene (HIPS), thermoplastic polyurethane (TPU), polyamides (PA), and polyetheretherketone (PEEK). Recently, fiber and particle filled filaments for FFF enjoy large interest within the AM industry to manufacture parts with enhanced mechanical properties or functionalize the properties of the composite (e.g., to increase thermal conductivity, as shown in Figure 3.26 with PA6 filament filled with discontinuous copper fibers [9]). Short glass or carbon fiber-reinforced are very prominent because of the additional stiffness and strength added to the manufactured part. Carbon fibers are of particular interest in FFF, because the fibers reduce the thermal expansion coefficient of the composite, reducing the residual stresses created by the layerwise deposition of the beads. Overall, the addition of reinforcing fillers has great potential for pushing the envelope for FFF to a wider range of applications of functional parts.

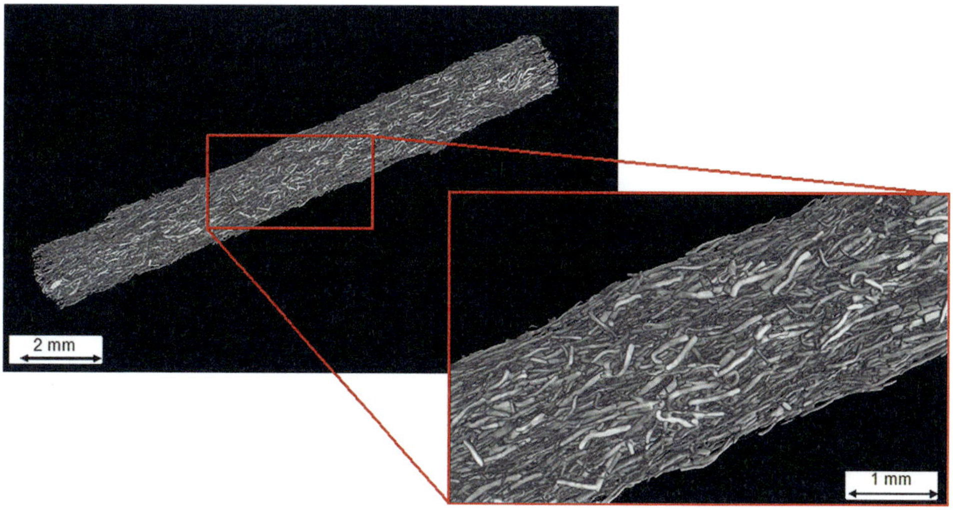

Figure 3.26 3D rendering of X-ray micro computed tomography scans of a discontinuous copper fiber-reinforced PA6 filament

3.4 Vacuum Bagging Techniques

This family of molding processes involves the use of vacuum to impregnate a fiber structure wedged between a vacuum bag and a rigid mold surface. Here, the pressure differential applied by the vacuum forces the resin through the fibrous structure, filling the air gaps between neighboring fibers with the polymer matrix. While most vacuum bagging processes involve continuous fiber structures such as fabrics and non-woven fiber mats, some include chopped fiber systems.

3.4.1 Fiber Spray-up Molding

In fiber spray-up molding, discontinuous fibers and thermoset resin aerosol are simultaneously sprayed onto a mold surface, as schematically depicted in Figure 3.27. Before the process begins, the mold surface has to be prepared. Initially, the mold is coated by a mold-release agent followed by a layer of gel-coat, which is used to provide a high-quality finish on the visible surface of the finished part. As soon as the gel coat hardens, the deposition of the compound begins. Glass fibers are fed into the spray from continuous glass fiber rovings, where the fibers are cut to lengths between 20 mm and 50 mm. In the spray gun, the fibers and the aerosol resin are mixed, which leads to a partial impregnation of the fibers. The mixture of fibers and resin is sprayed onto the gel coat layer and, depending on performance requirements, the thickness can be varied. The orientation of the fibers in the final part depends on the position and angle of the spraying guns, which has to be taken into account in the part design.

Once the surface is sprayed, rollers are used to consolidate the compound, remove entrapped air, and ensure good fiber wetting. Alternatively, the consolidation can be achieved by imposing a vacuum between the mold surface and a film, or vacuum bag, which is placed over the fiber and resin mixture. The resin is allowed to cure after consolidation before demolding.

Fiber spray-up molding has been a popular process in the manufacture of large surface parts such as truck wind-deflectors and boats. As with hand-layup techniques, fiber spray-up molding is being replaced by closed-mold systems such as RTM or vacuum-assisted resin infusion techniques to reduce volatile emissions.

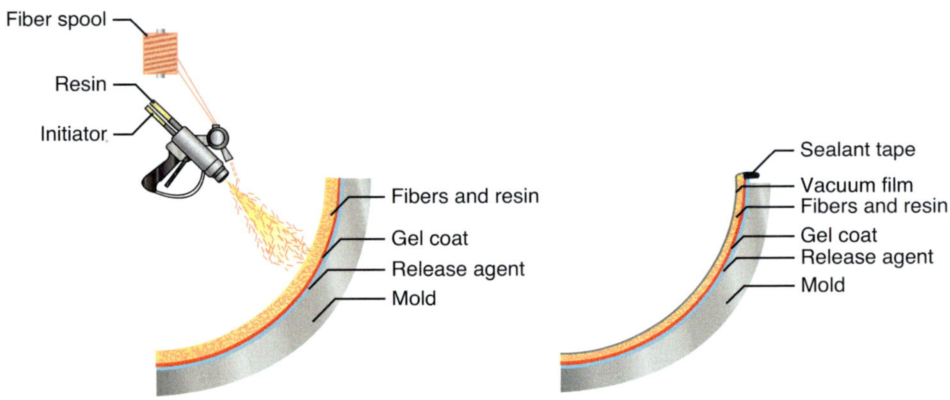

Figure 3.27 Spray-up molding process: schematic of the process (left) and construction of the layup (right)

3.4.2 Hand-Layup Molding

One of the oldest resin impregnation techniques that uses a vacuum bag is hand-layup molding. There are numerous continuous non-woven fiber mat structures as well as woven textiles made of glass and carbon fibers available for hand-layup. For hand-layup molding, the preparation of the mold is crucial for process robustness and final surface quality, and an airtight system is mandatory for the allowable pressure drop when under vacuum. A mold with a smooth surface will yield a homogenous outer layer in the finished part.

The process begins by applying a mold release agent to the mold surface to allow an easy release of the finished part. Next, a gel coat is applied to provide a high-quality finish on the visible surface of the finished part. The fiber layers are cut to predetermined dimensions, oriented, and stacked on a concave or convex mold. The fiber layers follow a layup schedule composed of different types of fiber weaves and mats depending on the specific application and forces experienced by the final part. The finished fiber layup is referred to as a preform. Once the preforms are in place, the resin is applied over the fiber mats using a roller. To ensure full impregnation of the fiber preforms, a film or vacuum bag is placed over the preform and resin. After sealing the film around the edges, a vacuum is applied between the mold surface and the vacuum bag, forcing the resin into the preform and so maximizing the impregnation. In some cases, the vacuum-bagging step is skipped and a roller is used to consolidate the resin into the fiber mats. This variation leads to inferior resin impregnation, which exhibits more air entrapment and voiding. Figure 3.28 shows the basic setup in a hand-layup molding process.

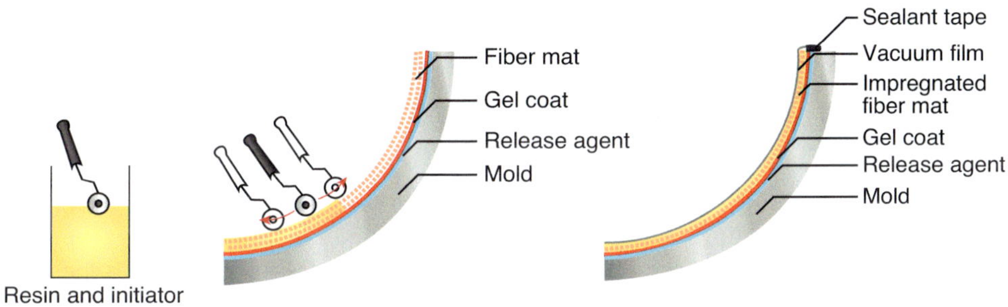

Figure 3.28 Hand-layup molding process: schematic of the process (left) and construction of the layup (right)

3.4.3 Vacuum-Assisted Resin Infusion

Vacuum-assisted resin infusion (VARI) techniques offer cost advantages over layup techniques that process prepregs, a semi-finished material that consists of fiber layups pre-impregnated with resin. In the VARI technique, a continuous fiber reinforcement, or a chopped fiber system, is impregnated with a low viscosity liquid thermoset resin by forcing the resin into a fibrous structure trapped in a vacuum formed between the mold surface and flexible plastic film or vacuum bag. Similar to the hand layup molding technique, the mold surface preparation is important to create a perfect seal and to render a part with a high quality surface. In the VARI molding technique, the fiber preform is cut and placed in layers over the mold surface. The fiber layers are then covered with three additional layers: i) a treated nylon release layer, sometimes referred to as peel-ply, ii) a flow media that increases the permeability of the resin through the fibers, and iii) a vacuum bag (Figure 3.29).

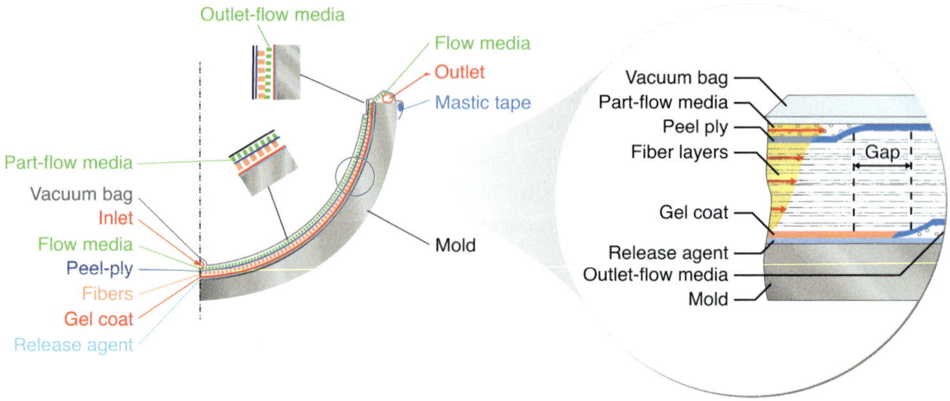

Figure 3.29 Vacuum-assisted resin infusion schematic

Due to the high permeability of the flow media, velocity profiles in VARI exhibit an incline through the thickness of the layers; as the resin streams through the flow media, it advances faster near the vacuum bag and slower near the mold surface. To compensate for this ramp effect, two independent flow media layers are used. The first layer, or the outlet-flow media, is placed between the fiber layup and the mold surface in the last region the resin reaches, often in an area that is discarded after the part cures. The second layer, or part-flow media, is placed over the peel-ply layer. The part-flow media covers the entire mold surface, except for the region that contains the outlet-flow media. A small gap is left between the outlet-flow media and the part-flow media to create a resistance in the flow and ensure an even resin front, synchronizing the flow pattern across the thickness of the preform layers (see Figure 3.30). The gap forces the flow on the top of the preform to slow down, allowing the flow front near the mold surface to catch up. This technique is required to prevent dry spots near the mold surface in the final part.

After properly setting up the flow media layers, the vacuum bag is sealed around the edges and a vacuum is applied to at least one outlet on the mold. This allows air to be drawn from the system, creating a vacuum between the vacuum bag and the mold surface. Once the system is found free of leakage, or when an acceptable leakage rate is achieved, the catalyzed thermoset resin is degassed to remove trapped air bubbles that occur during the mixing process. The resin is then allowed to flow into the mold through one or multiple inlets. Upon full saturation, the resin supply is sealed off and the part remains under vacuum until the resin cures.

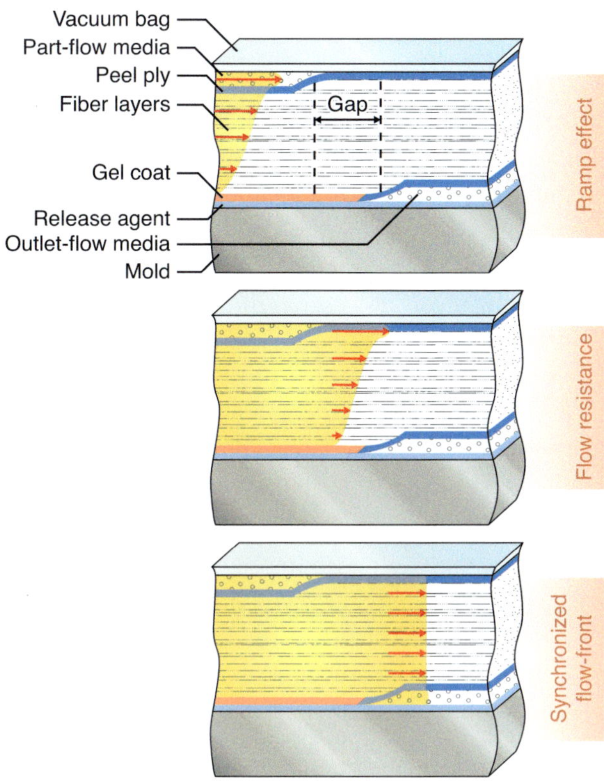

Figure 3.30 Illustration of the ramp effect

Compared to other open mold processes, vacuum infusion minimizes air voids in the part; it generates fiber-to-resin content of up to 60% and reduces volatile organic compound (VOC) emissions [11]. A major disadvantage associated with this process is the resulting waste: the peel-ply, flow media, and vacuum bag material are single-use. Advancements that reduce waste incorporate a reusable silicon vacuum bag or integral flow media that remains in the final part. Another disadvantage is the difficulty to control the amount of resin in the final part due to the flexible vacuum bag that can bound the top layer. Only highly controlled vacuum systems will render repeatable parts. VARI has long cycle times due to the manual operations necessary, making it a technology for small batch productions rather than applications that require high production rates.

3.4.4 Resin Transfer Molding Techniques

In contrast to the VARI techniques that use a vacuum and only one mold half, resin transfer molding (RTM) and structural reaction injection molding (S-RIM) make use of two mold halves with a controlled thickness and controlled mold surface on both sides. RTM and S-RIM are two similar liquid composite molding (LCM) processes that are well suited to manufacture medium-to-large, complex, lightweight, and high-performance composite components primarily for the aerospace and automotive industries. In these processes, the fiber mat preform is placed in a closed mold to be impregnated by a low viscosity, reactive liquid resin in a transfer or injection process. The process characteristics of RTM and S-RIM differ in the resins used, mixing and injection set up, mold requirement, cycle time, fiber volume fraction, and suitable production volume. However, both LCM processes are recognized as the most feasible and structurally efficient approach to mass-produce lightweight, high-strength, and low-cost structural composite components.

The preform is an assembly of dry (unimpregnated) reinforcement media, preshaped through a shaping or draping process and assembled with urethane-foamed cores into a three-dimensional skeleton of the actual part. A wide variety of reinforcements are available, such as woven and non-woven fiber products, die-cut continuous strand mats (CSM), random mats consisting of continuous or discontinuous fibers laid randomly by a binder adhesive, knit braiding or two- and three-dimensionally braided products, or hybrid preforms made of layers of different types of media. The selection of the preform architecture depends on the desired structural performance, processability, long-term durability, and cost.

The classic RTM process uses a wide range of thermally activated resins, but due to their low cost, polyesters are the most commonly used resins in RTM. Other resins used include epoxies, vinyl esters, acrylic/polyester hybrids, acrylamide resins, and methylmethacrylate vinyl esters. These thermally activated resin systems do not react appreciably at the initial resin storage temperature and rely on the heated mold walls to accelerate the chemical reaction. As a result, the filling times for RTM can be as long as 15 minutes, and the cycle time can be of an hour or longer, depending on the resin and application. Slow injection and reaction rates of some RTM systems can be used to achieve excellent impregnation of complex shaped parts. In addition, the low viscosity of the resin and the slow injection rate result in low injection-pressure and clamp-force requirements. As a result, RTM allows the use of so-called soft tools (e.g., wood backing epoxy molds or aluminum molds), which require significantly lower investments compared to competitive manufacturing systems. Because of the relatively long cycle time, RTM is generally limited to a low-volume production (i.e., less than 10,000 parts). Figure 3.31 shows a resin transfer molding setup.

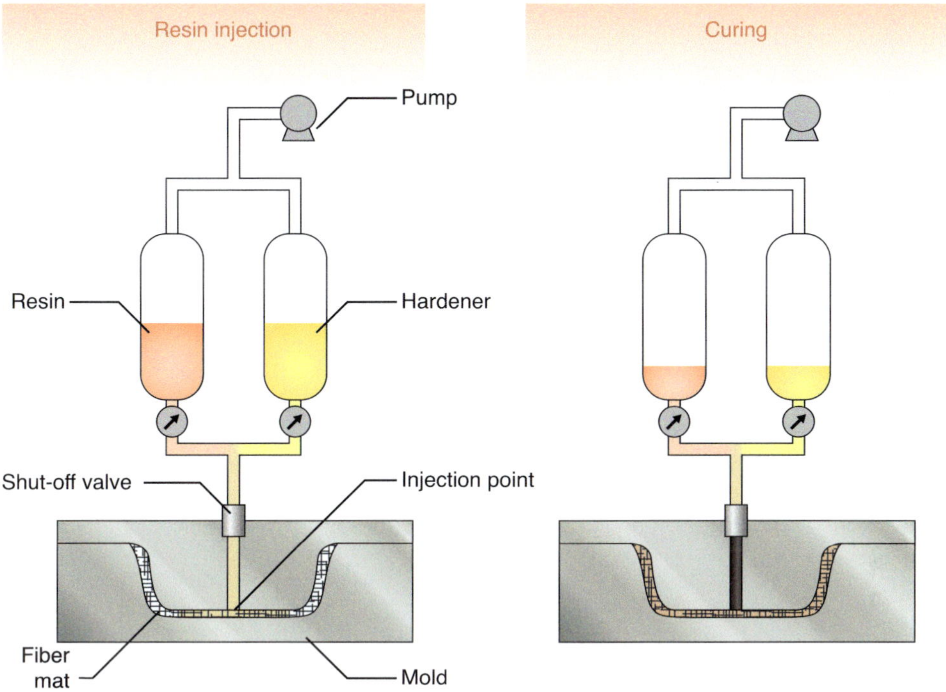

Figure 3.31 Resin transfer molding process

The main challenges with RTM and S-RIM technologies are the control of resin flow and fiber impregnation. Poor fiber impregnation can lead to voids between fiber and matrix, resulting in localized stress concentrations and possible crack initiation sites. One way to overcome these challenges is to use an assisting vacuum to force the resin through the network of fibers. This technique, schematically depicted in Figure 3.32, is referred to as vacuum-assisted resin transfer molding (VA-RTM).

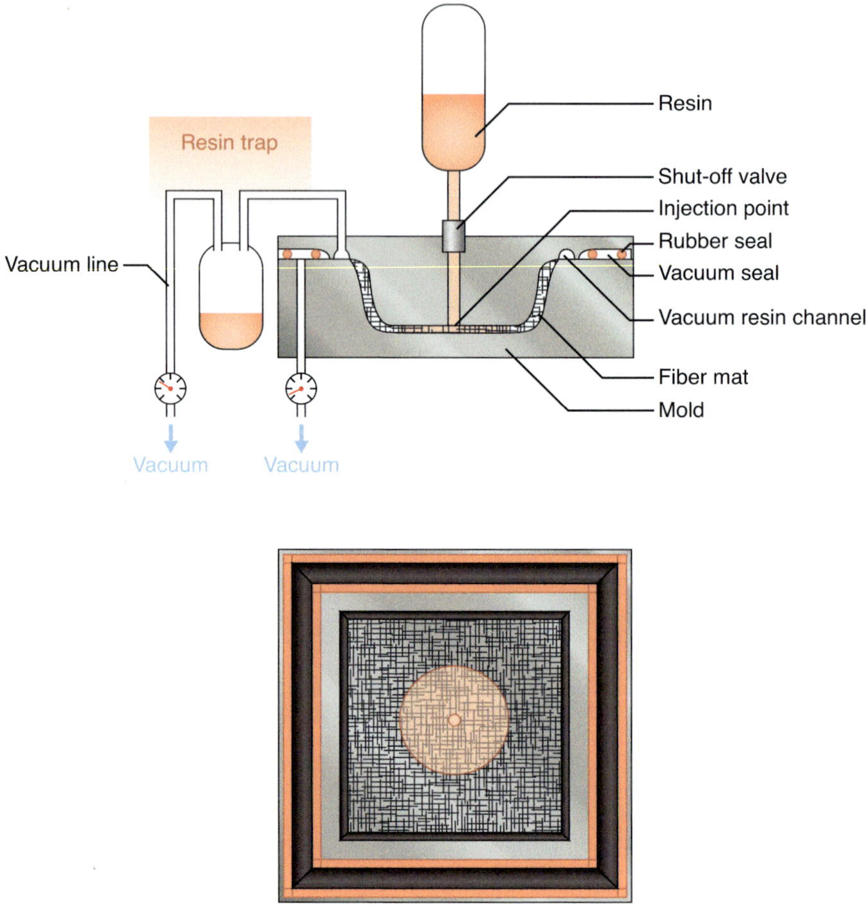

Figure 3.32 Vacuum-assisted resin transfer molding process

In contrast to the classic RTM and VA-RTM, compression resin transfer molding (C-RTM) uses a partially opened mold. The mold halves in C-RTM have a gap between them, which is sufficiently large to inject the resin into the cavity. Once a specific amount of resin is injected, the mold gap is closed, forcing the resin into the porous preform structure. Once the mold is closed, the preform is fully impregnated, and the resin is allowed to cure before demolding occurs. The main advantage of this process is the reduced time for resin injection and impregnation, allowing usage of fast curing systems. Figure 3.33 shows the compression resin transfer molding process.

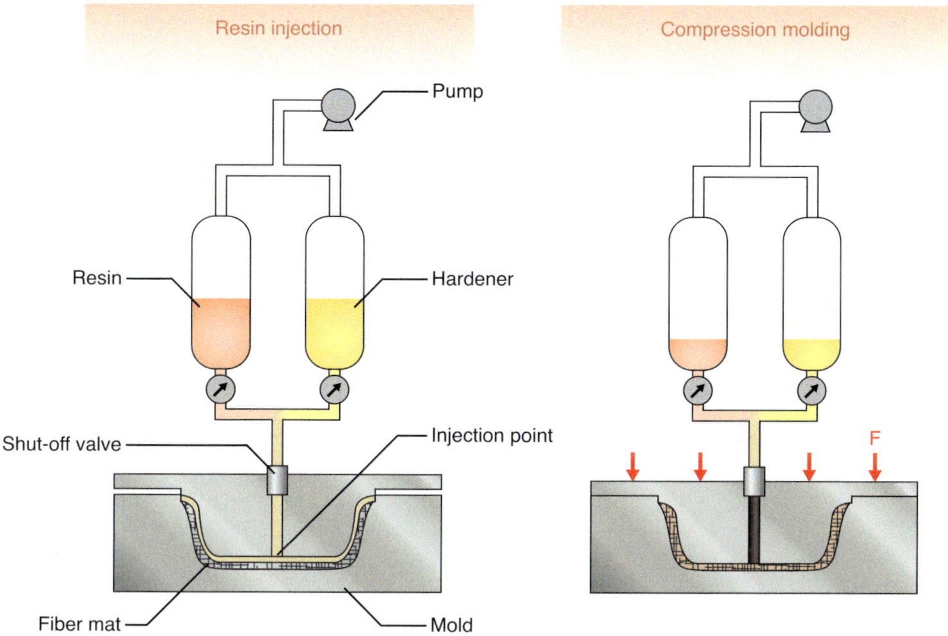

Figure 3.33 Compression resin transfer molding process

Finally, a variation of a resin transfer molding technique is the structural reaction injection molding (S-RIM) technique. The resins used in S-RIM are mixing activated. Common resins used for S-RIM include urethane, acrylamate, and dicyclopentadiene. S-RIM derives its name from the RIM process, from which its resin chemistry and injection techniques are adapted. The chemical reaction is activated by impingement mixing of two highly reactive components in a special mixing head under high pressure. Upon mixing, the mixture is subsequently injected into the mold at a lower pressure. The resin starts to cure as it impregnates the preform and forms the matrix of the composite. Due to the fast reaction rate and rapid buildup of viscosity from curing, the cavity has to be filled within a few seconds. In addition, the cycle time is as short as 1 minute.

Flow distances for typical S-RIM applications are therefore limited to 0.6 m to 0.9 m from the inlet gate. Furthermore, the volume fraction and construction of the reinforcement have to be selected carefully to facilitate fast, complete filling before gelation occurs. Because of the high injection rate and short cycle time, S-RIM generally uses steel tools and is suitable for medium- to high-volume production (10,000 to 100,000 parts). Figure 3.34 shows a schematic of the S-RIM process.

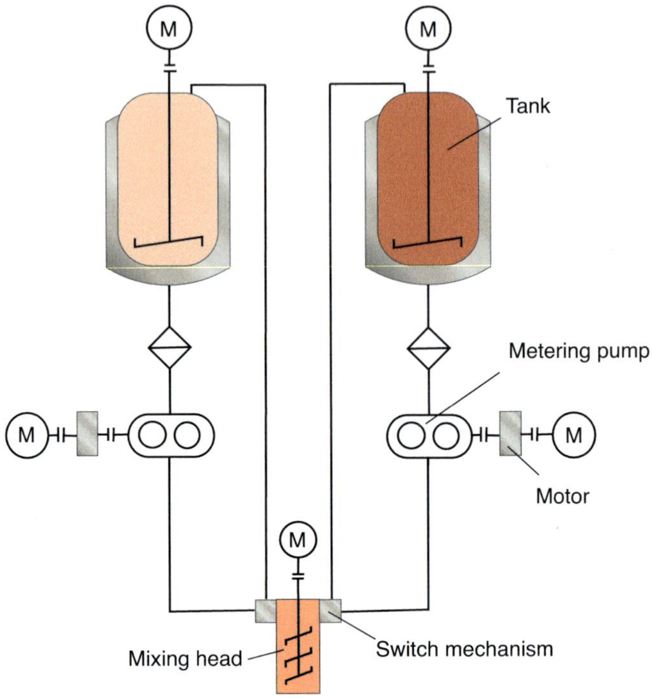

Figure 3.34 Structural reaction injection molding process

Parts made of discontinuous fiber-reinforced composites, such as LFT and SMC, are generally not used for structural applications due to lower mechanical performances. On the other side of the spectrum, composite manufacturing technologies such as prepreg layup can be used to produce high-strength lightweight parts used in the aerospace industry. However, the prepreg layup technology is slow, expensive, and labor-intensive. RTM and S-RIM offer viable options for producing lightweight, high-performance structural components for both aerospace and automotive industries. In addition, low injection pressure is another process advantage of both RTM and S-RIM. The injection pressure varies with the permeability of the fiber mat, part geometry, and the injection rate. A low injection rate requires inje tion pressures between 70 kPa and 140 kPa, and a fiber content of 10% to 20% by volume. A rapid mold filling requires injection pressures between 700 kPa and 1400 kPa, and higher fiber contents can be reached, varying from 30% to 50% by volume. Furthermore, the RTM and S-RIM processes employ closed molds, which reduce or eliminate the emission of hazardous vapor. Other advantages include more repeatable part thickness and minimal trimming and de-flashing of the final part.

3.5 Processes for Hybrid Structures

The development of hybrid composites has been a focus of research efforts in order to further improve the performance of conventional fiber-reinforced composites through combination of material classes and processes. Hybrid composites contain more than one type of reinforcing phase, creating an advanced mixture of materials with unique features. The purpose of hybridization is to improve properties such as cost, weight, failure behavior, and fatigue characteristics. Hybrid composites can be tailored so that the part meets the design requirements in a more cost-effective way than conventional composites. Several different types of hybrid composites exist and are distinguished by the arrangement of the reinforcing constituent. Interply composites consist of alternating layers of two or more materials that are stacked in a regular mode. Sandwich hybrids, also known as core-shell composites, consist of a core material that is sandwiched between two layers of another material. More hybrid composite classes are emerging for niche applications, such composites that are reinforced with a mix of continuous metal and glass fibers.

The most prominent class of hybrid composites consists of fiber-reinforced composites and metal inserts. These plastic-metal hybrid composites are made by preforming a metal sheet, which is over-molded with the polymer to add ribs, attachment points, and other features. This type of hybrid combines the good characteristics of metals such as ductility, impact, and damage tolerances with the benefits of fiber-reinforced composites such as high specific strength, high specific stiffness, and excellent corrosion and fatigue resistance. Plastic-metal hybrid composites are used to replace heavy structural components, particularly for automotive or aerospace applications. One of the first plastic-metal hybrid structures was the front end of the Audi A6 in 1998, which lowered its cost by 10% and weight by 15% [10]. Ford introduced the hybrid front end to their Focus in 1999, making it 20% cheaper and 40% lighter than the steel part it replaced. This front end integrated steel inserts over-molded to create ribs and other features using polyamide 6 reinforced with 30% glass fiber content by weight. The front end is depicted in Figure 3.35.

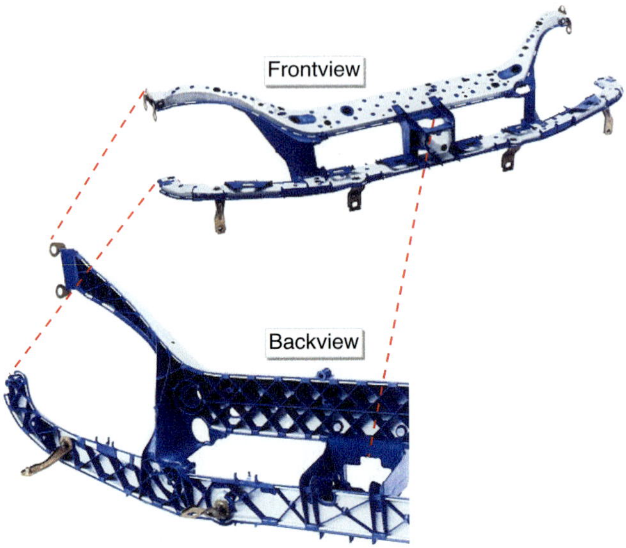

Figure 3.35 Ford Focus hybrid front-end structure made of steel inserts over-molded with polyamide 6 reinforced with glass fibers 30% by weight (courtesy Dynamit Nobel Kunststoffe)

■ References

[1] Osswald, T. A. and Menges, G., *Materials Science of Polymers for Engineers*, 3rd ed., Hanser, Munich (2012)

[2] Leibfried, D., Untersuchungen zum Werkzeugfüllvorgang beim Spritzgiessen von thermoplastischen Kunststoffen, Ph. D. Thesis, RWTH Aachen University, Aachen, Germany (1970)

[3] Tadmor, Z., Molecular orientation in injection molding, *J. Appl. Polym. Sci.*, vol. 18, no. 6, pp. 1753–1772 (1974)

[4] Lafranche, E., Injection moulding of long glass fibre reinforced polyamide 6-6: guidelines to improve flexural properties, *EXPRESS Polym. Lett.*, vol. 1, no. 7, pp. 456–466 (2007)

[5] El Barche, N., Candal, M. V., Yanev, A., and Osswald, T. A., Experimental Study of Fiber Attrition within a Long Fiber Glass-Reinforced PP under Controlled Conditions, in Proceedings of the Annual Technical Conference, Las Vegas, NV (2014)

[6] Goris, S., Back, T., Yanev, A., Brands, D., Drummer, D., and Osswald, T. A., A novel fiber length measurement technique for discontinuous fiber-reinforced composites: A comparative study with existing methods, *Polym. Compos.* (2017)

[7] Osswald, T. A., *Understanding polymer processing: processes and governing equations*, 2nd ed., Hanser, Munich (2017)

[8] Blanther, J. E., Manufacture of contour relief maps, US473901 (1982)

[9] Mulholland, T., Goris, S., Boxleitner, J., Osswald, T. A., and Rudolph, N., Fiber Orientation Effects in Fused Filament Fabrication of Air-Cooled Heat Exchangers, *JOM*, vol. 70, no. 3, pp. 298–302 (2018)

[10] Al-Sheyyab, A., Light-Weight Hybrid Structures - Process Integration and Optimized Performance, Ph.D. Thesis, University of Erlangen-Nuremberg, Erlangen (2008)

[11] Brouwer, W. D., van Herpt, E. C. F. C. and Labordus, M., Vacuum injection moulding for large structural applications, *Composites Part A: Applied Science and Manufacturing*, 34 (6): 551-8 (2003)

4 Microstructure in Discontinuous Fiber-Reinforced Composites

The performance of a discontinuous fiber-filled composite part is directly linked to the microstructure of the fibers. This microstructure not only affects mechanical behavior, such as stiffness and strength, but also properties such as electrical and thermal conductivity as well as the rheological behavior of a fiber-filled melt. This chapter provides an overview of the present understanding of the process-induced fiber microstructure for discontinuous fiber-reinforced composites. The chapter addresses the three main microstructural property changes, namely, fiber attrition, fiber orientation development, and changes in fiber concentration during processing. First, the underlying physics behind microstructure development during processing is discussed, followed by techniques recently developed to characterize the microstructure development during the process and within a discontinuous fiber-filled composite part.

■ 4.1 Process-Induced Microstructure in Discontinuous Fiber-Reinforced Composites

Fibers embedded within a polymer matrix during processing, such as injection or compression molding, not only tend to rotate, which affects the overall fiber orientation distribution, but also experience stresses that will cause it to deform, break, and segregate, profoundly affecting the mechanical behavior of the molded composite part. This section will present fiber attrition, fiber alignment or orientation, and fiber matrix separation individually, and help the reader to gain some understanding of the fundamental aspects of microstructure development during processing and help raise awareness where the underlying physics are not fully understood.

4.1.1 Fiber Attrition

The superior mechanical performance of fiber-reinforced composites is achieved by transferring the applied load from the matrix to the reinforcing fibers (for more details see Chapter 5). Consequently, the interfacial adhesion as well as the length of the fibers embedded in the matrix determine the maximum transferable load. Preserving a fiber length that is sufficiently long to transfer the load, while maintaining economical process cycles, is a major research objective in academia and industry [1–3]. Fiber attrition refers to the fact that fibers break during processing and that the length of the fibers is inevitably reduced from its initial value. While the initial length of the fibers may be uniform and up to 30 mm for long fiber thermoplastic (LFT) materials, fiber attrition results in a distribution of fiber lengths in the finished part ranging from powder-like fibers (< 0.1 mm) to a small fraction of fibers retaining their initial length. In general, the degree of fiber attrition is influenced by the process, processing parameters, material selection, and mold design [4]. Since mechanical properties depend on the fiber length, it is desired to preserve the length of the fibers as much as possible by using processing conditions that treat the fibers more gently. Along the entire process line, the fibers are subjected to severe stresses, including hydrodynamic effects, fiber–fiber interactions, and fiber–equipment interactions, which can reduce the average fiber length even for LFT materials to only 0.5–1.0 mm in the finished part [1, 2, 5, 6]. Fiber breakage can be attributed to the following mechanisms [4]:

- *Melting effects:* This damage mechanism appears at the beginning of polymer processing when the pellets are partially molten, and the interaction between fibers and pellets leads to bending and shear stresses in the fibers, as illustrated in Figure 4.1 (right). This effect is considered to cause a large percentage of fiber attrition overall and only occurs in the melting zone of the screw [7].
- *Fiber–fiber interaction:* In highly filled material combinations, the fibers collide excessively during processing, which results in length reduction due to fiber bending and friction between fibers, as shown in Figure 4.1 (left).
- *Fiber–melt interaction:* Fibers are embedded in the thermoplastic matrix, which introduces large stresses to the fibers due to the hydrodynamic forces during the entire process; see Figure 4.1 (left).
- *Fiber–wall interaction:* Fibers collide and interact with the solid walls of the equipment. Abrasion and friction between the fibers and the walls lead to stress concentration within the fibers, demonstrated by Figure 4.1 (left).

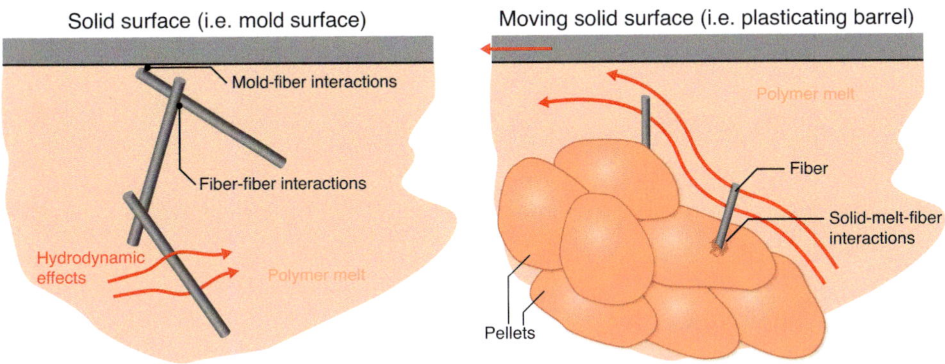

Figure 4.1 Mechanisms of fiber breakage during processing: mold–fiber interactions, fiber–fiber interactions, and hydrodynamic effects

The length of each fiber may vary from 0 to L_{max} and the overall sample consists of N total fibers. The entire population can be represented by a histogram with n bins, where each bin has N_i fibers with the length l_i. The total length of the population $\mathcal{L}$ is defined as

$$\mathcal{L} = \sum_{i=1}^{n} (N_i l_i) \tag{4.1}$$

The obtained distribution is conventionally summarized by calculating representative average values, such as the number-average length and the weight-average length, in order to obtain a concise understanding of the fiber population as a whole. The number-average fiber length, L_N, is calculated as

$$L_N = \frac{\sum_{i=1}^{n} (N_i l_i)}{\sum_{i=1}^{n} (N_i)} \tag{4.2}$$

The weight-average (or length-average) fiber length, L_W, is a second characteristic average of the fiber length distribution, emphasizing the proportion of long fibers in the distribution. This is of particular interest for molded LFT samples since they typically show a length distribution with a distinct peak followed by a long tail decay [8]. It is calculated using

$$L_W = \frac{\sum_{i=1}^{n} (N_i l_i^2)}{\sum_{i=1}^{n} (N_i l_i)} \tag{4.3}$$

The length distribution can be represented by a discrete probability density function to capture the measured data, which could also be fitted to defined continuous probability density functions, such as the Weibull [9] or Burr distribution functions [10]. The discrete probability density function $\mathfrak{n}(l)$ can be defined as

$$\mathfrak{n}(l) = \frac{N_i}{N} \tag{4.4}$$

This distribution function can also be defined in terms of the length-weighted distribution $\mathfrak{w}(l)$ using

$$\mathfrak{w}(l) = \frac{N_i l_i}{\mathcal{L}} \tag{4.5}$$

By integrating $\mathfrak{n}$ over length, the cumulative length distribution $\mathfrak{N}$ can be obtained, which describes the fraction of fibers that are smaller than, or equal to, a given length l_j:

$$\mathfrak{N}(l_i) = Pr(l \leq l_j) = \frac{\sum_{i=1}^{j} N_i}{N} \tag{4.6}$$

Similarly, the length-weighted cumulative distribution $\mathfrak{W}$ is calculated integrating $\mathfrak{w}$ over all lengths using

$$\mathfrak{W}(l_i) = Pr(l \leq l_j) = \frac{\sum_{i=1}^{j} N_i l_i}{\mathcal{L}} \tag{4.7}$$

The cumulative distributions $\mathfrak{N}$ and $\mathfrak{W}$ offer a consistent way to graphically compare different data sets without the need of computed average values. Figure 4.2 illustrates $\mathfrak{n}$ and $\mathfrak{N}$ (top), and $\mathfrak{w}$ and $\mathfrak{W}$ (bottom) for the same data set. The graphs indicate the corresponding average values as well as the median values from the number distribution and length-weighted distribution ($L_{N,50}$ and $L_{W,50}$).

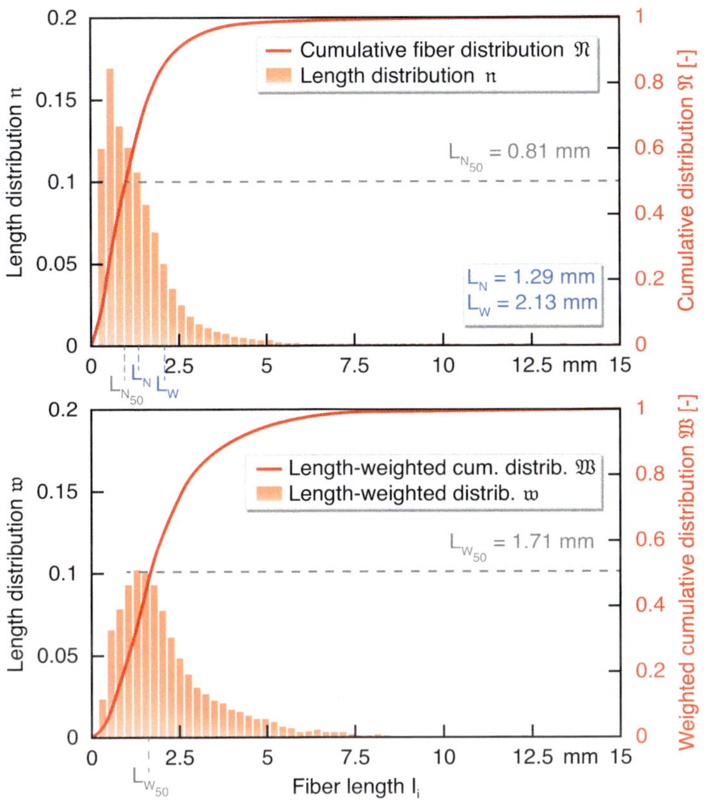

Figure 4.2 Example of a density function 𝔫 and cumulative fiber distribution 𝔑 (top), and a weighted length distribution 𝔴 and cumulative length-weighted distribution 𝔚 (bottom). Both distributions represent the same data set

The majority of fiber breakage is associated with the plasticizing phase and different research groups have studied fiber breakage along the different zones of the screw. Turkovich and Erwin [11] found that most fiber length degradation occurs in the melting zone with only minor attrition in the feed or conveying zone. This might be attributed to partially embedded fibers in the solid pellets, which are bent and sheared off during plasticizing, as previously shown in Figure 4.1. In 2005, Lafranche et al. [12] quantified the fiber length degradation in the plasticizing unit for long glass fiber-reinforced polyamide 6. They measured the fiber length at the different zones of the screw and summarized the results in Figure 4.3. Their results suggest that the remaining fiber length at the nozzle is less than 50% of the initial length (9 mm). Inoue et al. [7] found that the screw design in the compression zone has a substantial impact on fiber breakage. They found that an optimized screw design (low shear screw with a variable pitch) can reduce the fiber breakage compared to a standard screw.

Rohde et al. [13] conducted a comprehensive study on the influence of processing conditions on the residual fiber length in the molded part in LFT injection molding. They designed a full factorial experimental plan and varied the processing parameters, including injection speed, back pressure, holding pressure, and screw speed. The outcome of their analysis suggests that back pressure has a strong influence on fiber attrition. A change from 50 to 80 bar back pressure reduced the residual fiber length in the molded part by approximately 30%. Injection speed, holding pressure, and screw speed were not found to be significant in their molding trials. Only the combination of increased back pressure and holding pressure was significant in their statistical analysis. Nevertheless, the authors state that it is difficult to isolate the mechanisms that cause fiber attrition due to the complex phase change and shear history that is present in the injection molding process. In contrast, Lafranche et al. [12] conducted a similar experimental study but reported that the screw speed has a larger impact on fiber breakage than back pressure, and they suggest that the fiber-wall interactions due to screw movement are more important than the fiber-melt interaction caused by the back pressure.

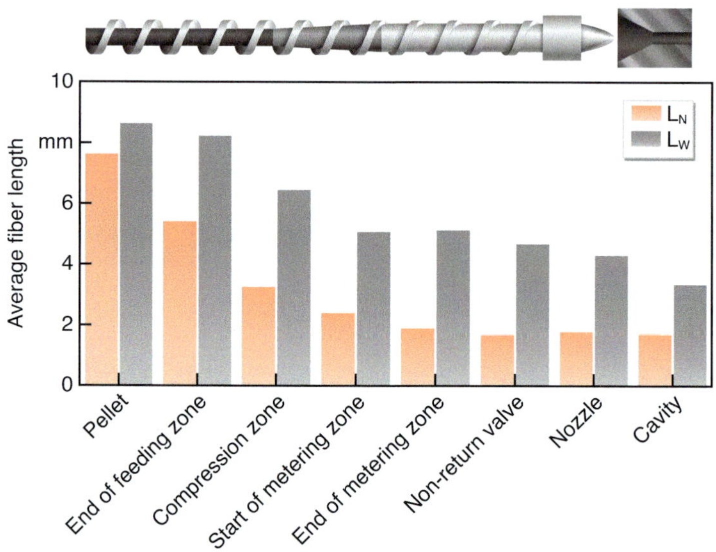

Figure 4.3 Fiber breakage during the plasticizing stage. Adapted from [12]

Metten and Cremer [14] found that increased holding pressure decreased the fiber length in the finished part, suggesting that increased holding pressure promotes fiber breakage during LFT injection molding. While their measurements show an average fiber length of 4.8 mm at 200 bar holding pressure, the fiber length drops below 3 mm at 450 bar holding pressure. This increase in fiber breakages might be caused by the additional motion of the molten core at elevated holding pressure.

Partially embedded fibers shear off with the motion of the core when the holding pressure is applied [14].

In summary, fiber length in the finished part is highly dependent on the attrition during processing. Plasticizing is the most severe phase and can reduce the fiber length by 50 percent or more when measured using weight averaging. Further attrition occurs in the mold cavity, which can depend on the mold design and flow management. An extensive literature review indicates that fiber attrition during plasticizing can be reduced by optimizing process parameters such as screw speed, pressure, and temperature. Also, screw design such as variable pitch and geometry can also help reduce fiber attrition. In most cases, it is a balancing act between production throughput versus fiber attrition reduction.

4.1.2 Fiber Alignment

Discontinuous fibers embedded in the molten thermoplastic matrix rotate and translate because of the deformation of the suspension during processing. Internal shear and elongation stresses induce a fiber orientation distribution that causes anisotropic properties in the molded part [14–16]. The evolution of fiber orientation in a highly-filled material is determined by processing conditions, cavity design, and material properties.

In general, a distinct orientation pattern through the thickness can be observed in injection molded parts, referred to as the core-shell-skin structure, as shown in Figure 4.4. The core layer in the center of the part consists of fibers that are predominately oriented in the cross-flow direction, whereas the fibers in the shell layers are aligned with the flow direction [18]. The skin layer is very thin and usually shows a random fiber orientation [4]. This specific orientation pattern emerges due to the *fountain flow* effect (see Figure 4.5), which was first studied by Tadmor and describes the reorientation of the polymer molecules (or fibers) at the advancing flow front [19].

The material at the flow front is forced to move from the center of the cavity outwards onto the mold surfaces. The material freezes upon contact with the cold mold, forming a solidified layer. As additional material enters the cavity, the advancing flow front is continuously stretched and rolls onto the cool wall, where it freezes instantly. The high-shear region at the solidified layer and the stretching in the advancing molten core cause a strong in-flow fiber alignment in the shell layers [4]. As briefly mentioned in Chapter 3, seven layers can be identified, and each layer has its own characteristic fiber orientation distribution, as schematically illustrated in Figure 4.4. The thick core layer consists of fibers predominately oriented in cross-flow direction. Two transition layers with a random fiber orientation surround the core layer. The fibers in the thick shell layers are mainly aligned along the direction of flow. Two thin outer layers (or skin layers) at the mold surfaces show a random fiber orientation.

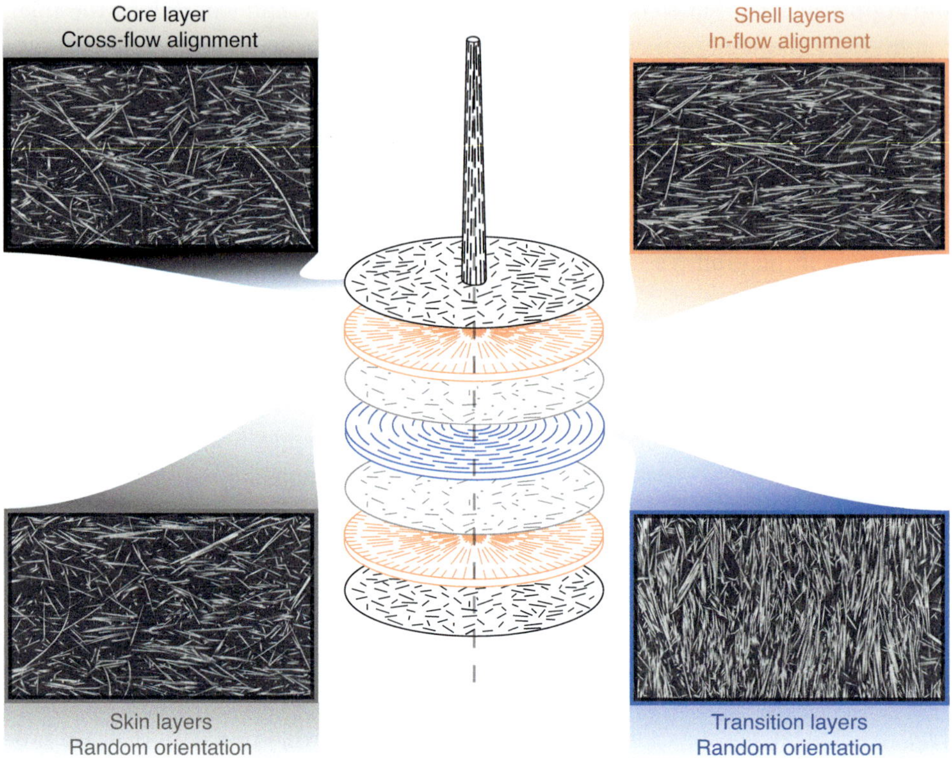

Figure 4.4 Illustration of the typical fiber orientation distribution through the thickness of an injection molded part, including μCT images of a 40 wt % glass fiber-filled PP sample

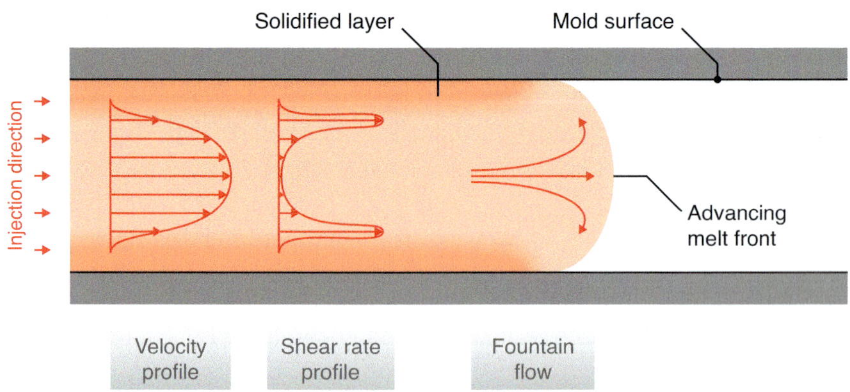

Figure 4.5 Fountain flow effect: flow pattern in injection molding at the advancing flow front

From a mathematical perspective, the orientation of a single rigid fiber in a three-dimensional (3D) space is described by the angle pair (φ, θ), or by the unit vector $\boldsymbol{p}(\varphi, \theta)$ directed along the fiber axis as shown in Figure 4.6. The unit vector $\boldsymbol{p}(\varphi, \theta)$ is described by

$$\boldsymbol{p} = \begin{Bmatrix} p_1 \\ p_2 \\ p_3 \end{Bmatrix} = \begin{Bmatrix} \cos\varphi \sin\theta \\ \sin\varphi \sin\theta \\ \cos\theta \end{Bmatrix} \tag{4.8}$$

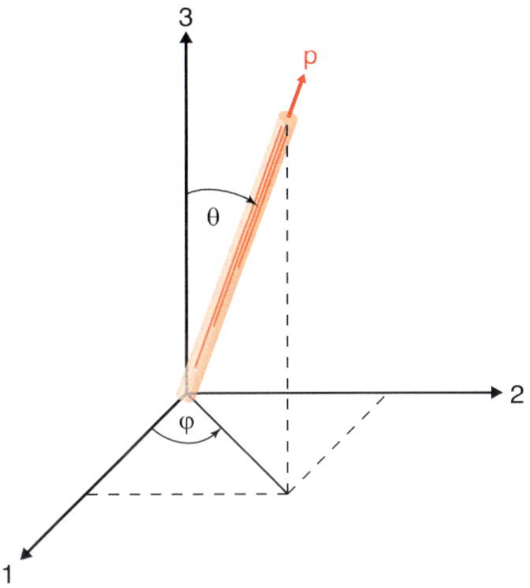

Figure 4.6 Schematic of the orientation of a single fiber in 3D space

Even in a small volume, adjacent fibers are not aligned in the same direction and instead have a distinct orientation distribution. For a network of fibers in a defined volume, a complete description of the orientation is given by a probability density function $\psi(\boldsymbol{p})$, which describes the probability of a fiber oriented between the angles θ_i and $(\theta_i + d\theta)$, and between φ_i and $(\varphi_i + d\varphi)$ [4]. However, calculating the probability density function $\psi(\boldsymbol{p})$ is too computationally intensive to be implemented in simulation applications. Advani and Tucker [20] proposed a tensorial representation of the fiber orientation, which ultimately emerged as the conventional representation in the field of discontinuous fiber-reinforced composites. The Advani-Tucker *orientation tensor* A_{ij} considers an average orientation property of all fibers within a discrete volume. The volume is assumed to be large enough to contain a sufficient amount of fiber and still small enough to ensure a uniform orientation distribution of the fibers within the volume. The components of the symmetric second-order orientation tensor are defined as [20]:

$$A_{ij} = \oint p_i p_j \, \psi(\boldsymbol{p}) \, d\boldsymbol{p} \tag{4.9}$$

where $p_i p_j$ is the dyadic or tensor product of the fiber orientation vector $\boldsymbol{p}$ with itself, and $\oint(\ldots)\psi(\boldsymbol{p})\, d\boldsymbol{p}$ denotes an integral over all possible fiber orientations. In terms of the orientation distribution and the angle pair (φ, θ), the tensor components can be calculated as follows [3]:

$$
\begin{aligned}
A_{11} &= \langle \sin^2 \varphi \cos^2 \theta \rangle & A_{12} &= \langle \sin^2 \varphi \cos \theta \sin \theta \rangle \\
A_{22} &= \langle \sin^2 \varphi \sin^2 \theta \rangle & A_{13} &= \langle \sin \varphi \cos \theta \cos \theta \rangle \\
A_{33} &= \langle \cos^2 \theta \rangle & A_{23} &= \langle \sin \varphi \cos \theta \sin \theta \rangle
\end{aligned} \tag{4.10}
$$

where the angle brackets $\langle\ \rangle$ indicate the average of all fibers in a volume. Numerical implementations for process simulation and structural analysis typically use the tensorial representation [21, 22]. Experimental data is also conventionally illustrated using the orientation tensor form. The diagonal components of the second order orientation tensor (A_{11}, A_{22}, and A_{33}) describe the degree of orientation with respect to the defined coordinate system. Conventionally, the reference coordinates are defined so that the 1-direction represents the in-flow direction, the 2-direction is the cross-flow direction, and the 3-direction is the thickness direction. The off- diagonal components of the orientation tensor show the tilt of the orientation tensor from the coordinate axes. Hence, they are zero only if the coordinate axes align with the principal directions of the orientation tensor [20]. The physical interpretation of the tensor components focuses mainly on the diagonal components of the tensor, illustrated in Figure 4.7. A completely random orientation (left) can be described by the diagonal components $A_{11} = A_{22} = A_{33} = \frac{1}{3}$ while perfectly aligned fibers (Figure 4.7, center), for example, in the 2-direction, can be described by $A_{22} = 1$ and $A_{11} = A_{33} = 0$. An example for a biaxial, random in-plane orientation is illustrated in Figure 4.7 (left) and the corresponding diagonal components of the orientation tensor are $A_{11} = A_{22} = 0.5$ and $A_{33} = 0$.

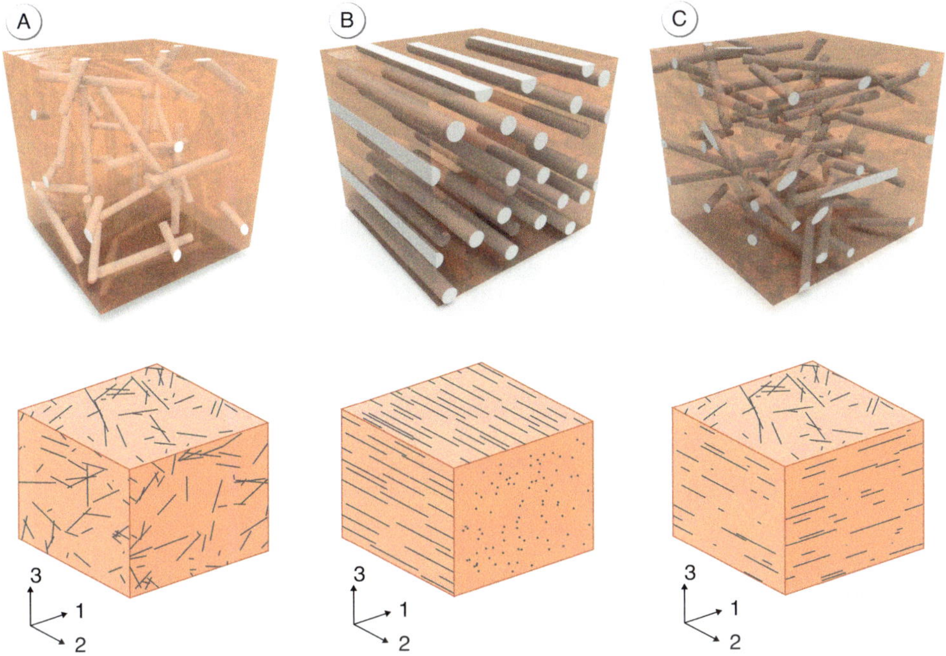

Figure 4.7 Orientation of a fiber population within a volume: randomly oriented fibers (A), fibers aligned in the 2-direction (B), and random in-plane fiber orientation (C). The figures in the top row illustrate the 3D fiber configuration and the bottom row depicts the 2D projected to corresponding plane

Process-induced fiber alignment has been a focus of numerous experimental and numerical studies in the field of discontinuous fiber-reinforced composites. Most work focuses on the fiber orientation pattern that emerges due to the fountain flow effect and the characteristics of the shell-core structure. Figure 4.8 shows an example of a measured fiber orientation distribution plotted along the part thickness using the conventional orientation tensor representation. The correlation between processing conditions and fiber orientation was studied by various research groups with a focus on experimental and numerical studies. Bailey and Rzepka [23] published a comprehensive experimental analysis of the fiber orientation in injection molding of 50 wt % glass fiber-filled polyamide 6 (PA6). The outcome of their analysis characterized the impact of processing conditions on morphology of the shell-core structure in the molded part. In terms of the degree of fiber alignment, their results show that injection speed and holding pressure are the most influential processing parameters. At low injection molding speeds, the relative core thickness reduced to 58% of the core thickness at standard injection speed. Their overall conclusion is that decreases in the core layer width are caused by any change in mold filling that induces higher stresses. Additionally, Bailey and Rzepka found that the core layer is wider for longer fibers than for shorter fibers. Their hypothesis is that the length of the fibers influences the mobility of fibers to rearrange. Consequently, longer fibers are limited in their aptitude to orient. Toll and Andersson conducted a similar study and came to the same conclusions as Bailey and Rzepka in terms of impact of fiber length and core layer thickness. Additionally, Toll and Andersson [18] report a higher in-plane orientation for a longer fiber compared to short fiber-filled compounds. Vincent et al. [24] studied the influence of cavity thickness on the fiber orientation by injection molding simple plates with thicknesses varying from 1.1 to 5.0 mm. Their results suggest a strong correlation of the shell-core structure and the part thickness. The thicker the part, the more distinct the core-shell structure becomes, as shown in Figure 4.8.

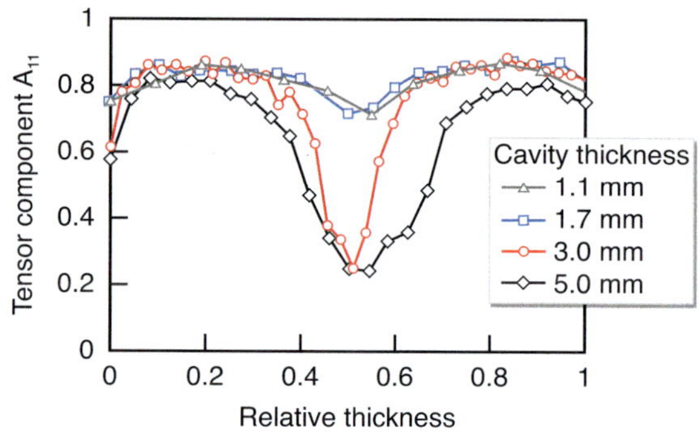

Figure 4.8 Fiber orientation measured for varying cavity thicknesses. Adapted from [24]

4.1.3 Fiber–Matrix Separation

The reinforcing effect of fibers is directly related to the concentration of the fibers in a material. The complex flow characteristics present in injection molding during mold filling can cause spatial differences in the fiber concentration within the finished part. The phenomenon of fiber–matrix separation describes the process-induced variation of fiber concentration. Earlier work showed that fiber migration during processing occurs on two scales: variation in *the local through-thickness fiber concentration* of a molded part and the *global fiber concentration gradients* along the flow path.

Previous studies have addressed the process-induced concentration gradients throughout molded parts for filled thermoplastics. Toll and Andersson [18] published results showing an increase in fiber concentration from nominal 30 wt % in the raw material to 36 wt % at the end of the flow path of a film-gated plate for glass fiber-reinforced polyamide 66 (PA66). In a similar study, O'Regan et al. [25] observed an increase of 3 to 4 vol % at the end of the flow path in injection molding of fiber-reinforced PA66. Lafrance et al. [12] showed an increase from 39 wt % at the gate to 51 wt % at the end of the flow path for a fan-gated plate. Using a spiral mold, Kubat and Szalanczio [26] studied the filler migration effects of a low-density polyethylene filled with glass spheres. Along a total length of 1950 mm, they measured the filler concentration and found substantial migration of the particles towards the end of the flow path. Their results suggest a relative increase of up to 25% filler concentration at the tip of the spiral for glass spheres with diameters between 53 and 105 μm.

Hegler and Menning [27] published work on the filler separation effects during injection molding of glass bead- and glass fiber-filled thermoplastics using dumbbell and rectangular box specimens. Their results show that the separation effects were more pronounced with beads than with fibers, indicating an influence of the shape and dimensions of the fillers. Furthermore, they conclude that mold geometry and filler concentration are key parameters determining the degree of filler migration in injection molding. Other processing parameters, such as mold temperature, injection speed, matrix material, and screw speed did not substantially impact the filler-matrix separation phenomena in their experiments.

Papathanasiou and Ogadhoh [28] studied particle migration during mold filling of glass bead-filled polystyrene with an average bead size varying from 50 to 500 μm. Their results agree with previous studies and showed that the filler matrix separation increases substantially with larger beads. The largest relative increase was 15% more fillers at the end of the flow for glass beads with a diameter of 500 μm. Another major outcome of their work was the observation that particles appear to accumulate at the free surface. For the largest particle sizes used in their study, the concentration of particles near the free surface was almost double than in the feed material. Papathanasiou and Ogadhoh's results strongly indicate that the filler migration depends on the particle size and nominal concentration in the feed material. In 2011, Kovacs published a similar study detailing the segregation effect of glass beads along the flow for injection molded plates and its impact on volumetric shrinkage [29]. His investigation into filler migration again indicated a dependency on particle size and nominal concentration, confirming Papathanasiou and Ogadhoh's study [28]. Figure 4.9 shows Kovacs's results of the filler separation along the flow length for glass bead diameters varying from 11 μm to 203 μm (nominal glass bead concentration of 40 wt %).

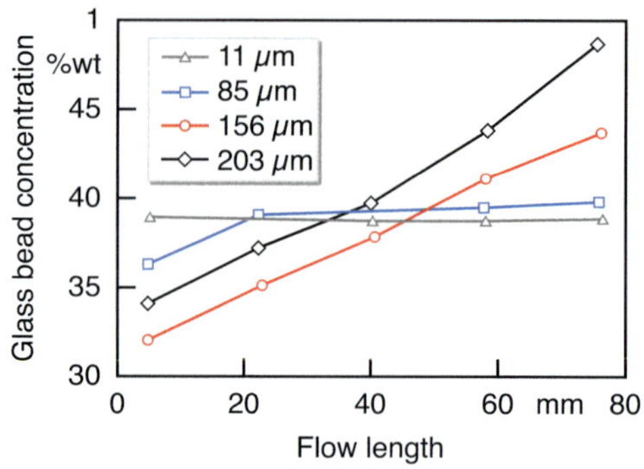

Figure 4.9 Glass bead migration along the flow path for various glass bead sizes (nominal bead concentration: 40 wt %). Adapted from [28]

Mondy et al. [30] conducted fiber migration experiments in a wide-gap Couette rheometer for low aspect ratio nylon fibers in glycol. Their measurements show a migration of fibers to the low-shear regions. While the authors did not find an impact of aspect ratio on the migration, they found an influence of fiber concentration suggesting that the effect of shear-migration increases at elevated concentrations.

Recent experimental work on fiber migration was done by Ramzy et al. [31] for hemp and sisal fiber-filled polypropylene. Similar to Kuba and Szalanczio, they used a spiral cavity and measured the concentration variation along the flow path. As reported in previous studies, the results showed an increase in filler concentration along the spiral flow length. The fiber agglomeration at the tip of the spiral showed a relative increase of up to 30%.

All these studies share the conclusion that the fiber concentration increases towards the end of the flow path, which suggests that the last filled location carries an accumulated amount of filler. Furthermore, the processing conditions appear to play a much smaller role than the filler properties when evaluating the degree of fiber–matrix separation. The influencing factors are reported to be the concentration of the filler and the shape of the filler, as the migration effect is more pronounced in general for longer fibers at higher concentrations than it is for shorter fibers at low concentrations [18].

The through-thickness filler concentration gradient in injection molding has been explored much less frequently. A comprehensive study was conducted by Toll and Andersson [18] for 30 wt % glass fiber-reinforced PA66 and a simple plate geometry. They found an agglomeration of fibers in the core layer using a sectioning and microscopic measurement protocol. Their results suggest almost 40% more fibers in the core layer than in the shell for their long fiber PA66 (initial fiber length of 10 mm). For the short fiber grade (initial fiber length of 0.6 mm), the fiber agglomeration is less pronounced with only 10% more fibers in the core than in the shell region.

While other publications also mention the phenomenon of fiber agglomerating in the core of an injection molded part, the observations are mostly side-effects in these studies and no comprehensive analysis exists nor have theories been formulated to explain the underlying mechanism. Velez-Garcia et al. [32] focused on fiber orientation measurements for short fiber-reinforced polybutylene terephthalate (PBT) by applying a newly developed sectioning procedure. From the obtained micrographs, they also reported a fiber concentration gradient through the thickness of the molded center-gated disk. Recently, Sun et al. [33] published an experimental study on using μCT to quantify the fiber orientation of an LFT injection molded instrument panel. While also focusing on orientation measurements, they incidentally analyzed the pixel fraction from μCT scans indicating a strong agglomeration of fibers in the core layer of the molded part.

At this point, no comprehensive explanation of the fiber–matrix separation phenomena has been derived and validated. A common hypothesis for the observed phenomena is that shear-induced particle migration causes the variation in fiber concentration. Shear migration describes the irreversible accumulation of particles due to a shear stress gradient [34]. In a simple Poiseuille flow, it was shown that particles tend to migrate from the high shear region at the wall towards the low shear region at the centerline [34, 35]. Similar migration was found in a wide-gap Couette rheometer flow [30]. Overall, shear migration is attributed to be a main cause for concentration gradients in molding discontinuous fiber-reinforced materials and result in process-induced fiber–matrix separation. However, the underlying physics have not been proven yet and additional studies to clarify the mechanisms are needed.

Fiber–matrix separation can also be affected by part design and features of the part. The separation effect in rib-filling has been studied by a few research groups and describes the phenomenon of fibers accumulating in the base of the rib while the tip of the rib remains almost fiber-free [36, 37]. The phenomena of fiber jamming in rib-like features is shown in Figure 4.10 and is known for unexpected failure caused by the low fiber concentration in the ribs.

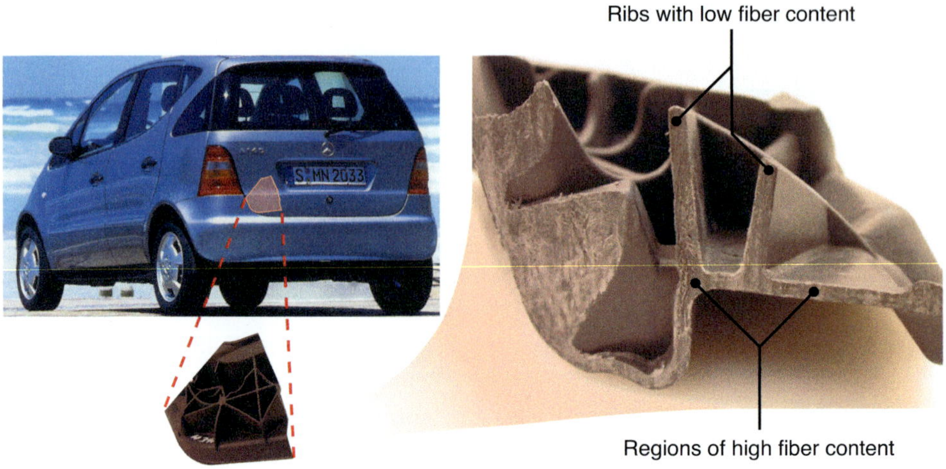

Figure 4.10 Fiber jamming in narrow part features resulting in premature failure due to large fiber concentration gradient

4.1.4 Fiber–Matrix Bonding

As discussed in Chapter 2 of this book, the bonding between the fiber and the matrix has a profound effect on the overall mechanical behavior of the composite material. The bonding properties are reflected by the *interfacial shear strength* between fiber and matrix.

The interfacial shear strength depends on the chemistry of the *surface treatment*, or *sizing*, that is applied on the fiber's surface. As will be discussed in more detail in Chapter 5, in a composite material the load is transferred from the matrix to the fiber through shear stresses between the matrix and fiber interface. If the bonding strength between the fiber and matrix is low, the weakest link in the composite can be the interface. In such situation, once the fiber and the matrix de-bond, the load distribution on the fibers in the vicinity increases, triggering failure of fibers. Hence, pushing up the binding strength will have a profound effect on the overall strength of the composite. A cursory analysis of the numbers presented in Chapter 2 shows how a 35% increase in interfacial shear strength by adding a sizing to untreated fibers will result in a 60% increase in the overall strength of the composite.

Bonding strength and sizing remain very important areas of research that have found a new application with the advent of hybrid structures. In particular, hybridization between sheet metal geometries, over molded with fiber-filled polymers, has brought a whole new area where bonding strength becomes very critical.

4.1.5 Microstructure–Property Relationship

The final configuration of the fibers, i. e., orientation, length, and dispersion, in the finished part depends on the process parameters used. As the reinforcing phase in the compound, the state of the fibers determines the mechanical performance of the finished product as illustrated in the Figure 4.11. The anisotropy, which is introduced by the alignment of the fibers, has a substantial influence on the mechanical properties. The tensile strength is higher in the direction of fiber alignment and lower in the cross direction. Hence, anisotropy is a key aspect of designing LFT parts since process-induced fiber alignment is inevitable in LFT injection molding. Similarly, the concentration and length distribution of fibers also play an important role in the microstructure–property relationship. Longer fibers tend to increase the strength and stiffness of the final part. But there is also cross play: longer fibers tend to interact with each other, resulting in resistance to alignment along the flow, which means less anisotropy and reduced warpage. Similarly, an increase in concentration of fibers increases the mechanical properties; however, if processing conditions are not optimal, fiber bundles may be found within the part. Bundling is the state where the fiber clusters are not fully dispersed during melting. Bundles often show reduction in mechanical properties. Dispersion of the fibers remains an area of research and is not well understood.

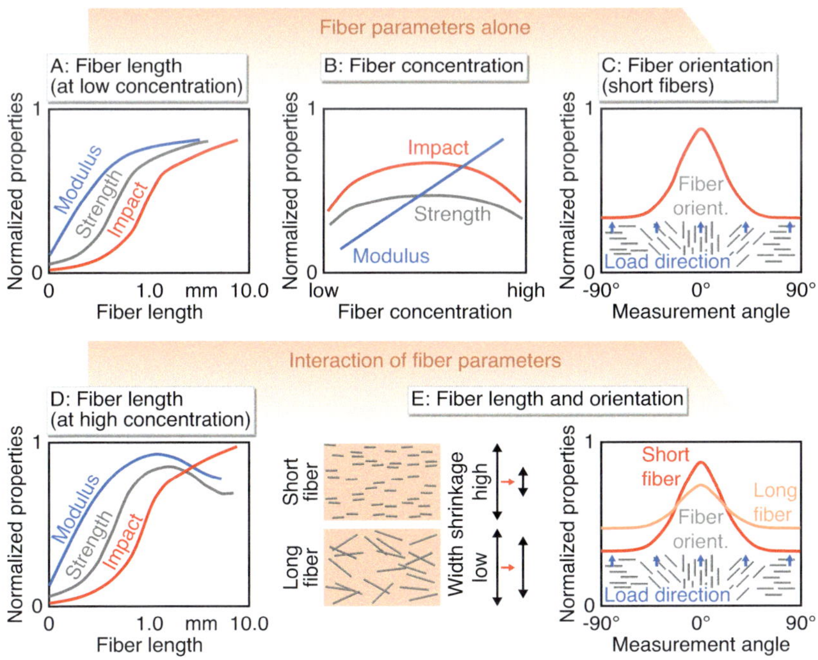

Figure 4.11 Summary of the process-microstructure-property relationship for discontinuous fiber-reinforced thermoplastics: impact of individual fiber parameters (top) and commonly observed effects from interactions of the fiber parameters (bottom)

Thomason et al. conducted the most comprehensive work to date on the structure-property relationship of discontinuous fiber-reinforced thermoplastics and published several papers from 1996 [38] to 2009 [39]. He and his research group investigated the mechanical properties of glass fiber-reinforced PP with fiber concentrations up to 73 wt % and initial fiber lengths varying from 0.1 mm to 50 mm. For all material combinations, specimens were manufactured by either a wet-molding process or injection molding. Comprehensive mechanical testing, including tensile, flexural, and impact tests, were conducted at ambient temperature as well as at −30 °C. Also, their work included characterization of the local fiber orientation and fiber length. Ultimately, they were able to draw conclusions about the relationship between the fiber configuration and the mechanical performance of the material. The key conclusions of their work are:

- A linear increase in tensile strength with fiber concentration was found for wet-molded polypropylene (PP) samples. The tensile strength was also shown to depend on the average fiber length, up to 3 mm. However, the increase in strength leveled off for all material combinations with fiber length greater than 3 mm, which corresponds to an aspect ratio of 270 (fiber diameter is 11.0 μm) [38].
- The modulus of fiber-reinforced PP increased linearly with the fiber concentration until 40 wt %, and the rate of increase was considerably less at higher concentrations. Thomason's results suggest that the modulus of the composite is independent of the fiber length after 0.5 mm (aspect ratio of 45) [40].
- The impact strength increased with increasing fiber concentration and the strength also improved with fiber length, but converged at lengths above 6 mm (aspect ratio of 540) [41].
- For injection molded samples that result in a fiber length distribution, as opposed to a fairly uniform fiber length in the wet-molded samples, the results showed that fiber length had a significant impact. Although the modulus did not show much difference between LFT and SFT, the tensile modulus, flexural modulus, and impact resistance all increased with longer fibers [42, 43].
- For injection molded PP samples, the strength and impact properties showed a maximum value at fiber concentrations of 40 to 50 wt % and, at higher concentrations, these properties decreased substantially and reached the values of unreinforced polypropylene at the highest fiber concentration (73 wt %). The modulus, however, increased linearly with fiber concentration over the entire range [3].
- Fiber attrition in the injection molded PP samples showed a linear dependence on fiber concentration. In injection molded PA6 samples, their results also indicated an increase of fiber attrition for fibers with lower diameters [39].

The main results of Thomason's studies on the relationship between fiber concentration and mechanical properties for injection molded samples can be summarized in a single graph, as shown in Figure 4.12. The graph shows the stiffness, strength, and impact performance as a function of fiber concentration [3]. The properties are each normalized to the values of the unfilled matrix material. While the modulus increased linearly with fiber concentration, the strength and impact performance show a maximum at 40 to 50 wt %. A further increase in fiber concentration resulted in decreased strength and impact properties. At highly filled levels above 60 wt %, Thomason's results suggest that these properties even decrease to the same level as unfilled PP.

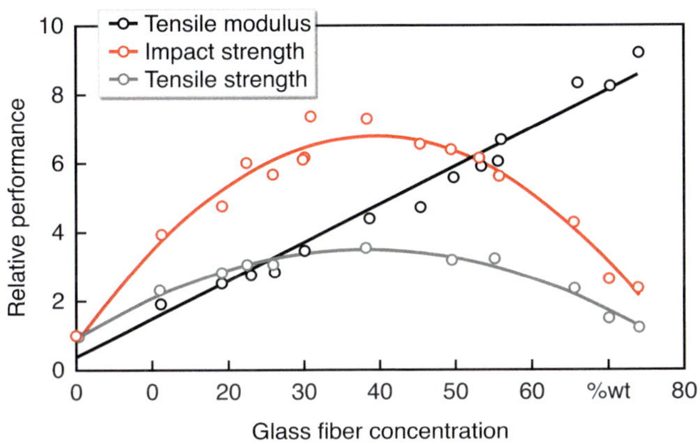

Figure 4.12 Normalized mechanical properties as function of fiber concentration (by weight), obtained experimentally: tensile modulus, tensile strength, and impact strength. Adapted from [3]

The effect of the residual length of the fibers on the mechanical properties of a molded part is illustrated in Figure 4.13. The figure is adapted from Schemme [44] and is based on the experimental work from Thomason et al. [38, 40, 41] discussed in the previous paragraphs. The graph visualizes the effect of the fiber aspect ratio on the tensile modulus, tensile strength, and impact strength. It can be seen that these properties saturate, after which point any increase of the fiber length would not further improve the mechanical properties. The figure also qualitatively illustrates the expected range of residual fiber aspect ratios for LFT injection molding, reported in the literature and measured experimentally in this work. This expected range would suggest a strong sensitivity for tensile strength and impact strength for injection molded LFT parts. While Schemme's graph offers a qualitative overview of the relationship between fiber aspect ratio and mechanical properties, the quantitative interpretation of this graph should be done carefully, because a universal validity has not been proven yet. For example, the underlying experimental data do not account for any variation of fiber orientation or local fiber concentration due to the change in fiber length.

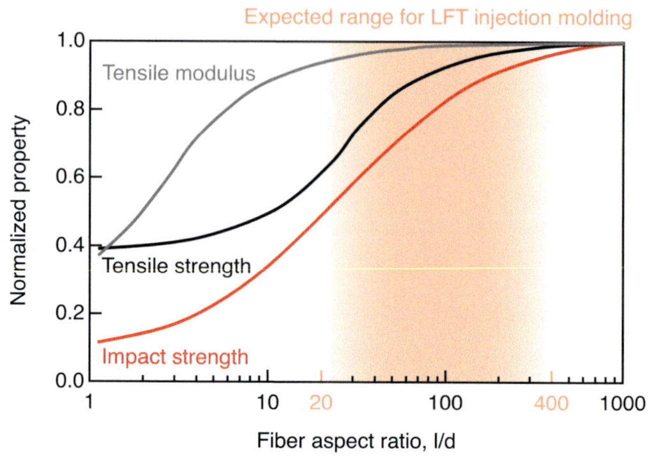

Figure 4.13 Normalized mechanical properties as function of fiber aspect ratio for tensile modulus, tensile strength, and impact strength. Adapted from Schemme et al. [44]. The highlighted range illustrates the qualitative range of fiber aspect ratios expected in LFT injection molding

As will be discussed in Chapter 7 of this book, mold filling simulation tools are widely accepted in industry to predict the microstructure development during processing. To illustrate this, the predicted anisotropy and its effect on the mechanical performance of injection molded parts are illustrated in Figure 4.14. The process simulation results show the effect of fiber alignment on the stress-strain behavior for a simple fan-gate plate geometry. For this plate design, the majority of the fibers show a fiber alignment in flow direction. Consequently, the 0° specimen shows a much higher stiffness than the 45° and 90° specimens.

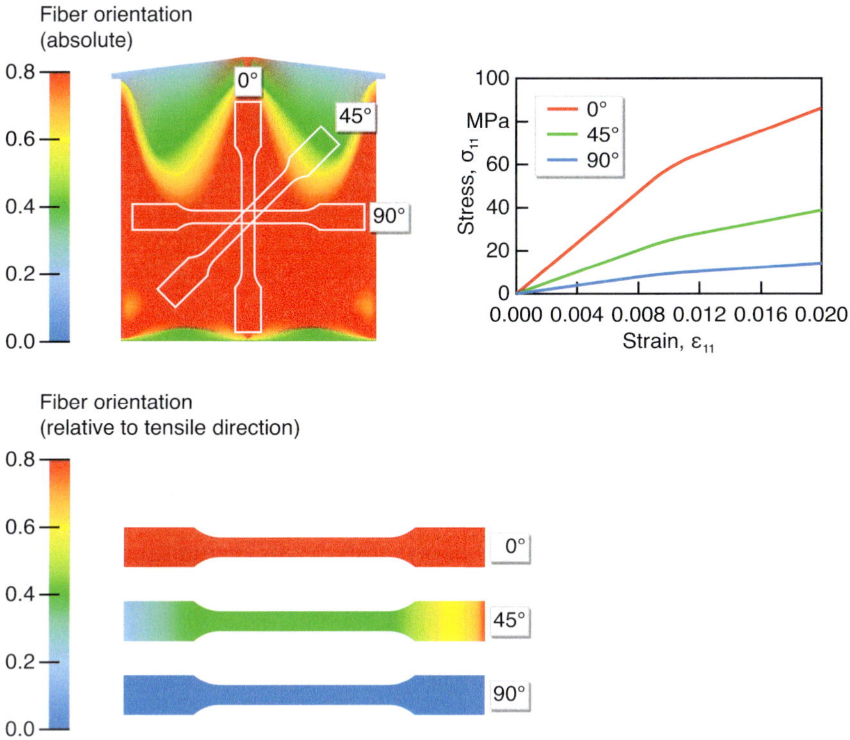

Figure 4.14 Illustration of the process-induced fiber orientation, obtained from a basic process simulation of a fan-gated plaque. The impact of fiber orientation, caused by the mold filling, is shown for specimens extracted at 0, 45, and 90 degrees

For a more detailed discussion on process simulation, the reader should consult Chapter 7 of this book. Furthermore, the effect of process-induced microstructure on the mechanical properties within a plaque is investigated in detail in Chapter 8.

4.2 Characterization Techniques for Fiber Microstructure Analysis

Quantifying the process-induced fiber configuration in molded samples is of utmost importance so that the underlying physics of the process–microstructure relationship can be studied experimentally. Furthermore, accurate and reliable experimental data are required to test and develop numerical models and predictive tools. At times, the lack of accurate experimental data limits the progress of validating modeling approaches [2, 6]. Nevertheless, the characterization of fiber configuration is a cumbersome task since even small samples comprise millions of fibers. A wide variety of measurement approaches exist, but no standard has been accepted nor has a uniform procedure been defined. While some measurement concepts share similarities, there are substantial differences in key aspects of the measurement techniques and the execution of the actual measurements. The following sections provide an in-depth review and discussion of measurement techniques used to quantify the orientation, length, and concentration in discontinuous fiber-reinforced polymers.

4.2.1 Measuring Fiber Orientation

For decades, fiber orientation measurements were performed manually by physically sectioning the sample and inspecting the cross sections using optical reflection microscopy [45–47], which is referred to as the method of ellipses (MoE). The fibers leave elliptical footprints on the polished cross sections, which are detected and analyzed for individual fibers. The fiber orientation can be quantified by analyzing the shapes of the ellipses, as illustrated in Figure 4.15. An image analysis system imports the captured micrographs and fits ellipses to determine the angles ϕ and θ individually for each fiber [47]. Although newer systems are automated to a degree, this technique still requires a lot of manual work to achieve accurate results. Furthermore, the destructive nature of this method does not allow additional testing of the sample, e.g., fiber length analysis. Moreover, the two-dimensional observation of the ellipses results in ambiguity with respect to the out-of-plane angle θ, because the fiber can be oriented at θ or $\theta+\pi$, as illustrated in Figure 4.15. However, this ambiguity can be resolved by an additional etching after each polishing step, so that a fiber's shadow is visible in the micrograph. By determining the location of the shadow relative to the corresponding fiber, the angle θ can be distinctly defined [32].

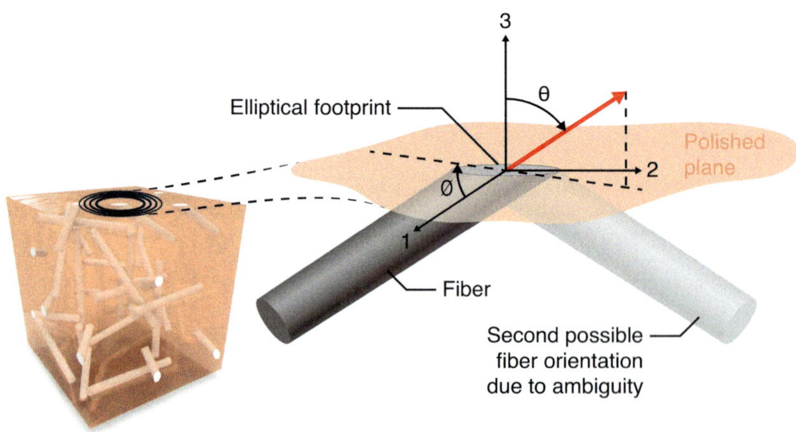

Figure 4.15 Basic principle of the method of ellipses and illustration of the ambiguity in the out-of-plane angle θ

A relatively new approach to measure fiber orientation distribution within fiber-reinforced plastic parts is applying the X-ray micro-computed-tomography (μCT) technology. This technique is a non-destructive testing (NDT) method, which enables a full 3D analysis of a specimen and the quantification of internal material structure [48–50]. The advantage of μCT is that it can achieve high resolution of the full 3D microstructure without destroying the specimen. Originally, μCT was solely used in medical applications, but it has now evolved to be a conventional characterization technique in the material sciences and in industrial applications. A basic schematic of the μCT process is illustrated in Figure 4.16.

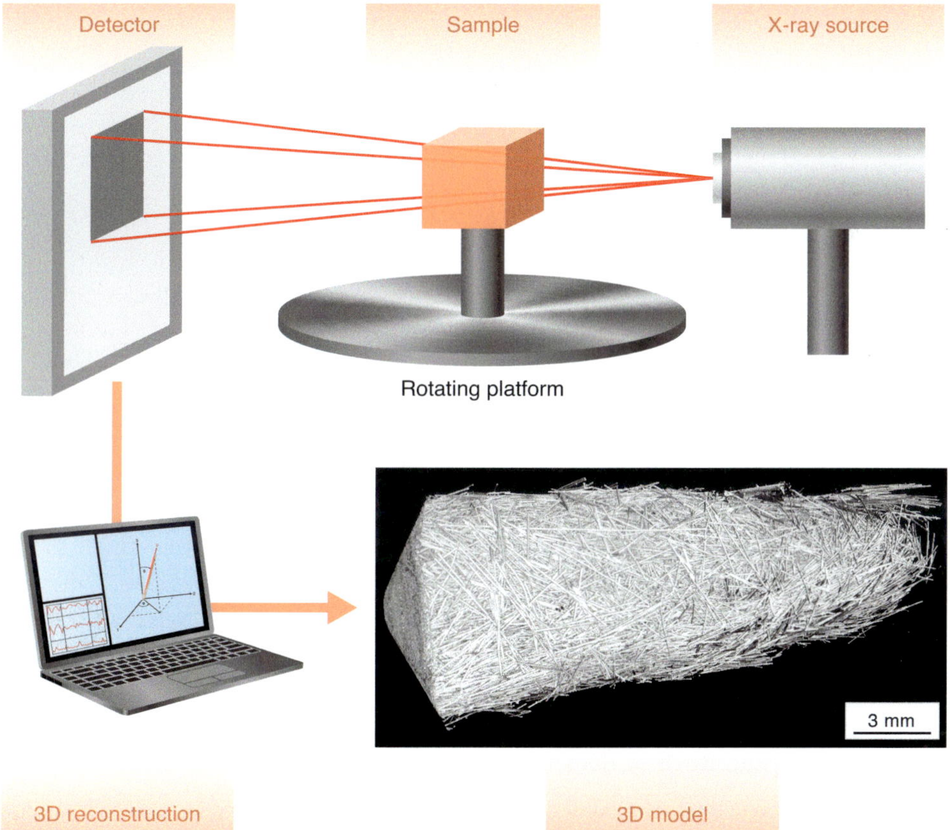

Figure 4.16 Illustration of the X-ray computed-tomography process: the μCT system comprising the detector, X-ray source, and rotating platform (top), the computer-assisted tomographic reconstruction (bottom left), and a 3D rendering of the data reconstruction (bottom right)

In general, the system consists of an X-ray source, a rotating platform and a detector. The basic principle of μCT is to irradiate a sample with penetrating X-rays, which are attenuated and captured downstream of the object with a detector system creating radiographs. At defined energy levels, the X-ray source irradiates the specimen, which is placed on a rotating platform to achieve a full scan of the sample. The detector records the attenuated X-rays as radiographs at incremental angles during the rotation of the sample. Each captured projection (radiograph) is a two-dimensional intensity distribution of the attenuated X-rays. The intensity distribution is directly related to the material's atomic density. A phase of high atomic density within the specimen absorbs more energy than low-density materials.

The X-rays interact with the sample and are attenuated in various ways. For the energy range of the X-rays used in material science, the most relevant mechanisms are Compton scattering and photoelectric effects [48]. Most fundamentally, the attenuation of an X-ray through a sample can be described by the change in intensity I as

$$dI = -\mu_{eff}(x)I(x)dx \tag{4.11}$$

where μ_{eff} is the effective linear attenuation coefficient and x is the location on the path. Integration of Equation (4.11) along the path C gives

$$-\ln\frac{I}{I_0} = \int_C \mu_{eff}(x)dx \tag{4.12}$$

which is also referred to as the Beer-Lambert law. Computing the correct value of μ_{eff} at each position along the path C is the central problem of computed tomography, called tomographic reconstruction. Several approaches and algorithms exist for the reconstruction and the reader is directed to [48] for more detailed information.

After the completion of the 3D reconstruction, the µCT data set can be further processed for qualitative and quantitative analysis of the actual specimen. Different software and algorithms exist to process the µCT images and analyze the fiber orientation distribution. A major challenge in using µCT for fiber orientation analysis is the amount of data that is generated, which has to be post-processed to quantify the orientation. For example, a scan of a 20 × 20 × 3 mm^3 sample at a resolution of 10 µm results into 8 GB of data. There are software packages from the medical field, e.g., Mimics® (Materialise NV, Belgium) that are adapted to be used for fiber orientation analysis [51]. Another software package for material science and industrial application is VG Studio MAX (Volume Graphics GmbH, Heidelberg, Germany), which also has a module for fiber orientation analysis and quantifies the fiber orientation using the structure tensor approach [52]. It was shown that VG Studio MAX provides fast and accurate µCT data analysis, making it the most commonly used software for fiber orientation measurements at the moment [53]. Additionally, research groups frequently develop their own image processing algorithms to quantify the fiber orientation from µCT scans.

The resolution (voxel size) of the scan is one of the most important parameters that determine the accuracy of the feature analysis of scanned samples. However, the trade-off between scan resolution and sample size needs to be considered. In general, finer scan resolutions require samples with smaller dimensions. Although a very fine resolution might be needed for an accurate analysis, a small sample might not fully represent the local orientation, especially for long fiber-filled materials. It has not been established what the minimum resolution might be to perform an accurate fiber orientation analysis for a specific fiber type. Furthermore, µCT scanning is limited in the case of composite materials with constituents that have similar atomic densities (e.g., carbon fiber-reinforced polyamide) since the contrast of the constituents in the radiograph is low and identifying the fibers can become challenging [80].

There is a wide variety of analysis resolution and sample size used in the microstructure analysis of fiber-reinforced samples throughout reported studies. The sample dimensions range from 3 µm voxel size resolution (sample size: 4 mm disk with a thickness of 4 mm) [54] up to a voxel size of 40 µm (sample size: 40 mm × 40 mm × 30 mm) [55]. To provide the reader an overview of recently published studies that apply fiber orientation measurements, Table 4.1 summarizes the imaging technique, sample size, and resolution used in their work.

Table 4.1 Overview and Summary of Fiber Orientation Measurement Techniques Applied in Recently Published Studies

Material	Imaging Technique	Sample Dimensions	Resolution	Reference
PP/GF	µCT	$3 \times 3 \times 3$ mm^3	8.0 µm	[56]
PA66/GF	µCT	$3 \times 3 \times 3$ mm^3	2.5 µm	[57]
PP/GF	µCT	$16 \times 16 \times 3$ mm^3	6.0 µm	[33]
PA6/GF	µCT	$4 \times 4 \times 2$ mm^3	1.4 µm	[49]
PA66/GF	µCT	$10 \times 10 \times 2$ mm^3	6.5 µm	[58]
PA6/GF	µCT	$3.6 \times 3.1 \times 3.9$ mm^3	9.0 µm	[59]
Natural Fibers	µCT	$6 \times 6 \times 3.5$ mm^3	7.8 µm	[60]
PP/GF	µCT	n/a	8.1 µm	[61]
PA66/GF	µCT	$5 \times 20 \times$ (1 to 3) mm^3	6.0 µm	[62]
PP/GF	µCT	Ø 4 mm and 4 mm thickness	3.3 µm	[54]
PA6/GF	µCT	Ø 4 mm and 4 mm thickness	1.8 µm	[63]
PA/GF	µCT	Ø 4 mm and 4 mm thickness	0.7 µm	[64]
PA6/GF	Cross-sectional polishing and µCT	$3.2 \times 4 \times 15$ mm^3	9.0 µm	[65]
PP/GF	Cross-sectional polishing	$1.5 \times 2 \times 1.7$ mm^3	0.24 µm	[66]
PP/GF	Cross-sectional polishing	$2 \times 2 \times 0.7$ mm^3	n/a	[67]

Bernasconi et al. and Gandhi et al. conducted a comparison between the MoE approach and μCT analysis techniques. Both groups showed strong agreement between these two methods [51, 65]. The MoE requires a simpler experimental setup but is destructive and time-consuming. μCT offers the great benefit of being nondestructive and providing fast results, but it is a larger capital investment and also requires advanced computing power for post-processing. In the next section, two case studies are provided that compare MoE and μCT measurement approaches for fiber orientation analysis.

Case Studies: Comparison of MoE and X-Ray μCT Analysis

In the first case, the fiber orientation analysis is described and discussed for injection molded and compression molded samples. The injection molded sample consists of short and straight fibers while the compression molded sample has longer and curved fibers. Samples from a large plaque at locations adjacent to each other were selected for this study. These two samples are selected in such a way that the fiber orientation for each sample along one face is the same.

Key steps used for MoE were discussed in the previous section and an illustration of the used equipment is presented in Figure 4.17. Important to note is that any metallographic technique is acceptable as long as fiber fracturing and fiber pull-out is minimized, providing a smooth surface for etching. To increase the contrast to allow for automated fiber orientation measurement, the samples were etched to generate a rougher and less reflective matrix. Oxygen plasma etching was used to retain good control over the process. Automated microscopy was used to collect an image for a large cross section, as shown in Figure 4.17. The fiber orientation is estimated from digital image processing. These MoE method-based measurements were conducted at Oakridge National Laboratory [51].

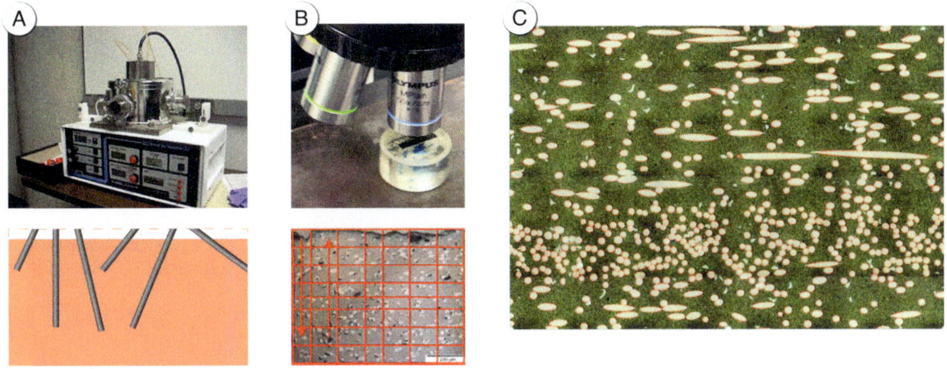

Figure 4.17 Details of the sample preparation and data collection for the method of ellipses: a) a sample is embedded in epoxy puck, finely ground, and then plasma-etched to offset the resin area by a few microns compared to the fibers, so we can get good contrast in the image; b) magnified (× 50) digital images are collected using motorized stage for the whole cross section; c) each image is analyzed to estimate fiber orientations (image obtained with an Olympus BX40 with × 50 N Plan). The study was done at Oakridge National Laboratory, USA [51]

The μCT scan was carried out on the adjacent samples, with a resolution of 3 μm, which results in a voxel size of 3 μm. Since the glass fibers are 14–17 μm in diameter, a resolution of 3 μm is considered sufficient to accurately capture a single fiber in the image. The μCT data sets are processed using Mimics® Innovation Suite (Materialise NV, Belgium) to obtain the fiber centerlines, which are used to compute the fiber orientation distribution and orientation tensor. Figure 4.18 summarizes the key steps in analysis of the μCT scan using Mimics® Innovation Suite.

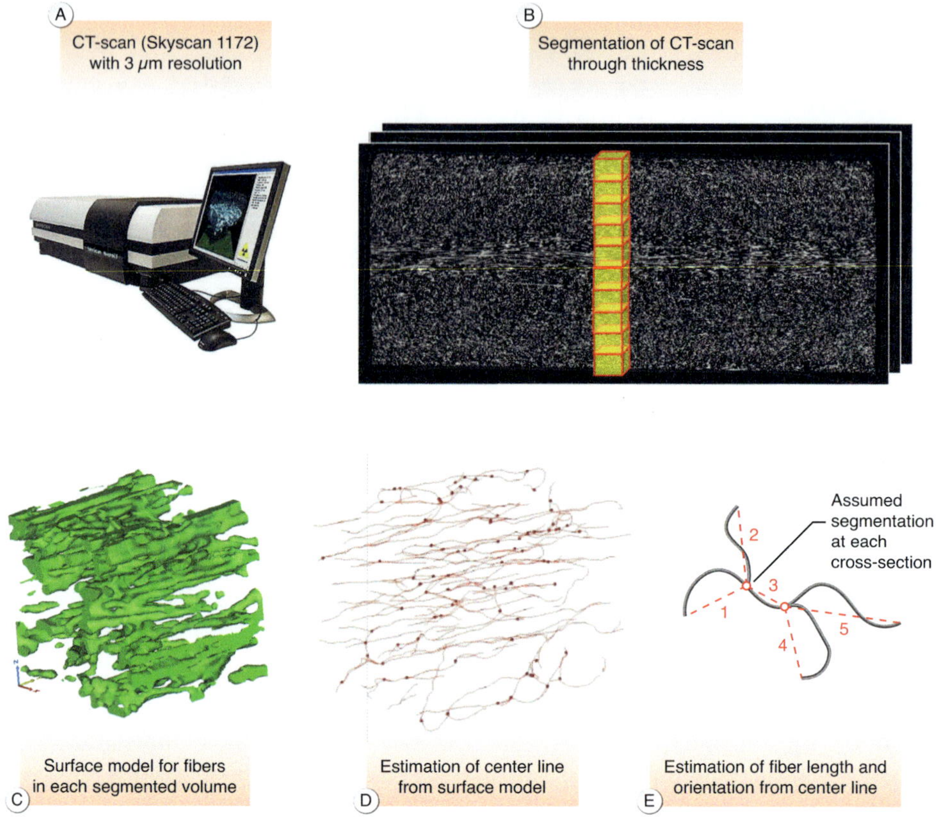

Figure 4.18 Overview of the μCT measurement approach and the processing using Mimics®

The μCT data is very large, commonly exceeding several gigabytes of data. The key challenge is how to identify single fibers to quantify the fiber orientations from such large-size data. To address this computational challenge, first the scanned volume is divided into a number of regions along the thickness as shown in Figure 4.18(B). Next, the selected regions are imported into Mimics®, where the fibers' volumes are extracted using gray value threshold, resulting in the 3D geometry of the fibers in the selected region; see Figure 4.18(C). Next, which is a key to success for a quick calculation, is creating a centerline through the 3D fiber geometry using the "Fit Centerline" function in Mimics®, as shown in Figure 4.18(D). This step greatly helps because a centerline representation requires much less data compared to the 3D fiber surface. Finally, the centerlines from Mimics® are exported and separated into non-connected segments. The orientation and length of each fiber segment is captured, as shown in Figure 4.18(E). The final output of the analysis is the local orientation tensor, which can be exported and plotted.

For this case study, plaques were injection molded and compression molded. From the plaque, samples with the dimensions of 10 × 5 × 2 mm^3 were extracted and analyzed. The results of fiber orientation measurements are presented in Figure 4.19, which shows the A_{11} tensor component across the normalized thickness of the sample for both-MoE and µCT-methods. The results suggest that there is a very good match between the two measurement methods for the injection molded sample, Figure 4.19(a). However, for the compression molded sample, Figure 4.19(b), we see similar trends but not a match as good as with the injection molded sample.

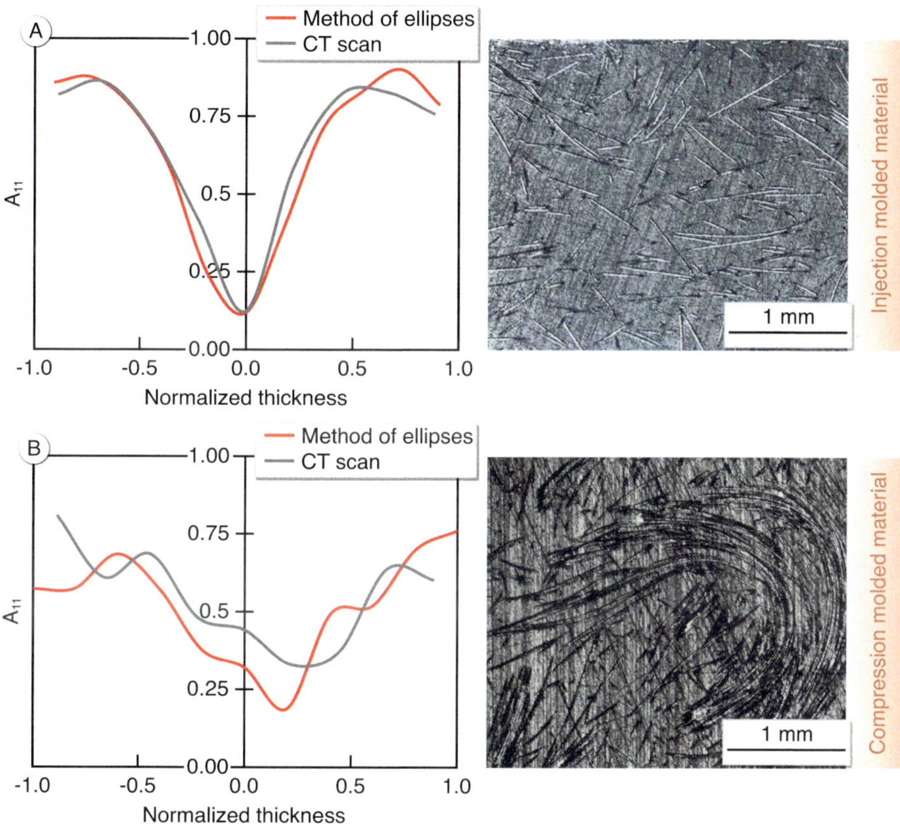

Figure 4.19 Comparison of fiber orientation from µCT scan and method of ellipses (MoE) for injection molded samples: a) injection molded material; b) compression molded material

The main reason for the difference is the morphology of the fibers. The fibers in the injection molded part are short and straight. For the straight fibers, the method of ellipses and the µCT scan approach are dealing with the same geometry, and therefore both match well, as illustrated in Figure 4.20 (left). However, fibers in the compression molded samples are longer and curved, which means that the fiber orientation measurement based on an elliptical cross section on surface (MoE) and the fiber orientation based on two end points of the fiber (μCT) are different, as shown in Figure 4.20. In general, we have observed that the fibers tend to start curving with higher aspect ratios.

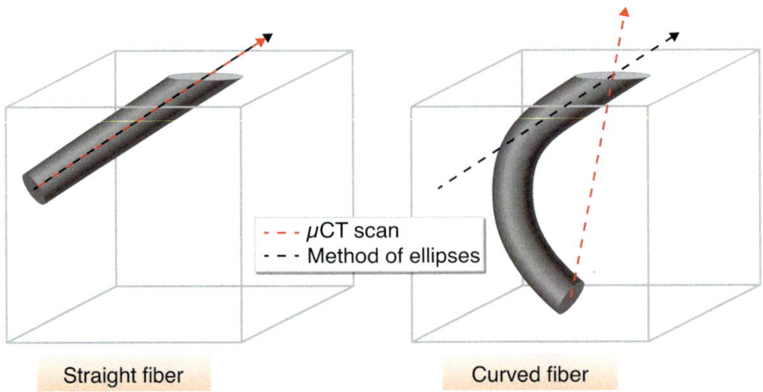

Figure 4.20 Fiber orientation measurement: key difference between the method of ellipses (MoE) and the μCT approach used here. For the straight fiber, both estimates will be the same, as shown on the left; however, for the curved fiber, there will be a difference, as shown on the right

For curved fibers, the question of orientation is more complex to describe, because the orientation (tangent) changes at each point along the length. The orientation angle for the MoE depends on the location of the measurement. For the μCT scan approach, the orientation depends on the location and size of the small volume selected for the calculations. One approach to address this can be to discretize the fibers into small straight segments, as shown in Figure 4.21, and then use the μCT scan on each of the small volumes. Since the interest in measuring the fiber orientation is to estimate the material properties, the idea of breaking the μCT scan image into smaller segments and then using the orientation in each segment to estimate effective material properties can be practical in addressing the curved fibers. Such approach also aligns well with the multiscale modeling discussed in Chapter 6, where the material properties at microscale based on the microstructure are considered.

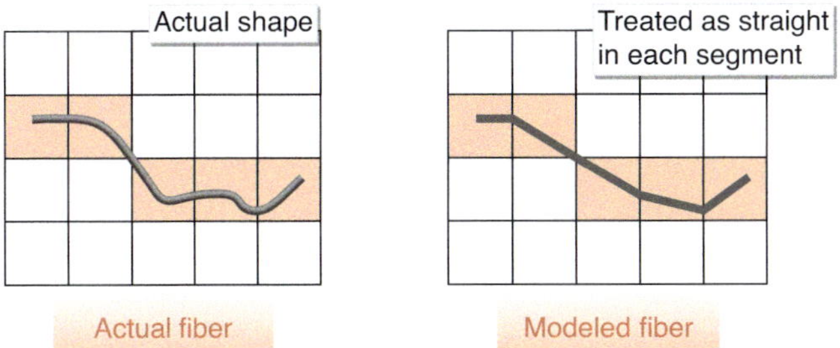

Figure 4.21 Long curved fibers can be considered as piecewise linear. In this case the fiber will be part of all the shaded volume segments (five) and contribute towards the orientation tensor in each of the five-volume segments

A second comparative study of fiber orientation measurement techniques was performed comparing the conventional method of ellipses (MoE; see Section 4.2.1) and μCT analyses for injection molded samples. Two samples (Samples A and B) with dimensions of 5 × 5 × 2.85 mm^3 were extracted from the center of an injection molded plate (40 wt % glass fiber-reinforced PP). The samples were first scanned using X-Ray μCT at 5 μm scan resolution and exactly the same samples were then analyzed using the MoE [68].

The μCT data sets were analyzed with three different image-processing algorithms, which were described in the previous chapter:

- **Mimics®:** A customized module [68] for the image-processing software package *Mimics Innovation Suite*. The module identifies single fibers in the 3D μCT data set and fits centerlines along the axis of each fiber. From the fitted centerlines, Mimics computes the fiber orientation distribution and the orientation tensor. This approach requires a very fine voxel size since individual fibers in the scan need to be segmented. Although a 5 μm scan resolution provides suffiently high resolution to separate the fibers, the developer suggests a resolution of less than one sixth of the fiber diameter.
- **VG Studio MAX™:** A commercially available software package for μCT data processing by Volume Graphics GmbH, Heidelberg, Germany. The *Fiber Composite Material Analysis Module* provides a quantitative fiber orientation analysis using the structure tensor approach [78], which does not rely on resolving the orientation distributions of single fibers but the distribution of the average orientations obtained from the direction of local gradients in grayscale values.
- **Slit Method (SM) Algorithm:** A new image-processing method that uses basic 2D projections to compute the fiber orientation distribution from 2D image stacks, developed by Goris [68, 69]. This image-processing technique does not require resolving single fibers but computes an intensity distribution that directly relates to the fiber orientation.

The main characteristics of the four analysis methodologies are summarized in Table 4.2. Figure 4.22 shows the obtained fiber orientation results for Sample A (the tensor components A_{11} and A_{22} are shown in two separate graphs for clarity). The results suggest that all measurement techniques show good agreement, particularly in the transition region from the core to the shell layers.

Table 4.2 Summary of Fiber Orientation Measurement Techniques for the Comparative Study

	MoE	Mimics®	VG Studio MAX™	SM Algorithm
Imaging Technique	Cross-sectional polishing and microscopy	µCT (3D)	µCT (3D)	µCT (2D image stacks)
Fiber Detection	Single fiber (elliptical cross section)	Single fiber (centerline fitting)	Fiber population (structure tensor [78])	Fiber population (numerical projections [96])
Destructive Testing	Yes	No	No	No
Required Resolution	Very fine (< 1 µm)	Less than 1/6 of fiber diameter	Less than fiber diameter	Less than fiber diameter

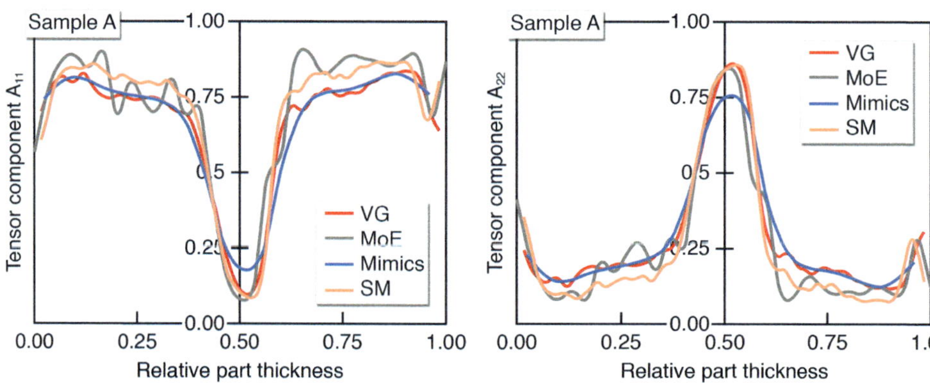

Figure 4.22 Results of the fiber orientation analyses of Sample A for all four measurement techniques: tensor components A_{11} (left) and A_{22} (right)

Figure 4.23 shows the fiber orientation results of Sample B. Similar to Sample A, the results suggest the measurement techniques match well. Some discrepancies can be seen in results obtained with the MoE technique in the core layer, indicating some irregularities in the measurements at a relative thickness of 0.45. These irregularities might be attributed to the fact that the increments in polishing and cross section inspection are performed in coarser steps compared to the µCT analyses. Also, the surface edging required to overcome the ambiguity of the angle detection [32] might not fully work for fiber dense core layer.

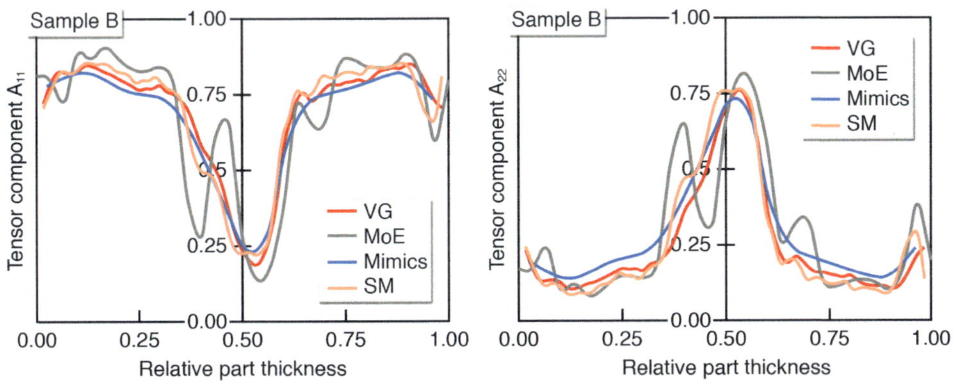

Figure 4.23 Results of the fiber orientation analyses of Sample B for all four measurement techniques: tensor components A_{11} (left) and A_{22} (right)

The outcome of the comparative study is that the measurement techniques show strong agreement in the results. In particular, the different analysis tools for µCT scans quantify matching fiber orientation distributions. Although the actual scan resolution for Mimics® is larger than the recommended value (1/6 of fiber diameter ≈ 3.2 µm), the results agree well with VG Studio MAX™ 3.0 and the new SM algorithm. Slight differences in the MoE technique might be caused by the coarser through-thickness resolution. Hence, fewer data points through the thickness are available due to the more time-consuming steps of polishing, sample edging, and image capturing of this technique.

Overall, this comparative study shows the advantages of µCT as a non-destructive and time-efficient methodology for fiber orientation analysis. However, the processing of the µCT data is critical and not trivial. While the analysis tools in comparative study agree well, the outcome of the fiber orientation analysis is dependent on the appropriate selection of the image-processing algorithm.

The resolution and the spot size of the scan is one of the most important parameters, limiting the scale of features that can be analyzed. As described in the previous section, scan resolution and sample size represent an important trade-off for fiber orientation analysis. Although a very fine resolution might be needed for an accurate analysis, a small sample might not fully represent the local orientation, especially for long fiber-filled materials.

A sensitivity analysis of the scan resolution on the results on the fiber orientation analysis was performed, using the VG Studio Max™ analysis method. The objective was to determine the largest voxel size and largest sample size without affecting the results of the fiber orientation analysis. An identical sample molded with dimensions of 10 × 10 × 3.2 mm³ was scanned at resolutions varying from 5 µm to 30 µm and the fiber orientation was analyzed. The injection molded sample comprises a 40% glass fiber-filled PP and fibers with a diameter of 19 ± 1 μm. The results of the sensitivity analysis are illustrated in Figure 4.24.

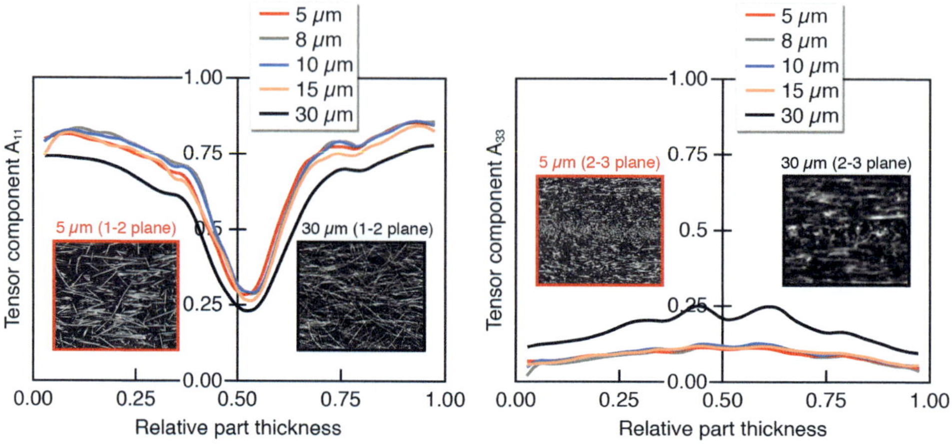

Figure 4.24 Results of the sensitivity analysis showing the influence of voxel size on the results of the fiber orientation analysis for an identical sample: tensor components A_{11} (left) and A_{33} (right)

The outcome of this sensitivity analysis suggests that the same fiber orientation distribution can be obtained for larger voxel sizes up to 15 μm. While the orientation can still be captured qualitatively at a 30 μm voxel size, there is a clear error due to the lack of resolution. The fiber orientation analyses for the scans at 5 to 15 μm lead to similar results with discrepancies of less than 1%. The scan at 30 μm shows qualitatively the same through-thickness orientation, but the values are skewed. The value for A_{11} at 30 μm is 0.1 less than the results at smaller voxel sizes and the A_{33} value is on average 0.12 larger.

It can be concluded that the voxel size impacts the outcome of fiber orientation analyses. While the sensitivity depends on the material combination of the composite and the fiber size, the results of this limited study suggest that accurate fiber orientation can be obtained at a voxel size less than the diameter of the fiber (19 ± 1 μm). Further reducing the voxel size does not improve the outcome of the fiber orientation analysis. Hence, it might not be necessary to scan at the smallest possible voxel size to quantify fiber orientation accurately and larger samples would allow a better representation of the local properties. In particular, larger sample sizes can be essential for a robust microstructure analysis, if local irregularities such as voids or non-dispersed fiber bundles are expected. Figure 4.25 illustrates the impact of a non-dispersed fiber bundle on the orientation analysis. The sample contains an irregularly orientated bundle and was scanned at a voxel size of 10 μm. The fiber orientation for the sample was done separately for the volume that contains the fiber bundle illustrated by quadrant (a) in Figure 4.25 and the remaining volume. The irregularly aligned fiber bundle has a significant impact on the results of fiber orientation analysis. The outcome of the fiber orientation analysis for quadrant (a) shows a largely skewed orientation at the location of the bundle compared to the analysis of the remaining volume.

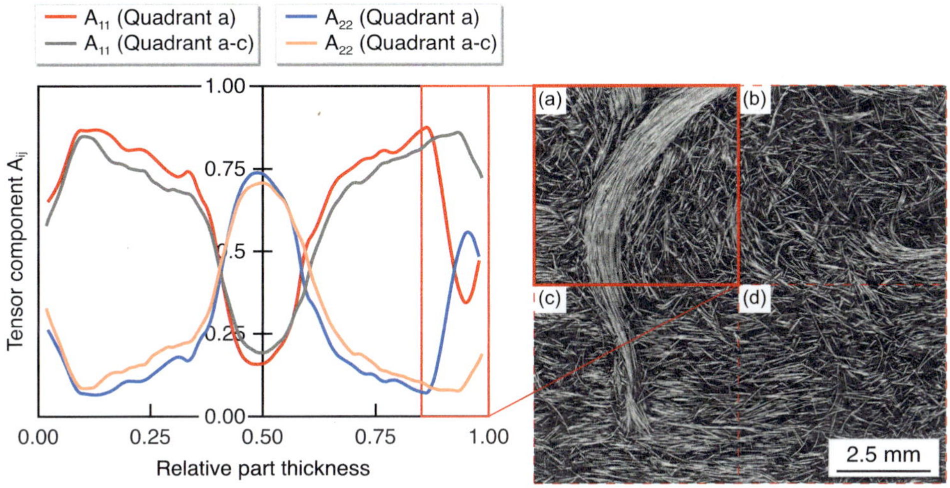

Figure 4.25 Illustration of the impact of an irregular fiber bundle on the fiber orientation analysis for an injection molding sample (40 wt % glass fiber-reinforced PP, average fiber diameter of 19 ± 1 μm)

With the advancement in computers and image-processing technology, current trends show that the methods based on μCT scanning are becoming faster and easier. Also, it is expected that the μCT scan-based method may be used to address the long and curved fibers; for example, to break the long curve fiber into linear segments and measure the orientation tensor for each segment.

4.2.2 Review of Fiber Length Measurement Methods

This section reviews the current approaches for characterizing fiber length in fiber-reinforced polymer samples. There are many different methods to measure fiber length, which will be described and discussed throughout this chapter. Figure 4.26 shows an overview and classification of the most common measurement techniques.

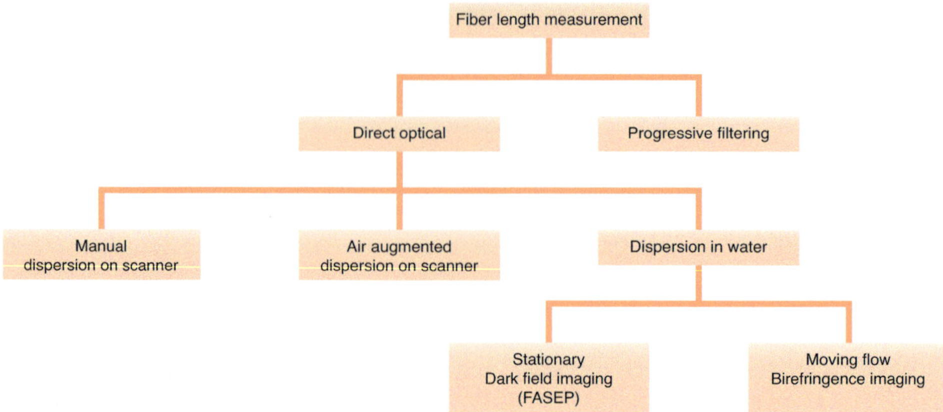

Figure 4.26 Classification of the main measurement approaches to determine the fiber length

The progressive filtering methodology is a classical approach, well known in material sciences, to determine the particle size distribution by sieving particles through discrete mesh sizes, as shown in Figure 4.27. After removing the matrix material by thermal or chemical degradation, the remaining fibers are collected. From the residual fibers captured in each mesh, a discrete fiber length distribution can be obtained. One of the major sources of error with such an approach is caused by the longer fiber tending to trap smaller fibers as shown in Figure 4.28, resulting in higher weight for the longer fibers. This approach is fast and robust but can lack accuracy; therefore, it can be used for quality control of parts in a production environment and can be a useful tool to provide a relation between the process settings and residual fiber length in the part.

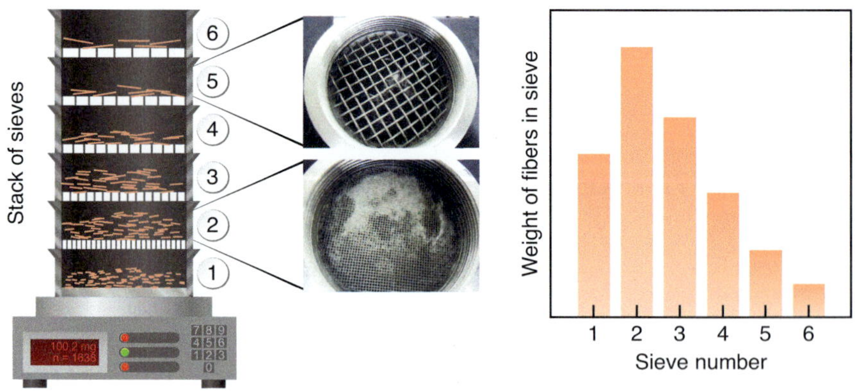

Figure 4.27 Progressive filtering to determine fiber length of samples

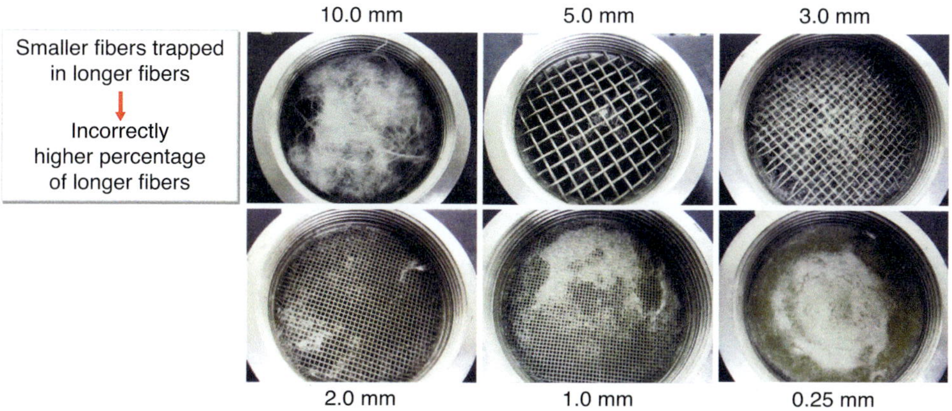

Figure 4.28 Example of progressive filtering for a glass fiber-reinforced sample after injection molding

The direct optical methods (see Figure 4.26) are used for more accurate measurements of fiber length and to study the fundamental relationship of processing conditions, fiber length, and the mechanical properties of molded parts. Table 4.3 summarizes the optical measurement techniques from recently published studies. The majority of the approaches share four main steps. First, a sufficiently large sample is extracted from the part and the matrix material is removed by pyrolysis or chemical digestion. Second, a subset of the fibers is selected and prepared to be scanned (down-sampling and fiber dispersion). In the third step, a digital image is created using either a microscope or an optical document scanner (digital imaging). Finally, the digital image is analyzed using image-processing software to determine the length of individual fibers. The detection of the fibers in the digital image is done either manually by clicking the endpoints of a fiber or with the support of (semi-) automated image-processing algorithms. In most cases the practical approach is hybrid, the algorithm is used to measure as many fibers as possible. After that, the remaining fibers that are often significantly overlapping or have poorer image are measured manually selecting the end points. Based on many years of experience, the authors prefer a sample size of 1000 to 2000 fibers for a good measurement.

Table 4.3 Overview of Fiber Length Measurement Techniques Used in Recently Published Studies

Material	Correction for Down-sampling	Fiber Dispersion	Fiber Detection	Fibers per Sample	Reference
SFT	No	Diluted suspension	Manual	360 ± 60	[70]
PP/GF	No	Manual dispersion	Manual	100	[71]
PP&PA12/GF	No	Manual dispersion	Manual	300	[72]
PP/GF	No	Manual dispersion	Manual	500	[73]
PP/GF	No	Manual dispersion	Manual	500	[74]
PP/GF	No	Manual dispersion	Manual	500	[43]
PP/GF	No	Manual dispersion	Manual	500	[75]
PP/GF&CF	Yes	Manual dispersion	Manual	2000	[5]
PP/GF&CF	Yes	Manual dispersion	Manual	2000	[76]
PA6/GF	No	Diluted suspension	Manual	400	[77]
PA6/GF	No	Diluted suspension	Manual	400	[78]
PA6/GF	No	Diluted suspension	Manual	1000	[12]
PP/GF	No	Diluted suspension	Manual	1000	[7]
GF	No	Manual dispersion	Semi-automatic	1000	[79]
PP/GF	No	Diluted suspension	Semi-automatic	800	[80]
PP/GF	No	Diluted suspension	Semi-automatic	2000	[81]
PP/GF	No	Diluted suspension	Semi-automatic	2000	[82]
PP/GF	Not needed	Diluted suspension	Semi-automatic	2000	[1]
PP/GF	Not needed	Manual dispersion	Semi-automatic	3000	[83]
PP/GF	No	Manual dispersion	Automatic	1000	[84]

For short fiber-filled materials, the International Organization for Standardization (ISO) has defined a standard for fiber length measurements (ISO:22314) [70]. The matrix removal is done by pyrolysis at 625 °C for 1.5 hours and the fiber dispersion is performed by mixing the fibers in water using ultrasound instead of any mechanical action. After completely drying the suspension, 300 ± 60 randomly picked fibers are measured manually with a microscope or stereoscope. The standard states that this measurement technique is only applicable to short fiber-reinforced materials. Although this standard exists, only one publication was found that specifically analyzed the fiber length according to this standard [85]. Nevertheless, most measurement procedures in the field are similar to the ISO:22314 standard with some minor variation such as heating temperature and number of fibers measured.

The practice of using a liquid to disentangle and disperse the fibers after removing the matrix is adopted by the majority of measurement techniques (see Table 4.3).

The fibers are placed in a liquid and mixed either manually [83] or using ultrasound [1, 7]. After mixing and dispersing the fibers, a fraction of the suspension is separated and used to generate digital images. The most common approach is to use water as a liquid with detergent [86] or glycerin [77] as additive to promote fiber dispersion. Ultrasonic vibrators are also used to cause low energy stirring of the mixture to disperse the fibers in the water.

To some degree, the fiber detection step can be supported by image-processing software. However, only a few programs can perform a fully automatic recognition of fibers due to intersecting or curved fibers, which is a particular challenge for longer fibers. Semi-automatic image-processing algorithms are capable of detecting the length of isolated fibers with no or little curvature [1, 84, 87]. Nevertheless, semi-automatic image-processing algorithms still require manual input from the operator to isolate single fibers in a cluster. In the majority of the reported studies, the fiber detection is still done completely manually by clicking the endpoints of fibers.

Almost all the published work on fiber length measurements applies a *down-sampling step* to decrease the amount of fiber that needs to be measured from more than a million fibers in the original sample to less than 3000 fibers. However, the down-sampling step is one of the major factors influencing the accuracy of the obtained results. The cause of this bias comes from the preferential selection of longer fibers, as suggested by Kunc and Nguyen [5, 76].

Kunc and Nguyen [5, 76] described the impact of down-sampling on skewing the obtained fiber length distribution and they suggest a defined down-sampling concept that ensures statistically representative results for a subset of the fibers. After removing the matrix, an epoxy resin is injected in the center of the dry fibers using a thin needle. The deposited epoxy forms a cylindrically-shaped plug through the entire thickness of the fiber bed. The epoxy is cured, and all fibers embedded within the so-called epoxy plug will be measured. The remainder of the fibers that are not bound to the epoxy plug are gently removed by a brush and discarded. A second pyrolysis is necessary to remove the epoxy plug and free the fibers for the dispersion step.

It is important to note that any down-sampling skews the obtained results, including the epoxy plug procedure. Nevertheless, this bias in the obtained results can be accounted for if a defined down-sampling procedure is used by applying a geometrical correction equation proposed by Kunc and Nguyen [5, 76]. The proposed correction accounts for the fact that relatively more long fibers are selected during down-sampling than short fibers due to the small area of the epoxy plug (down-sampling area) [5]. The *Kunc Correction* is illustrated in Figure 4.29.

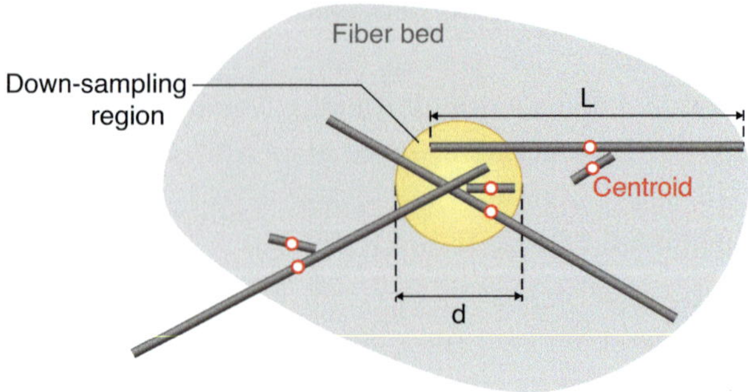

Figure 4.29 Illustration of the *Kunc Correction* procedure, adapted after [76]

A population of three short and three long fibers is down-sampled in a circular region with the diameter d. Although the centroid positions of the long and short fibers are similar, the down-sampling results in the collection of only one short fiber, but all the long fibers. It can be assumed that the fibers are oriented in the 2D plane and the correction can be formulated by applying a geometric correction for the centroid of a fiber L_i passing through the down-sampling area [76]. The corrected frequency $N(L_i)$ of fibers with the length L_i is obtained by:

$$N(L_i) = \theta(L_i)\left(1 + \frac{4L_i}{\pi d}\right) \tag{4.13}$$

where $\theta(L_i)$ is the raw frequency of fibers of length L_i obtained after down-sampling and d is the diameter of the down-sampling area.

FASEP® (Karg Industrietechnik, Krailling, Germany) is a commercially available fiber length measurement system, which can also be used for LFT materials [87]. After the pyrolysis step, the fibers are mixed in a solution of water and glycerin. The suspension is then placed in an ultrasonic bath to disentangle and disperse the fibers. Subsequently, a set of diluting steps is performed by adding more water and glycerin. Combined with gentle stirring, the fibers are uniformly dispersed in a highly diluted mixture. Finally, small portions of the liquid are extracted randomly with a pipette into a Petri dish to create a digital image of the sample. A customized dark-field imaging approach is used to get high-quality digital images. The digital images are analyzed with a customized module of the image-analyzing software Image-Pro Plus (Media Cybernetics, Inc., Rockville, USA), which can detect most fibers semi-automatically. The supplier recommends measuring at least 2000 fibers by repeating the extraction of additional samples from the diluted suspension. This approach offers two distinct advantages: first the fibers are highly diluted so when an image is taken in petri dish, most fibers are dispersed and, second, the dark-filed imaging fixtures result in high-quality images that are easy to analyze. Because of these two advantages, the automated measurement of fibers from digital images is highly successful.

Although not every publication in Table 4.3 describes the measurement approach in detail, there are significant differences in the fiber dispersion and fiber detection steps. Also, the amount of measured fibers varies from 100 to 3000. The selection of the subsample is mostly done manually by collecting random fibers using tweezers. The fiber dispersion is achieved by either mixing the dry fibers into a suspension, similar to ISO:22314, or by manually spreading the fibers using brushes or tweezers. In the majority of the publications, the detection of the fibers is done by hand and only some report an automatic solution using image-processing algorithms. Most studies focused on fiber breakage in glass fiber (GF) filled composites, but a few also measured the length distribution in carbon fiber (CF) filled materials [5, 76].

The large variation in important aspects of the measurement techniques raises questions about the comparability, repeatability, and accuracy of the different studies. Reported results and measured fiber lengths can even differ by a factor of two due to the measurement technique, which was pointed out by the European Alliance for Thermoplastic Composites (EATC) in a call for a new standardized fiber length measurement technique [88]. The fiber length distribution obtained from the collection and down-sampling of fibers is necessarily skewed and does not accurately represent the local fiber population. Additionally, too much manual handling of the fibers with tweezers, spatulas, or similar tools will result in the loss of very short fibers and/or the breakage of longer fibers. Lastly, the number of fibers in a sample can vary by more than an order of magnitude from one study to another, and it is questionable whether a set of 1000 or fewer fibers can be statistically representative of the local fiber length distribution of the sample. Lastly, repeated measurements are rarely performed or reported. Hence, little-to-no information about the measurement variation is available.

A new measurement approach was developed at the Polymer Engineering Center (PEC) at the University of Wisconsin-Madison. Goris et al. [6] proposed a method combining elements from existing procedures with many added improvements to make the measurement faster and robust. This new method is called *PEC method*. Figure 4.30 shows the main steps of the PEC measurement technique.

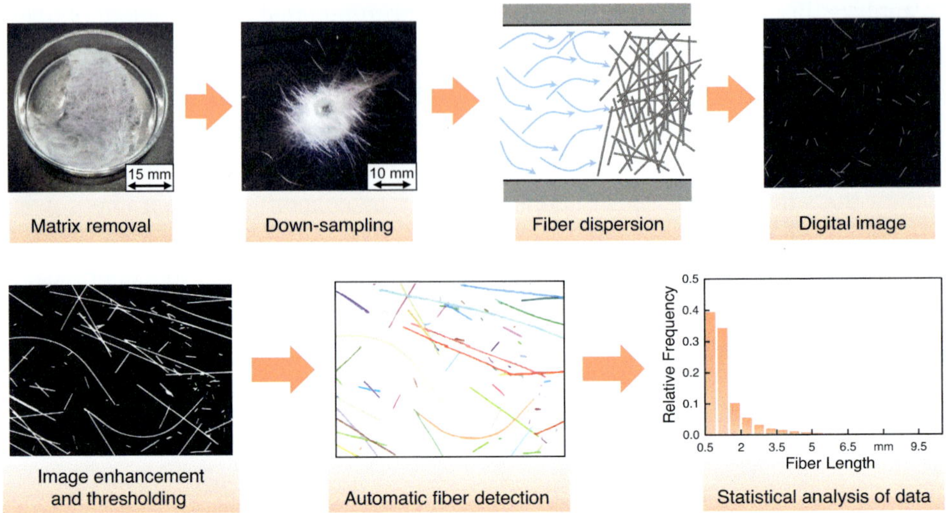

Figure 4.30 Overview of the steps of the PEC fiber length measurement technique

In the first step of the PEC method, a sample with the dimensions of at least twice the initial fiber length is extracted to avoid measuring fibers that crossed the cutting plane during extraction (e.g., an initial fiber length of 15 mm results into extracting round samples with a diameter of 30 mm). Next, the sample is placed in a brass sample holder and constrained by a brass lattice to ensure that the fibers stay in place when the matrix is removed. The pyrolysis is performed in a laboratory furnace at 500 °C for 2.0 h. Subsequently, a representative down-sampling step is performed, adapted from the epoxy-plug method described by Nguyen and Kunc [5, 76]. All fibers in the defined down-sampling region are collected by injecting an epoxy resin through the thickness of the fiber bed using a hypodermic needle. The down-sampling step allows the collection of a controlled and representative subset of fibers that statistically resemble the fiber length distribution of the initial sample. Furthermore, it removes all fibers that were cut at the edges of the sample when it was extracted from the molded part. Lastly, a repeatable amount of fibers for the actual measurement can be collected for each measurement by controlling the injected volume of epoxy. The down-sampling step with the resin injection is an important step to reduce the overall sample size. The presented measurement technique makes use of the epoxy plug method proposed by Kunc et al. [5, 76]. All fibers in a defined down-sampling region are collected by injecting a resin through the thickness of the fiber bed using a needle. The outcome of the down-sampling is shown in Figure 4.31. The overall objectives of this PEC procedure are:

- To collect a representative population of fibers that statistically resembles the fiber length distribution in the initial sample.

- To remove all fibers that were cut at the edges of the sample during extraction from the molded part.
- To obtain a defined amount of fibers for the actual measurement.

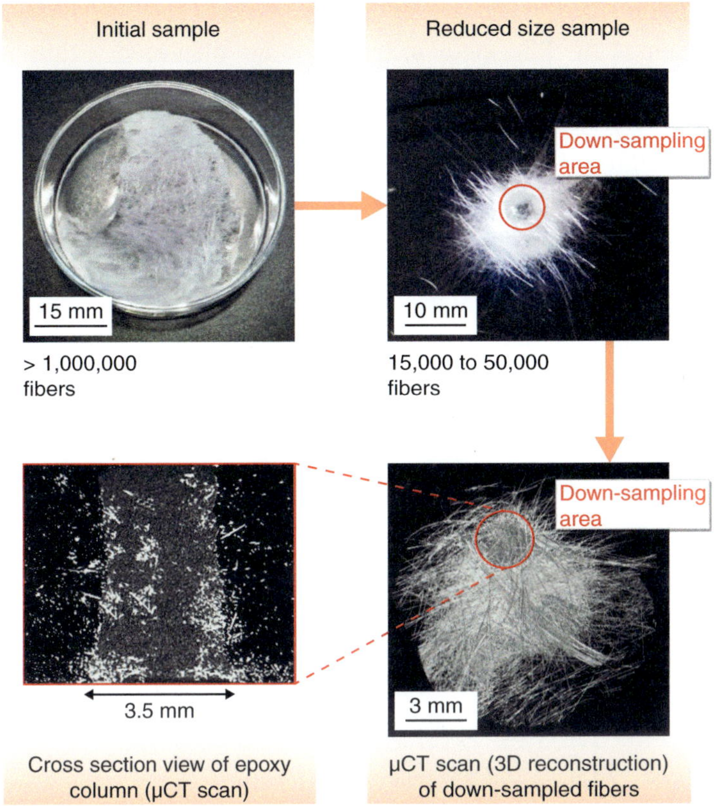

Figure 4.31 Down-sampling step: initial sample after pyrolysis (top left image), subsample after resin injection (top right), and µCT reconstruction of the sample (bottom)

In contrast to the epoxy system applied by Nguyen and Kunc, the PEC method presented here uses an ultraviolet (UV) curable epoxy (BONDIC™ Liquid-glue B0027N07MM, Aurora, Canada). This UV curable epoxy showed optimal flow behavior for the down-sampling step since the viscosity is sufficiently low to be easily injected through a fine hypodermic needle (22-gauge size), but still high enough to form a plug with uniform thickness. Figure 4.31 illustrates the down-sampling procedure and shows a µCT scan of the epoxy plug. The UV curable epoxy is a Newtonian fluid and the viscosity at ambient temperature was measured to be 7.33 Pa·s. Furthermore, the curing time is less than 5 seconds compared to several hours for the mixing-activated epoxy used by Nguyen and Kunc [5, 76].

After the epoxy plug is cured, all surrounding fibers are removed carefully using a brush. The fibers embedded within the epoxy are recovered by a second pyrolysis step at 500 °C for 1 h. The heat ramp is reduced to 5 K/min to slow down the thermal degradation of the epoxy slightly. This initiates an initial disentanglement of the fibers, which is particularly helpful at higher fiber concentrations.

In the next step of the PEC method, the fibers are dispersed within an enclosed chamber system using bursts of slightly compressed air. This is a quite unique feature as the air turbulence causes the fibers to disentangle and disperse uniformly. A few small air bursts (less than 0.5 seconds each) at a pressure of approximately 1 bar are sufficient to disperse the fibers uniformly. It takes less than 15 seconds for all fibers to settle. The fibers fall on a thin glass sheet, which is used to move the sample onto a film scanner. Overall, the fiber dispersion step in an enclosed chamber system using turbulent air flow allows the uniform dispersion of a large amount of fibers in less than 30 seconds. A schematic of the design of the dispersion chamber is shown in Figure 4.32. The enclosed chamber is made of stainless steel and has overall dimensions of 380 mm (height) × 210 mm (width) × 255 mm (depth). The glass sheet is a laboratory-grade float glass and its dimensions are 210 × 260 × 1 mm^3. The inner surface of the dispersion chamber is polished to ensure a smooth surface and coated with a thin layer of dry graphite to prevent fibers from sticking to the surface of the chamber. Pre-trial experiments showed that less than 1% of the glass fibers are lost in the dispersion process.

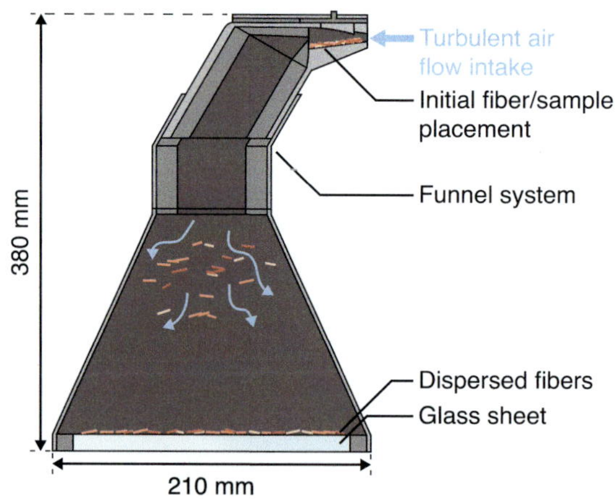

Figure 4.32 Sketch of the dispersion chamber for the preparation of the fiber sample for the PEC method

The dispersed fibers are scanned between 1200 dpi and 2400 dpi to generate a digital image using a flatbed scanner with a transparency unit (Epson Perfection V750 Pro by Seiko Epson Corporation, Suwa, Japan). All fibers in the image are automatically detected by an image-processing algorithm developed by Perez et al. [55] and based on the work of Wang [89]. Even bent and intersecting fibers are detected by this algorithm automatically. This ability is particularly important for accurate results when examining LFT materials because longer fibers can be substantially curved, even after pyrolysis [86]. The current version of the algorithm is coded in MATLAB R2016a and takes between 5–10 min to process the images of each sample.

To ensure the measured fiber length distribution is statistically representative, between 15,000 and 50,000 fibers are recommended for each sample. The measurement time for the PEC method is observed to be approximately 60 min per sample, excluding furnace time.

4.2.2.1 Comparative Study of Measurement Techniques

A comparative study for fiber length measurements was performed for injection molded LFT samples, molded with long glass fiber-reinforced polypropylene (PP) with 30 wt % (PPGF30) and 40 wt % (PPGF40) fiber concentration. The material is supplied as coated pellets with a nominal fiber length of 15 mm. The material was processed on a 320-ton Arburg 720S injection molding machine equipped with a general-purpose screw and diameter of 52 mm. The mold geometry for this study is a simple fan-gated plaque with a thickness of 2 mm as illustrated in Figure 4.29. The processing parameters are listed in Table 4.4, which follow the general processing guidelines suggested by the material supplier [90]. Two different molding conditions were defined: *standard* and *gentle*. The standard processing conditions use a higher screw speed (35 rpm) and back pressure (50 bar) than the gentle conditions, which process the material at 27 rpm and 13 bar.

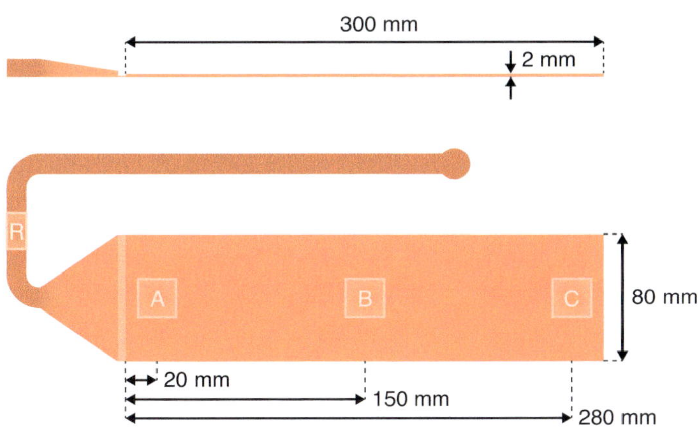

Figure 4.33 Part dimensions and sample locations for fiber length analysis

Table 4.4 Processing Conditions for the Injection Molding Trials (Standard vs. Gentle Conditions)

Molding Parameter	Standard Conditions	Gentle Conditions
Melt Temperature [°C]	250	250
Mold Temperature [°C]	50	50
Back Pressure [bar]	50	13
Screw Speed [rpm]	35	27
Injection Time [s]	2.5	2.5

Samples from location B at plaques for both 30% glass fiber PP and 40% glass fibers PP material are selected for both the standard and gentle processing conditions. The fiber lengths are measured using four different procedures for comparison. The fiber measurement procedures used in this comparative study are i) a baseline for the conventional fiber measurement technique, ii) FASEP measurement system, iii) Full Analysis, and iv) the PEC Method (see previous section for details). Table 4.5 summarizes the main characteristics of the compared measurement approaches.

Table 4.5 Overview of Measurement Techniques for the Comparative Fiber Length Analysis Study

	FASEP	Full Analysis	PEC Method
Fiber Dispersion	Diluted suspension	Manual	Turbulent air chamber
Sample Size	Representative (random) fraction	All fibers	Representative down-sampling
Measured Fibers	2000 to 15,000	350,000 to 750,000	15,000 to 50,000
Image Processing	Semi-automatically	Fully automatic	Fully automatic
Measurement Duration	2 hours per sample	8 hours per sample	Less than 1 hour per sample

As discussed, the majority of current measurement procedures manually select a random subset of fibers from a larger sample after the matrix was removed. These results can be skewed by both the amount of measured fibers and the bias in down-sampling. In order to generate a baseline for the comparative study of this work, ten identical samples (PPGF40, gentle conditions, position B) were measured according to the reported conventional measurement procedures, referred to as manual measurements. The manual measurement procedure was performed as follows.

Samples with the dimensions of 30 × 30 × 2 mm^3 were extracted from the molded part and the matrix was removed by pyrolysis at 500 °C for 2.0 h. From each sample, a subset of fibers was carefully removed, randomly, using plastic tweezers, following the procedures described in published studies. The measured amount of fibers varied from 282 to 4612 fibers, which is within the range of the reported studies. The fibers were manually dispersed using fine brushes and a digital image was created. Each fiber in the digital image was manually detected by clicking the endpoints using a developed MATLAB tool. The digital images were generated at 2400 dpi, translated to a binary image, and the endpoints of each individual fiber were selected manually. The length of curved fibers was determined by splitting the fiber in segments, fitting circular arcs to any curved segments and adding the length of all segments. This protocol of manual measurements is assumed to reflect the capabilities of the conventional measurement techniques.

The results of this evaluative study are summarized in Figure 4.34, which shows the obtained weight-average length and the number of fibers in each repetition. The manual technique shows a substantial variation in the measured fiber length distributions with the weight-average fiber length varying from 1.03 mm to 3.96 mm (mean of 2.37 mm and standard deviation of 0.95 mm). The large variation in the measurements confirms the claim by EATC that the variation in conventional measurement techniques might differ by a factor of two [88].

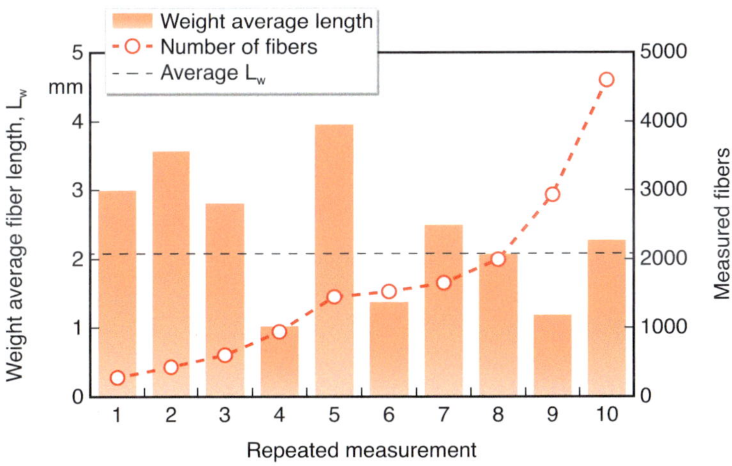

Figure 4.34 Results obtained from current procedure using manual measurements. Ten individual measurements of samples extracted from PPGF40 plates and molded at gentle processing conditions. This is used as the baseline to compare with other methods

Full Analysis

The Full Analysis is a measurement system based on the work of Krasteva [91] and described in [6]. The Full Analysis measures the entire fiber population in a sample and does not perform any down-sampling step. The size of the sample is 12 ×

12 × 3 mm^3 comprising approximately 350,000 to 750,000 fibers for a single measurement. The sample size was determined as the optimal trade-off between the amount of fibers that need to measured and the error introduced by cutting fibers when extracting the sample from the molded part [91]. After pyrolysis, all fibers are dispersed manually in subsets of 40,000 to 50,000 fibers. A digital image of each subset is generated and processed with an automated image-processing algorithm. Due to the amount of the overall fibers in a single sample, the manual dispersion is time-consuming.

Two different sets of molded samples were defined for the comparative study: PPGF30 (gentle conditions, location B) and PPGF40 (gentle conditions, location B); for sample location, refer to Figure 4.34. Measurements were repeated five times at all locations with the present method and the results are shown as average values, with error bars indicating the standard deviations. Two repeated measurements were performed with the FASEP system and one repetition for the Full Analysis. The results for the PPGF40 (gentle conditions, location B) samples are summarized in Figure 4.35 as average length values and the cumulative length distribution. Figure 4.36 summarizes the obtained average values for the PPGF30 (gentle conditions, location B) samples.

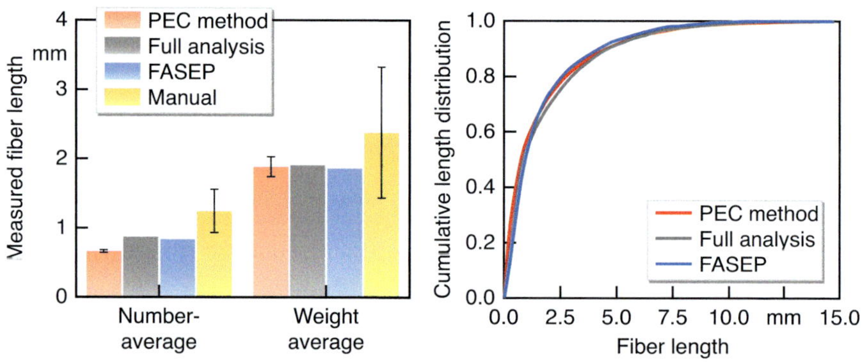

Figure 4.35 Comparison of the outcome of the fiber length measurements for all measurement protocols (PPGF40, gentle conditions, location B): average fiber length values, standard deviations indicated by error bars (left), and cumulative fiber length distribution (right)

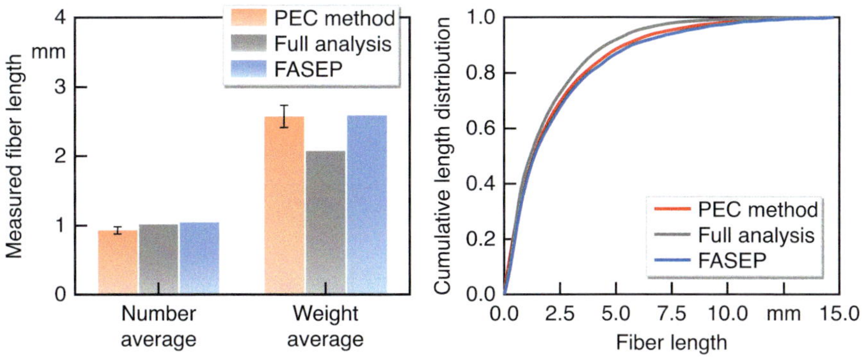

Figure 4.36 Comparison of the outcome of the fiber length measurements for all measurement protocols (PPGF30, gentle conditions, location B): average fiber length values, standard deviations indicated by error bars (left), and cumulative fiber length distribution (right)

While the FASEP measurements barely show any deviation between the two repeated measurements (± 0.1 mm), the measurements from the proposed method show a slightly higher deviation. The Full Analysis suggests a lower weight average length for the GF30, whereas the results for GF40 show a discrepancy of less than 2% between FASEP, the Full Analysis, and the present method. However, the results from the manual measurement suggest a 27% higher value for the weight-average fiber length and the standard deviation over the ten repeated samples is 0.95 mm (or 40.0% of the mean).

The fiber length distribution was measured for all process conditions and sample locations using the proposed PEC method and the Full Analysis for a further comparison between the two measurement techniques. Five repetitions for the PEC method and one repetition for the Full Analysis were performed for each location. Figure 4.37 shows the weight-average fiber length of the PPGF30 trials, for standard conditions (left) and gentle conditions (right). The results from the PPGF40 trials are summarized in Figure 4.38 for the standard conditions (left) and gentle conditions (right). The results show a strong agreement between the measurement techniques. While some samples show a slightly larger discrepancy of around 20% (PPGF40 gentle conditions, location A and PPGF30 gentle conditions, location B), the difference for all other locations is on average less than 5%.

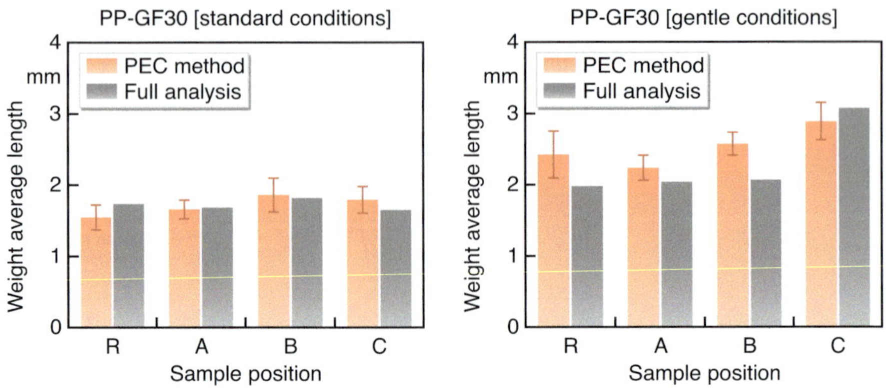

Figure 4.37 Comparison of Full Analysis and PEC method for all locations: PPGF30 at standard conditions (left) and PPGF30 at gentle conditions (right)

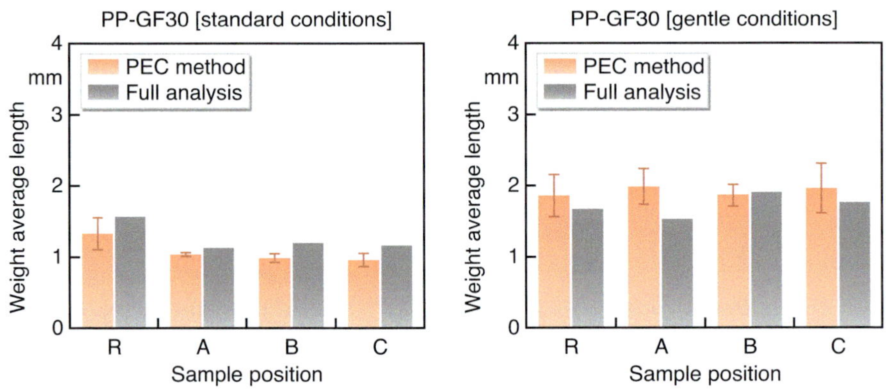

Figure 4.38 Comparison of Full Analysis and PEC method for all locations: PPGF40 at standard conditions (left) and PPGF30 at gentle conditions (right)

The results of the comparative study show the importance of a standardized measurement technique for fiber length analysis. The conventional method of manually selecting a subset of fibers and measuring a small amount of fibers per sample can result in substantial inaccuracies and large variations across repeated measurements. As reported in recent studies [5, 6, 76], there is a bias towards longer fibers from manually collecting and analyzing fiber samples. While measuring a small amount of fibers is certainly less time-consuming regardless of the actual measurement approach, it is questionable if fewer than 1000 fibers can be statistically representative of the local fiber length distribution of the sample. The Full Analysis approach of a sufficiently large sample might be the ideal procedure since it generates a statistically robust data set and no correction for down-sampling is necessary. However, the time-consuming sample preparation (fiber dispersion and generation of the digital image) might limit the application of this approach. For example, larger experimental studies on fiber breakage could require repeated measurements and analyzing 500,000 fibers per sample without automated fiber dispersion is very challenging.

Because it dilutes an entire fiber sample in a liquid, the FASEP system does not require any correction due to down-sampling. However, randomizing the fibers in the suspension and diluting the mixture requires time-consuming manual effort to be performed correctly. Also, a complete randomization of fibers within a highly-diluted suspension might be difficult to achieve for LFT samples due to the wide range of fibers in the samples. Larger fibers settle faster than smaller ones. Hence, it is challenging to ensure that a truly randomized sample is extracted from the diluted mixture. Nevertheless, the FASEP system uses a measurement protocol with defined steps resulting in repeatable measurements. At the time of this work, it is the only commercially available measurement system for fiber length analysis.

The outcome of the comparative study shows that the PEC method can be used to generate accurate results with a high level of repeatability, while minimizing the manual effort required to handle, prepare, and analyze a fiber sample. It improves upon the shortcomings of previously proposed characterization methods and the fiber length of large samples to be accurately measured in a timely manner to obtain statistically meaningful data.

The proposed PEC methodology also works for other fiber types. Initial tests with carbon fibers and natural (sisal and hemp) fibers showed that the same methodology can be applied. Only the matrix removal steps need to be modified accordingly since pyrolysis is not possible for natural fiber-reinforced composites and a chemical matrix removal is necessary. Carbon fiber-reinforced materials require a nitrogen atmosphere and carefully monitored pyrolysis to prevent oxidation of the fibers. Moreover, the scan resolution needs to be increased to at least 4800 dpi for carbon fibers due to the smaller fiber diameter.

4.2.3 Measuring Fiber Concentration

The average fiber concentration of a sample is commonly measured by removing the thermoplastic matrix, either chemically or by pyrolysis, and calculating the weight fraction from the remaining fibers. Due to its simplicity, most authors suggest removing the matrix for glass fiber-reinforced materials by thermal degradation (pyrolysis). Rohde et al. studied the impact of matrix removal by pyrolysis or chemical decomposition on the morphology of single fibers [1]. The results show that performing pyrolysis at 500 °C for 2 hours is optimal for a PP sample. Their work indicates that the properties of the glass fibers can change at higher temperatures and the fibers become more brittle. As Rohde et al. pointed out, the increased brittleness can cause additional fiber breakage in the handling of the sample. On the other hand, the time needed to completely remove the matrix at lower temperatures is too long to ensure a timely analysis. In other studies, scientists used higher temperatures and different incineration times that ranged from 800 °C and 1 h [77] to 530 °C and 5 h [12].

The fiber concentration variation in the thickness direction of samples can also be measured by pyrolysis. This approach requires the collection of the material of several layers through the thickness. For each data point along the thickness direction, the sample needs to be prepared by grinding and burning to determine the fiber density. This measurement procedure is not only cumbersome, but also suffers limitations in terms of sample size, because the part thickness for common injection molded parts ranges between 1 to 4 mm. To obtain sufficient data points (> 10), the ground samples need a thickness between 0.1 and 0.4 mm. On this scale, fiber pull out by the grinding process can cause errors in analysis. Recently, through-thickness analyses were published that applied μCT to determine the fiber density [33]. The authors determined the fiber density by calculating the fraction of pixels that represent fibers to the total amount of pixels. The authors found the expected fiber agglomeration in the center of the part, but their values of fiber density (average of 4 vol %) are actually off by a factor of two when compared to the expected nominal fiber volume fraction (9 vol %).

Here we are presenting a new approach, called μCT procedure, to apply μCT scanning and quantify the through-thickness fiber concentration using image processing. In this approach the scanned digital μCT data sets are processed to determine the change of fiber concentration through the sample thickness by applying an image-processing algorithm. The process flow chart of the procedure is illustrated in Figure 4.39. First, the raw μCT data set is aligned and registered using VG Studio MAX™ 3.0. Subsequently, the data set is exported as an image stack (2D slices) oriented normal to the thickness direction. The 2D slices comprising grayscale images are imported into MATLAB and the fiber volume fraction through the thickness of the sample is calculated using the newly developed algorithm described in the following paragraphs.

The grayscale images are transformed into binary images by segmentation, which requires selecting a reasonable threshold value to separate each image into black (matrix) and white (fibers) pixels as shown in Figure 4.40 (left). In the grayscale 2D slice, each pixel with the coordinates (x, y) has a distinct grayscale value $f(x, y)$ which depends on the bit depth (e.g., a 16-bit grayscale image has 65,536 tonal levels).

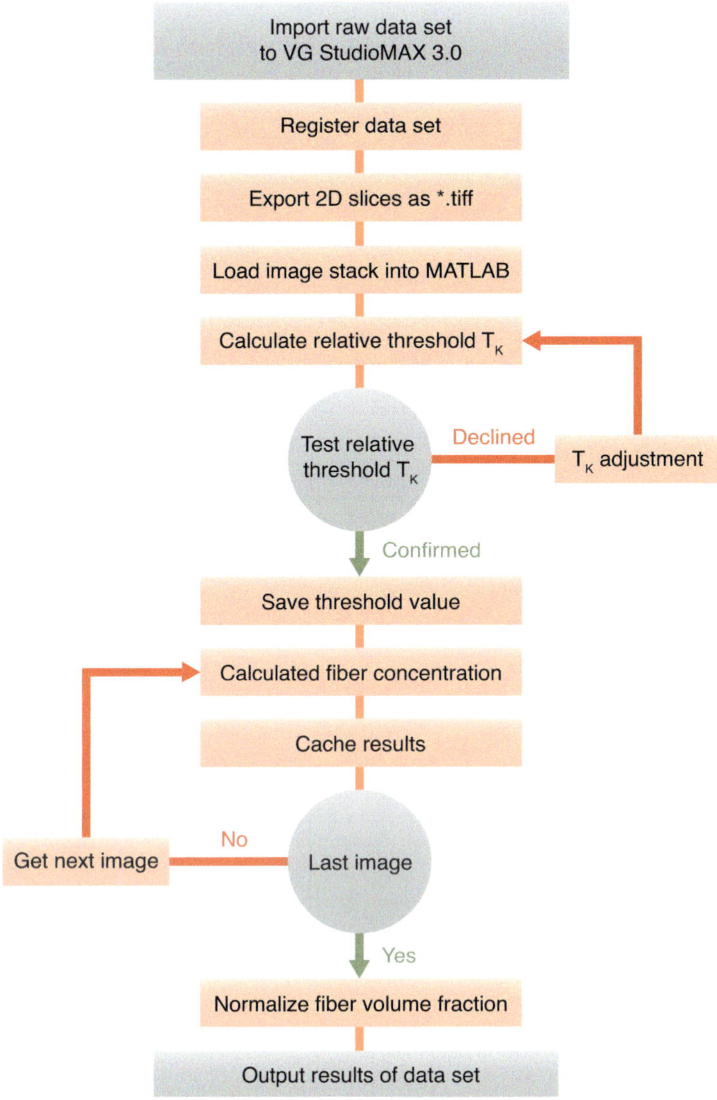

Figure 4.39 Flow chart of the procedure for the through-thickness concentration analysis

A basic thresholding approach is used to convert the grayscale image into a binary image by setting a global threshold value. The value for each pixel in the binary image is calculated by

$$g(x,y)=\begin{cases}1\ ,\ if\ f(x,y)>T_k\\0\ ,\ if\ f(x,y)<T_k\end{cases} \tag{4.14}$$

where T_k is the relative grayscale value for thresholding. For each 2D slice z, the fiber volume fraction $\varnothing_f(z)$ is calculated by the fraction of white pixels in the entire image $N \times M$:

$$\varnothing_f(z) = \frac{\sum_{i=1}^{n} \sum_{j=1}^{m} g(x_i y_j)}{N \times M} \tag{4.15}$$

The true T_k is unknown due to the gradual change in the grayscale value between the two phases as illustrated in Figure 4.40 (left). The choice of the threshold directly determines the size of the segmented phases and, thus, the fiber volume concentration. Even at very fine resolutions, the true threshold value cannot be detected directly from the μCT data set. A heuristic procedure for the segmentation is proposed by calculating the threshold value as the midpoint between the mean value representing the fibers and the mean value of the background. This selection might not result in the true threshold value, but any uncertainty around the value would merely shift the obtained fiber volume concentration as shown in Figure 4.40 (top right). Hence, selecting a single value for T_k and performing a normalization step for the entire image stack (μCT data set) resolves the ambiguity in selecting the true threshold value.

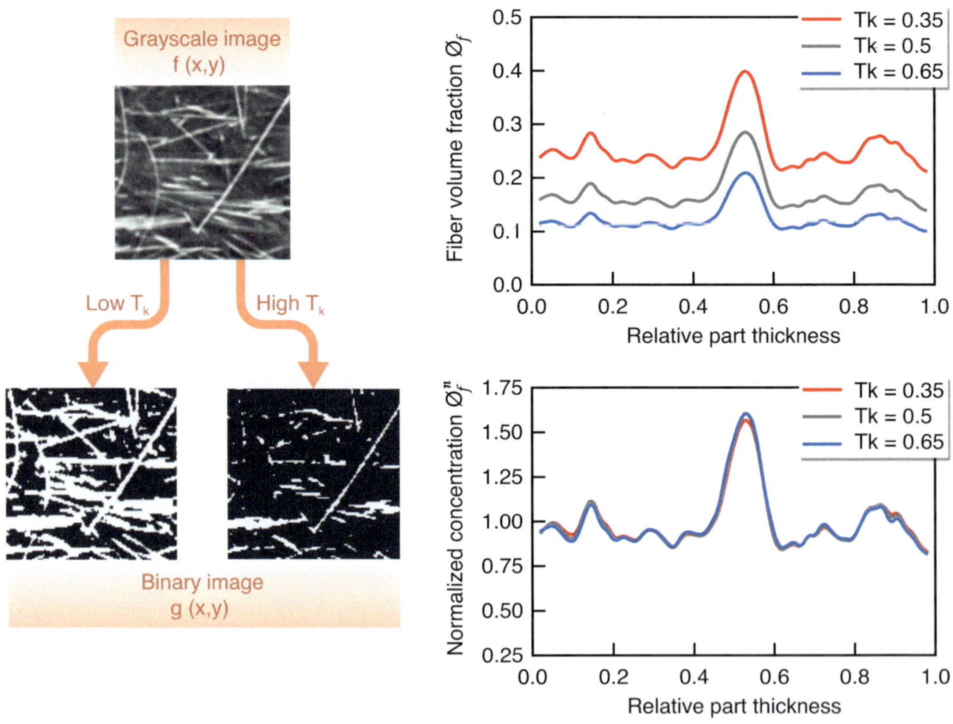

Figure 4.40 Transforming a grayscale image into a binary image (left), the obtained fiber volume fraction for varying relative threshold values (top right), and the normalized fiber concentration distribution (bottom right)

The average fiber concentration of the entire μCT data set $\bar{\varnothing}_f$ is calculated from the individual fiber volume concentration $\varnothing_f(z)$ of each slice. The normalized fiber concentration distribution $\varnothing_f^n(z)$ can be obtained for each image as follows

$$\varnothing_f^n(z) = \frac{\varnothing_f(z)}{\bar{\varnothing}_f} \qquad (4.16)$$

For a wide range of values for T_k, the normalized fiber concentration distribution is the same, as shown in Figure 4.40 (right). Only at extreme values, where $T_k < 0.3$ or $T_k > 0.7$, skewed distributions are seen. With this approach, it is possible to accurately obtain the through-thickness fiber concentration from μCT data. If needed, absolute values for the fiber concentration distribution can be calculated by measuring the local fiber concentration of the entire sample through pyrolysis.

The proposed measurement protocol for the through-thickness fiber concentration analysis using μCT was validated by performing a milling and pyrolysis procedure. A simple plate geometry was used to mold samples at 20 wt % and 40 wt % as shown in Figure 4.41. Samples at the center location of the plate were extracted and measured. The procedure was repeated three times for each sample.

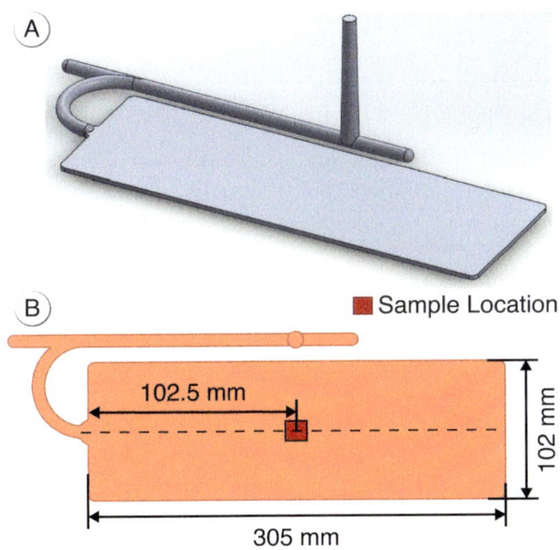

Figure 4.41 Sketch of the plate geometry (a) and illustration of the sample location for the microstructure analysis (b)

After the μCT scan of a sample, the identical sample was milled down in defined increments of 0.2 mm along the thickness. The shavings were carefully collected using a vacuum setup and a 25 μm mesh to ensure capturing all material during milling. The fiber weight concentration for each layer was determined by measuring the sample weight before and after pyrolysis on a high precision scale Explorer

(Ohaus, Parsippany, USA) with an accuracy of ± 0.01 mg. After converting the obtained measurements to volume fraction, the results from the μCT measurements and from the pyrolysis can be compared. The fiber diameter of the material is 19 μm and single fiber filaments can be clearly identified in the scans at resolutions below this value. Pre-trial experiments on the impact of resolution on the results showed that finer resolutions do not change the outcome. Table 4.6 summarizes the scan parameters used for all subsequent measurements. The sample size at 5.25 μm is 10 × 15 × 2.85 mm^3.

Table 4.6 μCT Scan Parameters of the Zeiss Metrotom 800 Used for this Study

Parameter	Value
Voltage [kV]	50
Current [μA]	80
Integration Time [ms]	1000
Gain [-]	8.0
Spot Size [μm]	5.0
Voxel Size [μm]	5.25

The outcome of the comparison is shown in Figure 4.42. The results show strong agreement between the two measurement procedures. Discrepancies between the results from the milling procedure and the μCT analysis are less than 2.5% for each of the six samples. Hence, the proposed μCT analysis protocol is suitable to accurately and efficiently quantify the fiber concentration through the thickness of the part. Not only is the μCT approach non-destructive, it is a faster analysis of the through-thickness fiber concentration and it also allows additional fiber orientation analyses of the same μCT data set.

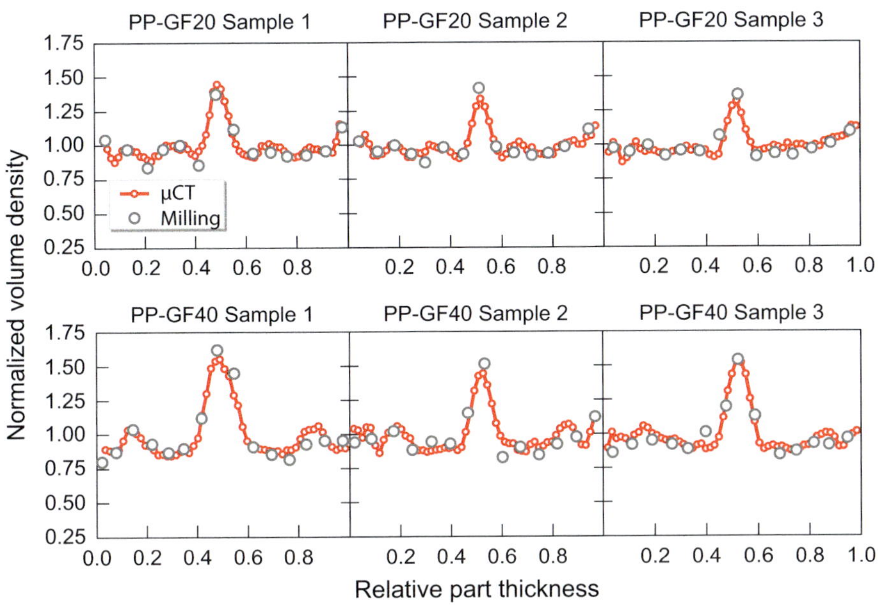

Figure 4.42 Comparison of the milling procedure and μCT procedure for the through-thickness fiber concentration measurements

4.2.4 Fiber Microstructure Characterization Conclusions

The analysis of the microstructural properties of discontinuous fiber-reinforced composites is a challenging and cumbersome task, but it is a necessary step to obtain a fundamental understanding of the phenomena present in LFT processing. This chapter highlights novel characterization techniques that use optical measurement systems and image processing to quantify the microstructural properties. While accuracy is the most important objective in developing a measurement procedure, the time required to take measurements also plays an important role.

The comparative studies show that the comparability of reported results across different studies using different measurement methodologies is limited due to the influence of the measurement approach causing skewed results. This was particularly evident in the fiber length analysis, where the conventional measurement approach yielded results that were not repeatable. Hence, any comparison of reported results across different studies and research groups has to be done with care due to the discrepancies in measurement techniques. Overall, the outcome highlights the need for a more standardized approach in characterizing the fiber length of molded samples to allow a fair and accurate comparison between experimental studies. As it overcomes the shortcomings of conventional measurement techniques, the developed fiber length measurement technique is a first step to standardization.

Analyzing the through-thickness fiber orientation and fiber concentration distribution was made possible using the μCT procedure and it proved to be a mighty tool for the fiber microstructure analysis. A measurement protocol using μCT and image processing was developed and validated for fiber orientation and fiber concentration analyses. While the investment costs of a μCT system limits the application of μCT for microstructure analysis of fiber-reinforced samples, it shows clear advantages over conventional measurement techniques. In fact, μCT allows the analysis of both fiber concentration and fiber orientation from the same data set. Additionally, image-processing algorithms for fiber length analysis using μCT are being successfully applied to short fiber-reinforced composites [53]. The 3D analytical capabilities and non-destructive nature of X-ray μCT is particularly useful to connect the process-microstructure relationship with the microstructure-property relationship, because the fiber microstructure of test samples can be characterized before mechanical testing. Overall, it was shown that the size of the sample and number of repeated measurements are important to reliably obtain the local information of the microstructural property. A single sample can carry local heterogeneities or irregularities, which might skew the outcome of a measurement, if the sample size is small with respect to the heterogeneities. In particular, the longer fibers in LFT materials can result in non-dispersed fiber bundles, as shown in Figure 4.43.

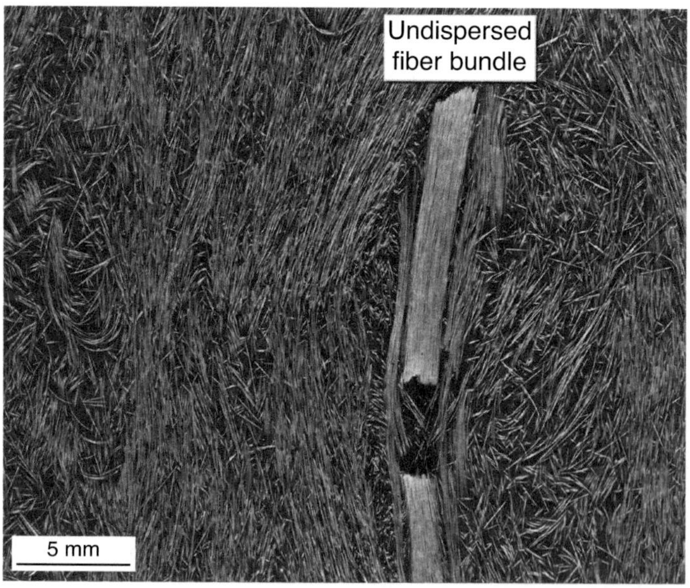

Figure 4.43 Illustration of an undispersed fiber bundle in an injection molded part (40 wt % glass fiber-reinforced PP)

4.3 Case Study: LFT Injection Molding of a Simple Plaque Geometry

The analysis of the microstructural properties of discontinuous fiber-reinforced composites and the control of the macroscale properties is a challenging task. While the characterization is a cumbersome step, it is crucial for a fundamental understanding of the phenomena present in LFT processing. This section shows a selected case study to illustrate the highly heterogeneous fiber configuration caused during processing.

A simple plate geometry was molded and the microstructural properties were analyzed. The part geometry used in this study is a simple plate with dimensions of 102 × 305 × 2.85 mm^3. The cavity is filled through a 20 mm edge-gate with the same thickness as the plate and is fed through a 17 mm full-round runner, as illustrated in Figure 4.44. The material used in this study is a commercially available long glass fiber-reinforced polypropylene (SABIC STAMAX™ LFT), which is supplied as 15 mm long pellets. Table 4.7 summarizes the main material properties of the glass fiber-reinforced polypropylene (PPGF).

Table 4.7 SABIC STAMAX™ LFT Material Properties According to the Material Supplier [92]

Material Property	Value
Nominal Fiber Length [mm]	15.0
Fiber Diameter [µm]	19.0 ± 1
Density of Fibers [g/cm^3]	2.55
Density of Matrix [g/cm^3]	0.905
Young's Modulus of Single Fibers [GPa]	73
Ultimate Strength of Single Fibers [MPa]	2600

The design-of-experiments (DoE) consists of nominal fiber concentrations varying from 5 wt % to 60 wt %. Table 4.8 describes the fiber volume and fiber weight concentration, as well as the feed material used to achieve the respective nominal concentrations for each trial. PPGF20, PPGF30, PPGF40, and PPGF60 are provided as compounded pellets by the material supplier (coated long fiber pellets). PPGF05, PPGF10, and PPGF50 were achieved by mixing higher fiber concentrations with neat PP (SABIC™ PP 579S) in a cement mixer before feeding it into the hopper of the injection molding machine. The neat PP is the same as the matrix material of the coated long fiber STAMAX™ pellets.

Table 4.8 Outline of the Injection Molding Trials: Nominal Fiber Concentration and Raw Materials

Trial ID	Fiber Weight Concentration	Fiber Volume Concentration	Raw Materials
PPGF05	5 wt %	1.8 vol %	25% PPGF20 and 75% neat PP
PPGF10	10 wt %	3.8 vol %	50% PPGF20 and 50% neat PP
PPGF20	20 wt %	8.2 vol %	STAMAX™ 20YM240
PPGF30	30 wt %	13.2 vol %	STAMAX™ 30YM240
PPGF40	40 wt %	19.1 vol %	STAMAX™ 40YM240
PPGF50	50 wt %	26.2 vol %	83% PPGF60 and 17% neat PP
PPGF60	60 wt %	34.7 vol %	STAMAX™ 60YM240

The parts were molded on a 130-ton Supermac Machinery SM-130 injection molding machine. The processing settings followed the general processing guidelines provided by the supplier (SABIC™) and are summarized in Table 4.9. The processing conditions are kept constant unless noted otherwise.

Table 4.9 Summary of Processing Conditions for the Injection Molding Trials

Material Property	Value
Melt Temperature [°C]	250
Mold Temperature [°C]	50
Back Pressure [bar]	5
Injection Time [s]	2.5
Holding Pressure [bar]	300
Holding Time [s]	22
Ultimate Strength of Single Fibers [MPa]	2600

The characterization of the fiber configuration is focused on the following three microstructural properties:

- **Fiber Length Distribution:** The local fiber length and its change along the flow path within the molded plate, which was obtained by measuring the fiber length at Location 1 *(gate region)*, Location 2 *(center of plate)*, and Location 3 *(end of the flow)*, as illustrated in Figure 4.44(b). Additionally, the fiber length after plasticating was measured to understand how much additional fiber length reduction occurs during mold filling. The measurements were performed according to the PEC method, as described in the previous section.
- **Fiber Orientation Distribution:** The through-thickness fiber orientation is measured at the same three locations along the centerline of the plate. The mea-

surements were done using mCT and using the VGStudio MAX™ for the fiber orientation analysis.

- **Fiber Concentration Distribution:** The through-thickness fiber concentration change in fiber concentration was obtained at Locations 1, 2, and 3 by applying the characterization process described in the previous section. The change in global fiber concentration throughout the entire plate was measured by pyrolysis of 13 × 26 × 2.85 mm^3 samples extracted from the plate, as illustrated in Figure 4.44(c).

Additionally, the fiber length and fiber concentration measurements were performed on purged material samples, which were obtained by air shots from a retracted injection unit at 10% of the injection speed.

The results of this experimental study are presented and discussed in the following sections focusing on the three microstructural properties individually before concluding the outcome of this study and discussing the correlation between fiber length, fiber orientation, and fiber concentration at the end of this chapter.

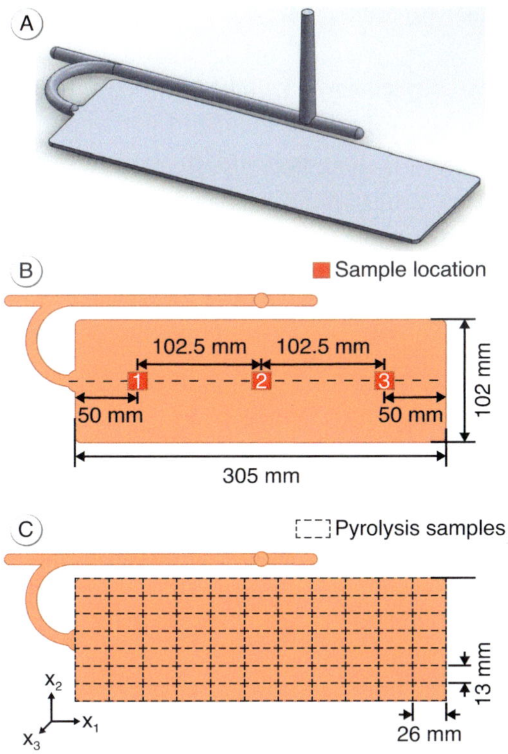

Figure 4.44 Sketch of the plate geometry and illustration of the sample locations for the microstructure analysis

4.3.1 Analysis of the Process-Induced Fiber Alignment

Figure 4.45 summarizes the measured fiber orientation at Locations 1, 2, and 3 for all trials (PPGF05 through PPGF60). The plots show the fiber orientation along the direction of flow, A_{11}, and in the cross-flow direction A_{22}. The fiber orientation in thickness orientation, A_{33}, is not shown for clarity ($A_{11} + A_{22} + A_{33} = 1$). Three samples were analyzed for each location and the low standard deviation of the measurement results suggests a high reproducibility in the molded samples.

The trials at higher concentrations (PPGF20 to PPGF60) show the expected core-shell-skin pattern. The thick core layer consists of fibers predominately aligned along the cross-flow direction (A_{22}) while the fibers in the shell layer are oriented along the direction of flow (A_{11}) due to the fountain flow effect. The orientation in thickness direction (A_{33}) is low with average values of less than 0.06 and is uniform through the thickness for all trials.

The measurements of the samples with 5 wt % and 10 wt % nominal fiber concentration (PPGF05 and PPGF10) indicate an alternating fiber orientation through the thickness of the sample, which is particularly distinctive at Location 2 (see Figure 4.45):

- Core layer: A thin core layer shows a slight preferential fiber alignment in cross-flow direction with an A_{22} value of 0.58 for PPGF05 and 0.63 for PPGF10.
- Layer with fiber orientation along flow direction: After a small transition region, two thick layers with a strong in-flow fiber alignment can be observed at a relative thickness of approximately 0.35 (0.65). The measurements suggest a strong fiber alignment in the direction of flow with average values for A_{11} of 0.78 for both PPGF05 and PPGF10.
- Cross flow alignment layer: After a second transition region, the measurements indicate a distinctive layer with fibers predominately oriented in the cross-flow direction at a relative layer thickness of approximately 0.15 (0.85). The PPGF05 trial shows a higher degree of orientation in this layer with an average value for A_{22} of 0.71 while the results for PPGF10 suggest a value of 0.59.
- After a third transition region, the thin skin layer close to the surface can be identified with preferential orientation in the direction of flow.

For all trials, the samples close to the gate (Location 1) show a wider core layer than those at Locations 2 and 3, while the degree of orientation in the shell layer is lower. This can be explained with the radial flow field close to the gate due to the edge gate.

The measurements for PPGF60 at Location 3 suggest a large core layer and an overall random fiber orientation, including an overall larger standard deviation for the repeated measurements. This indicates that the high fiber fraction of 60 wt %

might have caused a non-uniform flow front at the end of flow due to the high suspension viscosity.

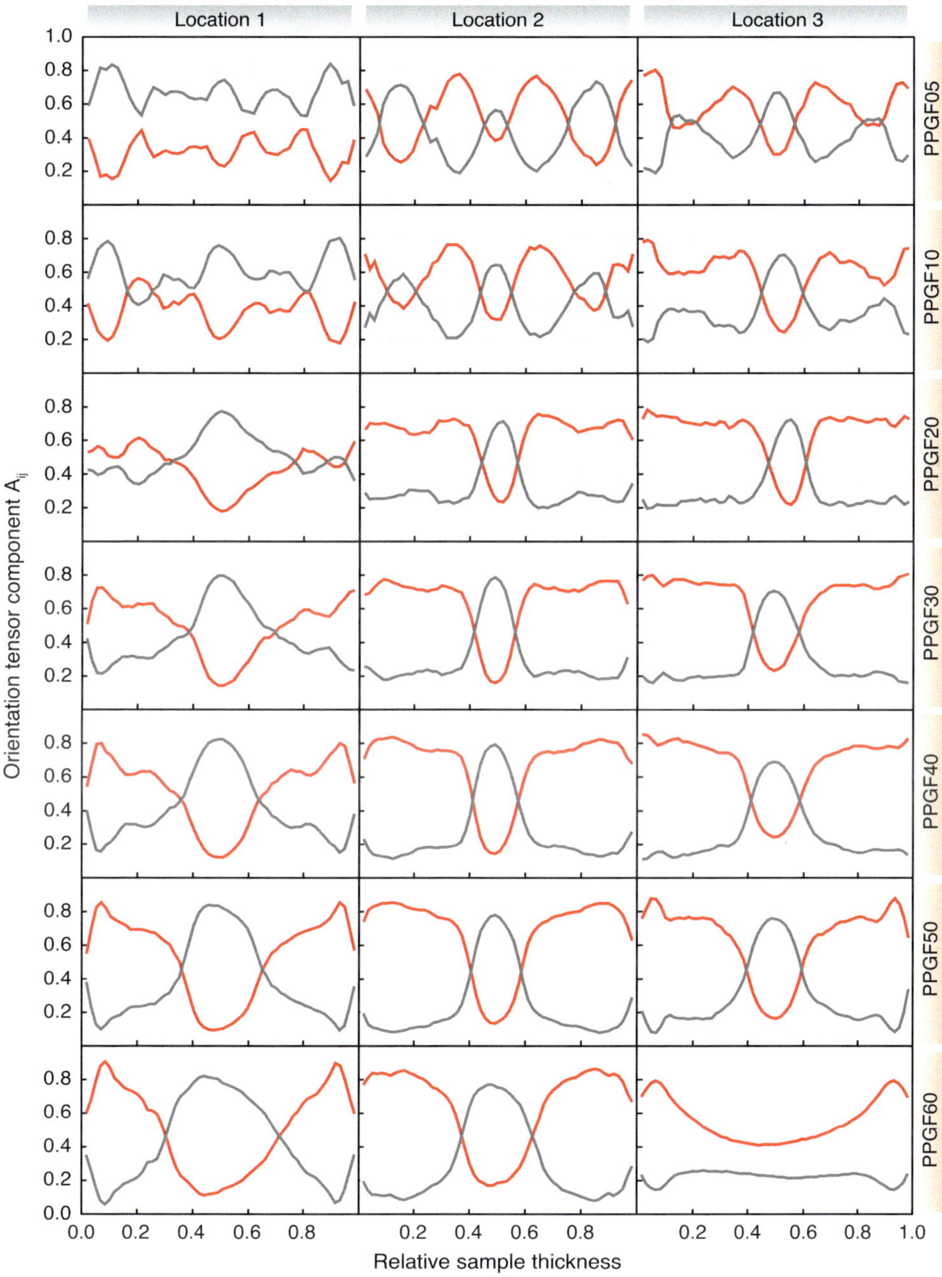

Figure 4.45 Fiber orientation measurements at Locations 1, 2, and 3 for all trials: graphs showing A_{11} and A_{22} plotted along the thickness (A_{33} is not shown for clarity)

The measurements show that the nominal fiber concentration affects the degree of orientation in the core and shell layers as well as the width of the core layer. Figure 4.46 shows the preferential orientation A_{11} in the shell layer (Figure 4.46 left) and the orientation in cross-flow direction A_{22} in the core along with the core layer width (Figure 4.46 right) for all trials. In the shell layer, the fiber alignment in the direction of flow (A_{11}) increases with increasing nominal fiber concentration; the strongest alignment occurs with 50 wt %, corresponding to a value of 0.85 for A_{11}. The cross-flow fiber orientation in the core also increases with increasing fiber concentration until 40 wt % and slightly reduces for higher concentrations. The maximum value for A_{22} in the core layer is 0.79 for PPGF40. The core width shows a steady increase from 0.09 for PPGF05 to 0.23 for PPGF60.

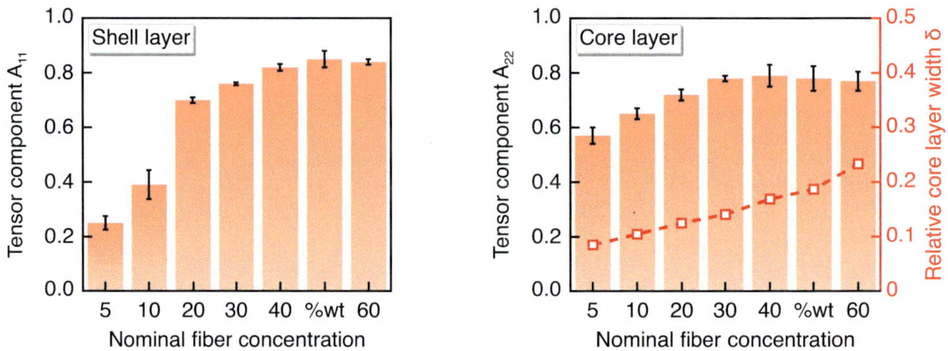

Figure 4.46 Measured fiber orientation in the shell layer showing degree of in-flow alignment A_{11} (left) and core layer showing cross-flow alignment A_{22} (right) at Location 2 for all trials

4.3.2 Analysis of the Process-Induced Fiber Breakage

The local fiber length was measured in the purged material and at three locations along the centerline of the molded plates. Figure 4.47 summarizes the obtained measurements represented by the number-average fiber length, L_N, and the weight-average fiber length, L_W, for all trials. Overall, the initial fiber length of 15 mm decreases substantially throughout the processing. The results show that increased nominal fiber concentration results in more fiber breakage and reduced fiber length in the molded plates. While additional fiber breakage occurs during mold filling, the majority of the length reduction occurs during plasticating, evident in the measured fiber length in the purged material. The longest weight-average fiber length, L_W, was 4.37 mm for PPGF10 (29.1% of the initial fiber length) and the shortest value for L_W was obtained for PPGF60 at 1.60 mm (10.7% of the initial fiber length). The measurements for PPGF05 and PPGF10 suggest a steady decrease in LW from the purged material and along the sample locations in the molded plates. In the purged material, L_W was 4.11 mm for PPGF05 and 4.30 mm for PPGF10 and, at Location 3, reduced to 3.09 mm (PPGF05) and 2.92 mm (PPGF10).

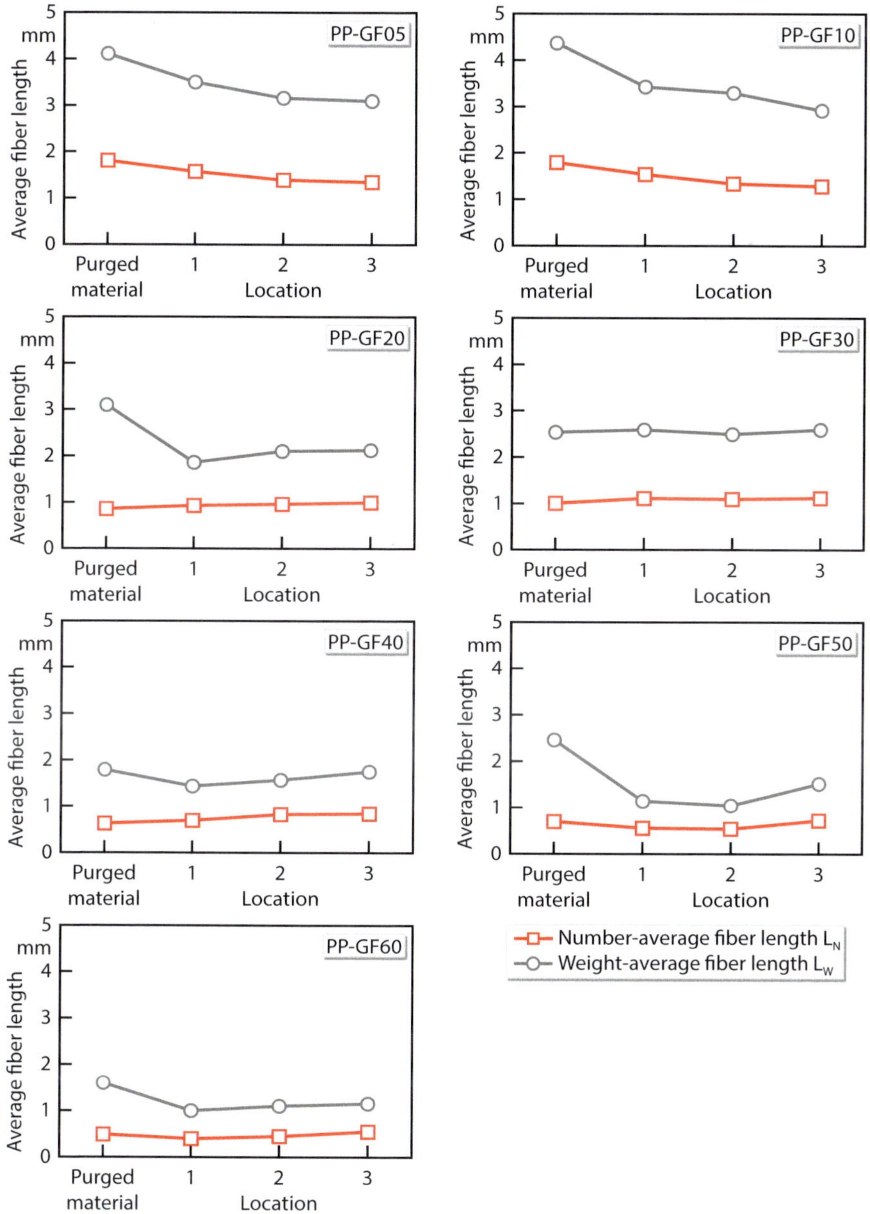

Figure 4.47 Results of the local fiber length measurements for all trials, showing the number-average fiber length, L_N, and the weight-average fiber length, L_W, at four locations: in the purged material, close to the gate (Location 1), at the center of the plate (Location 2), and at the end of the flow (Location 3)

At higher nominal fiber concentrations, the average fiber length increased from Location 1 to Location 3, suggesting that the last filled portion of the part carries longer fibers compared to the region close to the gate. The increase is most dominant for PPGF40 with a measured L_W value of 1.44 mm at Location 1 and 1.75 mm at Location 3, or a relative increase of 22%.

Figure 4.48 shows the fiber length in the plate (averaged for Locations 1 through 3) as a function of the nominal fiber volume concentration. Evidently, the fiber concentration has an impact on the residual fiber length. The effect is most distinct for weight-average fiber length, which decreases from 3.25 mm for PPGF05 to 1.08 mm for PPGF60. Overall, the measurements highlight the severe fiber breakage that is present during LFT injection molding. The measured weight-average in the plate is only 21.7% of the initial length (15 mm) for PPGF05 or less than 7.5% for PPGF60.

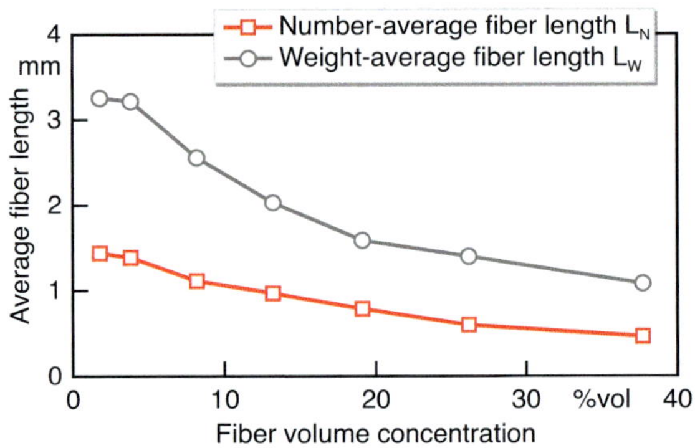

Figure 4.48 Average fiber length in the molded plate for all trials (averaged for Locations 1, 2, and 3)

4.3.3 Analysis of the Process-Induced Fiber Concentration

The global fiber concentration variation throughout the molded plates was determined by pyrolysis. The results for all trials are shown in Figure 4.49. The measurements show regions of low fiber concentration and elevated fiber concentration for each trial. PPGF05 has a peak of up to 20% higher fiber concentration along the center of the plate. PPGF10 to PPGF60 show a tendency of depleted fiber concentration close to the gate and increased fiber concentration to the end of the flow.

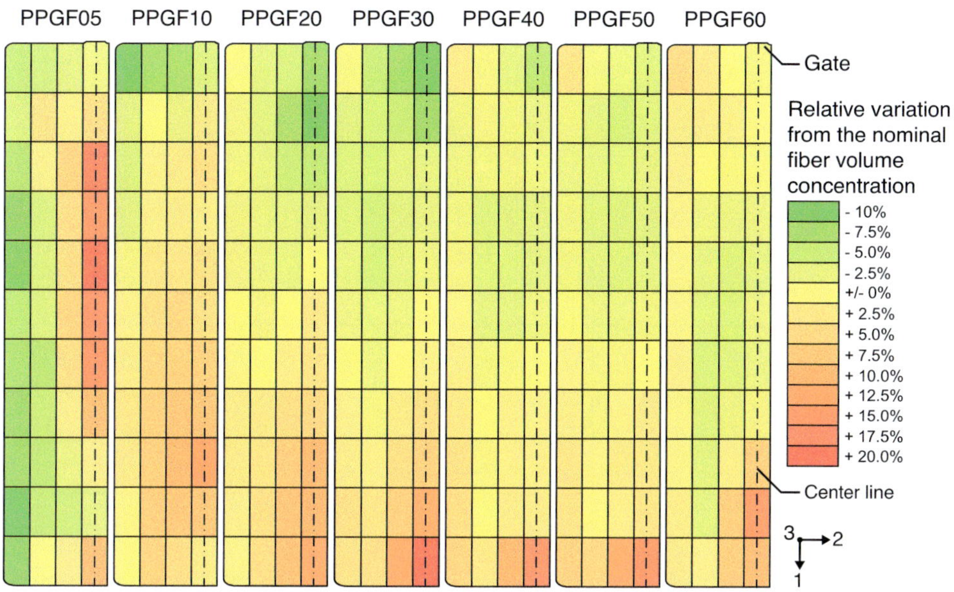

Figure 4.49 Results of the global concentration gradient analysis throughout the molded plate for all trials

The through-thickness fiber concentration was determined using X-ray mCT, as described in Section 4.2.3 of this chapter. Figure 4.50 shows the measured through-thickness fiber concentrations at Locations 1, 2, and 3 for all trials. The results suggest that the fiber concentration varies substantially in the thickness direction of all molded plates. The trials at higher concentrations (PPGF20 to PPGF60) show a common pattern in the measured concentration distribution. The measurements indicate a core layer with significantly higher fiber concentrations, which reaches values of up to 1.5 times the nominal concentration for PPGF40 (Location 2). The shell layers and surface regions have fewer fibers. For samples close to the gate (Location 1), the width of the core layer is wider than at Locations 2 and 3. There are also secondary concentration peaks close to the sample surfaces (relative thickness of 0.1), reaching values of 1.05 (PPGF20, Location 2) to 1.5 (PPGF40, Location 2). Only PPGF60 shows a fairly constant through-thickness concentration at the end of the flow path (Location 3).

The trials at diluted suspensions (PPGF05 and PPGF10) show a different trend in through-thickness fiber concentration. At Location 1, the core layers for both trials have a fiber depleted core with a normalized fiber concentration of 0.75 and these trials show a maximum of 1.25 in the transition region between shell and core at a relative thickness of approximately 0.3 (0.7). PPGF05 shows a similar pattern of a minimum in the core enclosed by two maxima at Locations 2 and 3 as well. On the other side, the PPGF10 results suggest a minor peak of approximately 1.1 in the core layer, which indicates a through-thickness pattern similar to these found at higher concentrations.

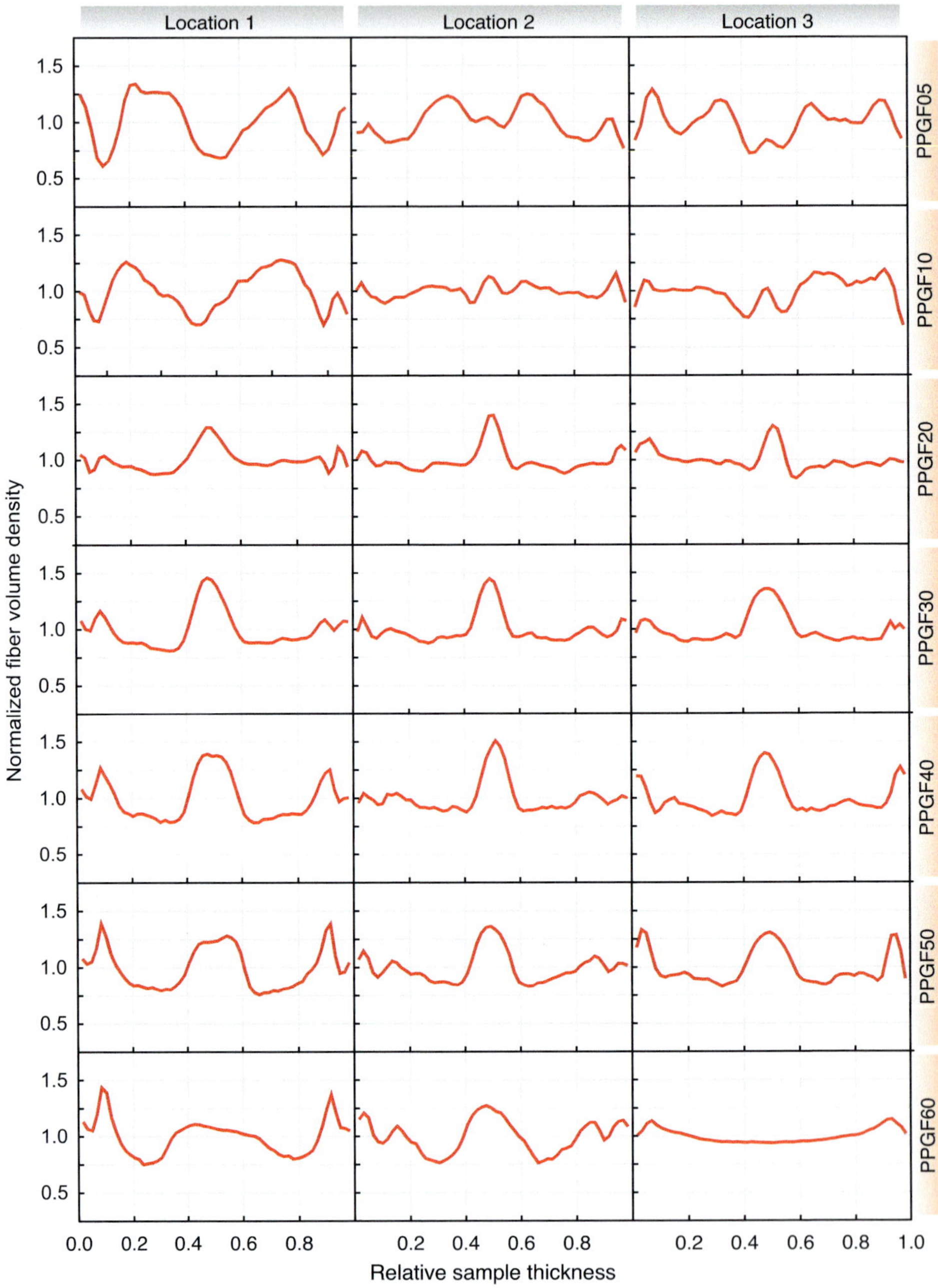

Figure 4.50 Results of the global concentration gradient analysis throughout the molded plate for all trials

For a more concise evaluation, two characteristic values can be defined based on the shape of the normalized fiber volume concentration: the core layer width δ and the maximum fiber concentration in the core layer ϕ_f^{max}, as illustrated in Figure 4.51 (left). The core layer width and maximum concentration measured at Location 2 for all trials is shown in Figure 4.51 (right). The relative core width increases at higher nominal fiber concentrations from 0.10 (PPGF05) to 0.26 (PPGF60). While the maximum fiber concentration in the core layer is 0.97 for PPGF05, it increases to 1.50 for PPGF40, which is the overall highest fiber concentration for all trials. For PPGF50 and PPGF60, it decreases to 1.36 and 1.27, respectively.

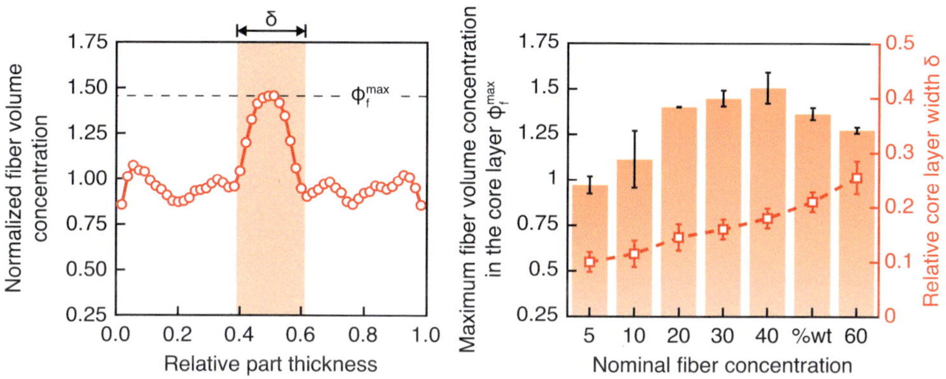

Figure 4.51 Illustration of core layer thickness δ and fiber concentration maximum ϕ_f^{max} (left) and obtained values at Location 2 for all trials (right)

4.3.3.1 Flow Front Analysis

Partial mold fillings were conducted to investigate the transient fiber concentration during mold filling and at the flow front. The partial mold fillings, or short shots, were molded at the same processing conditions without a packing phase for the PP with a nominal fiber concentration of 40 wt % (PPGF40). Short shots at 25%, 50%, 75%, and 90% fill were done and the fiber concentration was measured using µCT and pyrolysis. Figure 4.52 shows photographs of partially molded parts and the relative sample locations, which are 5 mm, 10 mm, and 17 mm from the advancing flow front. Additional measurements of the local fiber concentration through pyrolysis were performed to obtain the fiber concentration gradient along the flow length. Slices between 5 mm and 15 mm were cut from a 20 mm wide strip along the center line of the partial mold fillings.

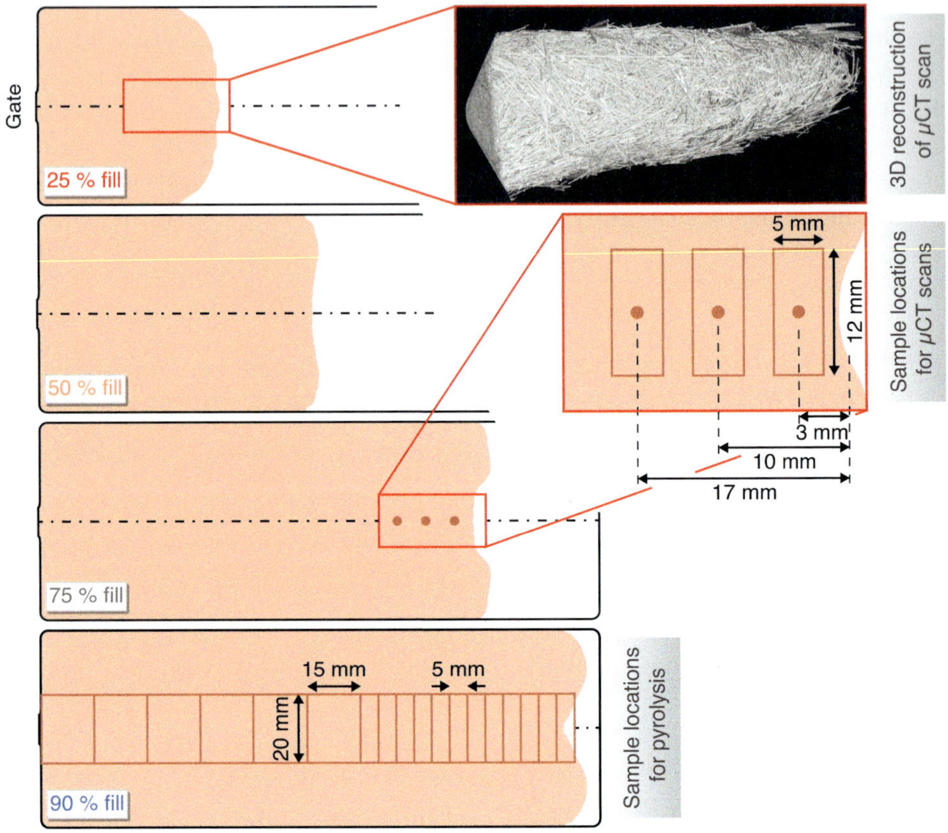

Figure 4.52 Flow front analysis: PPGF40 short shots (photos) at 25%, 50%, 75%, and 90% fill, 3D reconstruction of a μCT scan of the flow front, and sample locations for the measurements

Figure 4.53 shows the measured through-thickness fiber concentration for the partially filled moldings at a nominal fiber concentration of 40 wt % (PPGF40). The results suggest a fairly uniform fiber concentration through the thickness at the flow front without a distinct fiber agglomeration in the core layer for all fillings (Figure 4.53 left), but a decreasing fiber concentration towards the surface. At 13 mm from the flow front (Figure 4.53 center), the measurements suggest fiber agglomerating in the core. The peak in fiber concentration in the core decreases with increasing mold fill with a nominal fiber concentration of 1.5 at 25% fill and 1.1 for 90%. The results of the samples extracted at 17 mm from the flow front show a more distinct core-shell structure, as seen in Figure 4.53 (right). The peak fiber concentration in the core shows a maximum of approximately 1.4 for 50%, 75%, and 90% fill. For the 25% fill, the peak is larger at 1.6 times the nominal fiber concentration.

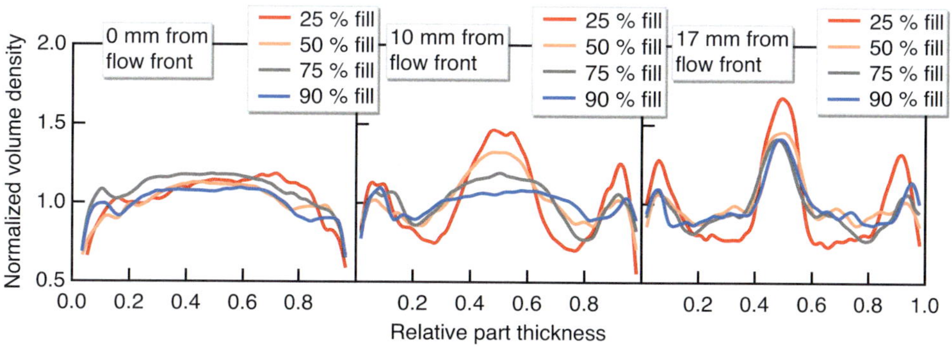

Figure 4.53 Measured through-thickness fiber concentration at the flow front for partially filled cavity molding (25%, 50%, 75%, and 90% fill) at a nominal fiber concentration of 40 wt % (PPGF40): at the flow front (left), 10 mm from the flow front (center), and 17 mm from the flow front (right)

Figure 4.54 summarizes the measured local fiber concentration with increasing distance from the melt front (three repeated measurements per location). The measurements suggest a substantial peak at the flow front itself can be observed, which reaches between 48 wt % for the 75% mold fill and up to 54.7 wt % for the 25% mold fill trial. Values above the nominal concentration (40 wt %) are measured up to 25 mm from the flow front. At distances above 25 mm, the measured fiber concentration is on average 38.5 wt %.

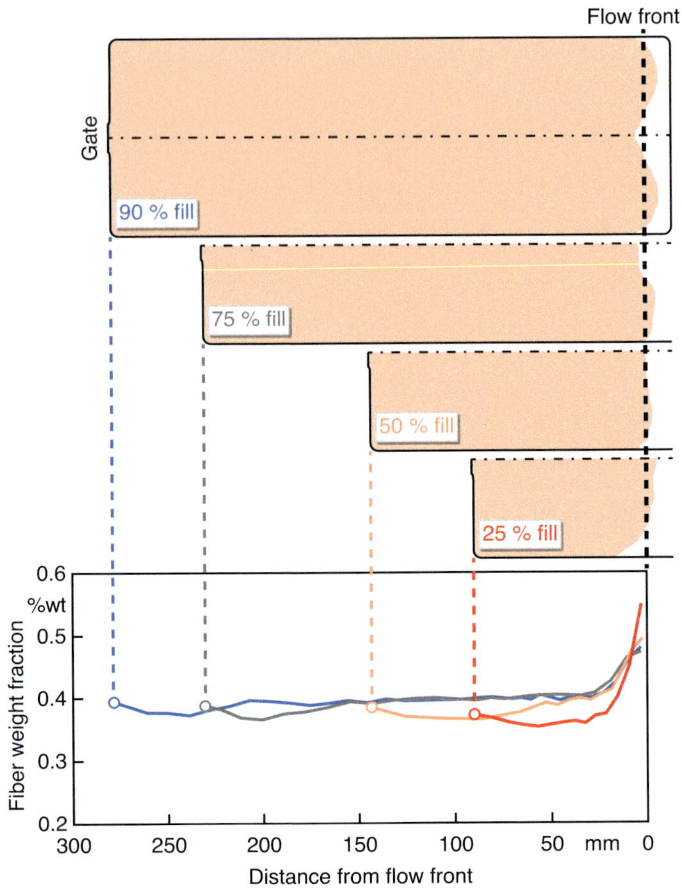

Figure 4.54 Measured fiber concentration with increasing distance to the flow front for partial mold fillings (25%, 50%, 75%, and 90% fill) at a nominal fiber concentration of 40 wt % (PPGF40). The illustration of the partial mold fillings is shown for reference (top)

4.3.4 Discussion and Conclusions

The results of this experimental study highlight the substantial impact of the process on the microstructure, inducing a heterogeneous fiber configuration along the flow path as well as through the thickness of the injection molded plates.

The through-thickness orientation at higher fiber concentrations (PPGF20 to PPGF60) shows the expected core-shell-skin pattern. The measurements suggest that the fiber alignment in both the core and shell slightly increases with increasing concentration. Throughout all trials, the core layer width increases steadily from a relative width of 0.09 for PPGF05 to 0.23 for PPGF60. The measurements at lower concentrations (PPGF05 and PPGF10) suggest an additional layer in the through-thickness fiber orientation. This could be attributed to the packing density of the fibers and the free volume of the fibers in the suspension. In general, a suspension of randomly oriented rods can be divided into three regimes, depending on the volume concentration, ϕ_f, and the aspect ratio, a_r, of the fibers [93]. In the *dilute regime* ($\phi_f < 1/a_r^2$), fibers are sufficiently far apart so that there are no contacts between neighboring fibers. The *semi-concentrated regime* is characterized by some restrictions of fiber movement due to interactions between adjacent fibers. The movement and rotation of fibers in the concentrated regime is highly restricted in all directions and the high packing density results in constant interactions between fibers. Evans and Gibson [94] found experimentally that the maximum volume concentration for randomly oriented fibers can be computed by $\phi_f^{max} = 5.3/a_r$. This relationship can be used to estimate the dividing line between *semi-concentrated* and *concentrated regimes* [95]. Figure 4.55 illustrates the three regimes and indicates the location of the measured fiber lengths from all trials in this study. The PPGF05 and PPGF10 are within the semi-concentrated regime, suggesting that the fibers have less restriction in movement as opposed to trials with higher concentrations. Hence, this might explain the additional orientation layer for the lower fiber concentrations as the fibers have room and more free volume to move and rotate during mold fill.

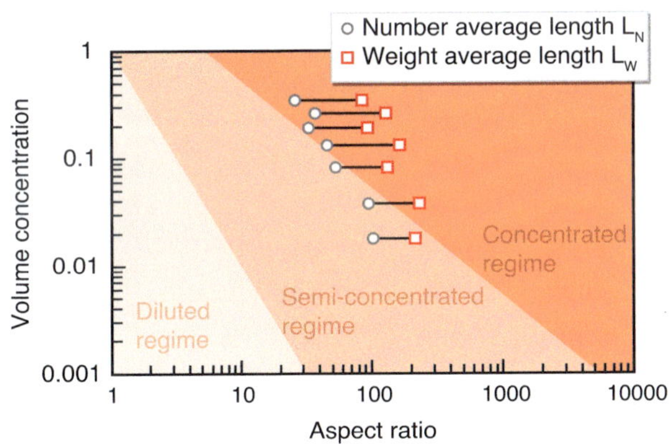

Figure 4.55 Illustration of the concentration regimes for fiber suspension and measured fiber aspect ratio from all trials in this study

The analysis of the through-thickness fiber concentration showed a substantial fiber agglomeration in the core layer and fewer fibers in the shell layers. Figure 4.56 shows a stitched set of low resolution µCT scans of a complete PPGF40 plate (left) and 2D slices of high resolution scans of the core layer and the shell layer respectively (right). This figure highlights qualitatively the fiber concentration gradient induced by injection molding process. At a nominal fiber concentration of 40 wt %, the measurements suggest 1.5 times the normal concentration of fibers in the core. The degree of fiber agglomeration can safely be assumed to have an impact on the other microstructural properties.

Figure 4.56 Qualitative illustration of the process-induced fiber–matrix separation in LFT injection molding for PPGF40: stitched set of low resolution µCT scans (left) and 2D slices of high resolution scans of the core layer and the shell layer (right)

The measured through-thickness fiber concentration shows a core-shell pattern similar to that found in the fiber orientation analysis. In fact, the measurements suggest a correlation between the through-thickness orientation and the concentration of the fibers. Figure 4.57 shows combined plots of the through-thickness analysis for PPGF05 (a), PPGF20 (b), PPGF40 (c), and PPGF60 (d) at Location 3. The fiber agglomeration in the core aligns with the fiber alignment in the core layer. Noticeably, the width of core layer for both the fiber concentration and the fiber orientation matches well. Particularly interesting is the correlation for PPGF05 with the two layers surrounding the core layer. In these two layers, the maximum fiber concentration coincides with the peak fiber alignment along the direction of flow (A_{11}).

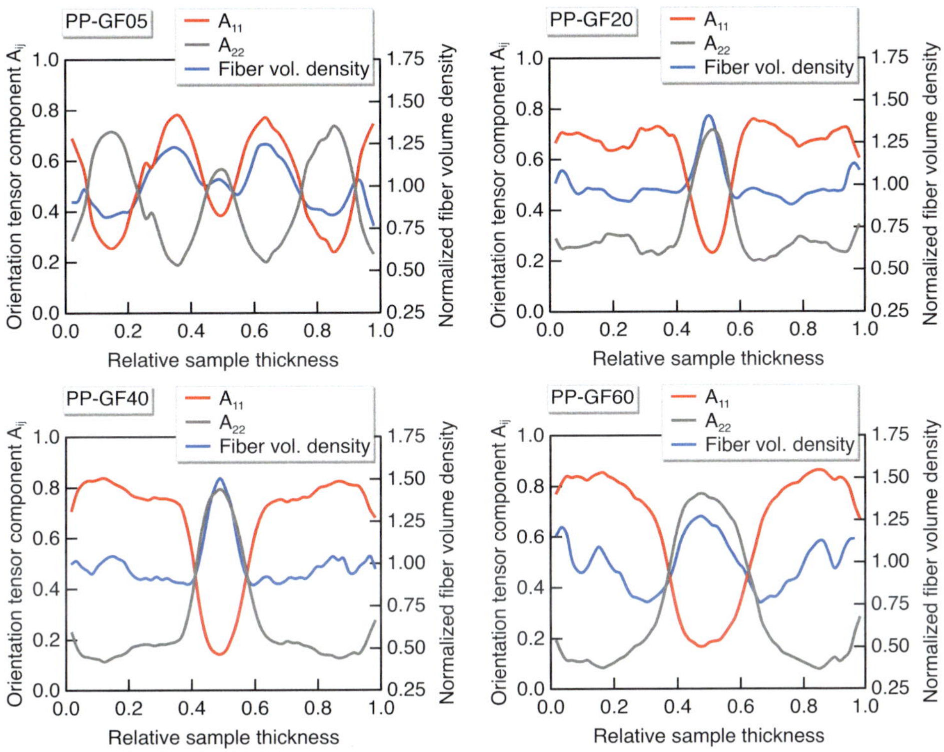

Figure 4.57 Correlation of the through-thickness fiber concentration and fiber orientation for PPGF05 (a), PPGF20 (b), PPGF40 (c), and PPGF60 (d) at Location 2

The emerging core-shell-skin pattern for the fiber orientation is attributed to the fountain flow effect. The observed through-thickness fiber concentration distribution might also be related to the fountain flow phenomenon, as the overall deformation at the advancing flow front causes substantial reorientation of the fibers. The degree of fiber alignment determines how closely the fibers can pack together, or vice versa.

The fiber concentration at any stage during mold filling is not low enough to allow a completely random fiber orientation. In fact, the fiber concentration in most of the trials is in the concentrated regime, as previously shown in Figure 4.55. Hence, the fibers need to be always aligned to some degree. The peak fiber concentration found at the advancing flow front in the PPGF40 partial mold fillings implies strong fiber alignment to accommodate the agglomerating fibers.

This high packing density at the flow front might not be able to carry through the reorientation process at the advancing flow front, at least not for all fibers at the flow front. Hence, the fiber concentration in the shell layers is lower than in the core. Furthermore, Andersson and Toll [40] suggest that the fibers in the core are compressed by the fibers that move through the fountain flow. Furthermore, they argue that stresses emerge acting on the core layer fibers due to the elasticity of the fibers going through the reorientation process, which further compresses the fibers in the core layer and packs them more densely. This hypothesis would also explain why the trials at diluted concentrations (PPGF05 and PPGF10) show different trends in the through-thickness concentration. The low fiber concentration in these trials allows for more random fiber orientation through the part and the alignment-concentration dependence is not as pronounced.

The analysis of fiber concentration through pyrolysis for the entire molded plate shows that there is also a global concentration gradient. The measurements show increasing fiber concentration along the flow path for all trials. While the concentration peak for PPGF05 is in the center of the plate, the measurements of the other trials suggest peaks at the end of the flow. Throughout all trials, the samples close to the gate have the lowest fiber concentration. The findings are aligned with what has been reported in other studies [12, 18, 25–28], which show similar trends.

Although no clear explanation for this phenomenon has been established, a reasonable hypothesis for the variation is the interaction between partially embedded fibers and the molten core at the interface of the solidified layer during mold filling. Two mechanisms can cause a change in fiber concentration, as shown in Figure 4.58. Partially embedded fibers in the solidified layers are exposed to deformation and stresses caused by the advancement of the molten core during cavity filling. Ultimately, the fibers can be either sheared off or pulled out and swept along with the molten core. Either mechanism can result in an increased fiber concentration at the last filled part of the cavity. Analyzing the fiber concentration of the partial mold fillings lends support to the theory because it shows a significant peak in fiber concentration at the melt front for all partial mold fillings. This peak suggests that an elevated concentration of fibers is carried along the flow front. Measurements of the raw pellet material and the plasticated material before the injection phase (purged material) confirmed that the fiber concentration is constant. Hence, the observed concentration gradients in the molded plates are due to the process-induced fiber–matrix separation during cavity fill.

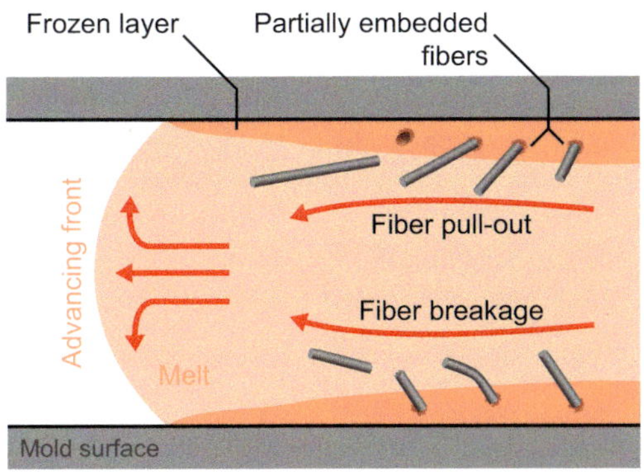

Figure 4.58 Schematic of the fiber pull-out and fiber breakage during cavity filling due to partly embedded fibers

The fiber length measurements suggest that substantial fiber breakage occurs throughout the entire process. The residual weight-average fiber length, L_W, in the molded plates reduces to approximately 3.25 mm for PPGF05 and 1.08 mm for PPGF60. Figure 4.59 shows a qualitative illustration of the expected effect on the mechanical properties using Schemme's graph [44] (see Section 4.1.5), highlighting the challenges that process-induced fiber breakage poses for the mechanical performance of the molded part. For all nominal fiber concentrations, Schemme's work suggests that the measured residual lengths are within a range where tensile strength and impact strength are very sensitive to the residual fiber length. Hence, even slight improvements in preventing fiber breakage could significantly increase the mechanical performance, while small decreases in fiber length could significantly degrade performance.

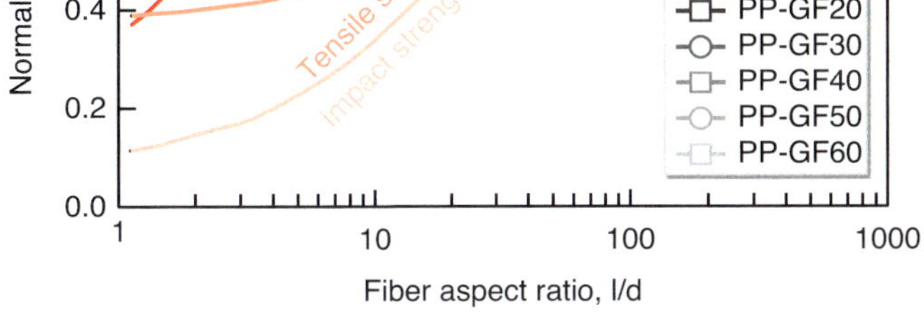

Figure 4.59 Illustration of the expected effect of the residual fiber aspect ratio on the normalized mechanical properties for the different PPGF grades

References

[1] Rohde, M., Ebel, A., Wolff-Fabris, F., and Altstädt, V., Influence of Processing Parameters on the Fiber Length and Impact Properties of Injection Molded Long Glass Fiber Reinforced Polypropylene, *Int. Polym. Process.*, vol. 26, no. 3, pp. 292–303 (2011)

[2] Phelps, J.H., Abd El-Rahman, A.I., Kunc, V., and Tucker, C.L., A model for fiber length attrition in injection-molded long-fiber composites, *Compos. Part Appl. Sci. Manuf.*, vol. 51, no. 0, pp. 11–21 (2013)

[3] Thomason, J.L., The influence of fibre length and concentration on the properties of glass fibre reinforced polypropylene: 7. Interface strength and fibre strain in injection moulded long fibre PP at high fibre content, *Compos. Part Appl. Sci. Manuf.*, vol. 38, no. 1, pp. 210–216 (2007)

[4] Osswald, T.A. and Menges, G., *Materials Science of Polymers for Engineers*, 3rd ed., Hanser, Munich (2012)

[5] Nguyen, B.N. et al., Fiber Length and Orientation in Long-Fiber Injection-Molded Thermoplastics – Part I: Modeling of Microstructure and Elastic Properties, *J. Compos. Mater.*, vol. 42, no. 10, pp. 1003–1029 (2008)

[6] Goris, S., Back, T., Yanev, A., Brands, D., Drummer, D., and Osswald, T.A., A novel fiber length measurement technique for discontinuous fiber-reinforced composites: A comparative study with existing methods, *Polym. Compos.* (2017)

[7] Inoue, A., Kazuya, M., Tanaka, T., Arao, Y., and Sawada, Y., Effect of screw design on fiber breakage and dispersion in injection-molded long glass-fiber-reinforced polypropylene, *J. Compos. Mater.*, vol. 49, no. 1 (2015)

[8] Masubuchi, Y., Terada, M., Yamanaka, A., Yamamoto, T., and Ishikawa, T., Distribution function of fiber length in thermoplastic composites, *Compos. Sci. Technol.*, vol. 134, pp. 43–48 (2016)

[9] Burr, I.W., Cumulative Frequency Functions, *Ann. Math. Stat.*, no. 2, pp. 215–232 (1942)

[10] Weibull, W., A Statistical Distribution Function of Wide Applicability, *J. Appl. Mech.*, vol. 18, pp. 293–297 (1951)

[11] Turkovich v., R. and Erwin, L., Fiber fracture in reinforced thermoplastic processing, *Polym. Eng. Sci.*, vol. 23, no. 13, pp. 743–749 (1983)

[12] Lafranche, E., Krawczak, P., Ciolczyk, J.-P., and Maugey, J., Injection moulding of long glass fiber reinforced polyamide 66: Processing conditions/microstructure/flexural properties relationship, *Adv. Polym. Technol.*, vol. 24, no. 2, pp. 114–131 (2005)

[13] Rohde, M., Direct Processing of Long Fiber Reinforced Thermoplastic Composites and their Mechanical Behavior under Static and Dynamic Load, PhD Thesis, University of Bayreuth (2015)

[14] Metten, M. and Cremer, M., Langfaserverstaerkte Thermoplaste spritzgiessen - Verfahrensparameter beeinflussen die Faserlaenge, Kunststoffe, vol. 90, no. 1 (2000)

[15] Folgar, F. and Tucker, C.L., Orientation Behavior of Fibers in Concentrated Suspensions, *J. Reinf. Plast. Compos.*, vol. 3, no. 2, pp. 98–119 (1984)

[16] Jeffery, G.B., The Motion of Ellipsoidal Particles Immersed in a Viscous Fluid, *Proc. R. Soc. Lond. Ser. A*, vol. 102, no. 715, pp. 161–179 (1922)

[17] Bay, R.S. and Tucker, C.L., Fiber orientation in simple injection moldings. Part II: Experimental results, *Polym. Compos.*, vol. 13, no. 4, pp. 332–341 (1992)

[18] Toll, S. and Andersson, P.-O., Microstructure of long- and short-fiber reinforced injection molded polyamide, *Polym. Compos.*, vol. 14, no. 2, pp. 116–125 (1993)

[19] Tadmor, Z., Molecular orientation in injection molding, *J. Appl. Polym. Sci.*, vol. 18, no. 6, pp. 1753–1772 (1974)

[20] Advani, S. G. and Tucker, C. L., The Use of Tensors to Describe and Predict Fiber Orientation in Short Fiber Composites, *J. Rheol.* 1978-Present, vol. 31, no. 8, pp. 751–784 (1987)

[21] Phelps, J. H. and Tucker, C. L., An anisotropic rotary diffusion model for fiber orientation in short- and long-fiber thermoplastics, *J. Non-Newton. Fluid Mech.*, vol. 156, no. 3, pp. 165–176 (2009)

[22] Tseng, H. C., Chang, R.-Y., and Hsu, C. H., An objective tensor to predict anisotropic fiber orientation in concentrated suspensions, *J. Rheol.*, vol. 60, no. 2, pp. 215–224 (2016)

[23] Bailey, R. and Rzepka, B., Fibre Orientation Mechanisms for Injection Molding of Long Fibre Composites, *Int. Polym. Process.*, vol. 6, no. 1, pp. 35–41 (1991)

[24] Vincent, M., Giroud, T., Clarke, A., and Eberhardt, C., Description and modeling of fiber orientation in injection molding of fiber reinforced thermoplastics, *Polym. Blends Compos. Hybrid Polym. Mater.* MACRO 2004, vol. 46, no. 17, pp. 6719–6725 (2005)

[25] O'Regan, D. and Akay, M., The distribution of fibre lengths in injection moulded polyamide composite components, *J. Mater. Process. Technol.*, vol. 56, no. 1–4, pp. 282–291 (1996)

[26] Kubat, J. and Syalanczi, A., Polymer-Glass Separation in the Spiral Mold Test, *Polym. Sci. Eng.*, vol. 14, no. 12 (1974)

[27] Hegler, R. P. and Mennig, G., Phase separation effects in processing of glass-bead- and glass-fiber-filled thermoplastics by injection molding, *Polym. Eng. Sci.*, vol. 25, no. 7, pp. 395–405 (1985)

[28] Ogadhoh, S. O. and Papathanasiou, T. D., Particle rearrangement during processing of glass-reinforced polystyrene by injection moulding, *Compos. Part Appl. Sci. Manuf.*, vol. 27, no. 1, pp. 57–63 (1996)

[29] Kovács, J. G., Shrinkage alteration induced by segregation of glass beads in injection molded PA6: Experimental analysis and modeling, *Polym. Eng. Sci.*, vol. 51, no. 12, pp. 2517–2525 (2011)

[30] Mondy, L. A., Brenner, H., Altobelli, S. A., Abbott, J. R., and Graham, A. L., Shear-induced particle migration in suspensions of rods, *J. Rheol.*, vol. 38, no. 2, pp. 444–452 (1994)

[31] Ramzy, A. Y., El-Sabbagh, A. M., Steuernagel, L., Ziegmann, G., and Meiners, D., Rheology of natural fibers thermoplastic compounds: Flow length and fiber distribution, *J. Appl. Polym. Sci.*, vol. 131, no. 3, p. n/a-n/a (2014)

[32] Vélez-García, G. M., Wapperom, P., Baird, D. G., Aning, A. O., and Kunc, V., Unambiguous orientation in short fiber composites over small sampling area in a center-gated disk, *Compos. Part Appl. Sci. Manuf.*, vol. 43, no. 1, pp. 104–113 (2012)

[33] Sun, X., Lasecki, J., Zeng, D., Gan, Y., Su, X., and Tao, J., Measurement and quantitative analysis of fiber orientation distribution in long fiber reinforced part by injection molding, *Polym. Test.*, vol. 42, no. 0, pp. 168–174, Apr. (2015)

[34] Guazzelli, E. and Morris, J., *A Physical Introduction to Suspension Dynamics*, 1st ed., Cambridge University Press (2012)

[35] Nott, P. R. and Brady, J. F., Pressure-driven flow of suspensions: simulation and theory, *J. Fluid Mech.*, vol. 275, pp. 157–199 (1994)

[36] Kuhn, C., Walter, I., Taeger, O., and Osswald, T., Experimental and Numerical Analysis of Fiber Matrix Separation during Compression Molding of Long Fiber Reinforced Thermoplastics, *J. Compos. Sci.*, vol. 1, no. 2 (2017)

[37] Osswald, T. A., Numerical Methods for Compression Mold Filling Simulation, University of Illinois at Urbana - Champaign (1987)

[38] Thomason, J. L. and Vlug, M. A., Influence of fibre length and concentration on the properties of glass fibre-reinforced polypropylene: 1. Tensile and flexural modulus, *Compos. Part Appl. Sci. Manuf.*, vol. 27, no. 6, pp. 477–484 (1996)

[39] Thomason, J. L., The influence of fibre length, diameter and concentration on the impact performance of long glass-fibre reinforced polyamide 6,6, *Compos. Part Appl. Sci. Manuf.*, vol. 40, no. 2, pp. 114–124 (2009)

[40] Thomason, J.L., Vlug, M.A., Schipper, G., and Krikor, H.G. L.T., Influence of fibre length and concentration on the properties of glass fibre-reinforced polypropylene: Part 3. Strength and strain at failure, *Compos. Part Appl. Sci. Manuf.*, vol. 27, no. 11, pp. 1075–1084 (1996)

[41] Thomason, J.L. and Vlug, M.A., Influence of fibre length and concentration on the properties of glass fibre-reinforced polypropylene: 4. Impact properties, *Compos. Part Appl. Sci. Manuf.*, vol. 28, no. 3, pp. 277–288 (1997)

[42] Thomason, J.L., The influence of fibre length and concentration on the properties of glass fibre reinforced polypropylene: 5. Injection moulded long and short fibre PP, *Compos. Part Appl. Sci. Manuf.*, vol. 33, no. 12, pp. 1641–1652 (2002)

[43] Thomason, J.L., The influence of fibre length and concentration on the properties of glass fibre reinforced polypropylene. 6. The properties of injection moulded long fibre PP at high fibre content, *Compos. Part Appl. Sci. Manuf.*, vol. 36, no. 7, pp. 995–1003 (2005)

[44] Schemme, M., Langfaserverstaerkte Thermoplaste, *Kunststoffe*, vol. 93, pp. 106–109 (2003)

[45] Bay, R.S., Fiber orientation in injection-molded composites: A comparison of theory and experiment, Ph.D., University of Illinois at Urbana-Champaign (1991)

[46] Velez-Garcia, G., Experimental Evaluation and Simulations of Fiber Orientation in Injection Molding of Polymers Containing Short Glass Fibers, Virginia Polytechnic Institute and State University (2012)

[47] Hine, P.J., Davidson, N., Duckett, R.A., and Ward, I.M., Measuring the fibre orientation and modelling the elastic properties of injection-moulded long-glass-fibre-reinforced nylon, *Compos. Sci. Technol.*, vol. 53, no. 2, pp. 125–131 (1995)

[48] Stock, S.R., *MicroComputed Tomography: Methodology and Applications*, Boca Raton: CRC Press (2009)

[49] Nguyen Thi, T.B., Morioka, M., Yokoyama, A., Hamanaka, S., Yamashita, K., and Nonomura, C., Measurement of fiber orientation distribution in injection-molded short-glass-fiber composites using X-ray computed tomography, *J. Mater. Process. Technol.*, vol. 219, pp. 1–9 (2015)

[50] Goris, S. and Osswald, T.A., Process-induced fiber matrix separation in long fiber-reinforced thermoplastics, *Compos. Part Appl. Sci. Manuf.*, vol. 105, pp. 321–333 (2018)

[51] Gandhi, U., De Boodt, S., Kunc, V., and Song, Y.Y., Method to measure orientation of discontinuous fiber embedded in the polymer matrix from computerized tomography scan data, *J. Thermoplast. Compos. Mater.* (2015)

[52] Krause, M., Hausherr, J.M., Burgeth, B., Herrmann, C., and Krenkel, W., Determination of the fibre orientation in composites using the structure tensor and local X-ray transform, *J. Mater. Sci.*, vol. 45, no. 4, p. 888, Nov. (2009)

[53] Salaberger, D., Kannappan, K.A., Kastner, J., Reussner, J., and Auinger, T., Evaluation of Computed Tomography Data from Fibre Reinforced Polymers to Determine Fibre Length Distribution, *Int. Polym. Process.*, vol. 2011/03, pp. 283–291 (2011)

[54] Buck, F. et al., Two-scale structural mechanical modeling of long fiber reinforced thermoplastics, *Compos. Sci. Technol.*, vol. 117, pp. 159–167 (2015)

[55] Perez, C., Goris, S., and Osswald, T.A., Study on the Fiber Properties of a LFT Strand, SPE Automot. Compos. Conf. Exhib. ACCE, vol. 13th Annual (2013)

[56] Fliegener, S., Luke, M., and Gumbsch, P., 3D microstructure modeling of long fiber reinforced thermoplastics, *Compos. Sci. Technol.*, vol. 104, pp. 136–145 (2014)

[57] Ayadi, A., Nouri, H., Guessasma, S., and Roger, F., Determination of orthotropic properties of glass fibre reinforced thermoplastics using X-ray tomography and multiscale finite element computation, *Compos. Struct.*, vol. 136, pp. 635–649 (2016)

[58] Hannesschläger, C., Revol, V., Plank, B., Salaberger, D., and Kastner, J., Fibre structure characterisation of injection moulded short fibre-reinforced polymers by X-ray scatter dark field tomography, *Case Stud. Nondestruct. Test. Eval.*, vol. 3, pp. 34–41 (2015)

[59] Bernasconi, A., Cosmi, F., and Dreossi, D., Local anisotropy analysis of injection moulded fibre reinforced polymer composites, *Deform. Fract. Compos. Anal. Numer. Exp. Tech. Regul. Pap.*, vol. 68, no. 12, pp. 2574–2581 (2008)

[60] Graupner, N., Beckmann, F., Wilde, F., and Müssig, J., Using synchrotron radiation-based micro-computer tomography (SR μ-CT) for the measurement of fibre orientations in cellulose fibre-reinforced polylactide (PLA) composites, *J. Mater. Sci.*, vol. 49, no. 1, pp. 450–460 (2014)

[61] Köpplmayr, T. et al., Influence of fiber orientation and length distribution on the rheological characterization of glass-fiber-filled polypropylene, Polym. Test., vol. 32, no. 3, pp. 535–544 (2013)

[62] De Monte, M., Moosbrugger, E., and Quaresimin, M., Influence of temperature and thickness on the off-axis behaviour of short glass fibre reinforced polyamide 6.6 – Quasi-static loading, *Compos. Part Appl. Sci. Manuf.*, vol. 41, no. 7, pp. 859–871 (2010)

[63] Moennich, S., Gloeckner, R., and Becker, F., Analysis of Fibre Orientation using CT Data, (2011)

[64] Rolland, H., Saintier, N., Mareau, C., and Robert, G., Fatigue Mechanisms in Short Glass Fibre Reinforced Thermoplastic: In Situ X-Ray Microtomography Observations and Microstructure Modelling, in Proceedings of the 7th European Conference on Composite Materials, Munich, Germany (2016)

[65] Bernasconi, A., Cosmi, F., and Hine, P.J., Analysis of fibre orientation distribution in short fibre reinforced polymers: A comparison between optical and tomographic methods, *Compos. Sci. Technol.*, vol. 72, no. 16, pp. 2002–2008 (2012)

[66] Cieslinski, M.J., Baird, D.G., and Wapperom, P., Obtaining repeatable initial fiber orientation for the transient rheology of fiber suspensions in simple shear flow, *J. Rheol.*, vol. 60, no. 1, pp. 161–174 (2016)

[67] Meyer, K.J., Hofmann, J.T., and Baird, D.G., Initial conditions for simulating glass fiber orientation in the filling of center-gated disks, *Compos. Part Appl. Sci. Manuf.*, vol. 49, pp. 192–202 (2013)

[68] Goris, S., Characterization of the Process-Induced Fiber Configuration of Long Glass Fiber-Reinforced Thermoplastics, Ph.D. Thesis, University of Wisconsin-Madison, Madison, WI (2018)

[69] Goris, S. and Osswald, T.A., Fiber Orientation Measurements Using a Novel Image Processing Algorithm for Micro-Computed Tomography Scans, in Proceedings of the 15th Automotive Composites Conference and Exhibition, Novi, MI (2015)

[70] International Organization for Standardization, ISO 22314: Plastics – Glass-fibre-reinforced products – Determination of fibre length (2006)

[71] Bajracharya, R.M., Manalo, A.C., Karunasena, W., and Lau, K., Experimental and theoretical studies on the properties of injection moulded glass fibre reinforced mixed plastics composites, *Compos. Part Appl. Sci. Manuf.*, vol. 84, pp. 393–405 (2016)

[72] Bumm, S.H., White, J.L., and Isayev, A.I., Glass fiber breakup in corotating twin screw extruder: Simulation and experiment, *Polym. Compos.*, vol. 33, no. 12, pp. 2147–2158 (2012)

[73] Priebe, M. and Schledjewski, R., Processing and properties of glass/polypropylene in long fibre compounding extrusion, *Plast. Rubber Compos.*, vol. 40, no. 6–7, pp. 374–379 (2011)

[74] Wang, J. et al., Shear induced fiber orientation, fiber breakage and matrix molecular orientation in long glass fiber reinforced polypropylene composites, *Mater. Sci. Eng. A*, vol. 528, no. 7–8, pp. 3169–3176 (2011)

[75] Jin, G., Lin, X., Tian, G., Zhang, S., Wang, M., and Wang, X., Entrance flow of long glass fiber reinforced polypropylene through contraction die, *J. Reinf. Plast. Compos.*, vol. 35, no. 2, pp. 111–123 (2016)

[76] Kunc, V., Tucker, C.L., Velez-Garcia, G., Frame, B., and Nguyen, B.N., Fiber length distribution measurement for long glass and carbon fiber reinforced injection molded thermoplastics, in Proceedings of the 7th Annual Automotive Composites Conference and Exhibition, Novi, MI (2007)

[77] Inceoglu, F. et al., Correlation Between Processing Conditions and Fiber Breakage During Compounding of Glass Fiber-Reinforced Polyamide, *Polym. Compos.*, vol. 32, no. 11, pp. 1842–1850 (2011)

[78] Yilmazer, U. and Cansever, M., Effects of processing conditions on the fiber length distribution and mechanical properties of glass fiber reinforced nylon-6, *Polym. Compos.*, vol. 23, no. 1, pp. 61–71 (2002)

[79] Huq, A.M. A. and Azaiez, J., Effects of length distribution on the steady shear viscosity of semiconcentrated polymer-fiber suspensions, *Polym. Eng. Sci.*, vol. 45, no. 10, pp. 1357–1368 (2005)

[80] Zhuang, H., Ren, P., Zong, Y., and Dai, G.C., Relationship between fiber degradation and residence time distribution in the processing of long fiber reinforced thermoplastics, *EXPRESS Polym. Lett.*, vol. 2, no. 8, pp. 560–568 (2008)

[81] Teixeira, D., Giovanela, M., Gonella, L.B., and Crespo, J.S., Influence of injection molding on the flexural strength and surface quality of long glass fiber-reinforced polyamide 6.6 composites, *Mater. Des.*, vol. 85, pp. 695–706 (2015)

[82] Ren, P. and Dai, G., Fiber dispersion and breakage in deep screw channel during processing of long fiber-reinforced polypropylene, *Fibers Polym.*, vol. 15, no. 7, pp. 1507–1516 (2014)

[83] Stengler, R., Sandau, K., Hartwich, M.R., Hoehn, N., and Mayr, H., FASEP Ultra - Neuartige Prozessplanung und -steuerung bei der Verarbeitung von Langfaserverstärkten Thermoplasten (LFT) mittels einer neuen bildanalytischen Bestimmung von Faserlängenverteilungen, Querschnitt - Reports from Research and Development, vol. 23 (2009)

[84] Giusti, R., Dubrovich, I., and Lucchetta, G., Rapid and accurate image analysis procedure for fiber length measurements, SPE Plastics Research Online (2015)

[85] Kleindel, S., Salaberger, D., Eder, R., Schretter, H., and Hochenauer, C., Measurement and Numerical Simulation of Void and Warpage in Glass Fiber Reinforced Molded Chunky Parts, *Int. Polym. Process.*, vol. 30, no. 1, pp. 100–112 (2015)

[86] Hine, P., Parveen, B., Brands, D., and Caton-Rose, F., Validation of the modified rule of mixtures using a combination of fibre orientation and fibre length measurements, *Compos. Part Appl. Sci. Manuf.*, vol. 64, pp. 70–78 (2014)

[87] Hartwich, M.R., Mayr, H., and Stengler, R., FASEP ULTRA - Automated Analysis of fibre length distribution in glass-fibre-reinforced products, in Proceedings of Optical Measurement Systems for Industrial Inspection, vol. 2009 (2009)

[88] Schijve, W., Eigenschaftsermittlung und Pruefung von langfaserverstaerkten Thermoplasten, EATC (2008)

[89] Wang, H., Fiber Property Characterization by Image Processing, Master Thesis, Texas Tech University, Lubbock, TX (2007)

[90] SABIC, Processing Guides: SABIC Stamax, Official Company Website (2016) [Online]. Available: https://www.sabic.com/europe/en/products-services/plastics/technical/Processing-guides–SABIC-STAMAX

[91] Krasteva, D.L., Integrated Prediction of Processing and Thermomechanical Behavior of Long Fiber Thermoplastic Composites, PhD Thesis, University of Minho, Portugal (2009)

[92] SABIC, SABIC STAMAX Material Datasheet (2017)

[93] Tucker, C.L., Flow regimes for fiber suspensions in narrow gaps, *J. Non-Newton. Fluid Mech.*, vol. 39, no. 3, pp. 239–268 (1991)

[94] Evans, K.E. and Gibson, A.G., Prediction of the maximum packing fraction achievable in randomly oriented short-fibre composites, *Compos. Sci. Technol.*, vol. 25, no. 2, pp. 149–162 (1986)

[95] El-Rahman, A.I.A. and C.L. T. III, Mechanics of random discontinuous long-fiber thermoplastics. Part II: Direct simulation of uniaxial compression, *J. Rheol.*, vol. 57, no. 5, pp. 1463–1489 (2013)

[96] Goris, S. and Osswald, T.A., Fiber Orientation Measurements Using a Novel Image Processing Algorithm for Micro-Computed Tomography Scans, 15th Annual SPE Automot. Compos. Conf. Exhib., ACCE (2015)

5 Mechanics of Composites

Paul V. Osswald[1] *and Tim A. Osswald*

When referring to filled composite materials, we refer to materials that have fillers that are intentionally placed in the polymer matrix to make them stronger, lighter, electrically conductive, thermally conductive, or cheaper. Any filler will affect the mechanical behavior of a polymeric material. For example, long fibers will make it stiffer and stronger but usually denser, whereas foaming will make it more compliant but much lighter. On the other hand, a filler such as calcium carbonate will decrease the polymer's toughness while making it considerably cheaper. Figure 5.1 [1] shows a schematic plot of the change in stiffness as a function of volume fraction for several types of filler materials.

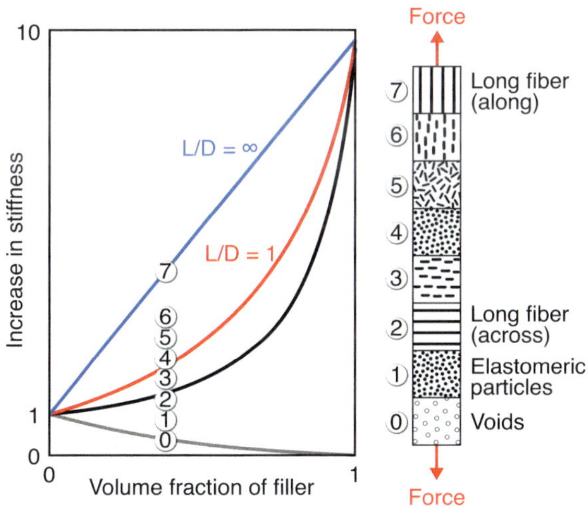

Figure 5.1 Relation between stiffness and filler type and orientation in polymeric materials

The most common composite materials are fiber-reinforced polymer composites, which can be either *continuous* and *discontinuous* or *chopped fiber composites*. The

[1] BMW AG, Munich, Germany

mechanics of long and short fiber composites are very similar as long as the chopped fibers are long enough to accomplish full transfer of stresses from the matrix to the fibers. This section will first discuss the mechanics of continuous fiber-reinforced composite materials and is followed by a discussion on what makes a short fiber system an appropriate composite material.

5.1 Anisotropic Strain–Stress Relation

Polymers filled with fibers are anisotropic, and traditional stress–strain relations are no longer valid. The three-dimensional anisotropic strain–stress relation where a local 1, 2, and 3 coordinate system has been chosen can be written as

$$\varepsilon_{11} = \frac{1}{E_{11}}\sigma_{11} - \frac{\nu_{21}}{E_{22}}\sigma_{22} - \frac{\nu_{31}}{E_{33}}\sigma_{33} \tag{5.1}$$

$$\varepsilon_{22} = -\frac{\nu_{12}}{E_{11}}\sigma_{11} + \frac{1}{E_{22}}\sigma_{22} - \frac{\nu_{32}}{E_{33}}\sigma_{33} \tag{5.2}$$

$$\varepsilon_{33} = -\frac{\nu_{13}}{E_{11}}\sigma_{11} - \frac{\nu_{23}}{E_{22}}\sigma_{22} + \frac{1}{E_{33}}\sigma_{33} \tag{5.3}$$

$$\gamma_{12} = -\frac{1}{G_{12}}\tau_{12} \tag{5.4}$$

$$\gamma_{23} = -\frac{1}{G_{23}}\tau_{23} \tag{5.5}$$

$$\gamma_{31} = -\frac{1}{G_{31}}\tau_{31} \tag{5.6}$$

where the remaining terms are symmetric. While strain and stress are second order tensors, due to symmetry they can be represented by a vector, such as shown for strain in the equation below:

$$\underline{\underline{\varepsilon}} = \begin{bmatrix} \varepsilon_{11} & \gamma_{12} & \gamma_{13} \\ \gamma_{21} & \varepsilon_{22} & \gamma_{23} \\ \gamma_{31} & \gamma_{32} & \varepsilon_{33} \end{bmatrix} \rightarrow \begin{Bmatrix} \varepsilon_{11} \\ \varepsilon_{22} \\ \varepsilon_{33} \\ \gamma_{12} \\ \gamma_{23} \\ \gamma_{31} \end{Bmatrix} \text{or} \begin{Bmatrix} \varepsilon_{1} \\ \varepsilon_{2} \\ \varepsilon_{3} \\ \varepsilon_{4} \\ \varepsilon_{5} \\ \varepsilon_{6} \end{Bmatrix} \tag{5.7}$$

Hence, the strain-stress relationship can be represented using a 4th order compliance tensor representing the constitutive material matrices as

$$\begin{Bmatrix} \varepsilon_{11} \\ \varepsilon_{22} \\ \varepsilon_{33} \\ \gamma_{12} \\ \gamma_{23} \\ \gamma_{31} \end{Bmatrix} = \begin{bmatrix} S_{11} & S_{12} & S_{13} & S_{14} & S_{15} & S_{16} \\ S_{21} & S_{22} & S_{23} & S_{24} & S_{25} & S_{26} \\ S_{31} & S_{32} & S_{33} & S_{34} & S_{35} & S_{36} \\ S_{41} & S_{42} & S_{43} & S_{44} & S_{45} & S_{46} \\ S_{51} & S_{52} & S_{53} & S_{54} & S_{55} & S_{56} \\ S_{61} & S_{62} & S_{63} & S_{64} & S_{65} & S_{66} \end{bmatrix} \begin{Bmatrix} \sigma_{11} \\ \sigma_{22} \\ \sigma_{33} \\ \tau_{12} \\ \tau_{23} \\ \tau_{31} \end{Bmatrix} \tag{5.8}$$

5.2 Laminated Composite

The most common composite structure is an aligned fiber-reinforced laminated composite, such as that shown schematically in Figure 5.2. Here, the fiber direction is defined as the longitudinal direction, 1, and the direction out of plane and perpendicular to the fibers as the transverse direction, 2.

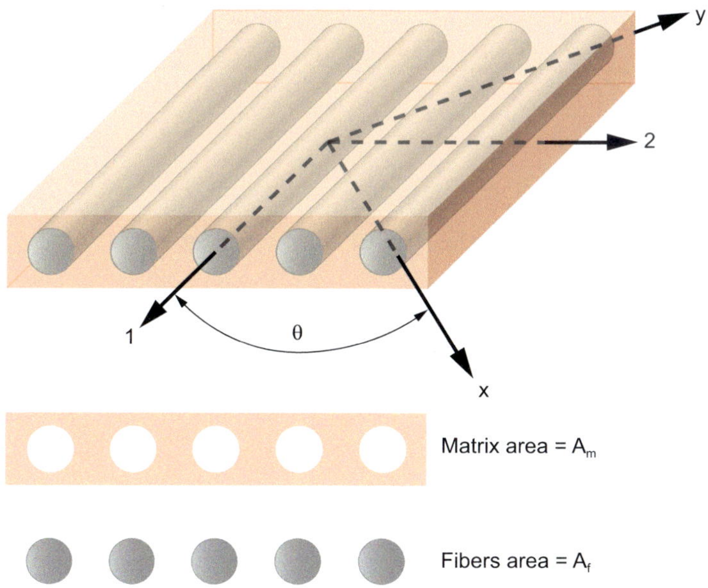

Figure 5.2 Schematic diagram of unidirectional continuous fiber-reinforced laminated structure

The cross-sectional area of the laminate is made up of fibers and matrix, such that

$$A_c = A_m + A_f \tag{5.9}$$

where c denotes composite, m matrix, and f fiber. Since the cross section of the laminate in Figure 5.2 remains constant along the length of the fibers, we can relate the volume fraction of fibers, ϕ, to the areas

$$\phi = \frac{V_f}{V_c} = \frac{A_f L}{A_c L} = \frac{A_f}{A_c} \tag{5.10}$$

where V denotes volume and A area. The same is true for the volume fraction of the matrix, $(1-\phi)$

$$(1-\phi) = \frac{A_m}{A_c} \tag{5.11}$$

Using the above information, the mass or weight fraction of fibers, ψ, is defined as

$$\psi = \frac{m_f}{m_c} = \frac{\rho_f V_f}{\rho_c V_c} = \frac{\rho_f}{\rho_c}\phi \tag{5.12}$$

In most situations it is easier to define the mass fractions in a composite material. However, when performing calculations we usually use volume fractions and we are often required to translate between the two in terms of ρ_f and ρ_m; we can use

$$\phi = \frac{1}{1 + \frac{1-\psi}{\psi} \cdot \frac{\rho_f}{\rho_m}} \tag{5.13}$$

5.2.1 Equilibrium, Deformation, and Constitutive Equations

The resultant force in a composite structure is the sum of the forces on the fibers and the force acting on the matrix

$$F_c = F_m + F_f \tag{5.14}$$

Since the fibers and the matrix are bonded to each other, their strain or deformation must be the same

$$\epsilon_c = \epsilon_m = \epsilon_f \tag{5.15}$$

Furthermore, we can say that the fibers, the matrix, and the composite have their respective constitutive behavior, which for simplicity we can say follows the behavior of a linear elastic solid

$$\sigma_c = E_c \epsilon_c \tag{5.16}$$

$$\sigma_m = E_m \epsilon_m \tag{5.17}$$

$$\sigma_f = E_f \epsilon_f \tag{5.18}$$

Since the axial stress applied to the laminate, σ, is defined by

$$\sigma = F / A \tag{5.19}$$

we can rewrite Eq. (5.14) as

$$E_c \epsilon_c A_c = E_m \epsilon_m A_m + E_f \epsilon_f A_f \tag{5.20}$$

Since the strain is the same throughout the cross section (Eq. (5.15)), the above equation reduces to

$$E_c = E_m \frac{A_m}{A_c} + E_f \frac{A_f}{A_c} \tag{5.21}$$

or

$$E_c = E_m (1 - \phi) + E_f \phi \tag{5.22}$$

Equation (5.22) is the well-known *rule of mixtures*, widely used when performing calculations with composite materials. It is easy to understand that Eq. (5.22) is used to compute a property that pertains to the longitudinal direction of the laminate and can be rewritten with the appropriate subscript

$$E_{11} = E_m (1 - \phi) + E_f \phi \tag{5.23}$$

where 1 denotes the longitudinal direction.

Similarly, the ultimate strength of the composite in the longitudinal direction of the fiber can be computed using the rule of mixtures.

5.2.2 Transverse Properties—The Halpin-Tsai Equation

At this point, we recognize that it is not as easy to calculate the transverse properties of the composite. Perhaps the most widely accepted model used today is the *Halpin-Tsai model* [2], which also takes into account the length to diameter ratio of the fiber, $\left(\frac{L}{D}\right)$, important for chopped or short fiber-reinforced composite materials

$$E_{11} = E_m \left(\frac{1 + \xi \eta \phi}{1 - \eta \phi} \right) \tag{5.24}$$

which approaches Eq. (5.23) for continuous fibers. The transverse modulus is computed using

$$E_{22} = E_m \left(\frac{1 + \xi \eta \phi}{1 - \eta \phi} \right) \tag{5.25}$$

with the appropriate value of ξ given below. The shear modulus, with

$$G_{12} = G_m \left(\frac{1 + \lambda \phi}{1 - \lambda \phi} \right) \tag{5.26}$$

In the above equations

$$\eta = \frac{\left(\frac{E_f}{E_m} - 1 \right)}{\left(\frac{E_f}{E_m} + \xi \right)} \tag{5.27}$$

$$\lambda = \frac{\left(\frac{G_f}{G_m} - 1 \right)}{\left(\frac{G_f}{G_m} + 1 \right)} \tag{5.28}$$

The parameter ξ is an empirical factor that is given by

$$\xi = 2 \left(\frac{L}{D} \right) (\text{Longitudinal direction}) \tag{5.29}$$

$$\xi = 2 \, (\text{Transverse direction}) \tag{5.30}$$

In the above equations, the subscripts f and m represent the fiber and matrix, respectively; L, the fiber length; D, the fiber diameter; and ϕ, the volume fiber fraction. It is important to mention that some fiber materials such as carbon have a longitudinal modulus, E_{f1}, and a transverse modulus, E_{f2}. For those materials, the term η in the above equations must be computed using the corresponding values of E_{f1} when calculating E_{11}, and E_{f2} when calculating E_{22}. The values of Poisson's ratio are computed using

$$\nu_{21} = \nu_m (1 - \phi) + \nu_{f21} \phi \tag{5.31}$$

for a contraction in the 2 direction resulting from a load in the 1 direction, and

$$\nu_{12} = \nu_{21} \frac{E_{22}}{E_{11}} \tag{5.32}$$

for a contraction in the 1 direction, resulting from a load in the 2 direction.

In addition to the Halpin-Tsai model, there are several other models in use today to predict the elastic properties of aligned fiber-reinforced laminates [1, 3]. Most models predict the longitudinal modulus quite accurately, as shown in Figure 5.3 [1], which compares measured values to computed values using the *mixing rule*. However, differences do exist between the models when predicting the transverse modulus, as shown in Figure 5.4 [1]. From Eq. (5.24) it is clear that the Halpin-Tsai model reduces to the mixing rule Eq. (5.23) for a continuous fiber system where a self-evident $\left(\frac{L}{D}\right) = \infty$ can be considered.

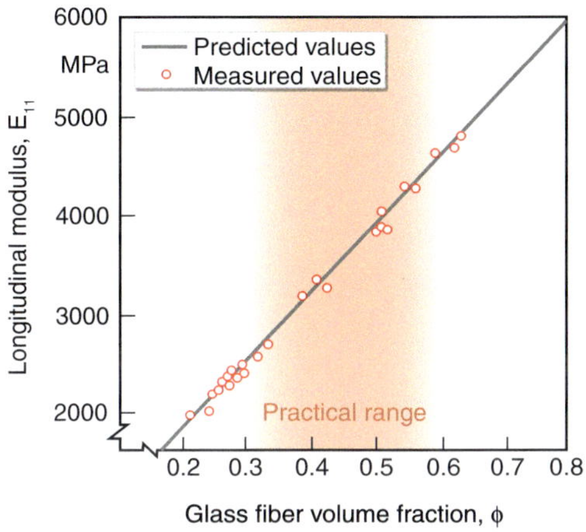

Figure 5.3 Measured and predicted longitudinal modulus for an unsaturated polyester/aligned glass fiber composite laminate as a function of volume fraction of glass content

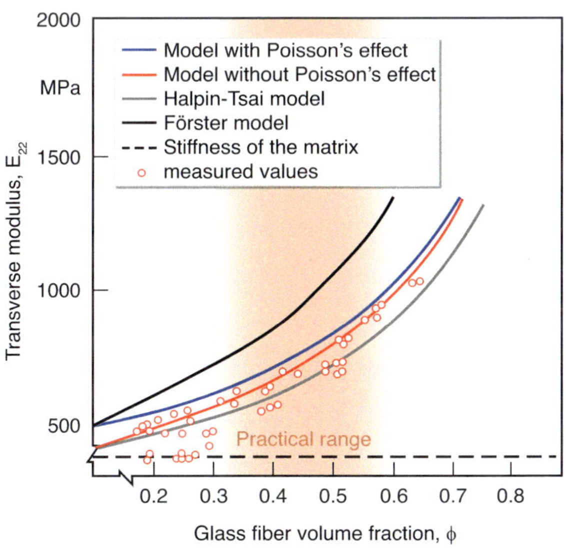

Figure 5.4 Measured and predicted transverse modulus for an unsaturated polyester/aligned glass fiber composite laminate as a function of volume fraction of glass content

Once the longitudinal and transverse terms are determined, we can easily define the strain-stress relation for the laminate in Figure 5.2 as

$$\varepsilon_{11} = \frac{1}{E_{11}}\sigma_{11} - \frac{v_{21}}{E_{11}}\sigma_{22} \tag{5.33}$$

$$\varepsilon_{22} = -\frac{v_{12}}{E_{11}}\sigma_{11} + \frac{1}{E_{22}}\sigma_{22} \tag{5.34}$$

$$\gamma_{12} = \frac{1}{G_{12}}\tau_{12} \tag{5.35}$$

which in matrix form is written as

$$\begin{Bmatrix} \varepsilon_{11} \\ \varepsilon_{22} \\ \gamma_{12} \end{Bmatrix} = \begin{bmatrix} \dfrac{1}{E_{11}} & -\dfrac{v_{21}}{E_{22}} & 0 \\ -\dfrac{v_{12}}{E_{11}} & \dfrac{1}{E_{22}} & 0 \\ 0 & 0 & \dfrac{1}{G_{12}} \end{bmatrix} \begin{Bmatrix} \sigma_{11} \\ \sigma_{22} \\ \tau_{12} \end{Bmatrix} \tag{5.36}$$

or

$$\begin{Bmatrix} \varepsilon_{11} \\ \varepsilon_{22} \\ \gamma_{12} \end{Bmatrix} = \begin{bmatrix} S_{1111} & S_{1122} & 0 \\ S_{2211} & S_{2222} & 0 \\ 0 & 0 & S_{1212} \end{bmatrix} \begin{Bmatrix} \sigma_{11} \\ \sigma_{22} \\ \tau_{12} \end{Bmatrix} \tag{5.37}$$

or

$$\{\bar{\boldsymbol{\varepsilon}}\} = [\bar{\boldsymbol{S}}]\{\bar{\boldsymbol{\sigma}}\} \tag{5.38}$$

where the subscripts 1 and 2 define the longitudinal and transverse directions, respectively, as described in Figure 5.2, $\{\bar{\boldsymbol{\varepsilon}}\}$ and $\{\bar{\boldsymbol{\sigma}}\}$ are strains and stresses when the coordinate system is aligned with the longitudinal fiber direction, and $[\bar{\boldsymbol{S}}]$ is the corresponding *compliance tensor.*

5.2.3 Transformation of Fiber-Reinforced Composite Laminate Properties

It is clear that the loads in the laminate in Figure 5.2 are not always aligned with the longitudinal direction of the laminate, but may be rotated by an angle θ. This is especially true when multi-layered laminated structures are made, where each layer has their own orientation. It is therefore necessary to rotate the laminate and its properties by an angle θ. Figure 5.2 depicts the laminate's material coordinate system $x-y$ and a rotated arbitrary coordinate system 1–2. If we rotate the axes from the 1–2 system to the $x-y$ system, we can transform the stress components using

$$\begin{Bmatrix} \sigma_{11} \\ \sigma_{22} \\ \tau_{12} \end{Bmatrix} = \begin{bmatrix} c^2 & s^2 & 2sc \\ s^2 & c^2 & -2sc \\ -sc & sc & \left(c^2 - s^2\right) \end{bmatrix} \begin{Bmatrix} \sigma_{xx} \\ \sigma_{yy} \\ \tau_{xy} \end{Bmatrix} \tag{5.39}$$

or

$$\{\bar{\boldsymbol{\sigma}}\} = [T_\sigma]\{\boldsymbol{\sigma}\} \tag{5.40}$$

where c represents $\cos\theta$ and s represents $\sin\theta$. The transformation of the strain components carry an extra 1/2 term for the shear strains and is written as

$$\begin{Bmatrix} \varepsilon_{11} \\ \varepsilon_{22} \\ \gamma_{12} \end{Bmatrix} = \begin{bmatrix} c^2 & s^2 & sc \\ s^2 & c^2 & -sc \\ -2sc & 2sc & \left(c^2 - s^2\right) \end{bmatrix} \begin{Bmatrix} \varepsilon_{xx} \\ \varepsilon_{yy} \\ \gamma_{xy} \end{Bmatrix} \tag{5.41}$$

or

$$\{\bar{\boldsymbol{\varepsilon}}\} = [T_\varepsilon]\{\boldsymbol{\varepsilon}\} \tag{5.42}$$

Combining the strain-stress relation in Eq. (5.37) with the above transformations, we can write

$$[T_\varepsilon]\{\boldsymbol{\varepsilon}\} = [\bar{\boldsymbol{S}}][T_\sigma]\{\boldsymbol{\sigma}\} \tag{5.43}$$

or

$$\{\boldsymbol{\varepsilon}\} = [T_\varepsilon]^{-1}[\bar{\boldsymbol{S}}][T_\sigma]\{\boldsymbol{\sigma}\} \tag{5.44}$$

The compliance matrix in the $1-2$ coordinate system has four independent components and the $x-y$ system has six. The inverse of $[T_\varepsilon]$ is equivalent to rotating the coordinates back by $-\theta$. This leads to

$$\begin{Bmatrix} S_{xxxx} \\ S_{yyyy} \\ S_{xxyy} \\ S_{xyxy} \\ S_{xxxy} \\ S_{yyxy} \end{Bmatrix} = \begin{bmatrix} c^4 & s^4 & 2c^2s^2 & c^2s^2 \\ s^4 & c^4 & 2s^2c^2 & c^2s^2 \\ c^2s^2 & c^2s^2 & \left(c^4 + s^4\right) & -c^2s^2 \\ 4c^2s^2 & 4c^2s^2 & -8c^2s^2 & \left(c^2 - s^2\right)^2 \\ 2c^3s & -2cs^3 & 2\left(cs^3 - c^3s\right) & \left(cs^3 - c^3s\right) \\ 2cs^3 & -2c^3s & 2\left(c^3s - cs^3\right) & \left(c^3s - cs^3\right) \end{bmatrix} \begin{Bmatrix} S_{1111} \\ S_{2222} \\ S_{1122} \\ S_{1212} \end{Bmatrix} \tag{5.45}$$

or

$$\{S\}=[R(\theta)]\{\bar{S}\} \tag{5.46}$$

With all the terms defined, we can now write a general strain-stress relation that in matrix form can be represented with

$$\begin{Bmatrix} \varepsilon_{xx} \\ \varepsilon_{yy} \\ \gamma_{xy} \end{Bmatrix} = \begin{bmatrix} S_{xxxx} & S_{xxyy} & S_{xxxy} \\ S_{yyxx} & S_{yyyy} & S_{yyxy} \\ S_{xyxx} & S_{xyyy} & S_{xyxy} \end{bmatrix} \begin{Bmatrix} \sigma_{xx} \\ \sigma_{yy} \\ \tau_{xy} \end{Bmatrix} \tag{5.47}$$

The engineering elastic constants in the x–y system can easily be computed using the above equations

$$E_{xx} = \frac{1}{S_{xxxx}} \tag{5.48}$$

$$E_{yy} = \frac{1}{S_{yyyy}} \tag{5.49}$$

$$G_{xy} = \frac{1}{S_{xyxy}} \tag{5.50}$$

$$\nu_{xy} = -\frac{S_{xxyy}}{S_{xxxx}} \tag{5.51}$$

$$\nu_{yx} = -\frac{S_{xxyy}}{S_{yyyy}} \tag{5.52}$$

Figure 5.5 [1] shows how the stiffness decreases as one rotates away from the longitudinal axis for an aligned fiber-reinforced composite with different volume fraction fiber contents. From the figure it is evident that for high volume fraction fiber contents only a slight misalignment of the fibers from the loading direction results in a drastic reduction of the properties. Along with the predicted stiffness properties, the figure also presents the stiffness for a composite with 0.56 volume fraction of fibers measured at various angles from the longitudinal axis of the composite. The measured and the predicted values agree quite well.

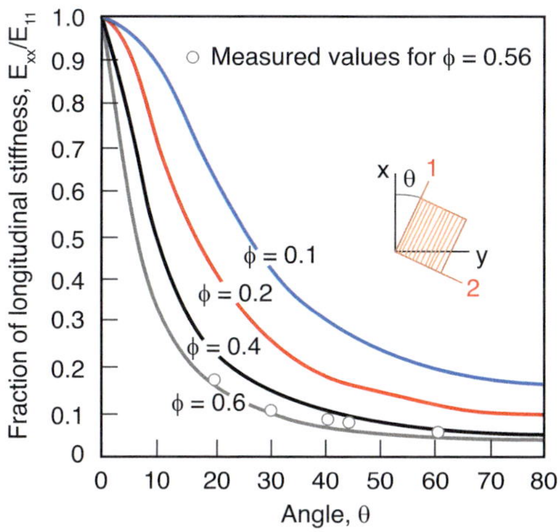

Figure 5.5 Measured and predicted elastic modulus in a unidirectional fiber-reinforced laminate as a function of angle between loading and fiber direction

5.2.4 Sample Application of a Laminated Composite

To illustrate laminated composite theory, the above system of equations were applied to a unidirectional, discontinuous epoxy-carbon fiber laminated composite with 50% fiber volume fraction and the resulting E_{xx}, E_{yy}, and G_{xy} were computed as a function of angular position θ. The cured epoxy matrix has a modulus of 3.7 GPa, and a modulus for the anisotropic carbon fiber is E_{f1} = 379 GPa and E_{f2} = 10 GPa. The modulus of rigidity for the carbon fiber is G_f = 15 GPa. The Poisson's ratios for the epoxy and the carbon are 0.4 and 0.2, respectively.

Using the above data in conjunction with the above equations, and assuming a very large L/D of 10,000, the following calculations were done

For the longitudinal direction: $\eta = \dfrac{\left(\dfrac{379}{3.7} - 1\right)}{\left(\dfrac{379}{3.7} + 20{,}000\right)} = 5.07 \times 10^{-4}$

For the transverse direction: $\eta = \dfrac{\left(\dfrac{10}{3.7} - 1\right)}{\left(\dfrac{10}{3.7} + 2\right)} = 0.505$

$$G_m = \frac{3.7}{2(1 + 0.4)} = 1.32\,\text{GPa}$$

$$\lambda = \frac{\left(\frac{15}{1.32} - 1\right)}{\left(\frac{15}{1.32} + 1\right)} = 0.9657$$

$$E_{11} = 3.7\left(\frac{1 + 20{,}000\left(5.07 \times 10^{-4}\right)(0.5)}{1 - \left(5.07 \times 10^{-4}\right)(0.5)}\right) = 191.3\ \text{GPa}$$

$$E_{22} = 3.7\left(\frac{1 + 2(0.505)(0.5)}{1 - (0.505)(0.5)}\right) = 7.44\ \text{GPa}$$

$$G_{12} = 1.32\left(\frac{1 + (0.9657)(0.5)}{1 - (0.9657)(0.5)}\right) = 3.79\ \text{GPa}$$

$$\nu_{21} = 0.4(1 - 0.5) + (0.2)(0.5) = 0.3$$

$$\nu_{12} = 0.3\frac{7.44}{191.3} = 0.012$$

Once E_{11}, E_{22}, G_{12}, ν_{21}, and ν_{12} are known, the terms in Eq. (5.37) can be computed and transformed with Eq. (5.45) to give E_{xx}, E_{yy}, and G_{xy} as a function of angular position θ. Figure 5.6 presents the plot for E_{xx}, E_{yy}, and G_{xy}, and Figure 5.7 shows G_{xy} with a different scale.

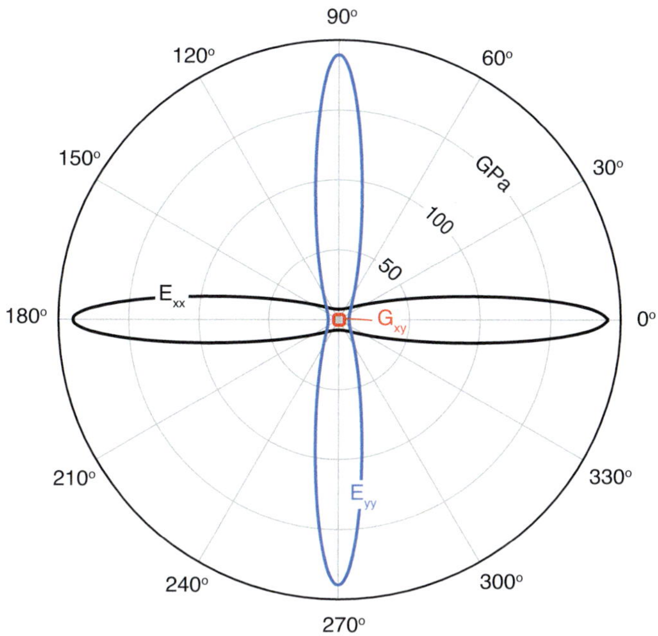

Figure 5.6 Predicted E_{xx}, E_{yy}, and G_{xy} for a unidirectional laminated epoxy-glass fiber composite as a function of θ

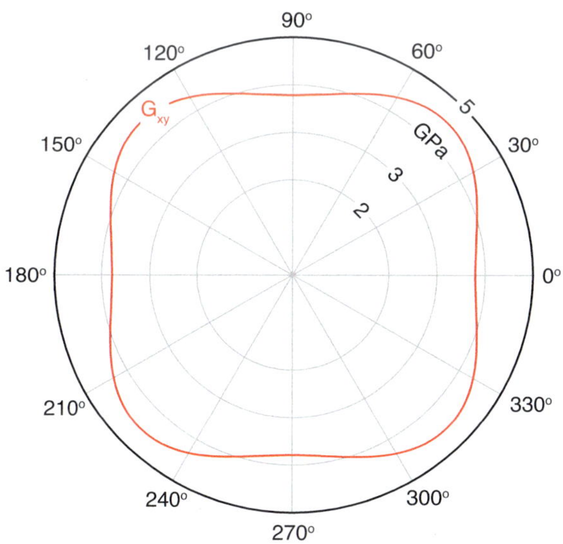

Figure 5.7 Predicted G_{xy} for a unidirectional laminated epoxy-glass fiber composite as a function of θ

Similarly, using the data given and computed above one can compute the moduli E_{11} and E_{22} for a cross-laminated $0° - 90°$ plate with $\phi = 0.5$ volume fraction of carbon fibers, where half of the fibers are oriented at 0° and the other half at 90°.

It may be intuitive that the properties E'_{11} and E'_{22} in a cross-laminated plate are the same. The procedure to calculate E'_{11} and E'_{22} in the cross-laminated plate is to take the results from the above calculations and perform a weighted average as

$$E'_{xx} = E'_{yy} = 0.5E_{xx} + 0.5E_{yy} \tag{5.53}$$

This results in the modulus shown in Figure 5.8.

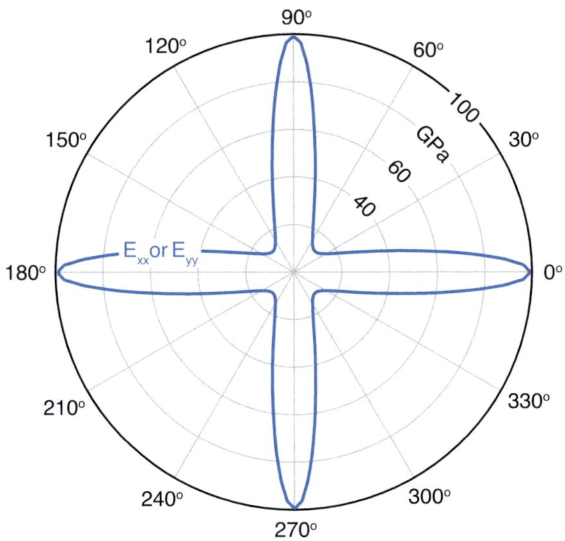

Figure 5.8 Predicted moduli for a cross-laminated $0° - 90°$ plate with $\phi = 0.5$ volume fraction of carbon fibers

■ 5.3 Discontinuous Fiber-Reinforced Composites

As was demonstrated in the previous section, discontinuous or chopped fiber composites require a somewhat different treatment, since the discontinuities can lead to incomplete load transfer between the matrix and the fibers and therefore a weakened composite structure. Hence, the object of a discontinuous fiber is to be long enough to maximize its load-bearing capacity. To illustrate this concept,

Figure 5.9 presents a schematic of a fiber embedded in a cylindrically shaped matrix. When the cylindrical body is loaded, the deformation within the matrix transmits the load to the fiber by shear. In the figure, we can see a high degree of shear deformation near the tip of the fiber and lower shear deformation as one moves away from the tip. The object is for the shear to reach zero before the center of the fiber is reached. At that point, a maximum normal stress is reached within the fiber. Hence, in order to transfer the load, the fiber must have a length L_t. This is schematically shown in Figure 5.10.

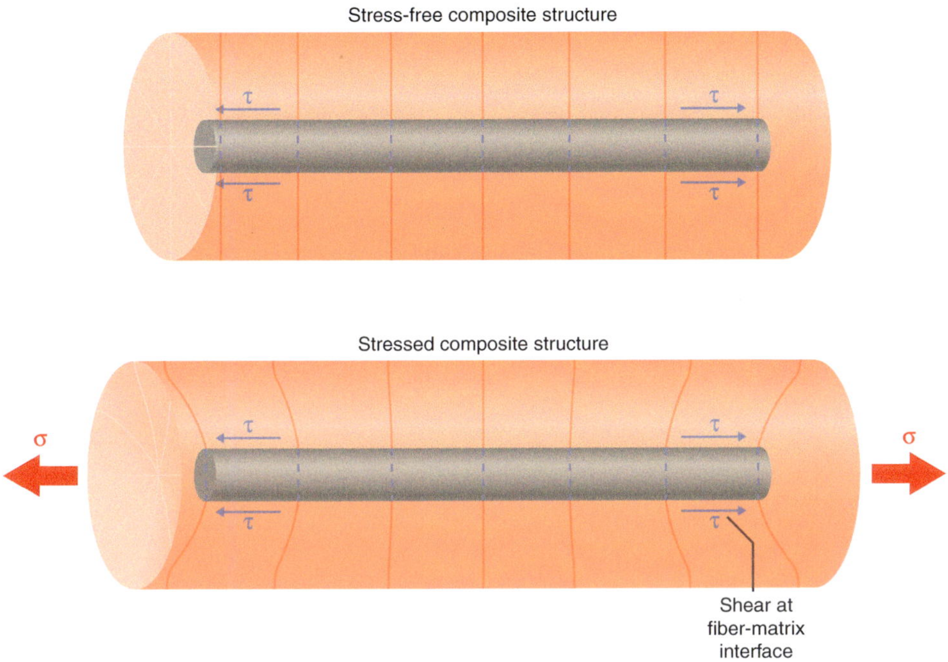

Figure 5.9 Schematic of matrix deformation with an embedded fiber

The shape of the shear stress at the fiber–matrix interface depends on the nature of the matrix. If the matrix behaves like an elastic material, such as is the case with most thermosetting resins, the shear stress has a sharp spike near the tip of the fiber, such as schematically depicted in the upper portion of Figure 5.10. With thermoplastics resins, which have a viscoelastic response to imposed stress fields, the sharp stresses relax rapidly, leading to a smoother stress distribution near the fiber tip, as shown in the lower part of Figure 5.10.

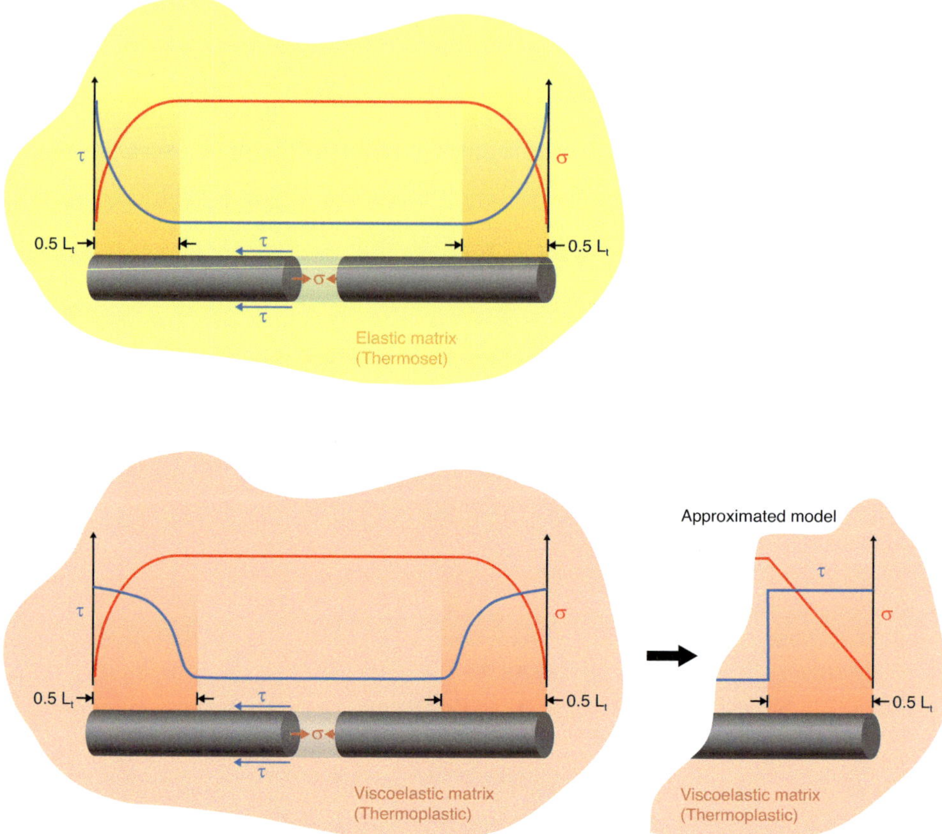

Figure 5.10 Schematic diagram of load transfer from matrix to fiber in a composite

If we use the simplified model presented for a thermoplastic matrix in Figure 5.10, we can calculate the force transmitted by shear stress at the interface using

$$F_{Shear} = \pi D L_t \tau / 2 \tag{5.54}$$

and the maximum force exerted by the fiber with

$$F_{Fiber} = \pi D^2 \sigma_f / 4 \tag{5.55}$$

For a load system aligned with the fibers, the maximum fiber stress, σ_f, is related to the stress applied to the system, σ, using Eq. (5.15)

$$\frac{\sigma_f}{E_f} = \frac{\sigma}{E_{11}} \tag{5.56}$$

Equating the force transmitted by shear to the maximum force exerted by the fiber, we can solve for the length required to transmit the load from the matrix to the fiber

$$L_t = \frac{\sigma_f D}{2\tau} = \frac{E_f}{E_{11}} \frac{\sigma D}{2\tau} \tag{5.57}$$

Failure of the composite structure occurs when either the matrix-fiber bonding strength is exceeded, or when the fiber breaks. Hence, in the optimal situation that maximizes the use of both matrix and fiber is when failure occurs simultaneously. For that situation we would like the shear stress to be the bonding shear strength, τ_b, and the fiber stress to be the ultimate strength of the fiber, σ_{fu}. With these values, we can rewrite Eq. (5.57) as

$$L_c = \frac{\sigma_{fu} D}{2\tau_b} \tag{5.58}$$

5.3.1 Sample Application of Critical Length for Load Transfer[2]

Glass fiber-reinforced polyamide 66 is a commonly used engineering resin to manufacture automotive structural components. Assuming the ultimate strength of the glass fiber is 4,585 MPa and the interfacial shear strength between the fiber and the matrix is 40 MPa, compute the critical fiber length for a 10 μm diameter fiber.

Using Eq. (5.58) we get

$$L_c = \frac{\sigma_{fu} D}{2\tau_b} = \frac{\left(4.585 \cdot 10^9 \text{ Pa}\right)\left(10^{-6}\text{m}\right)}{2\left(4.0 \cdot 10^9 \text{ Pa}\right)}$$

We can also approach this problem from a more complex point of view and model the system represented in Figure 5.9 using the finite element method. Since the system represented in Figure 5.9 is axis-symmetric, the problem reduces to a 2D problem such as represented in Figure 5.11. Here it is only necessary to model one quarter of the domain. A finite element output of the displacement field is presented in Figure 5.12 for the case where $\frac{L}{D} = 25$. The shear deformation in the matrix region that surrounds the fiber can be easily identified in this graph. Even the region near the center of the fiber reflects a visible amount of shear deformation, telling us that $\frac{L}{D} = 25$ is not sufficient to transfer the load from the matrix onto the fiber.

---

[2] Analysis courtesy of Dr. John Puentes.

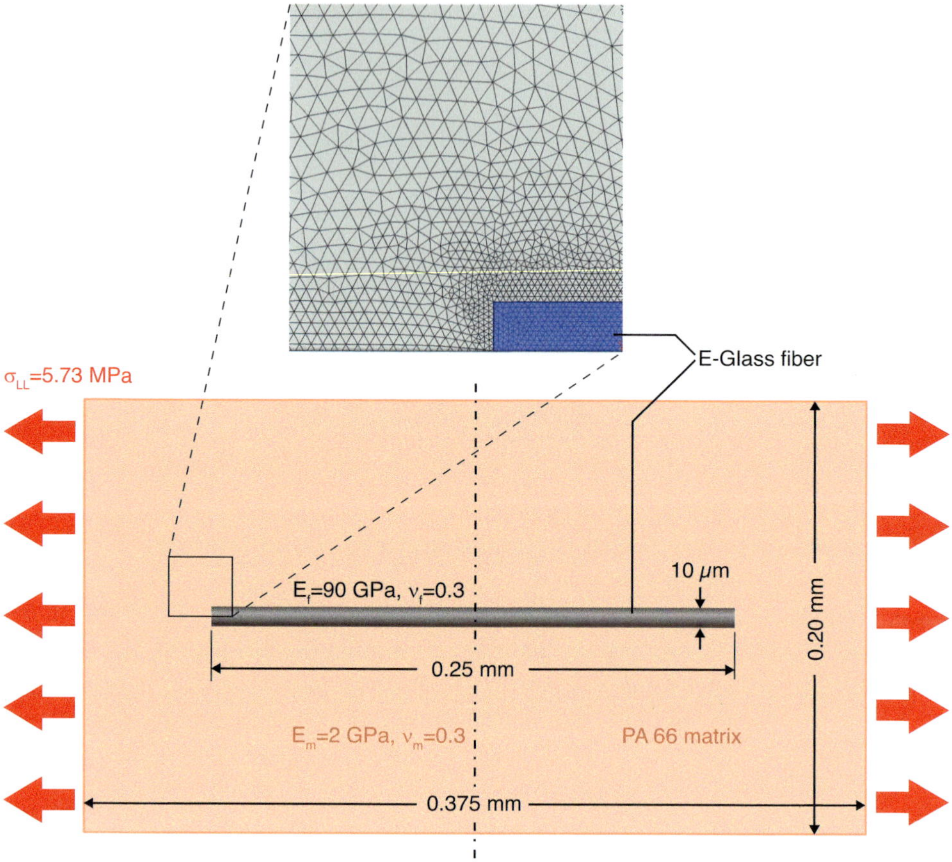

Figure 5.11 Model of a PA66 matrix cylinder with an embedded glass fiber

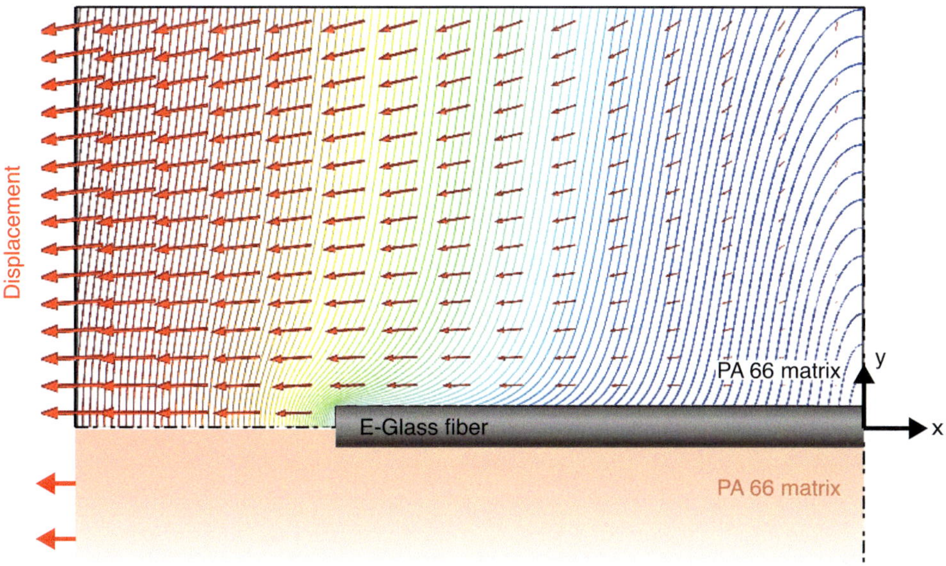

Figure 5.12 Displacement field of a PA66 matrix cylinder with an embedded E-glass fiber

A succession of runs were performed using the finite element method for different cases with $\frac{L}{D} = 25, 50, 75, 100, 150$, and 200. Figure 5.13 presents the stress within the fiber and the shear stress on the matrix-fiber interface for all six cases. The upper portion of the graph shows the axial fiber stress and the lower portion of the figure presents the shear stress on the surface of the fiber. It is evident that the fibers with $\frac{L}{D} = 25$ and 50 are not sufficiently long to achieve the maximum possible stress of 235 MPa. A fiber with $\frac{L}{D} = 25$ only achieves a maximum stress of 195 MPa, and a fiber with $\frac{L}{D} = 50$ achieves a stress of about 225 MPa. According to this linear-elastic finite element model, a fiber with an $\frac{L}{D}$ of at least 75 is required to accomplish a full load transfer.

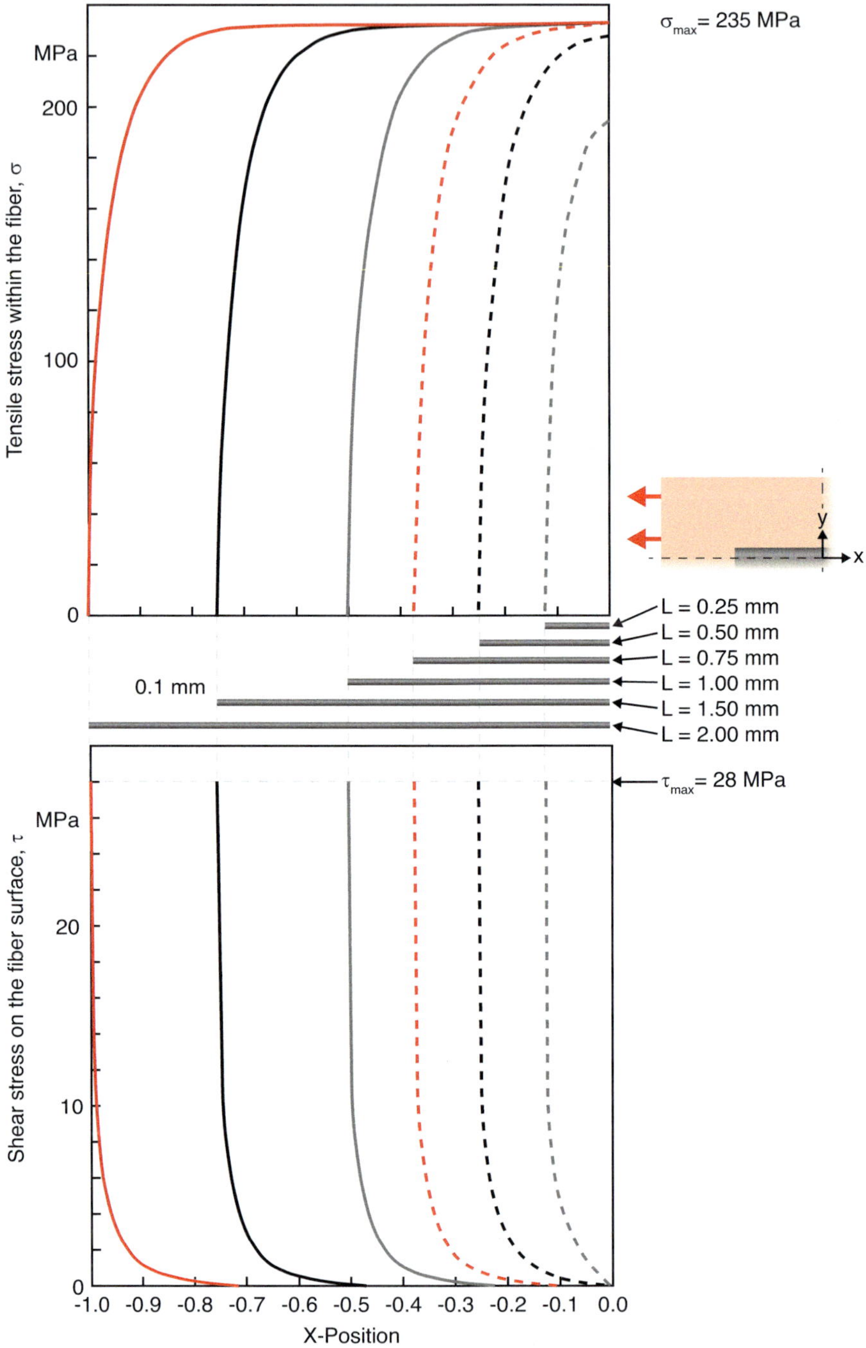

Figure 5.13 Displacement field of a PA66 matrix cylinder with an embedded E-glass fiber for various $\frac{L}{D}$ ratios

5.3.2 Reinforced Composite Laminates with a Fiber Orientation Distribution Function

The above transformation can be used to compute the properties of planar systems with a fiber orientation distribution function. This is done by superposing aligned fiber laminates rotated away from the principal 1–2 coordinate system by an angle θ and with a volume fiber fraction given by $\psi(\theta)$. The transformation is written as

$$\{S\} = \int_{-\pi/2}^{\pi/2} \left([R(\theta)] \{\bar{S}\} \psi(\theta) \right) d\theta \tag{5.59}$$

which can be written in discrete terms to be used with a fiber orientation distribution function attained from computer simulation:

$$\{S\} = \sum_{i=1}^{N} \left([R(\theta_i)] \{\bar{S}\} \psi(\theta_i) \right) \Delta\theta \tag{5.60}$$

Using Eq. (5.60), one can easily predict the stiffness properties of a part with randomly oriented fibers, where $\psi(\theta) = 1/\pi$, using[3]

$$\frac{1}{E_{xx}} = \frac{1}{E_{yy}} = \frac{1}{E_{\text{Random}}} = \frac{3}{8}\frac{1}{E_{11}} + \frac{3}{8}\frac{1}{E_{22}} - \frac{2}{8}\frac{v_{12}}{E_{11}} + \frac{1}{8}\frac{1}{G_{12}} \tag{5.61}$$

Figure 5.14 shows the increase in dynamic shear modulus for polybutylene terephthalate with 10 and 30% glass fiber content. While the orientation in injection molded test samples used for the measurements shown in Figure 5.14 are considered isotropic (with a random orientation), as will be shown in Chapter 5, injection molded items exhibit a complex seven-layered laminated structure.

[3] The incorrect expression $E_{\text{Random}} = 3/8E_{11} + 5/8E_{22}$ is often successfully used for low fiber content to approximate the stiffness of the composite with randomly oriented fibers. However, using this equation for composites with large differences between E_{11} and E_{22} may lead to an overestimate of stiffness by 50%.

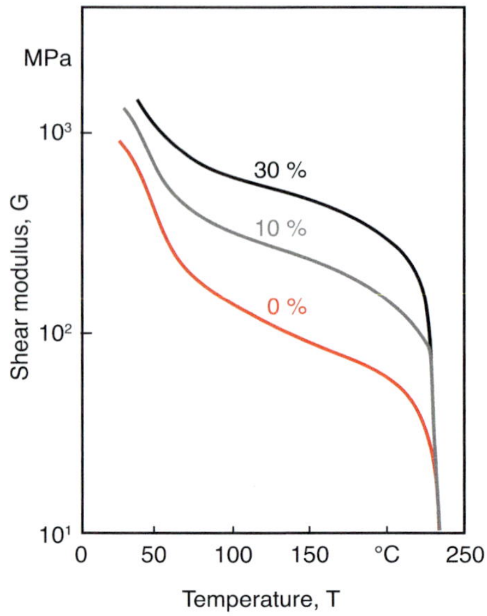

Figure 5.14 Shear modulus for a polybutylene terephthalate with various levels of glass fiber content by weight

5.4 Failure of Fiber-Reinforced Composites

5.4.1 Failure of an Axially Loaded Laminate

The failure of fiber filled composites should be controlled by the strength of the fibers. While fibers have a significantly higher ultimate strength, σ_{fu}, than the ultimate strength of the matrix, σ_{mu}, the fiber's maximum strain, $\epsilon_{f-\max}$, is significantly lower than the maximum strain of the matrix, $\epsilon_{m-\max}$, such as shown in Figure 5.15 for a carbon fiber and an epoxy matrix. Hence, in an axially loaded unidirectional fiber composite such as shown in Figure 5.2, the composite fails when $\epsilon_{f-\max}$ is exceeded. If we rewrite the force balance in Eq. (5.14) in terms of stress, we get

$$\sigma_c A_c = \sigma_m A_m + \sigma_f A_f \tag{5.62}$$

which results in the rule of mixtures for stress

$$\sigma_c = \sigma_m (1-\phi) + \sigma_f \tag{5.63}$$

If we assume that in the laminate failure occurs the moment the fibers rupture, we can rewrite the above equation as

$$X = \sigma'_{um}(1-\phi) + \sigma_{uf}\phi \tag{5.64}$$

where σ'_{um} is the stress in the matrix at the failure strain of the fiber (Figure 5.15). However, as can be seen from Eq. (5.64), the maximum stress the composite sustains can be lower than σ_{um} when $\phi = 0$, which is not possible. It is therefore necessary to correct the ultimate strength of the composite for those cases when the fiber volume fraction is so low that the failure is controlled by the matrix. For those cases, after the fibers break at $\epsilon_{f-\max}$, the system fails when the ultimate strength of the matrix is reached within the matrix itself. We can write

$$X = \sigma_{um}(1-\phi) \tag{5.65}$$

Equations (5.64) and (5.65) are combined to form the criterion presented in Figure 5.16. There we clearly see that, in order for the fibers to contribute to the strength of the composite, their volume fraction must be above a minimum value $\phi_{\min}$

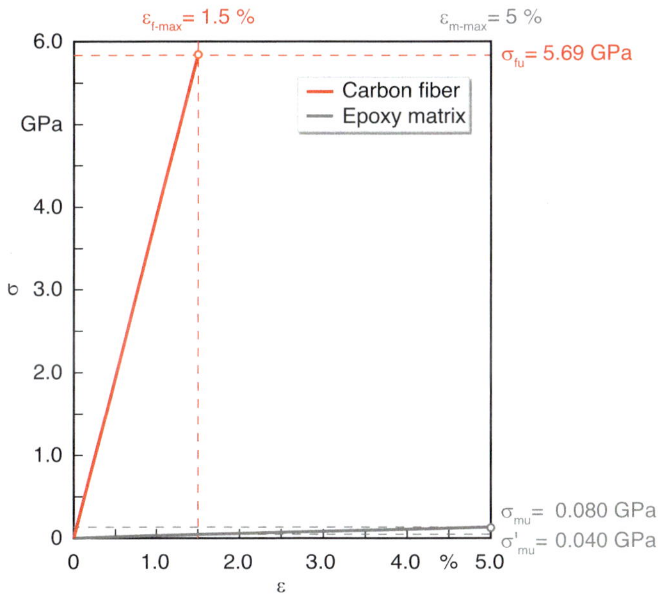

Figure 5.15 Stress-strain curves for a typical carbon fiber and epoxy resin

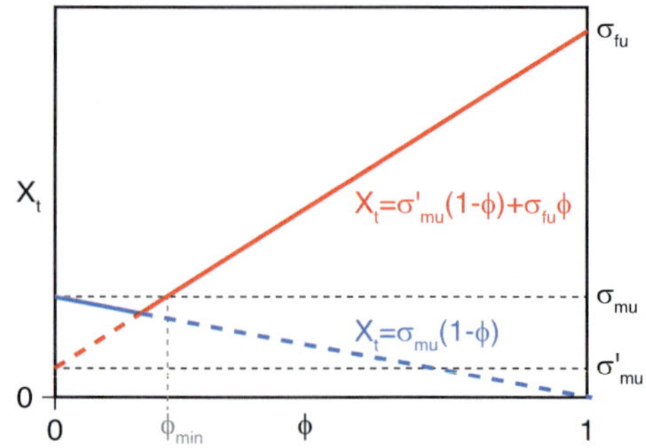

Figure 5.16 Composite strength as a function of fiber volume fraction

By inspecting Figure 5.16 one can see that there is a minimum volume fraction of carbon fibers necessary to reinforce epoxy resin in a unidirectional laminated composite. The minimum volume fraction, ϕ_{min}, is the point where the ultimate strength of the composite equals the ultimate strength of the unreinforced matrix,

$$\sigma_{um} = \sigma'_{um}\left(1-\phi_{min}\right)+\sigma_{uf}\phi_{min} \tag{5.66}$$

Solving for ϕ_{min} we get

$$\phi_{min} = \frac{\sigma_{um} - \sigma'_{um}}{\sigma_{uf} - \sigma'_{um}} = \frac{0.080\text{ GPa} - 0.040\text{ GPa}}{5.69\text{ GPa} - 0.040\text{ GPa}} = 0.0071 = 0.71\% \tag{5.67}$$

5.5 Failure Criteria for Composites with Complex Loads

Today, there are two general families of failure criteria. The first and oldest ones are the strength tensor based criteria, which represent the failure surface with a single scalar function, such as the Tsai-Hill [4, 5], which will be presented here, the Gol'denblat-Kopnov [6], the Malmeister [7] or Tsai-Wu [8], and the Theocaris [9] criteria. Furthermore, a new strength tensor based failure criterion recently developed by Osswald and Osswald [10], which includes all possible stress interactions, will be presented in this chapter. These models, such as the Malmeister model, often referred to as Tsai-Wu model, are widely used and have been incorporated into finite element analysis software to perform strength analysis of anisotropic parts. The second family of models includes those that incorporate physical aspects of fracture, often termed phenomenological or mechanistic models, such as the ones by Sun [11], Puck [12], Pinho [13], and Dávila [14]. These models do not include the interaction between longitudinal stresses and transverse stresses, but concentrate on the interaction between transverse stresses and transverse shear stresses, modeling quite well the shear strengthening effect during transverse compression. This second family of failure criteria was explicitly developed to predict the failure of UD-FRP. Since this book concentrates on discontinuous fiber filled composites, only strength tensor based failure criteria are presented. While these criteria are purely mathematical adaptations to experimental results, they can be easily implemented for the failure of any type of composite.

5.5.1 Maximum Stress Failure Criterion

The most basic, and often intuitive, failure criterion is the *maximum stress failure criterion.* From the analysis presented in the previous section, one can therefore say that the failure of the composite structure loaded in the 1 direction occurs perpendicular to that direction as depicted in Figure 5.17. Note that failures of composites are different depending on whether the stresses are tensile or compressive. In that case, the failure criteria is given by

$$\sigma_{11} \geq X_t \text{ or } X_c \tag{5.68}$$

where X_t or X_c represent the tensile and compressive strength in the 1 direction. As shown in Figure 5.17, other failure modes such as inter fiber failure exist, due to failure of the matrix:

$$\sigma_{22} \geq Y_t \text{ or } Y_c \tag{5.69}$$

$$\tau_{23} \geq S_T \tag{5.70}$$

$$\tau_{12} \geq S_L \tag{5.71}$$

where Y_t or Y_c represent the maximum stresses transverse to the fiber direction, and S_T and S_L represent the transverse and longitudinal shear strengths. For poorly bonded materials, failure occurs by delamination between fibers and matrix. For such cases one can write

$$\sigma_{22} \geq \sigma_B \tag{5.72}$$

$$\tau_{23} \geq \tau_B \tag{5.73}$$

$$\tau_{12} \geq \tau_B \tag{5.74}$$

where σ_B and τ_B are the bonding strengths between the matrix and the fiber.

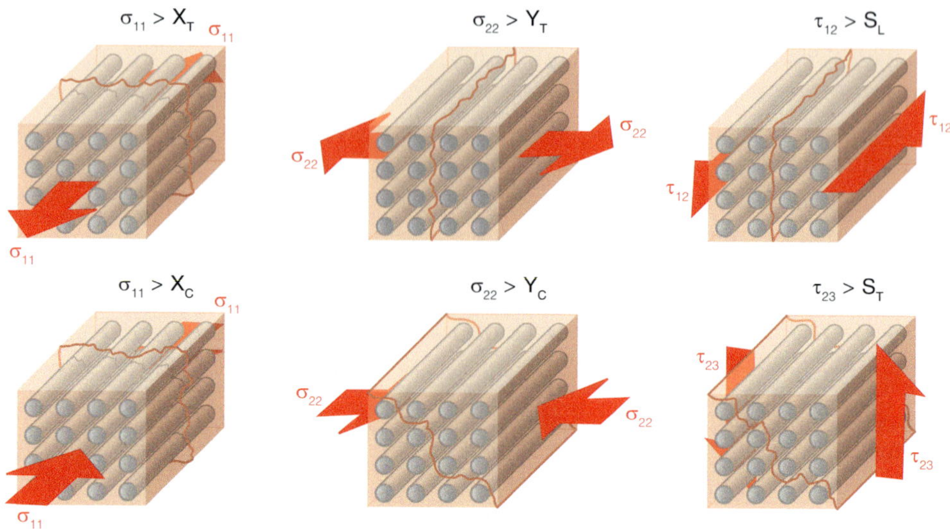

Figure 5.17 Failure modes in unidirectional composite laminates

To illustrate this failure criterion for a unidirectional carbon-epoxy laminated structure loaded in tension at any angle θ, with $\sigma_{22} = \tau_{12} = 0$, the following calculations and graphs will show the limiting stresses of the laminate using this oversimplified criterion. For this case, it is assumed that σ_{uc} = 3 GPa (0.5 volume fraction), X = 0.185 GPa, and S = 0.1 GPa.

If one takes the unidirectional laminate as depicted in Figure 5.2 and applies a tensile load σ_{11} at any angle θ, with $\sigma_{22} = \tau_{12} = 0$, one can rewrite Eq. (5.39) as

$$\sigma_{11} = \cos^2(\theta)\sigma_{xx} \tag{5.75}$$

$$\sigma_{22} = \sin^2(\theta)\sigma_{xx} \tag{5.76}$$

$$\tau_{12} = \sin(\theta)\cos(\theta)\sigma_{xx} \tag{5.77}$$

Using the above equations and the failure criteria, Eqs. (5.69)–(5.71), we can say that in order for the laminate not to fail the following must be true

$$\sigma_{xx} \leq \frac{X_T}{\cos^2(\theta)} \text{ and } \sigma_{xx} \leq \frac{Y_T}{\sin^2(\theta)} \text{ and } \sigma_{xx} \leq \frac{S_L}{\sin(\theta)\cos(\theta)} \tag{5.78}$$

Equation (5.78) is graphically depicted in Figure 5.18.

However, this failure criterion completely ignores that the strength in one direction actually depends on stresses that are not directly associated to the direction of that strength. In other words, there is a stress interaction in the various strengths. These strength interactions are included in the following two criteria presented below.

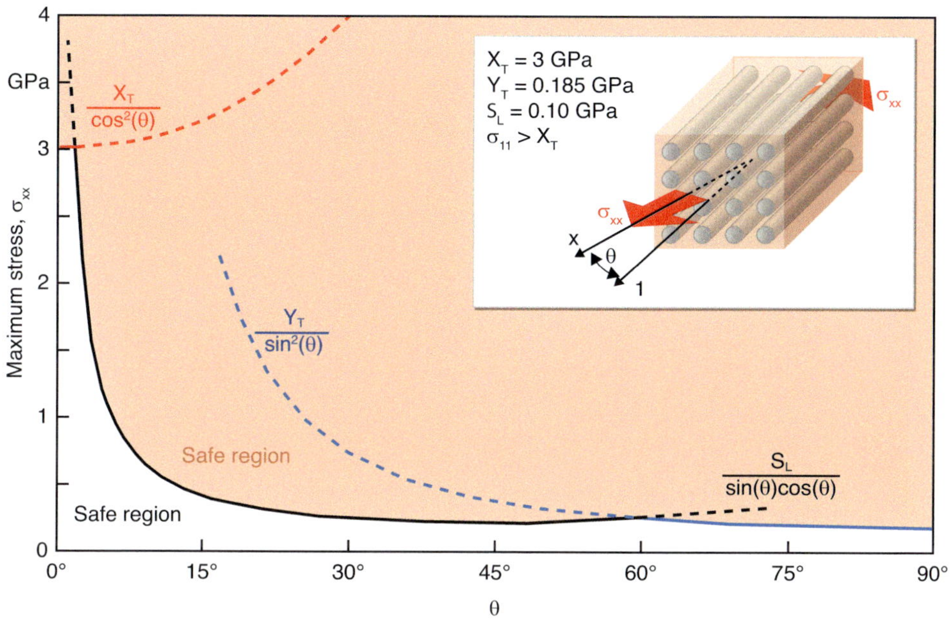

Figure 5.18 Failure modes in unidirectional composite laminates for an axial model

5.5.2 Tsai-Hill Failure Criterion

In the above analysis, the failure modes were independent from each other. However, there is always interaction between stresses. Laminates will fail more easily if, in addition to shear, a surface also feels a tensile normal stress. As mentioned above, there are many failure criteria to predict failure of composite laminated structures. The oldest one and one of the most widely accepted ones is the *Tsai-Hill criterion* given by

$$f = \left(\frac{\sigma_{11}}{X}\right)^2 + \left(\frac{\sigma_{22}}{Y}\right)^2 - \frac{\sigma_{11}\sigma_{22}}{X^2} + \left(\frac{\tau_{12}}{S_L}\right)^2 \tag{5.79}$$

Note that Tsai-Hill does not distinguish between tensile and compressive stresses. In Equation (5.79), f is a failure function: when it is greater or equal to 1, failure is eminent. The Tsai-Hill criterion describes a failure envelope within the stress field, and is more conservative than the stress criterion defined by the individual stress modes, as is demonstrated below. The third term in Equation (5.79) represents a stress interaction between σ_{11} and σ_{22}. However, no stress interactions between normal stresses and shear stresses are included.

When the Tsai-Hill failure criterion is implemented for a unidirectional carbon-epoxy laminated structure loaded in tension with a σ_{11} stress, but with $\sigma_{22} = \tau_{12} = 0$, the stress field as a function of θ can be written as

$$f = \left(\frac{\cos^2(\theta)\ \sigma_{xx}}{X}\right)^2 + \left(\frac{\sin^2(\theta)\ \sigma_{xx}}{Y}\right)^2 - \frac{\sin^2(\theta)\cos^2(\theta)\sigma_{xx}^{\ 2}}{X^2} + \left(\frac{\sin(\theta)\cos(\theta)\sigma_{xx}}{S_L}\right)^2 \tag{5.80}$$

Letting $f \geq 1$, the above equation can be rearranged as

$$\sigma_{xx} \leq \sqrt[-2]{\frac{\cos^4(\theta)}{X^2} + \frac{\sin^4(\theta)}{Y^2} - \frac{\cos^2(\theta)\sin^2(\theta)}{X^2} + \frac{\sin^2(\theta)\cos^2(\theta)}{S_L^{\ 2}}} \tag{5.81}$$

The above equation is plotted in Figure 5.19 together with the maximum stress failure criterion. As can be seen, the safe domain within the failure envelope has been somewhat reduced.

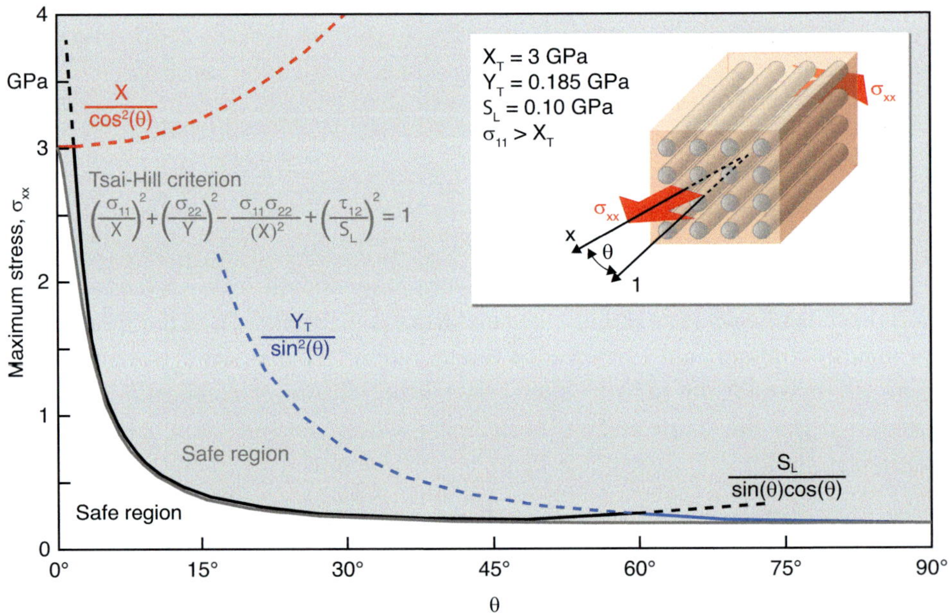

Figure 5.19 Tsai-Hill failure mode for a unidirectional composite laminate with an axial load

5.5.3 A New Strength Tensor Based Failure Criterion with Stress Interactions[4]

The new failure criterion presented here [10] is based on strength tensor based models such as the Gol'denblat-Kopnov and the Malmeister or Tsai-Wu models, which include the effect of different compressive and tensile strengths. As with the Tsai-Hill model, the model presented by Gol'denblat and Kopnov defines a scalar failure function as a function of strength tensors and stresses using

$$f = \left(F_{ij}\sigma_{ij}\right) + \left(F_{ijkl}\sigma_{ij}\sigma_{kl}\right)^{1/2} \tag{5.82}$$

where failure is expected when $f \geq 1$. The strength tensor components F_{ij} and F_{ijkl} are second and fourth order tensors, respectively, that depend on engineering strength parameters such as X_t, X_c, Y_t, Y_c, and S and satisfy the symmetry conditions, $F_{ij} = F_{ji}$ and $F_{ijkl} = F_{klij}$. By using an exponent of 1/2 in the second term of the equation, they achieved a linear criterion scalar function f. Malmeister [7] and Tsai and Wu [8] modified the Gol'denblat-Kopnov model by letting the exponent be equal to 1, eliminating the fractional exponents, but resulting in a function f that is quadratic with respect to stress.

Through this additive technique Gol'denblat and Kopnov were able to couple all the failure modes into one single function, where a separate treatment of compressive and tensile modes is not required. For example, for the plane stress case the Gol'denblat-Kopnov criterion becomes

$$\begin{aligned} f &= F_{11}\sigma_{11} + F_{22}\sigma_{22} + F_{12}\tau_{12} \\ &+ \left(F_{1111}\sigma_{11}^{\;2} + F_{2222}\sigma_{22}^{\;2} + F_{1212}\tau_{12}^{\;2} + 2F_{1122}\sigma_{11}\sigma_{22} + 2F_{1112}\sigma_{11}\tau_{12} + 2F_{2212}\sigma_{22}\tau_{12}\right)^{1/2} \end{aligned} \tag{5.83}$$

where the usual notation of σ_{ij} and τ_{ij}, for normal and shear stresses, respectively, was used. Gol'denblat and Kopnov used Equation (5.83), without the stress interaction terms F_{1112} and F_{2212}, to solve for most of the strength tensor components by assuming uniaxial conditions in the 1 and 2 directions or a shear condition in the $1-2$ plane. These are all listed in Table 5.1. In order to determine the $\sigma_{11} - \sigma_{22}$ interaction strength tensor component F_{1122}, Gol'denblat and Kopnov measured the positive and negative shear strengths, S_{45p} and S_{45n}, of a specimen where the fibers were oriented at 45°. The resulting F_{1122} stress interaction term is also listed in Table 5.1.

Figure 5.20 illustrates this effect by comparing both models to the first World Wide Failure Exercise (WWFE-I) data for the $\sigma_{22} - \tau_{12}$ failure plane. One can see that both models have the same failure surface, $f = 1$, but by delivering a failure function f that is linear with respect to the applied stress field, the Gol'denblat-Kopnov

[4] The reader should consult the paper by Osswald and Osswald [10] for the full derivation of this model.

is a more conservative approach. Furthermore, a linear function can be more easily implemented when using probabilistic failure analysis, as, for example, proposed in the Soviet Union in 1975 by Zaitsev et al. [15] and Thieme et al. [16] in Germany in 2014. Even when using a simple factor of safety approach, in the example depicted in Figure 5.20 the Gol'denblat-Kopnov criterion has no failures when using a safety factor above 1.15 ($f = 0.87$), whereas the Tsai-Wu approach would still predict a failure at a safety factor of 2.1 ($f = 0.47$).

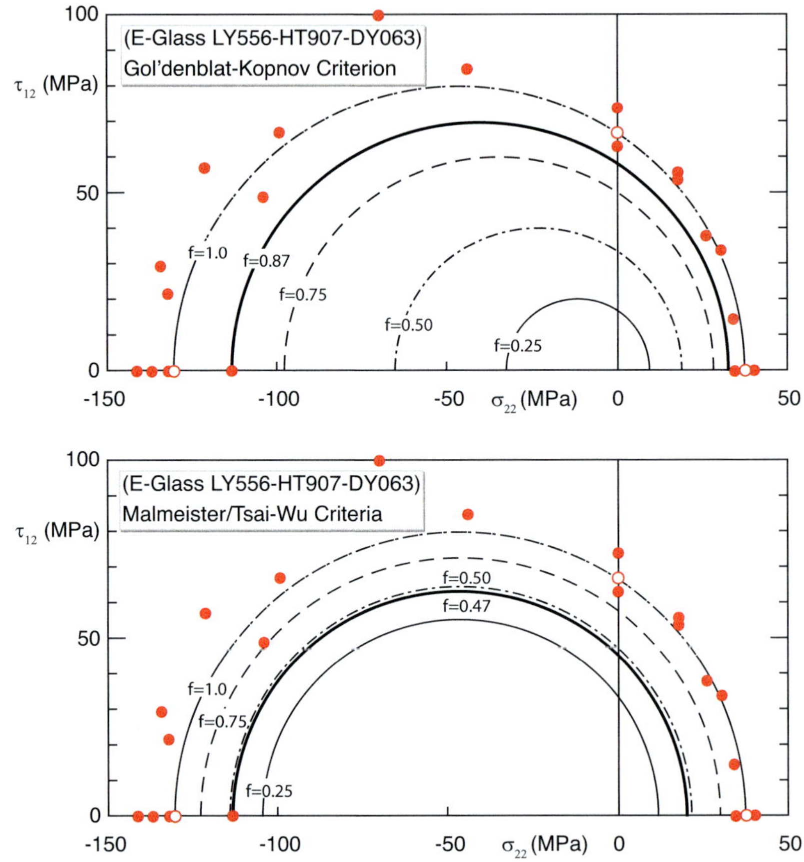

Figure 5.20 Comparing the Gol'denblat-Kopnov and Malmeister or Tsai-Wu failure criteria to WWFE-I data for the $\sigma_{22} - \tau_{12}$ failure plane

Table 5.1 presents the strength tensor components for both the Gol'denblat-Kopnov and the Malmeister or Tsai-Wu models. The $\sigma_{11} - \sigma_{22}$ interaction strength tensor component F_{1122} proposed by Tsai and Wu [8], listed in Table 5.1, depends on the factor f_{12}, which lies between −1 and 0. A popular approach is letting $f_{12} = -1/2$, because it leads to the classic von Mises theory for isotropic materials [17].

However, these models neglect the longitudinal and transverse interaction terms, F_{1112} and F_{2212}, which represent the $\sigma_{11}-\tau_{12}$ and $\sigma_{22}-\tau_{12}$ interactions. These interactions, as well as a new $\sigma_{11}-\sigma_{22}$ interaction, are proposed in this paper and are presented in the next section.

Table 5.1 Strength Tensor Components for the Gol'denblat-Kopnov and the Malmeister/Tsai-Wu Failure Criteria

Tensor Component	Gol'denblat-Kopnov	Malmeister/Tsai-Wu
F_{11}	$\frac{1}{2}\left(\frac{1}{X_t}-\frac{1}{X_c}\right)$	$\frac{1}{X_t}-\frac{1}{X_c}$
F_{1111}	$\frac{1}{4}\left(\frac{1}{X_t}+\frac{1}{X_c}\right)^2$	$\frac{1}{X_t X_c}$
F_{22}	$\frac{1}{2}\left(\frac{1}{Y_t}-\frac{1}{Y_c}\right)$	$\frac{1}{Y_t}-\frac{1}{Y_c}$
F_{2222}	$\frac{1}{4}\left(\frac{1}{Y_t}+\frac{1}{Y_c}\right)^2$	$\frac{1}{Y_t Y_c}$
F_{12}	0	0
F_{1212}	$\frac{1}{S^2}$	$\frac{1}{S^2}$
F_{1122}	$\frac{1}{8}\left[\left(\frac{1}{X_t}+\frac{1}{X_c}\right)^2+\left(\frac{1}{Y_t}+\frac{1}{Y_c}\right)^2-\left(\frac{1}{S_{45p}}+\frac{1}{S_{45n}}\right)^2\right]$	$\frac{f_{12}}{\sqrt{X_t X_c Y_t Y_c}}$

Osswald and Osswald proposed interaction strength tensor components, F_{ijkl}, based on the slopes of the failure surface ($f=1$) at any of the points where the engineering strength values are known within an arbitrary $\sigma_{ij}-\sigma_{kl}$ plane. With minimal experimental data gathered to compute the slope of the failure surface around any of the four strength values within a $\sigma_{ij}-\sigma_{kl}$ plane, the interaction strength tensor component F_{ijkl} can be evaluated. For this, we derive the interaction terms between normal stresses and the interaction between normal and shear stresses separately, as schematically depicted in Figure 5.21.

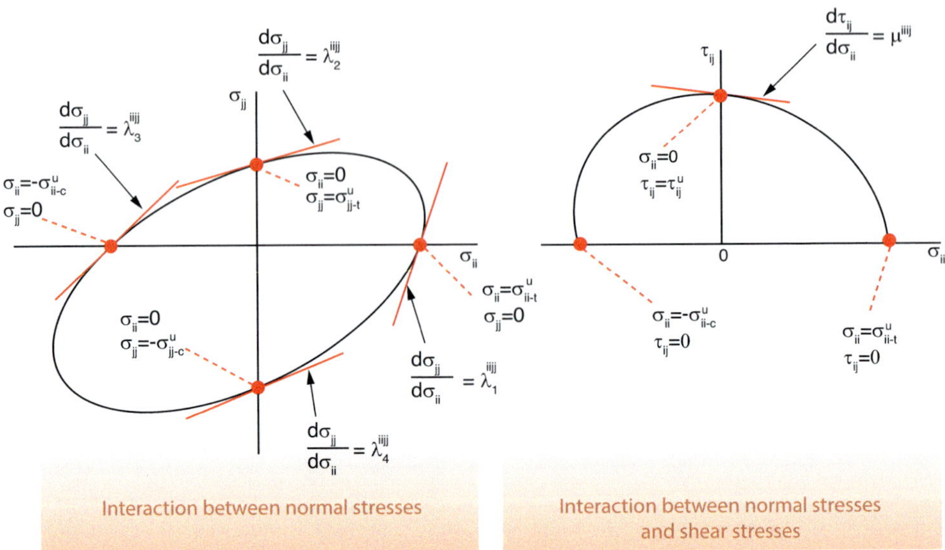

Figure 5.21 Locations on the failure surface within the $\sigma_{ii} - \sigma_{jj}$ (left) and $\sigma_{ii} - \tau_{ij}$ (right) planes where the interaction F_{iijj} and F_{iiij}, respectively, can be evaluated

To illustrate this, let us concentrate on the $\sigma_{11} - \sigma_{22}$ interaction (F_{1122}) point on the failure surface, where $\sigma_{11} = \sigma^{u}_{11-t} = X_t$ and $\sigma_{22} = 0$ and the failure surface has a slope of $\dfrac{d\sigma_{22}}{d\sigma_{11}} = \lambda_1^{1122}$. Taking the derivative of Equation (5.83) with respect to σ_{11}, at the failure surface where $f = 1$ and $\tau_{12} = 0$, results in

$$0 = F_{11} + F_{22}\lambda_1^{1122} + \frac{1}{2}\left(F_{1111}X_t^2\right)^{-1/2}\left(2F_{1111}X_t + 2F_{1122}X_t\lambda_1\right) \tag{5.84}$$

from which the unknown strength tensor component, F_{1122}, can be computed as

$$F_{1122} = -\frac{F_{1111}^{\frac{1}{2}}\left(F_{11} + F_{22}\lambda_1^{1122}\right) + F_{1111}}{\lambda_1^{1122}} \tag{5.85}$$

Finally, after substituting for F_{11}, F_{1111}, and F_{22}, the $\sigma_{11} - \sigma_{22}$ strength tensor interaction term can be written as

$$F_{1122} = -\frac{1}{4}\left[\frac{2}{\lambda_1^{1122}X_t^2} + \frac{2}{\lambda_1^{1122}X_tX_c} + \frac{1}{X_tY_t} - \frac{1}{X_tY_c} + \frac{1}{X_cY_t} - \frac{1}{X_cY_c}\right] \tag{5.86}$$

F_{1122} can be evaluated at any of the remaining three axes intersections of the failure surface within the $\sigma_{11}-\sigma_{22}$ plane, theoretically resulting in the same computed numerical value for the strength tensor interaction term. Table 5.2 presents the interaction strength tensor components for the more general case where the subscripts 1122 were replaced with *iijj*.

Table 5.2 Interaction Strength Tensor Components for the Gol'denblat-Kopnov and the Malmeister/Tsai-Wu Failure Criteria ($\sigma_{11-t}^u = X_t$, $\sigma_{11-c}^u = X_c$, $\tau_{12}^u = S$, etc.)

Tensor Component	Gol'denblat-Kopnov	Malmeister/Tsai-Wu
F_{iiij}	$-\dfrac{F_{ii}}{\tau_{ij}^u} - F_{ijij}\mu^{iiij}$	$F_{ii}F_{ijij}^{\ 1/2} - F_{ijij}\mu^{iiij}$
$F_{iijj}\left(\lambda_1^{iijj}\right)$	$-\dfrac{\left(F_{ii}+F_{jj}\lambda_1^{iijj}\right)F_{iiii}^{\ 1/2}+F_{iiii}}{\lambda_1^{iijj}}$	$-\dfrac{F_{ii}+F_{jj}\lambda_1^{iijj}+2F_{iiii}\sigma_{ii-t}^u}{2\sigma_{ii-t}^u\lambda_1^{iijj}}$
$F_{iijj}\left(\lambda_2^{iijj}\right)$	$-\left(F_{ii}+F_{jj}\lambda_2^{iijj}\right)F_{jjjj}^{\ 1/2} - F_{jjjj}\lambda_2^{iijj}$	$-\dfrac{F_{ii}+F_{jj}\lambda_2^{iijj}+2F_{jjjj}\sigma_{jj-t}^u\lambda_2^{iijj}}{2\sigma_{jj-t}^u}$
$F_{iijj}\left(\lambda_3^{iijj}\right)$	$\dfrac{\left(F_{ii}+F_{jj}\lambda_3^{iijj}\right)F_{iiii}^{\ 1/2}-F_{iiii}}{\lambda_3^{iijj}}$	$\dfrac{F_{ii}+F_{jj}\lambda_3^{iijj}-2F_{iiii}\sigma_{ii-c}^u}{2\sigma_{ii-c}^u\lambda_3^{iijj}}$
$F_{iijj}\left(\lambda_4^{iijj}\right)$	$\left(F_{ii}+F_{jj}\lambda_4^{iijj}\right)F_{jjjj}^{\ 1/2} - F_{jjjj}\lambda_4^{iijj}$	$\dfrac{F_{ii}+F_{jj}\lambda_4^{iijj}-2F_{jjjj}\sigma_{jj-c}^u\lambda_4^{iijj}}{2\sigma_{jj-c}^u}$

To represent the interaction between normal stresses and shear stresses, such as F_{2212}, one assumes symmetry (replacing τ_{ij} by $\left|\tau_{ij}\right|$). Osswald and Osswald proposed an interaction term F_{iiij} based on the slope of the failure surface, $\dfrac{d\tau_{ij}}{d\sigma_{ii}} = \mu^{iiij}$, at $\tau_{ij} = \tau_{ij}^u = S$ and $\sigma_{ii} = 0$, as schematically depicted in Figure 5.21. For example, when deriving the F_{2212} interaction, one can take the derivative of Eq. (5.83) with respect to σ_{22}, at the failure surface where $f = 1$ and $\sigma_{22} = 0$, which results in

$$0 = F_{22} + \left[F_{1212}S\left(-\mu^{2212}\right) + F_{2212}S\right] \tag{5.87}$$

where the unknown strength tensor component, F_{2212}, can be solved for

$$F_{2212} = -\frac{F_{22}}{S} + F_{1212}\mu \tag{5.88}$$

The $\sigma_{22} - \tau_{12}$ strength tensor interaction term can be written in terms of engineering strength values as

$$F_{2212} = \left[\frac{\mu}{S^2} - \frac{1}{2Y_t S} + \frac{1}{2Y_c S} \right] \tag{5.89}$$

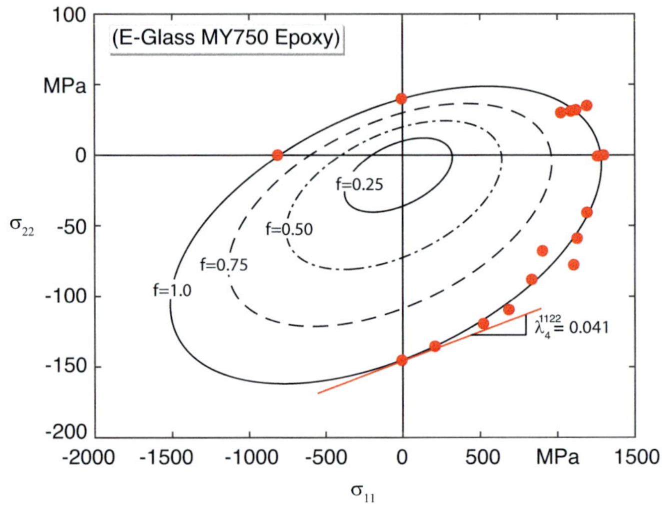

Figure 5.22 Comparing the present model using $\lambda_4^{1112} = 0.041$ to WWFE-I experimental data

This term is also presented in general form in Table 5.2 for both failure criteria.

To test this model with unidirectional composites in the $\sigma_{11} - \sigma_{22}$, $\tau_{12} - \sigma_{22}$, and $\tau_{12} - \sigma_{11}$ stress planes, it was compared to experiments performed with glass and carbon reinforced epoxy composites, taken from the WWFE-I [19].

WWFE-I fiber failure data for the $\sigma_{11} - \sigma_{22}$ plane was modeled using the new model with $\lambda_4^{1112} = 0.041$ and is presented in Figure 5.22. The resulting failure surface, $f = 1$, is identical to the failure surface predicted by the Malmeister or Tsai-Wu criteria using the interaction coefficient $f_{12} = -1/2$.

The model was also tested using inter-fiber failure data within the $\tau_{12} - \sigma_{22}$ stress plane resulting from an extensive experimental study performed by Cuntze et al. [20]. The new model is computed with two coefficients $\mu^{2212} = -0.57$ and $\mu^{2212} = -0.38$, where the first shows excellent agreement with the experimental data. The figure clearly shows how the Cuntze model captures the shear strength increase under compressive loads, but misses the compressive strength increase under shear loads. The experimental data exhibits a compressive transverse strength increase that is also reflected by the present model. It is not quite clear if the compressive transverse strength increase is real. While this "bulge" is present in many experiments, the literature remains silent on the reason why this occurs.

The third set of experimental data related to unidirectional FRPC materials is the longitudinal stress under shear stress in the $\tau_{12} - \sigma_{11}$. Using the new model with a parameter μ^{1112} = 0.054, the shear strengthening effect under tensile longitudinal loads is clearly observed in Figure 5.23.

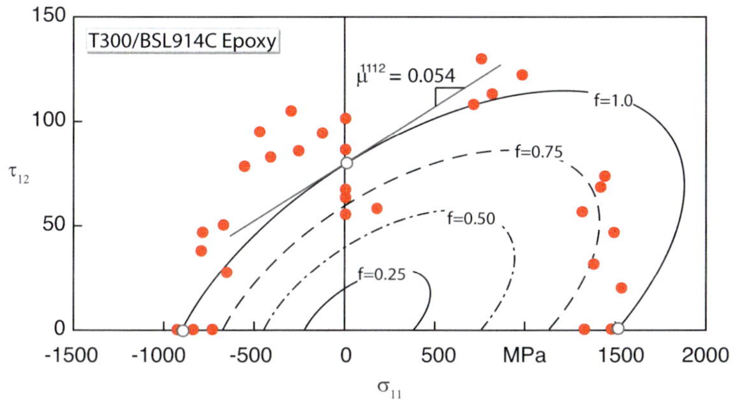

Figure 5.23 Comparing the present model to WWFE-I biaxial failure stress envelope data under longitudinal and shear stress loading results [19], using μ^{1112} = 0.054

To test the new model with an anisotropic material with a fiber orientation distribution, experiments performed on paperboard by Suhling et al. [21] were used. In their study, they found that when τ_{12} = 0, the Tsai-Wu criterion with $F_{1122} = -2.297 \cdot 10^{-4}$ (MPa)$^{-1}$ gave the best fit. However, when including all four levels of shear, τ_{12} = 0, τ_{12} = 6.9 MPa, τ_{12} = 10.3 MPa, and τ_{12} = 15.9 MPa, they had to drop the longitudinal-transverse stress interaction tensor component ($F_{1122} = 0$) to achieve an overall better fit. This compromised somewhat the results at τ_{12} = 0, because the tilt of the elliptical failure surface, observed in the experimental results, is lost when no stress interaction exists. On the other hand, in addition to including the stress interaction strength tensor F_{1122}, the new model is able to include both shear-stress-normal-stress interactions F_{1112} and F_{2212}, by adjusting λ_4^{1112}, μ^{1112}, and μ^{2212}, respectively. Figure 5.24 compares the strength data within the $\sigma_{11} - \sigma_{22}$ stress plane at the four different shear levels to the present model. It is clear that by including all three stress interaction strength tensor components, a very good match between model and experiments was achieved.

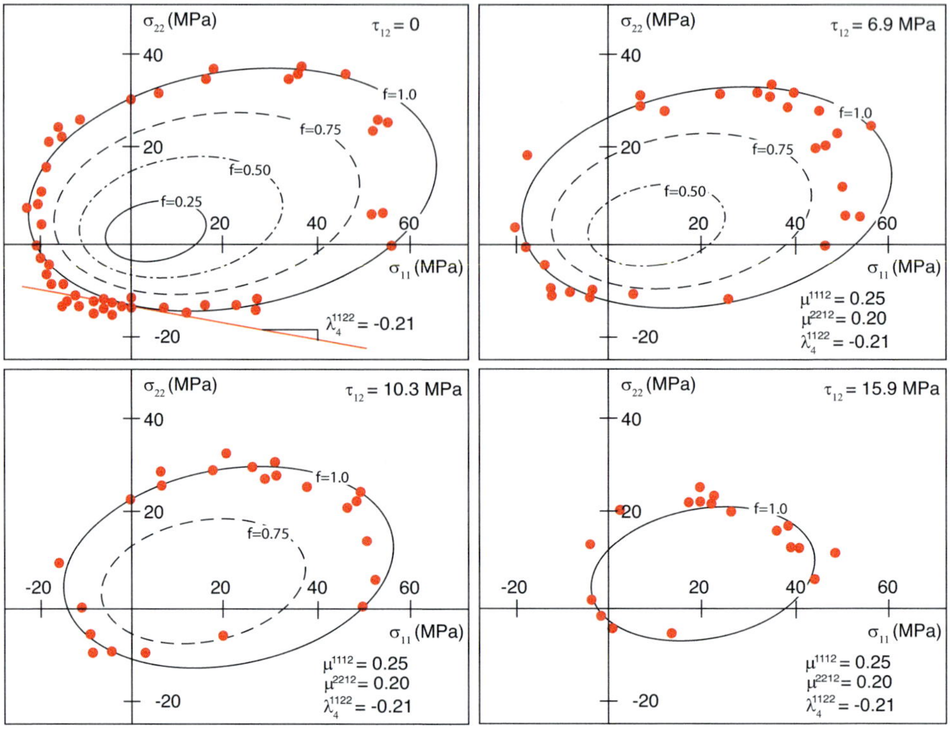

Figure 5.24 Comparing the present model to biaxial in-plane strength results for paperboard experimental results [21] using $\lambda_4^{1112} = -0.21$, $\mu^{1112} = 0.25$, and $\mu^{2212} = 0.20$

Finally, to validate the new model against failure data from woven fabric composite materials, experiments by Mallikarachchi and Pellegrino [22] were used. The Gol'denblat and Kopnov data presented sufficient information to show that the $\sigma_{11} - \sigma_{22}$ plane failure surface does not have a tilt and is therefore relatively circular. This is in agreement with the failure criterion developed in Greece by Theocaris [7] as well as his experimental work dealing with weaves [24]. The present model can fit the data used in the Gol'denblat and Kopnov paper very well, as can be seen in Figure 5.25. There is sufficient data available to evaluate F_{1122} at three locations in the Gol'denblat and Kopnov experimental data for a textile composite [4]. The authors computed the three slopes, which resulted in the same longitudinal-transverse stress interaction tensor component of approximately $F_{1122} = 1.8 \cdot 10^{-7}$ (MPa)$^{-1}$.

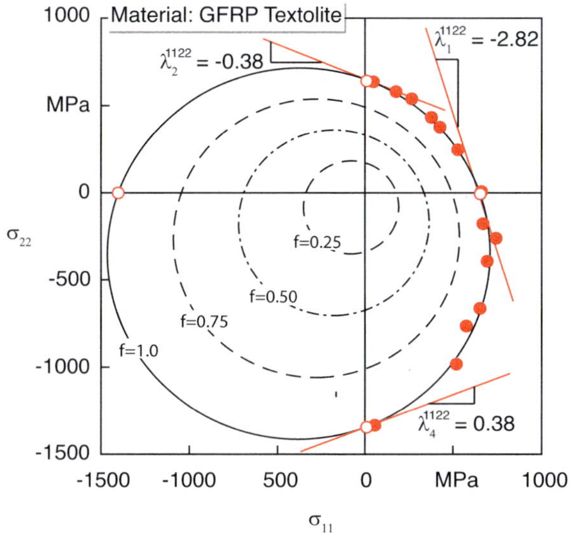

Figure 5.25 Comparing the present model to biaxial in-plane strength results for a weave GFRP laminate for $\tau_{12} = 0$ [6], with either $\lambda_1^{1112} = -2.82$, $\lambda_2^{1112} = -0.38$, or $\lambda_4^{1112} = 0.38$, which all give $F_{1122} = 1.8 \cdot 10^{-7}$ (MPa)$^{-1}$

Furthermore, to validate the new model against failure data from woven fabric composite materials with combined longitudinal, transverse, and planar shear stresses, experiments performed by Mallikarachchi and Pellegrino [22] on two-ply T300-1k/Hexcel 913 plain weave laminates were used. In their own analysis, Mallikarachchi and Pellegrino used the Tsai-Wu model with a four-fold symmetry about the third axis of the laminate, resulting in $F_{11} = F_{22}$, $F_{1112} = F_{2212}$, etc., and used Tsai-Wu's F_{1112} longitudinal-transverse stress interaction tensor component with $f_{12} = -\frac{1}{2}$. As a result, as shown in Figure 5.26, their failure surface was elliptical, contradicting the findings of Gol'denblat and Kopnov [6] as well as of Theocaris [23]. The present model was fit in such a way that it predicts a nearly circular failure surface by using $\lambda_4^{1112} = -0.35$. Furthermore, the parameters μ^{1112} and μ^{2212} were adjusted such that the two average data points within the $\sigma_{11} - \sigma_{22}$ failure plane, located at shear stress $\tau_{12} = 92.6$, fell on the failure surface. The values of $\mu^{1112} = \mu^{2212} = 0.102$ put the failure surface on top of the average values resulting in a perfect fit of the available data.

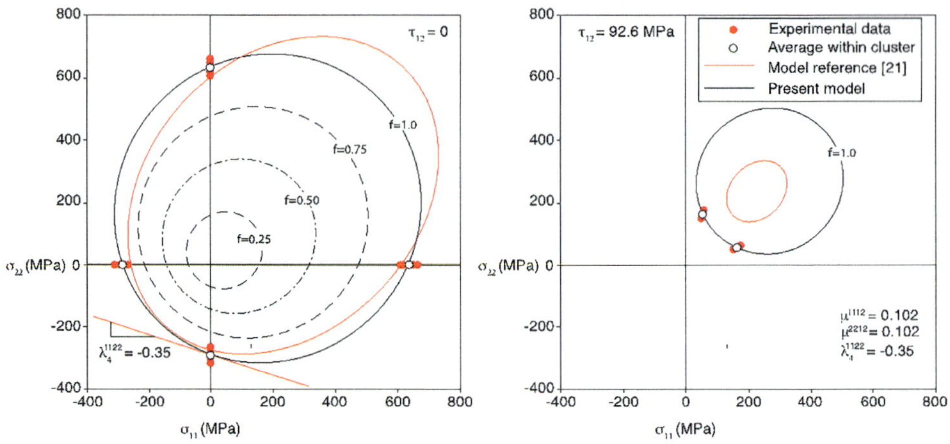

Figure 5.26 Comparing the present model to biaxial in-plane strength results for a two-ply plain weave CFRP laminate for $\tau_{12} = 0$ (left) and $\tau_{12} = 92.6$ MPa (right) [23], with $\lambda_4^{1112} = -0.35$ and $\mu^{1112} = 0.102$

References

[1] Osswald, T. A. and Menges, G., *Materials Science of Polymers for Engineers*, 3rd ed., Hanser, Munich (2012)

[2] Tsai, S. W., Halpin, J. C., and Pagano, N. J., *Composite Materials Workshop*, Technomic Publishing Co., Stamford (1968)

[3] Menges, G. and Bintrup, H., Chapter 4 in *Polymere Werkstoffe*, Georg Thieme Verlag, Stuttgart (1984)

[4] Hill, R. A., *Proc. Roy. Soc. A*, 193:281–297 (1948)

[5] Tsai, S. W., Strength characteristics of composite materials, NASA CR-224 (1965)

[6] Gol'denblat, I. I. and Kopnov, V. A., *Mekhanika Polimerov*, 1, pp. 70–78 (1965)

[7] Malmeister, A. K., *Mekhanika Polimerov*, 4, pp. 519–534 (1966)

[8] Tsai, S. W. and Wu, E. M., *J Comp Mater*, 5, pp. 58–80 (1971)

[9] Theocaris, P. S., *Acta Mechanica*, 79, pp. 53–79 (1989)

[10] Osswald, P. V. and Osswald, T. A., *Polymer Composites*, 39, pp. 2826–2834 (2018)

[11] Sun, B. J. and Quinn, D. W., Oplinger, Comparative Evaluation of Failure Analysis Methods for Composite Laminates. DOT/FAA/AR-95/109 (1996)

[12] Puck, A. and Schürmann, H., *Compos. Sci. Technol.*, 58, pp. 1045–1067 (1998)

[13] Pinho, S. T., Modeling failure of laminated composites using physically based failure models. Ph. D. Thesis, Department of Aeronautics, Imperial College London (2005)

[14] Dávila, C. G., Camanho, P. P., and Rose, C. A., *J. Comp. Mater.*, 39, pp. 323–345 (2005)

[15] Zaitsev, G. P., Pashkov, V. A., Strelyaev, V. S., *Moscow Aeronautical Tech. Inst.*, 8, pp. 3–9 (1973)

[16] Thieme, M., Boehm, R., Gude, M., and Hufenbach, W., *Compos. Sci. Technol.*, 90, pp. 25–31 (2014)

[17] Tsai, S. W. and Melo, J. D. D., *Compos. Sci. Technol.*, 123, pp. 71–78 (2016)

[18] Tsai, S. W. and Melo, J. D. D., An Invariant-Based Theory of Composites, *Compos. Sci. and Tech.*, DOI: 10.1016/j.compscitech.2014.06.017 (2014)

[19] Soden, P. D., Hinton, M. J., and Kaddour, A. S., *Compos. Sci. Technol.*, 62, pp. 1489–1514 (2002)

[20] Cuntze, R. G., Deska, R., Szelinski, B., Jeltsch-Fricker, R., Mechbach, S., Huybrechts, D., Kopp, J., Kroll, L., Gollwitzer, S., and Rackwitz, R, Neue Bruchkriterien und Festigkeitsnachweise für unidirektionalen Faserkunststoffverbund unter mehrachsiger Beanspruchung – Modellbildung und Experimente. VDI Fortschrittbericht. 5.506 (1997)

[21] Suhling, J. C., Rowlands, R. E., Johnson, M. W., and Gunderson, D. E., *Experim. Mech.*, 25, pp. 75–84 (1986)

[22] Mallikarachchi, H. M. Y. C. and Pellegrino, S., *Journal of Composite Materials*, 47, pp. 1357–1375 (2013)

[23] Theocaris, P. S., *Acta Mechanica*, 95, pp. 68–86 (1992)

6 Modeling and Simulation of Discontinuous Fiber-Reinforced Composites

6.1 Introduction to Finite Element Methods (FEM)

As an engineer designing a part, one is often interested in estimating the structural performance of the part, which is to predict how the designed part will perform during different potential loading conditions. Such capability can become very important in minimizing possibility of failure, which means a shorter design cycle. Also, such capability can offer opportunities to consider many design options and hence potential to optimize for weight and performance.

Typical design parameters needed to estimate the performance of a part are geometrical shape, how the part will be constrained, i.e., attachments or methods to secure the part, properties of the material used to build the part, and expected loading conditions. The goal of using modeling and simulation is to use this information and estimate the structural performance for various potential usage conditions accurately. The simplest analytical tool that can help achieve such goal starts with the fundamental equations of mechanics of materials or strength of materials. The fundamental equations basically deal with the stress and strain in the materials resulting from the applied load and deformations. Fundamental formulas describing stress and strain in fiber-reinforced composite materials were reviewed in Sections 5.1 and 5.2 of Chapter 5. The equations, describing the stress and strain state in the fiber-reinforced material represent a closed form solution based on the constitutive equations for fiber and matrix materials and geometrical details of the fibers, such as orientation, concentration, etc. While such fundamental equations are very important to model the basic behavior of the fiber-reinforced materials, they have limited appeal when used in design of industrial parts. This is because the closed form solutions emerging from the fundamental equations can be used only for very simple geometries, i.e., shapes where the boundary conditions can be described by mathematical functions.

It is necessary to address this limitation to analyze the structural performance of complex-shaped industrial parts. A numerical approach based on discretization, called finite element method (FEM), has been developed for this purpose. The key point of the finite element method is that the complex shape of a part is divided into a number of simpler shapes, called elements. The force deformation relationship for such element at selected points, which are called "nodes", is developed using a constitute equation representing the stress and strain relationship of the material and geometry of the element. Then, these elements and corresponding force and displacements are combined to form the stiffness matrix, i.e., the matrix relating force and deformation at each node within the whole part. Furthermore, the stiffness matrix is used to estimate structural performance for different loading and boundary conditions over the complex geometry of the part.

6.1.1 Basic Concept of FEM

There are many textbooks and publications available on finite element methods [1–3]. Most four-year engineering programs offer introductory courses in finite element methods. This book, on the other hand, will provide a basic understanding of the finite element method fundamentals at a concept level only. The detailed implementation is not discussed as it is assumed that it is already known or can be learned through other sources. This chapter focuses on addressing the challenges in using the finite element method for composite materials. The finite element method (FEM) is a numerical method to model structural and other engineering systems that can be represented by partial differential equations but cannot be solved due to the complex shape of the structure, as well as the complex boundary and loading conditions. To address these difficulties, attempts were made in the 1940s [4] to discretize and solve the partial differential equations. In the early 1950s, Turner's team at Boeing developed a discrete stiffness method [5] to address vibration for a complex aero structure shape, which can be considered as the foundation of the finite element methods. However, meaningful work with wide appeal was done in the 1960s and 1970s. J. H. Argyris [6] at the University of Stuttgart, R. W. Clough [7] at UC Berkeley, and O. C. Zienkiewicz [1] at Swansea University are considered early pioneers in the field. R. W. Clough is credited for coining the name "finite element method". This was followed by many other researchers around the world during the 1970s and 1980s developing stronger mathematical bases and algorithms that are in use by finite element software today [2, 3].

The basic idea behind the finite element method, stated in simple words, is to discretize a complex shape into many simpler shapes, called elements, for which the differential equation or partial differential equations can be solved. A complex part is discretized into thousands—and sometimes millions—of elements, which would make manual addition of such discrete systems impossible. However, through the development of computational equipment, such as by IBM in the 1960s and 1970s, when they developed mainframe computers that can handle larger amounts of data, such complex and cumbersome computational tasks became possible. Furthermore, as time passed, computing speed increased rapidly, almost doubling every two years. In parallel, the faster and efficient algorithms to represent the geometry and manipulate matrices took advantage of new computer architectures as they were being developed. Furthermore, during the same time there was a strong push to expand the technology to help the United States reach the moon. This resulted in rapid development in many areas. The finite element method also benefited and advanced significantly. One such effort was led by NASA and resulted in a general purpose structural analysis software program called NASA STRucture ANalysis (NASTRAN). The NASTRAN™ system was introduced to the public in 1971 [32]. Over time, it saw many improvements. Many commercial versions of the original NASA development, such as MSC NASTRAN™, NX Nastran™, NEi Nastran™, etc., are still in use in the industry today. Over the past six decades, significant human efforts have been expended in the development of FEM software suites. Researchers and mathematicians have developed improved algorithms for finite element method calculations and their efficient implementation into the computers.

Constitutive equations presented in Chapter 5, Equations (5.1) to (5.9), provide the relation between stress and strain for homogeneous anisotropic materials in terms of elastic constants and Poisson's ratio. The finite element method lends itself to implement these anisotropic material constitutive equations for structural analyses of geometrically intricate parts (Figure 6.1). Such parts can have complex load and boundary conditions.

Figure 6.1 Complex shapes can be analyzed using the finite element method

Due to the complex shape of a part, it becomes a challenge to develop mathematical expressions to predict stress, strain, and deformation fields. This would require solving differential equations with complex boundary conditions and constitutive relations. The finite element method (FEM) has been developed over the years to address this challenge. The key idea behind FEM is to break the complex shape of the structure into simple shapes called elements, where the governing equations as well as constitutive material laws are applied to solve for the force and deformation fields at selected points, called nodes, in the elements. Once such force deflection relations for each discrete element at selected points is known, all the elements in the structure of interest can be added to find the performance of the complete structure. Figure 6.2 presents a schematic diagram of this procedure.

Fundamentally, a structural analysis finite element (FE) software program solves Hooke's law in a complex three-dimensional field as

$$[K]\{d\}=\{F\} \tag{6.1}$$

where $[K]$ is the stiffness matrix of the three-dimensional structure, and $\{d\}$ and $\{F\}$ are the displacement and force vectors, respectively. The displacement and force vectors represent the boundary conditions, complex load system, and deformation field. As shown in Figure 6.2, the matrix and vectors in Eq. (6.1) result by adding individual finite element stiffness matrices and force and displacement vectors, defined by

$$[K^e]\{d^e\}=\{F^e\} \tag{6.2}$$

where the superscript e denotes "element". Depending on the complexity of the structure, different finite elements can be used, from 3 noded triangular elements to 20-noded bricks.

One of the goals of this chapter is to explain the general finite element method steps, which are also called finite element analysis (FEA). Hence, a simple 3-noded element will be used to illustrate the concept for a plane stress system. The key steps to develop Eq. (6.2) for an individual element are as follows:

- Approximate the displacement field within an element using the values at the nodes and interpolation functions called shape functions
- The strain field is expressed as a function of displacement and geometry within the selected element
- The stress field is expressed as a function of strain and the elasticity matrix as defined in Chapter 5
- Define the element stiffness matrix based on Galerkin's method

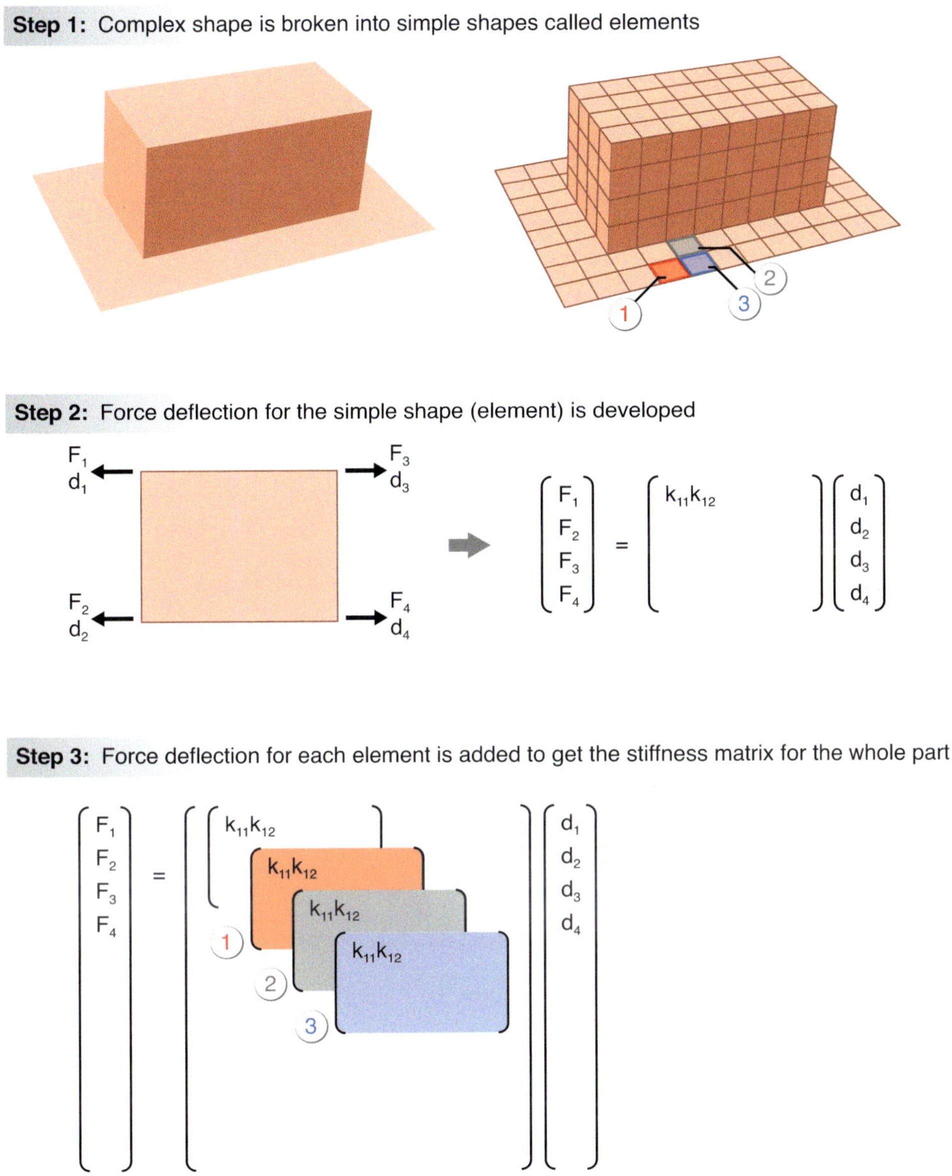

Figure 6.2 Finite element concept illustrated for a 3D part

Element Displacement Field Interpolation Functions

An arbitrary three-noded, two-dimensional element located within an *xy* coordinate system, as shown in Figure 6.3, is considered. This element is also often referred to as a *constant strain triangle*.

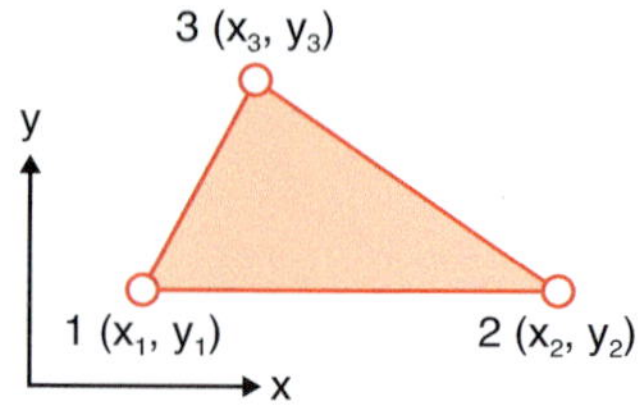

Figure 6.3 Three-noded element to illustrate the finite element concept

The *x*-displacement field, $u(x, y)$, can be approximated using a linear interpolation function as

$$u(x,y) = a_1 + a_2 x + a_3 y \tag{6.3}$$

The constants a_1, a_2, and a_3 can be evaluated by letting $u(x_1, y_1) = u_1$, $u(x_2, y_2) = u_2$, and $u(x_3, y_3) = u_3$, which in matrix form can be written as

$$\begin{Bmatrix} u_1 \\ u_2 \\ u_3 \end{Bmatrix} = \begin{bmatrix} 1 & x_1 & y_1 \\ 1 & x_2 & y_2 \\ 1 & x_3 & y_3 \end{bmatrix} \begin{Bmatrix} a_1 \\ a_2 \\ a_3 \end{Bmatrix} \tag{6.4}$$

This also can be written as

$$\begin{Bmatrix} a_1 \\ a_2 \\ a_3 \end{Bmatrix} = \begin{bmatrix} 1 & x_1 & y_1 \\ 1 & x_2 & y_2 \\ 1 & x_3 & y_3 \end{bmatrix}^{-1} \begin{Bmatrix} u_1 \\ u_2 \\ u_3 \end{Bmatrix} = [C]^{-1} \begin{Bmatrix} u_1 \\ u_2 \\ u_3 \end{Bmatrix} \tag{6.5}$$

which, substituted into Eq. (6.3), results in

$$u(x,y) = [1 \quad x \quad y][C]^{-1} \begin{Bmatrix} u_1 \\ u_2 \\ u_3 \end{Bmatrix} \tag{6.6}$$

Similarly, the *y*-displacement field, $v(x,y)$, can be approximated by

$$v(x,y) = [1 \quad x \quad y][C]^{-1} \begin{Bmatrix} v_1 \\ v_2 \\ v_3 \end{Bmatrix} \tag{6.7}$$

The above equations can also be written in a more generic form as

$$u(x,y)=\begin{bmatrix}N_1 & N_2 & N_3\end{bmatrix}\begin{Bmatrix}u_1\\u_2\\u_3\end{Bmatrix} \tag{6.8}$$

$$v(x,y)=\begin{bmatrix}N_1 & N_2 & N_3\end{bmatrix}\begin{Bmatrix}v_1\\v_2\\v_3\end{Bmatrix} \tag{6.9}$$

where N_1, N_2, and N_3 are often referred to as shape functions. The shape functions equal 1 when the coordinates of the corresponding node are used, and 0 when the coordinates of the other two nodes are used.

Stress and Strain Fields

Once the displacement field has been defined, one can easily approximate the strain field within the element using

$$\varepsilon_{xx}=\frac{\partial u(x,y)}{\partial x}=\begin{bmatrix}0 & 1 & 0\end{bmatrix}\begin{bmatrix}C\end{bmatrix}^{-1}\begin{Bmatrix}u_1\\u_2\\u_3\end{Bmatrix}=\begin{bmatrix}\frac{\partial N_1}{\partial x} & \frac{\partial N_2}{\partial x} & \frac{\partial N_3}{\partial x}\end{bmatrix}\begin{Bmatrix}u_1\\u_2\\u_3\end{Bmatrix} \tag{6.10}$$

$$\varepsilon_{yy}=\frac{\partial v(x,y)}{\partial y}=\begin{bmatrix}0 & 0 & 1\end{bmatrix}\begin{bmatrix}C\end{bmatrix}^{-1}\begin{Bmatrix}v_1\\v_2\\v_3\end{Bmatrix}=\begin{bmatrix}\frac{\partial N_1}{\partial y} & \frac{\partial N_2}{\partial y} & \frac{\partial N_3}{\partial y}\end{bmatrix}\begin{Bmatrix}v_1\\v_2\\v_3\end{Bmatrix} \tag{6.11}$$

$$\gamma_{xy}=\frac{\partial u(x,y)}{\partial y}+\frac{\partial v(x,y)}{\partial x}=\begin{bmatrix}0 & 0 & 1\end{bmatrix}\begin{bmatrix}C\end{bmatrix}^{-1}\begin{Bmatrix}u_1\\u_2\\u_3\end{Bmatrix}+\begin{bmatrix}0 & 1 & 0\end{bmatrix}\begin{bmatrix}C\end{bmatrix}^{-1}\begin{Bmatrix}v_1\\v_2\\v_3\end{Bmatrix} \tag{6.12}$$

Note that for this type of element, where the interpolation functions are linear, the strain is constant. Hence, this element is often referred to as *constant strain triangle*. Fully evaluating the above equations, the strain vector reduces to

$$\begin{Bmatrix}\varepsilon_{xx}\\\varepsilon_{yy}\\\gamma_{xy}\end{Bmatrix}=\begin{bmatrix}B\end{bmatrix}\begin{Bmatrix}u_1\\u_2\\u_3\\v_1\\v_2\\v_3\end{Bmatrix} \tag{6.13}$$

where $[B]$ is a derivative operator, or B-matrix, defined in terms of element geometry as

$$[B]=\frac{1}{\det[C]}\begin{bmatrix}(y_2-y_3) & 0 & (y_3-y_1) & 0 & (y_1-y_2) & 0\\ 0 & (x_3-x_2) & 0 & (x_1-x_3) & 0 & (x_2-x_1)\\ (x_3-x_2) & (y_2-y_3) & (x_1-x_3) & (y_3-y_1) & (x_2-x_1) & (y_1-y_2)\end{bmatrix} \quad (6.14)$$

where

$$\det[C]=2A=(x_2y_3-x_3y_2)-(x_1y_3-x_3y_1)+(x_1y_2-x_2y_1) \quad (6.15)$$

The stress field can be defined as a function of strain or displacement and the elasticity matrix as

$$\begin{Bmatrix}\sigma_{xx}\\ \sigma_{yy}\\ \tau_{xy}\end{Bmatrix}=[E]\begin{Bmatrix}\varepsilon_{xx}\\ \varepsilon_{yy}\\ \gamma_{xy}\end{Bmatrix}=[E][B]\begin{Bmatrix}u_1\\ u_2\\ u_3\\ v_1\\ v_2\\ v_3\end{Bmatrix} \quad (6.16)$$

where the elasticity matrix for a plane stress isotropic element is given by

$$[E]=\begin{bmatrix}\frac{E}{1-\nu^2} & \frac{E\nu}{1-\nu^2} & 0\\ \frac{E\nu}{1-\nu^2} & \frac{E}{1-\nu^2} & 0\\ 0 & 0 & G\end{bmatrix} \quad (6.17)$$

Element Stiffness Matrix

Having defined the displacement, strain, and stress fields throughout the finite element one can evaluate the element stiffness matrix using[1]

$$\left[K^e\right]=\int [B]^T[E][B]t\,dx\,dy \quad (6.18)$$

where t is the thickness of the plate with plane stress. $[B]$ and $[E]$ are defined in Eqs. (6.14) and (6.17).

[1] For the full derivation of a finite element stiffness matrix, the reader should consult the literature, such as Zienkiewicz [1], Bathe [2].

There are many different types of elements available to represent a structure's geometry. It is important to emphasize that the element formulation depends on the following two key items:

- Each element type has different interpolation or shape functions. For example, for an n-noded three-dimensional element, the displacement field is represented by

$$u(x,y)=\sum_{i=1}^{n} N_i u_i \tag{6.19}$$

$$v(x,y)=\sum_{i=1}^{n} N_i v_i \tag{6.20}$$

$$w(x,y)=\sum_{i=1}^{n} N_i w_i \tag{6.21}$$

  where u, v, and w are x, y, and z displacement fields, respectively. These shape functions, N_i, are used to derive the strain field for the complex three-dimensional geometry. The choice of N_i depends on the element geometry as well as the intended usage; this is one of the busiest research areas in the finite element development field. Commercial FEM software programs have a large library of elements, where each element is developed to support its intended usage, such as rods, plates, shells, and bricks with various numbers of nodes.
- Each material has its constitutive equations, which must be implemented into the element, that is, the stress–strain relation based on the material properties. For isotropic elastic materials, the linear elastic model based on Hooke's law is sufficient. This relation is relatively simple and well known. However, as discussed in Chapter 5, for fiber-reinforced materials the constitutive properties are much more complex and challenging. This is because the material is not homogeneous, i.e., there are distinct multiple phases in the material, with different properties. A method to develop stress–strain relations for the fiber-reinforced composite material from the individual fiber and matrix properties is one of the main interests in this book. Stress–strain expressions for fiber-reinforced materials in simple shapes, based on principles of mechanics, were presented in Chapter 5. However, such expressions must be developed for complex three-dimensional shapes and loading systems. Therefore, in the next section we will present the development of material properties for fiber-reinforced materials, and their implementation into finite element formulations.

A finite element can be one-dimensional, two-dimensional, or three-dimensional. The numbers of degrees of freedom for each node and element can also vary. Generally, depending on the shape of the part under consideration, the material used, and the type of loading and intended analysis, the element can be selected.

Reviewing the key steps and concepts in the development of finite element models may appear quite involved and overwhelming at first. However, the key advantage of the finite element approach is that it is a numerical approach and, therefore, most of the tasks involved can be automated and hence computerized. Over the past six decades, millions of hours have been invested in developing efficient algorithms, element formulations, and computational methods. As a result, many commercial software programs, such as ABAQUS™, LS-Dyna™, ANSYS™, MSC-NASTRAN™, COMSOL™, etc., which make finite element methods user-friendly and convenient to use, are available. Using such FEM software programs does not require deep programming knowledge or an in-depth understanding of the geometrical manipulations. However, it is important to understand the underlying assumptions and physics, as it helps the user to understand the limitations involved and hence properly interpret the results. Therefore, the approach in this chapter is to provide the reader with an essential understanding of the physics and assumptions and how they may influence the FEM predictions. For a deeper understanding, we recommend referring to other publications [1–3].

6.1.2 Status of FEM in the Industry

The key steps in applying the finite element approach on a complex 3D shaped part are presented in Figure 6.4. The first task will be to discretize the complex 3D shaped part geometry into finite elements. For a complex 3D shaped part, it can be quite a challenge to do such discretization manually. To address this challenge, pre-processors, such as Hypermesh, Patran, etc., with excellent interactive graphical user interfaces, have been developed and are available commercially. Typical pre-processors can start from a digital 3D part geometry and prepare the finite element model quickly and efficiently. Once the FEM analysis is completed the software can generate large amounts of numerical data, which could be stress and strain at the nodes or integration points for each element, or other relevant information such as nodal forces or displacements. Post-processors with a graphical interface can help analyze such large quantities of numerical output. The main strength of typical post-processing software is the data analysis and graphical interface that helps visualize the large amount of numerical data. Both the pre- and post-processing capabilities in the industry are well developed and readily available for use.

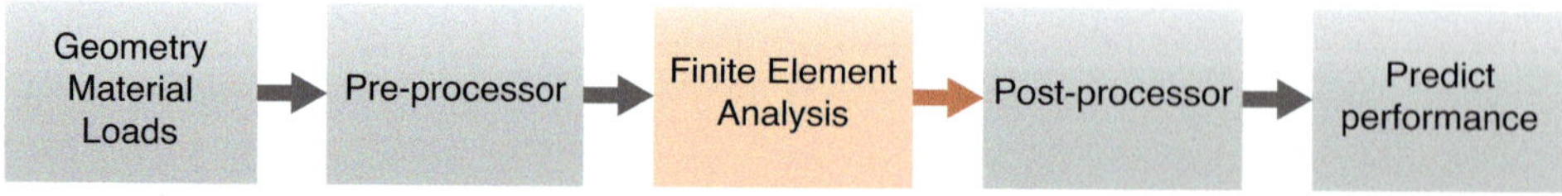

Figure 6.4 Key steps in finite element method for complex parts

Because of the improvements in the technology, simpler user interface, and increasing demand for the predictive tools, the usage of FEM in the industry has increased tremendously [8]. Many well-established and highly optimized commercial software programs, along with a suite of pre- and post-processors, are already available. Usage of the finite element method to model the structural performance, heat transfer, fluid flow, or electromagnetic systems is quite common. In general, any physical phenomenon that can be described using differential equations, or partial differential equations, can be modeled for complex shapes using the finite element method.

6.2 Key Challenges in Using FEM for Fiber-Reinforced Polymers

As mentioned before, the FEM implementation to study structural performance starts with the constitutive relation between stress and strain within the specific material. For the most common engineering materials, such as steel, aluminum, etc., the constitutive equations are linear, isotropic, and homogeneous. When dealing with fiber-reinforced polymers, one is faced with a complex system composed of multiple phases, the fibers, and the matrix. Furthermore, the fiber to matrix interface, as well as voids, constitute their own properties. Therefore, the assumptions of isotropy and homogeneity do not hold. To address this, new constitutive relations between stress and strain are required. In developing these relations, the main challenge is how to account for the multiple phases (fibers, polymers, their interface, and voids). Since each phase has its own unique properties, the composite material essentially is a non-homogeneous continuum with non-isotropic properties at the micro level. The goal in this chapter is to present the mathematical framework to describe the stress–strain relationship for such non-homogeneous materials and the process to integrate the constitutive relations for such materials into the FE formulation.

The different types of fiber materials and geometries used to reinforce the polymers have been discussed in Chapter 2. From the modeling perspective, as shown in Figure 6.5, the fibers in the polymer can be broadly classified based on their geometry:

- Continuous tapes and sheets
- Continuous fibers in fabric-form infused with polymer resin
- Discontinuous fibers

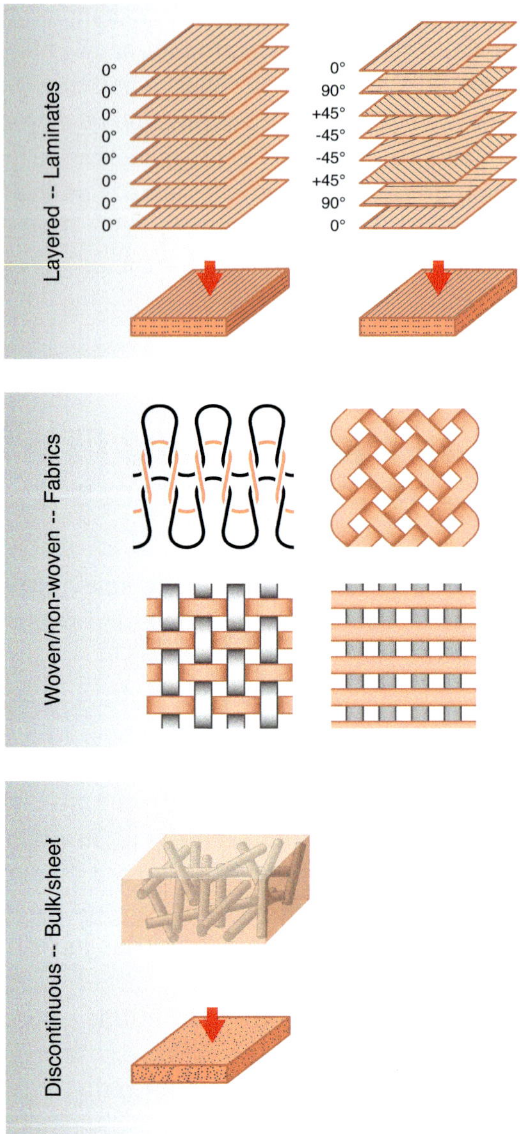

Figure 6.5 Different types of fiber-reinforced composite materials

The continuous layers of fiber-reinforced polymers are the most common form of composites in usage today. As discussed in Chapters 4 and 5, the structural properties for such tapes or sheets significantly vary with fiber orientation. It was also discussed that, with continuous glass or carbon fibers, the tapes and sheets can be layered with different orientations to achieve laminated structures with tailored anisotropic properties, which can be nearly isotropic properties. Continuous layered fiber composites offer excellent strength and stiffness per unit weight; hence, they have been in use in aerospace and other high-performance applications for many years. However, the material cost and processing time for such composites are quite high, which results in higher total cost. Consequently, their usage is limited to high-end applications. When developing material models for such materials, the fiber orientation in each layer is known based on the initial design and the manufacturing process used. The material models for such continuous fiber-reinforced polymers have been well developed over the past 50 years [10]. Also, the computational routines for such materials are available in most of the existing commercial FEA software programs. Methods to use material models to predict stiffness, strength, warpage, etc. have been well accepted and have been in usage for many years. However, damage predictions, or failure criteria, for such materials using fundamental mechanics are still under development. The main bottleneck is that the stress-strain distribution in an element is complex. Most of the current commercial methods largely depend on homogenized properties of the fiber-reinforced system, which does not account for stress-strain variation in the element accurately. Also the failure criteria for discontinuous fiber-reinforced polymers are not well developed yet. Efforts to estimate the failure and fatigue performance for such materials using tensor-based or phenomenological approaches have been demonstrated by Osswald and Osswald [9]. A summary of developments to improve prediction of damage and fatigue for continuous fiber-reinforced materials is available in [10, 31].

A fabric infused with polymer resins is the second form of material that needs consideration in developing the material model for use with finite element analysis software. The basic process involves infusing the fabric with the polymer resin and then solidifying to final shape by cooling for thermoplastic resin or curing for thermoset resin. As discussed in Chapter 3, many variations of these basic processes are being used in the industry today. These fabrics can be woven, braided, or knit. Generally, such fabrics have repeating patterns, which are known and, hence, the microstructure to represent fabric and resin can be developed using the repeating unit cells approach. The unit cells can represent the stress-strain distribution based on their repeating microstructure. Since the fibers are curved and interact with each other along with the surrounding resin material, modeling of such material structure even at unit cell level can be quite challenging. Wisetex, a software program [11] developed at KU-Leuven University in Belgium is one such tool that can help development of material models based on the repeating microstructure of the fabric. The software offers many unit cell templates for which the microstructure shape is predefined; however, the design parameters can be chosen based on actual measurements. Such customized unit cells help define the material properties, which can be coupled with the finite element models.

The third type of fiber-reinforced composite in common usage is discontinuous fibers. Typically, the injection or compression molding processes are used to make parts from such materials. In such processes the resin is heated to a melt state, compounded or mixed with fibers using extruders at high temperature, and compression molded, or the mixtures of resin and fibers are injection molded to fill the mold cavities. When the resin solidifies, the fibers can add strength and stiffness to the material. The most commonly used fibers in the industry are glass or carbon and the most common resin material is thermoplastic. Due to the high shear stresses generated during mold filling, particularly during injection molding, the fibers suffer significant breakage. The fibers that start at 15 mm in the initial pellet can break down to 0.25–1.5 mm within the final part. Such discontinuous fiber-reinforced thermoplastic materials, in conjunction with the injection or compression molding processes, result in a low-cost alternative when compared to the other two types of material discussed earlier. Furthermore, discontinuous fiber filled polymer materials offer design flexibility along with the possibility of large volume production. On the other hand, due to discontinuity of the fibers, as well as the relatively uncontrolled fiber orientation distributions, the overall improvement in the strength and stiffness within the final part can be limited when compared to their continuous fiber counterparts.

The development of material models for discontinuous fibers is a complex task. The main reason is that the fiber condition or microstructure for the injection and compression molded parts can vary significantly through the geometry of the part. Since the material properties depend on the fiber microstructure, it is essential to account for fiber orientation, fiber length, and fiber concentration while developing the material models. However, accounting for these is not trivial. Typically, the glass fiber diameter is between 14 and 20 μm and that of carbon fibers is between 5 and 7 μm. The aspect ratios can vary from 30 to higher than 400, depending on the processing parameters. As a result, there can be thousands of fibers even in a one-cubic-millimeter volume. Accounting for the complete 3D configuration of the fibers to generate a representative material model is a challenge. It is widely recognized that there is a strong need to address this challenge. It is observed that discontinuous fiber-reinforced polymers are significantly lower in both the material and manufacturing cost, and that they are suited for large volume production. This is of extreme importance to the automotive and other consumer applications industries. Hence, the next sections will concentrate on the material model development for discontinuous fiber systems.

While considering the development of the material models for discontinuous fiber-reinforced polymeric material for use in conjunction with FE structural analysis software, it is important to understand the difference in the physics compared to the isotropic materials. The material properties are a result of the mechanical interaction between two or more distinct phases at the micro level. The material model should be chosen such that this multi-phase interaction can be represented in the simulation. The main aspects to consider in the material model development for fiber-reinforced materials are discussed below. These are:

- **Geometrical condition of the fibers:** The fiber orientation, fiber length, and fiber concentration can vary throughout the injection or compression molded part. Therefore, the material properties will vary from one location to the other. Additionally, the shape of the fiber is quite important. For example, a curved fiber will be less effective than the straight fiber when it comes to stiffness and load carrying performance, but if intermingled with other fibers it can be more effective in load transfer. It is important to estimate and account for the fiber condition in the material model.
- **Properties of each phase:** In general, glass or carbon fibers used in the composite can be treated as linear elastic, whereas the polymer matrix, depending on the resin material, can be elastic, nonlinear elastic, elastic-plastic, or viscoelastic. The interface and void properties are not generally simple to measure; therefore, suitable assumptions may be made to include their influence.
- **Microstructure details:** If one is interested in finding the stiffness or force deflection performance, then the material properties can be homogenized at the elements level and used in the finite element model calculations. However, if one is interested in predicting damage to estimate failure load or fatigue life, it is important to include additional details in the microstructure model for the material. This is because crack initiation and propagation will depend on the localized stiffness as well as local stress and strain distribution within the typical element. In such a case, homogenization may not work well as the stress and strain distribution may not be accurate at a local level.
- **Loading conditions:** A typical fiber can take higher load in tension compared to compression because in compression it is likely to buckle and reduce its load capacity. Fiber-reinforced material performance in tension and compression are different and require different considerations in calculation as well as design.
- **Rate of loading:** The rate of loading can result in different strain rate through the composite at micro level. This is because each constituent material, the polymeric component, is likely to have different strain rate performance as well as strain rate dependence.

An expression for material properties for discontinuous fibers when all the fibers are aligned in one direction or when all the fibers are completely random was derived in Chapter 5, Section 5.3. For such relatively simple conditions the material properties such as elastic modulus, stress–strain distribution, etc. were estimated using the combination of principles of mechanics and observation of experimental results. In deriving these relations, it was assumed that the bulk material, i.e., fiber and polymer mixture is homogeneous. Such approach works well when estimating macro level properties, if all the fibers are aligned in the same directions. However, as briefly discussed in Chapter 5, when discontinuous fiber microstructure such as fiber length, orientation, and concentration vary throughout the part, as it occurs with injection or compression molded parts, the treatment is somewhat different. The microstructure for the discontinuous fiber composites in the finished part depends on the processing conditions. Details of the processing conditions and their effect on the fiber microstructure were discussed in Chapter 4. Also, it is well known that fiber microstructure is likely to be different from one location to another, which directly influences the material properties. This is a major a challenge with discontinuous fibers when developing material models that can be implemented into the finite element formulation presented earlier, because the finite element model discussed so far is based on the constitutive relations developed for homogeneous materials.

The goal of this chapter is to discuss how to address this challenge, that is, how to define the material properties for such parts for use in FE computation. This is a difficult task, where a perfect solution is not possible. However, sufficient know-how has been developed that helps in formulating a useful solution. The usefulness of the solutions presented here will be demonstrated with examples in the next chapters.

The other two issues that pose a modeling challenge when dealing with fiber-reinforced polymers are compressive stress behavior and deformation rate dependence. Both are important for impact loading applications. As mentioned before, when compressed, fibers tend to buckle. Taking such local fiber level buckling into calculations can be difficult. Furthermore, polymers have properties that are a function of the rate of loading or the rate of deformation, and are time dependent. These are very complex challenges and research to address them is in progress. Both these issues are only superficially addressed in this book.

■ 6.3 Modeling Discontinuous Fiber-Reinforced Materials for FEM Implementation

The material model to be implemented into a finite element model can be broadly classified into two categories, (1) macro scale or phenomenological models or (2) micro scale or multiscale models.

As the names imply, the main difference between the two is in the detail of how the effect of the constituent materials are accounted for. The selection of the method in the study is normally based on the availability of data, need for accuracy, and available resources. Details of each approach are discussed in the next sections.

6.3.1 Macro Scale or Phenomenological Modeling

Historically, macro scale or phenomenological modeling has been one of the most common methods used for the fiber-reinforced polymers in the industry. Essentially, the material is treated as a homogeneous, orthotropic material. The material properties such as stress and strain are measured through specimen, or coupon, testing along multiple directions, such as flow, cross flow, and a few additional directions in between; see Figure 6.6. It is assumed that the effect of the micro constituent is embedded in the measurements. The measurement is used to estimate the unknown constants in the phenomenological models using reverse engineering or curve-fitting methods. In industry, often, even simpler approaches are used. For example, the material is assumed to be isotropic with about 60 to 70 percent of the highest measured values.

Macro scale modeling is a significant simplification of the material properties. The constituent materials are considered at a global level only; the geometry and properties of each constituent material at the micro level are ignored. The fiber-reinforced polymer is considered as a single material with homogeneous properties, and the variation in the properties from location to location are ignored as well. The accuracy of the finite element model when using such gross material properties is quite limited. The main reason of its popularity is the simplicity of the model, the simple test methods required, and the need for almost no know-how. Engineers familiar with finite element analysis of traditional isotropic materials can easily adopt and use this type of material models. The measured data can be used to estimate parameters for the material constitutive laws being used in the finite element model.

Depending on the part being investigated and manufacturing processes employed, such approach may work well for the calculation of stiffness and force deformation estimates in many cases. Furthermore, when dealing with large complex assemblies, due to computational limitation, the macro scale is the only possible option. In industry, it is quite common to use such material models, where the constitutive material behavior and parameters needed are based on years of measurements and know-how.

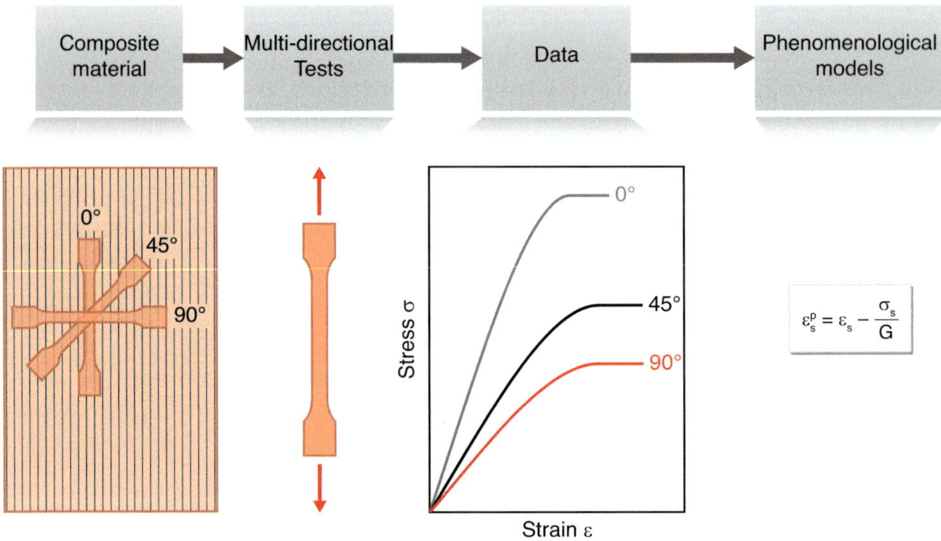

Figure 6.6 Macro scale approach to modeling

6.3.2 Micro Scale Modeling

A cross section of glass fiber-reinforced polymer composite magnified under a microscope is presented in Figure 6.7. As mentioned above, glass fibers are between 14 and 20 μm in diameter and carbon fibers are between 5 and 7 μm in diameter. In a small volume of one cubic millimeter, depending on length and volume concentration, one can expect thousands of fibers. Each of the constituents of the composite plays an important role in the final performance of that material. Particularly, for discontinuous fiber composites the fiber length, fiber alignment, fiber concentration, and fiber geometry vary from location to location and affect the final material properties within that space. If the fibers are random, evenly distributed, and are similar in length, the macro approach will work reasonably well for small deformations in the linear elastic regime. However, within the nonlinear behavior it would be very difficult to model the material behavior accurately. Also, it is imperative that the stress and strain field will be discontinuous due to the non-homogeneity of the material. This means that the damage initiation and failure estimates will be extremely difficult. Hence, microscopic models must be considered.

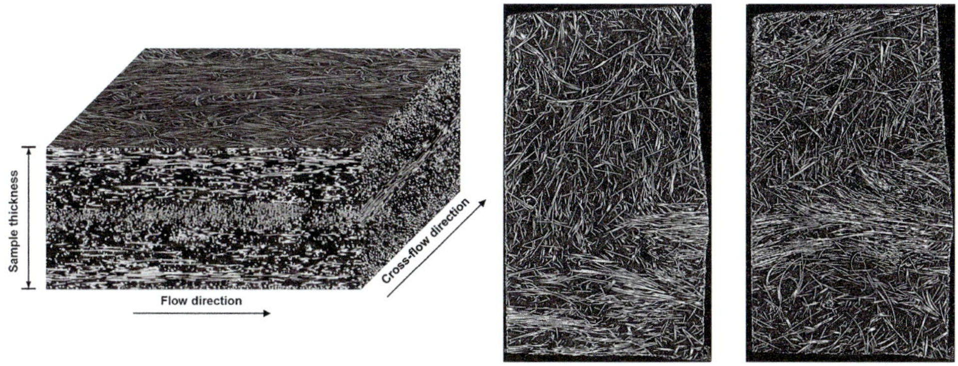

Figure 6.7 Glass fiber and polypropylene resin microstructure

With the micro scale model, ideally, one would like to account for each constituent of the composite material, i. e., fiber, resin, the interface, and their geometrical arrangements. A finite element model of the microstructure that includes all the fibers with their accurate geometry and matrix with voids can be considered. However, there are thousands of fibers even in one cubic millimeter. Modeling each fiber can result in a very large number of elements in the model, which brings computational challenges. Additionally, any constants such as interface properties are not well known. Such models, which we call a microscopic model in the true sense, are considered impractical for any useful engineering application. Many schemes to address these challenges and simplify the computational model development have evolved over time. These are called multiscale modeling. Multiscale modeling essentially incorporates the fiber details at a micro level (micrometer) and translates it into material models at macroscopic level (millimeter). There are many different schemes available to convert the microscopic details to macro level material data to be used in a finite element model.

The basic principle of multiscale modeling is presented in Figure 6.8. There, the concept of representative volume element (RVE) is used. RVE is the smallest volume which can accurately capture the structural behavior of the effective properties of the composite material. In nature, most of the structures are heterogeneous, and when dealing with such heterogeneous materials it is common to average out the effect of microstructure over a selected volume and call it effective properties. The size of such representative volume element can be debated. If the microstructure is periodic, such as with woven or layered continuous fibers, choosing the size of the RVE is easy. However, for discontinuous fibers it can be challenging. We would like the volume to be as small as possible, to capture the effect of material variation throughout the part; on the other hand, it should be large enough to average the effective composite properties as a continuum material. Generally, it is expected to be of an order of magnitude larger than the microstructure details being averaged [24].

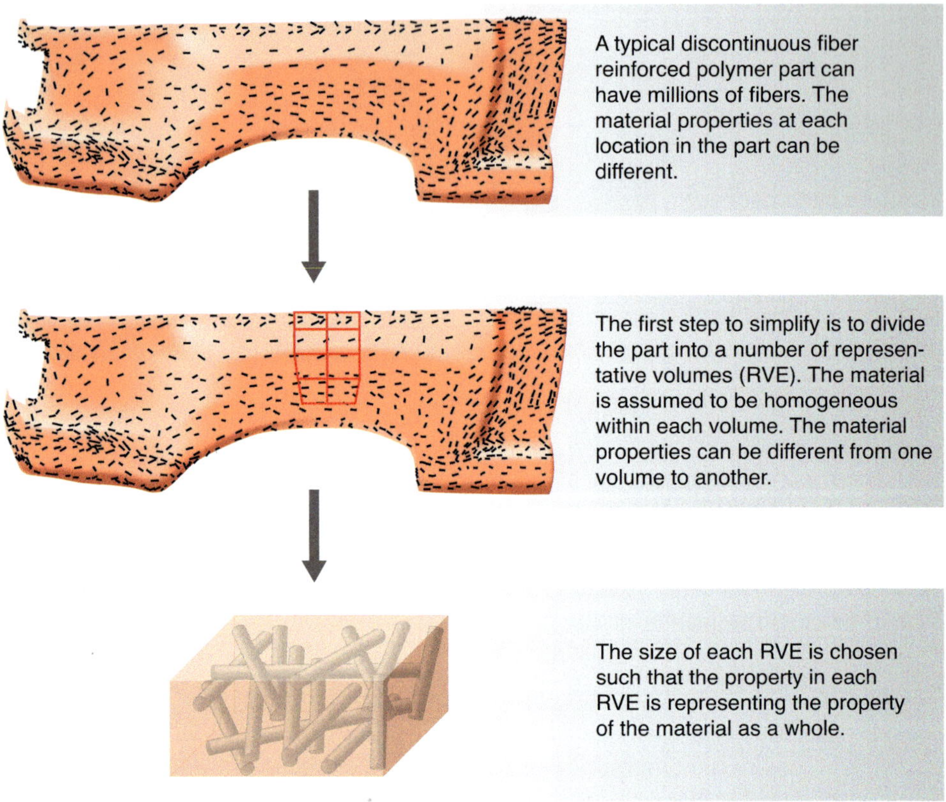

Figure 6.8 Representative volume element (RVE) concept: the part is divided into small volumes; each of them can represent the material model based on the microstructure. In FEM each such RVE can be a single finite element

The fiber-reinforced composite part can be divided into a number of representative volume elements (RVE). The constitutive material model for each RVE can be computed individually based on the fiber and resin properties and geometry details of the fibers such as fiber length, orientation, and concentration. There are many approaches to the computation of material properties in a RVE. The simplest computation is when the material properties for the RVE are assumed to be homogeneous and the averaged homogeneous material properties are used for the RVE in the finite element calculations. It is important to emphasize that, when the material properties for an RVE are assumed to be homogeneous, they can and are likely to be anisotropic for discontinuous fiber-reinforced composites. Hence, compared to the macroscopic approach discussed above, where the whole part was assumed to be homogeneous, with the RVE approach one breaks the part down into many smaller volumes at macro scale. While each of such volumes is assumed to be homogeneous, the properties from one volume to the other are expected to vary. The gross properties at whole part level can be quite different from the RVE (macro level) properties. Also, an RVE approach better represents the actual conditions in the material. Therefore, this can be a significant improvement in the overall material property representation in the FE model.

The representative volume element approach helps significantly in reducing the computation time as well as the modeling burden. The accuracy and computational cost of such an approach depends on how the representative volume is chosen and the method used to represent the stress and strain within that volume. Furthermore, once the finite element model is used for computation, the stress and strain estimated over the RVE must be interpreted. This is important since the microstructure which is assumed to be homogeneous over the RVE is heterogeneous, consisting of the fiber and polymer resin. Depending on the fiber geometry as well as the material properties of the fibers and resin, the stress and strain in the RVE will not be homogeneous but vary from location to location. The average stress based on the homogeneous material will not be accurate. Therefore, it will be important to be able to translate the average stress and strain to a meaningful stress and strain distribution within the RVE, based on the microstructure.

Since most of the materials found in nature are heterogeneous, with two or more distinct materials at a microstructure level, the concept of using a representative volume element is widely accepted in physics. As shown in Figure 6.9, there are many different schemes developed over the years to deal with the representative volume elements for fiber-reinforced materials. Each approach has its advantages and disadvantages. The selection of a representative volume element, for a given problem, is made based on the knowledge of the microstructure, available computing resources, and level of accuracy desired. In the following sections, various methods to define the RVE for fiber-reinforced materials will be discussed. To maintain the focus of this book, the key aspects and insights will be presented from a practical perspective. Detailed theoretical background and computational derivations can be found in many references [12, 13].

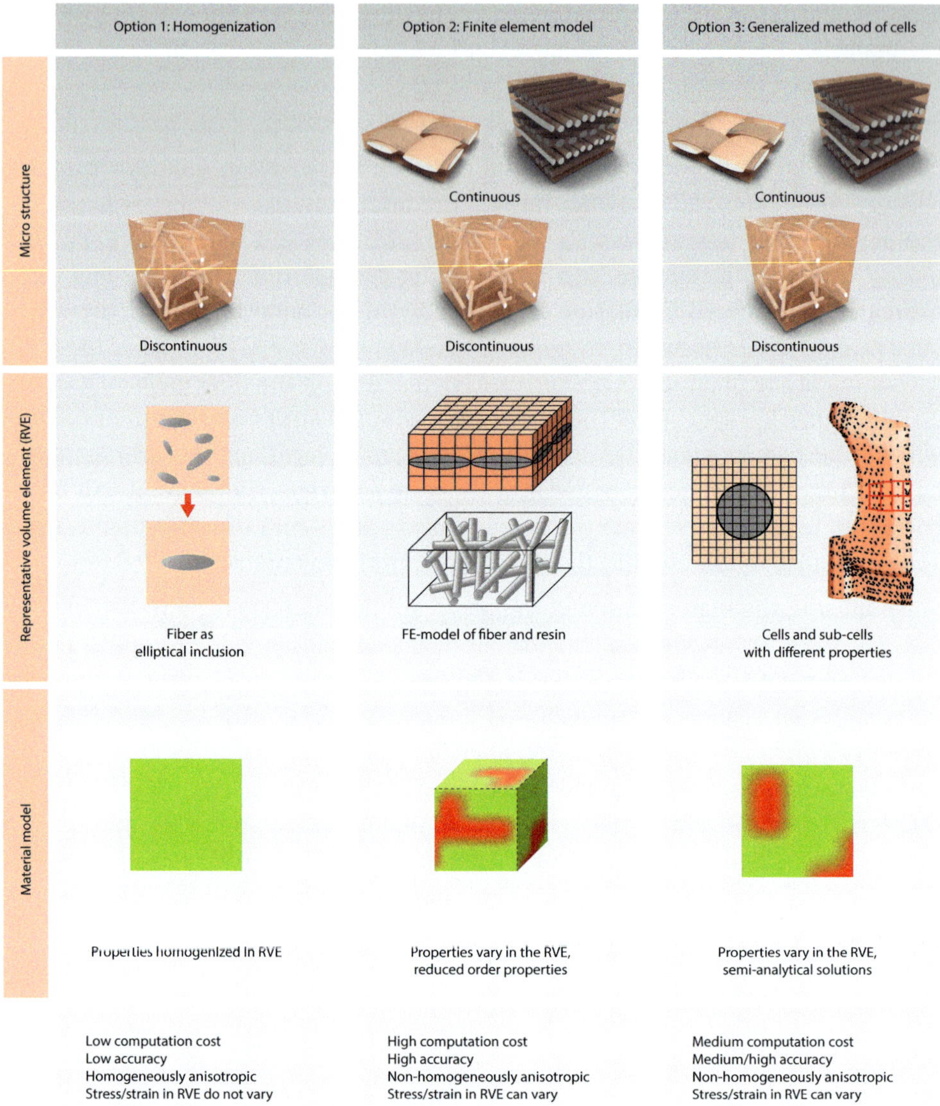

Figure 6.9 Methods to compute material properties based on microstructure for a representative volume element (RVE) for fiber-reinforced polymer composites

How a RVE can be used to generate microstructure based material properties of a fiber-reinforced composite for use with finite element analysis is presented in Figure 6.10. The key microstructure for the fiber-reinforced composite that affects the material properties are fiber length, fiber orientation, fiber concentration, fiber resin interface, and voids in the resin material. Such microstructure for discontinuous fibers is likely to vary from one location to other based on the manufacturing processes used. Therefore, it is important to know the fiber microstructure detail for each RVE. This is usually accomplished using process simulation. Process simulation can predict the fiber condition in the finished part. For each RVE within the part, the material properties are computed based on the microstructure details acquired from the process simulation and the method, selected from Figure 6.9, to model the microstructure material properties.

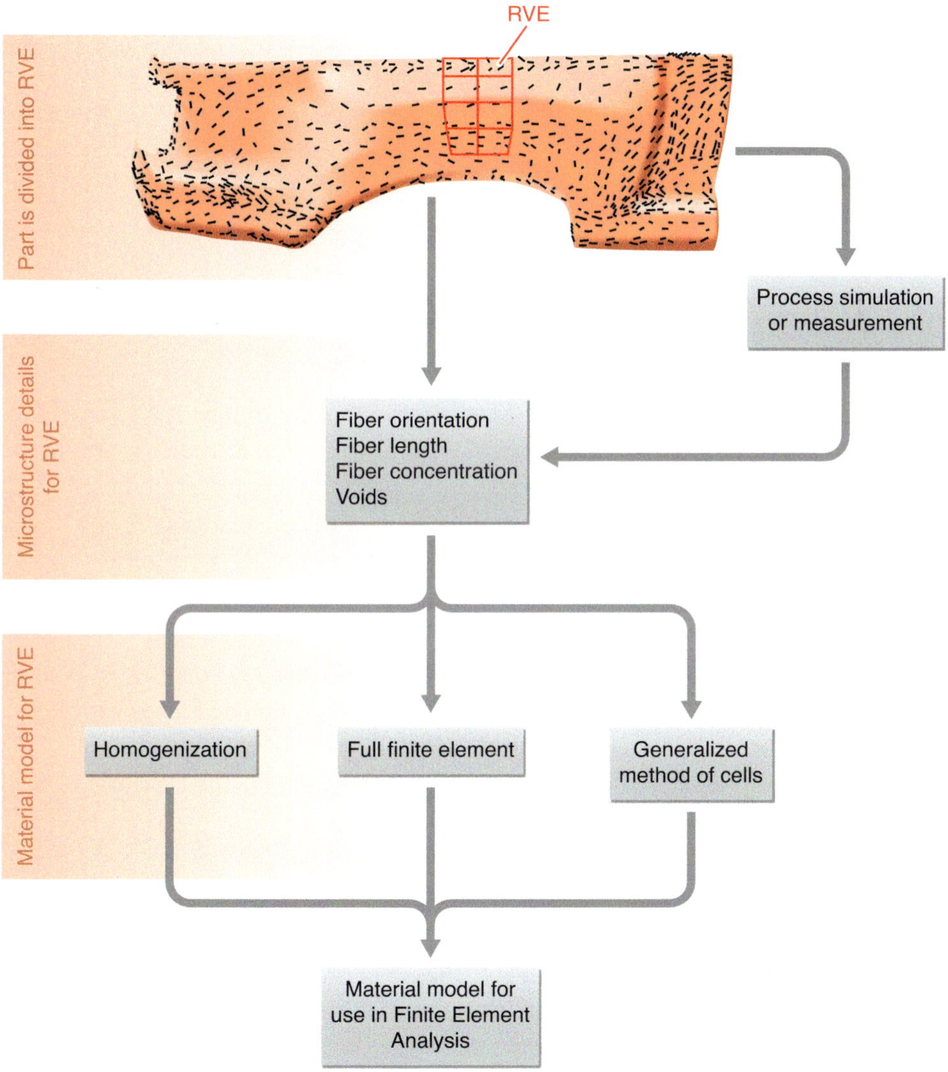

Figure 6.10 Approach to integrate material properties from RVE into the finite element analysis

6.3.2.1 Mean Field Homogenization

Mean field homogenization is a process of estimating average material properties for a selected volume that accounts for the microstructure of the composite. The mean field homogenization method [14, 15] was originally developed for the linear elastic region, and has since been expanded for plastic and nonlinear regions. These methods are computationally efficient and one of the most used homogenization approaches in practice today to model fiber-reinforced material in conjunction with FEA.

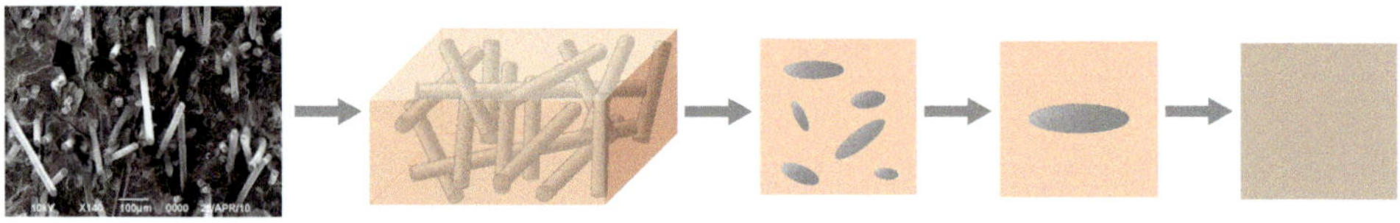

Figure 6.11 Mean field homogenization approach

The key goal and steps for such a method is as shown in Figure 6.11. The fiber-reinforced polymer representative volume element (RVE) is assumed to have a simplified geometry. Each fiber is assumed to be an ellipsoid embedded in the resin matrix. The aspect ratio[2] a of the ellipse is a parameter used to represent the ratio of fiber length to fiber diameter. It is assumed that the fibers are straight and well dispersed such that there is no interaction between fibers. The goal is to start with the simplified geometry and estimate the homogenized properties for the selected representative volume.

There has been strong interest in developing average properties for such composite material for a long time. The earliest efforts were developed by Voigt [16], who in 1889 developed the following simple expression for the homogeneous composite material stiffness E_c:

$$E_c = E_m(1-\phi) + E_f\phi \tag{6.22}$$

where the subscripts c, f, and m imply the composite, fiber (inclusion), and matrix, respectively. ϕ is the volume fraction of fibers. Here, it is assumed that the strain throughout the matrix and fiber is constant.

In 1929, Ruess [17, 18] assumed that the phases of the composite are subjected to a uniform stress, equal to the average stress, which results in an effective compliance (S) of the composite material as

$$S_c = S_m(1-\phi) + S_f\phi \tag{6.23}$$

---

[2] a is more commonly written as L / D; however, to keep Eshelby's notation, the authors chose to refer to it as a in this section.

Both of these approaches are not actually correct. The strain or the stress at the phase boundary for the two phases in the composite are not balanced. However, both are useful when estimating the lower and upper bound for the composite stiffness as

$$E_c^{\text{Ruess}} < E_c < E_c^{\text{Voigt}} \tag{6.24}$$

In 1957, Eshelby [19] developed an analytical solution for stress, strain, and displacement fields in a linear infinite elastic body containing an ellipsoidal inclusion. He investigated the case when the ellipsoidal inclusion embedded in the elastic body is deformed due to eigen-strain (uniform strain). Since the changes in shape and size for the inclusion are restricted because of the elastic material surrounding the inclusion, both the inclusion and the surrounding material remain in a stressed state of equilibrium. He also demonstrated that the strain in the infinite elastic body and the ellipsoidal inclusion are not the same. Eshelby derived a closed form solution for the strain and stress in the ellipsoidal inclusion and demonstrated it to be uniform throughout the inclusion. The uniform strain in the inclusion and the strain in the stress-free elastic continuous surrounding are related by

$$\varepsilon_{\text{inclusion}} = \mathbb{A}\varepsilon_m \tag{6.25}$$

where ε_m = eigen-strain. $\mathbb{A}$ is a fourth-order tensor relating the strain field in the elastic ellipsoid to the eigen-strain in the infinite continuum[3]. $\mathbb{A}$ is a complex property that is a function of the stiffness of the ellipsoid E_f and of the stiffness of the surrounding continuum E_m, expressed as

$$\mathbb{A} = \left[\mathbb{I} - \mathbb{S}E_m^{-1}(E_f - E_m)\right]^{-1} \tag{6.26}$$

where $\mathbb{A}$ is the fourth-order unit tensor and $\mathbb{S}$ is the fourth-order Eshelby tensor, whose components are defined as a function of $a = L/D$, aspect ratio of the elliptical inclusion, and ν_m, Poisson's ratio of the elastic body surrounding the inclusion (matrix); details of $\mathbb{S}$ are presented in the Appendix of this chapter.

The above expression is for dilute inclusions, i. e., it is assumed that the stress in the inclusion is caused by the matrix only, and not from interactions with other inclusions. In 1973, Mori and Tanaka [20] extended Eshelby's results for a single ellipsoidal inclusion and applied it to a two-phase system, such as a fiber and matrix composite, where the interaction between the inclusions is accounted for. The method was further advanced by many researchers, such as Benveniste [21] in 1987.

[3] Details are presented in the Appendix at the end of this chapter.

Mori and Tanaka [20] adapted the tensor $\mathbb{A}$ to a new fourth-order tensor $\mathbb{T}$, and an expression like Eshelby's is written as

$$\varepsilon_{\text{inclusion}} = \mathbb{T}\varepsilon_m \tag{6.27}$$

The tensor $\mathbb{T}$ can be expressed in terms of $\mathbb{A}$ and volume fractions ϕ of the inclusions as:

$$\mathbb{A} = \left[(1-\phi)\mathbb{I} + \phi\mathbb{T}\right]^{-1}\mathbb{T} \tag{6.28}$$

In this case, for dilute inclusion $\phi = 0$, $\mathbb{T} = \mathbb{A}$, the same as Eshlby's tensor. The elastic modulus for a composite can be expressed as

$$E_c = E_m + \phi(E_f - E_m)\ \mathbb{T} \tag{6.29}$$

The detailed derivation based on the Mori-Tanaka homogenization model is available in Aboudi [13]. Also, a practical expression for the Mori-Tanaka scheme, which can be implemented in computer programs, is available in Zhao [22] and Papathanasiou [23], and is also presented in the Appendix of this chapter for reference.

In addition to the Mori-Tanaka homogenization, many other variations of the mean field homogenization have been developed. An excellent review of different schemes is available in Klusemann [15] and Ortolano [24]. The key difference of the various approaches is how the stiffness of the composite material is assumed and the method used for estimating the average stress and strain in the material.

6.3.2.2 Representative Volume Element (RVE) Using a Finite Element Approach

The RVE can be defined, as shown in Figure 6.12, using the finite elements method. With such approach the geometry and material properties of the fibers, resin, inclusions, interface, and voids in the selected RVE can be better represented. Consequently, the details of the material microstructure as well as stress and strain variations in the microstructure will also improve. Since such RVE can offer the most accurate representation of the microstructure, such an approach results in the most accurate model for the fiber-reinforced structure.

The use of finite element RVE models for a discontinuous fiber-reinforced material for large parts can become very challenging, because many RVEs in the part will need to be modeled. Developing such model can be time-consuming and running an FE simulation would require large computing resources. To address this challenge many statistical approaches are developed that can make use of the knowledge of the statistical distribution of the microstructure features [25, 28] to reduce the number of degrees of freedom while working with the finite element models.

Full finite element models of the RVE are also often used, but their usage is limited to material research, where there is interest in learning detailed interactions among the constituents at microstructure level and models are often small at micro scale.

If the microstructure used in RVE is repeating in the part–for example, woven fabric–then the finite element based RVE can be used, because few RVE structures repeat themselves, which can help reduce computational cost. In general, such an RVE representing a woven loop can be modeled in detail with many degrees of freedom. To use such RVE as a source of material properties, the RVEs are often reduced to a smaller number of degrees of freedom, and then integrated as "call-in material property" into the finite element model.

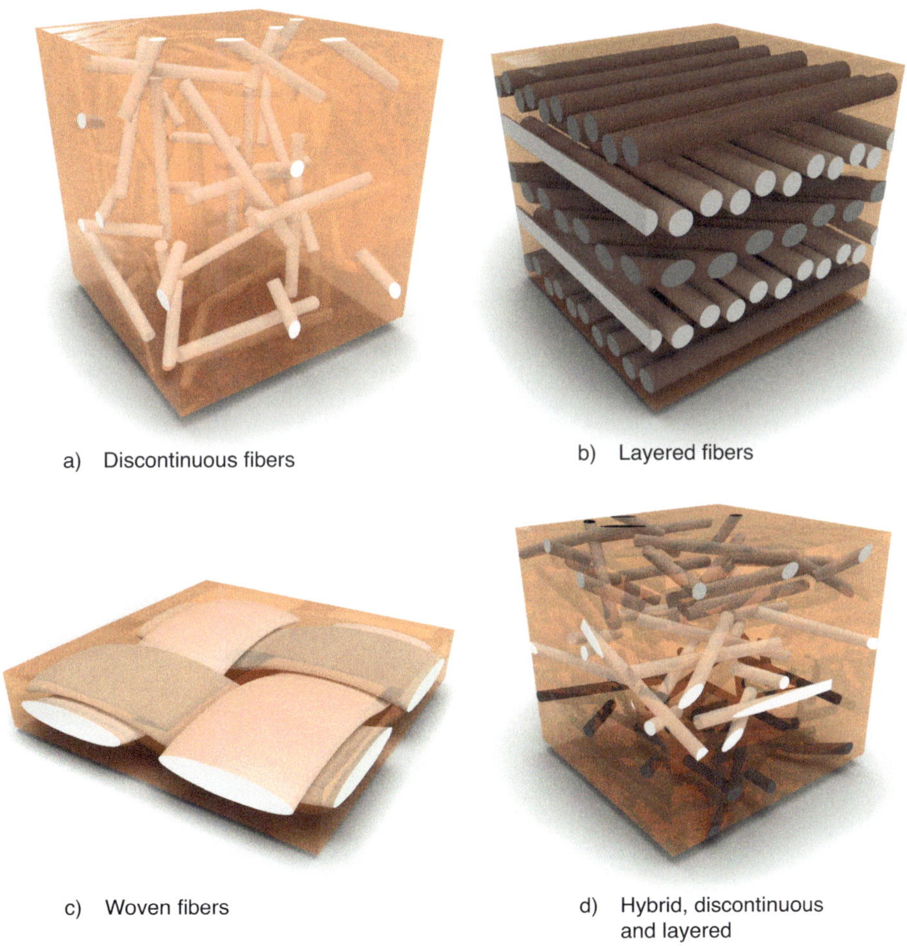

Figure 6.12 Representative volume element using finite element approach

6.3.2.3 Generalized Method of Cells

The *method of cells* was developed to represent the fiber-reinforced material for continuous fibers as well as periodic structures, using a discretization approach; see Figure 6.13. A cell can be divided into four sub-cells to represent the fibers that are periodically distributed in the resin. Each sub-cell can have different material properties. This improves the accuracy over homogenization as some variations in the stress and strain fields are possible within the cell. The method of cells often falls between the homogenization and the full finite element representation of the RVE in terms of accuracy and computing resources.

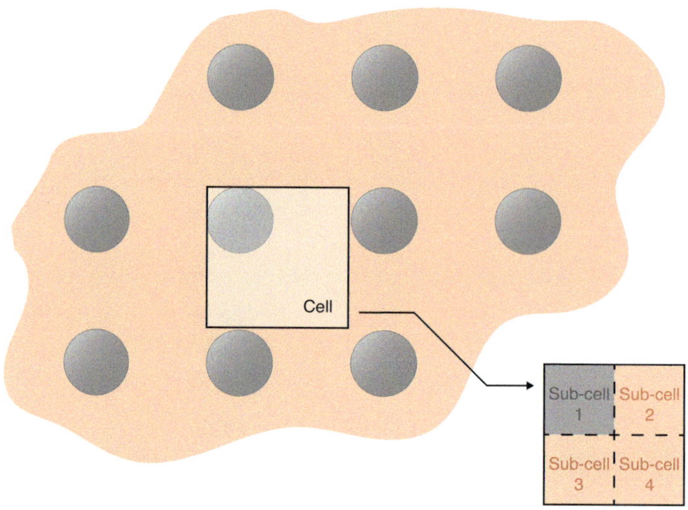

Figure 6.13 Method of cells for composite structure with periodic fiber

The method of cells was extended by Aboudi et al. [13] into the generalized method of cells. In the *generalized method of cells (GMC)* the sub-cells are typically periodic, they can repeat in all three directions, which is called triply periodic. GMC can be used for many different types of periodic fiber conditions such as continuous fibers in layered, braided, or woven condition.

Since GMC is a fully analytical, micromechanics model it provides significant improvement in computing cost while providing varying stress and strain fields in the RVE, which is a significant improvement over homogenization. The basic idea is to divide a repeating unit cell into an arbitrary number of generic cells. Each generic cell is made up of a number of sub-cells. The properties of each sub-cell are considered homogeneous within the sub-cell but properties are different from one sub-cell to the other sub-cell within a generic cell. Thus, choosing the appropriate material properties each generic cell can be defined to represent the targeted homogeneous structure. It is assumed that the displacement vector on each sub-cell varies linearly with the local coordinates. Also, the force and displacement are treated to be continuous from one sub-cell to another sub-cell within the generic cell. In general, the global parameters with GMC are acceptably accurate. The local stress–strain distributions improve due to homogenization, but some limitations are likely due to a built-in assumption of continuity.

The use of GMC for discontinuous fibers, as shown in Figure 6.14, was investigated by Pahr and Arnold [26]. They compared the GMC and the finite element model performance for the discontinuous fibers. The GMC model worked well for the elastic regime. However, they observed that for the elastic-plastic material, the regime between two short fibers shows softer response with GMC. This limitation was attributed to the lack of normal and shear component coupling in GMC. The same group at NASA has proposed the *high fidelity generalized method of cells* (Aboudi [27] and Arnold [30]), to address these limitations. The improvements do come at increased computational cost.

Software implementation for the generalized method of cells (GMC) which works with ABAQUS™ is available from NASA (National Aeronautics and Space Administration) for use within USA [29, 30]. Different variations of GMC, for which semi-analytical formulations are available, are also available in the software implementation.

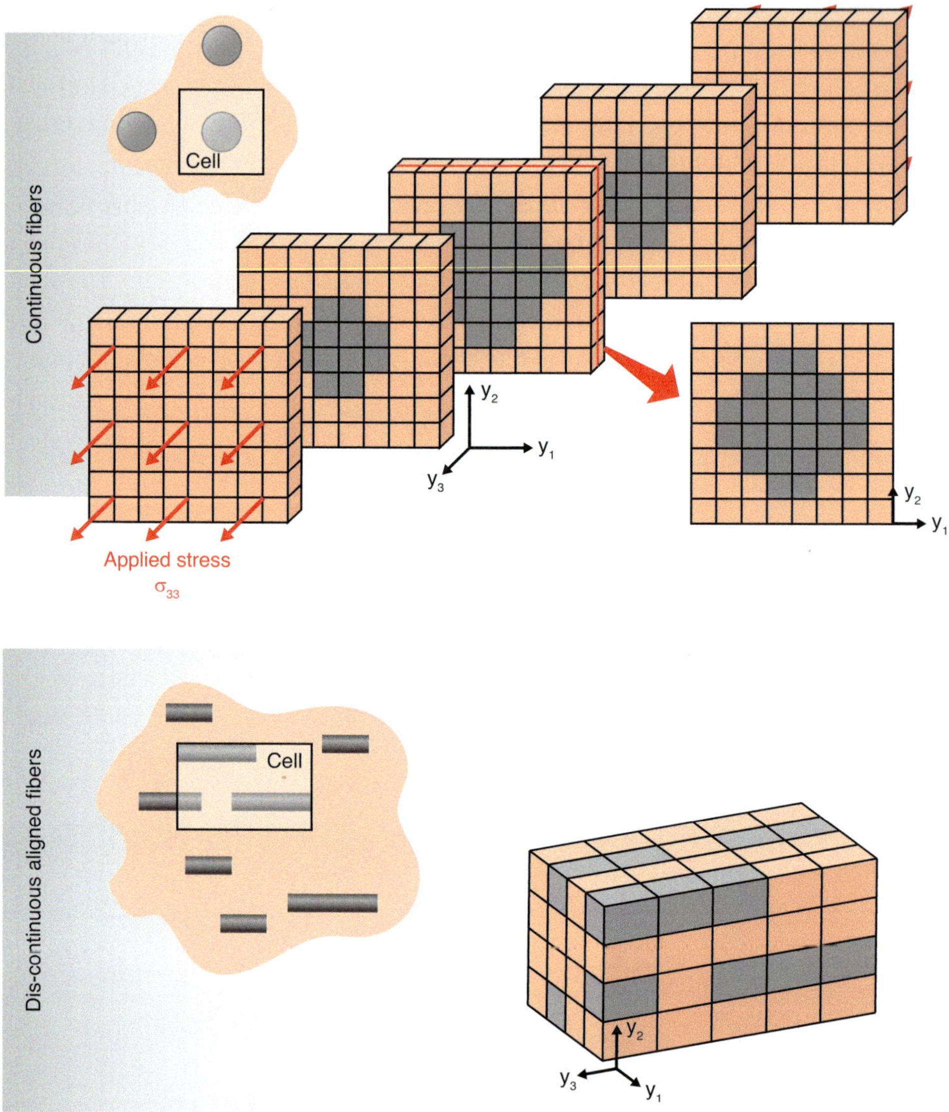

Figure 6.14 Generalized method of cells and sub-cells to represent the fiber condition: a) GMC for continuous fibers, b) GMC for discontinuous fibers

6.3.2.4 Repetitive Unit Cell (RUC)

When the microstructure repeats itself such as in woven, braided, or layered continuous fibers, they are often called repeating unit cells (RUC); Figure 6.15. Essentially, RUC is a specialized case of RVE, where the microstructure is periodic. Similar to RVE, the RUC can be coupled with the finite element model to generate microstructure based material properties. The RUC for the material can be modeled using the finite element model, GMC, or a fully analytical constitutive expression for the material properties in the repeating unit cell.

The main advantage of RUC is that the stress-strain distribution through the selected cell varies based on the known repeating microstructure. Often, fully analytical expressions for such structure are possible, which significantly reduces computing resources while improving accuracy. We have introduced them here as they are quite often referred for the woven and braided materials in many publications.

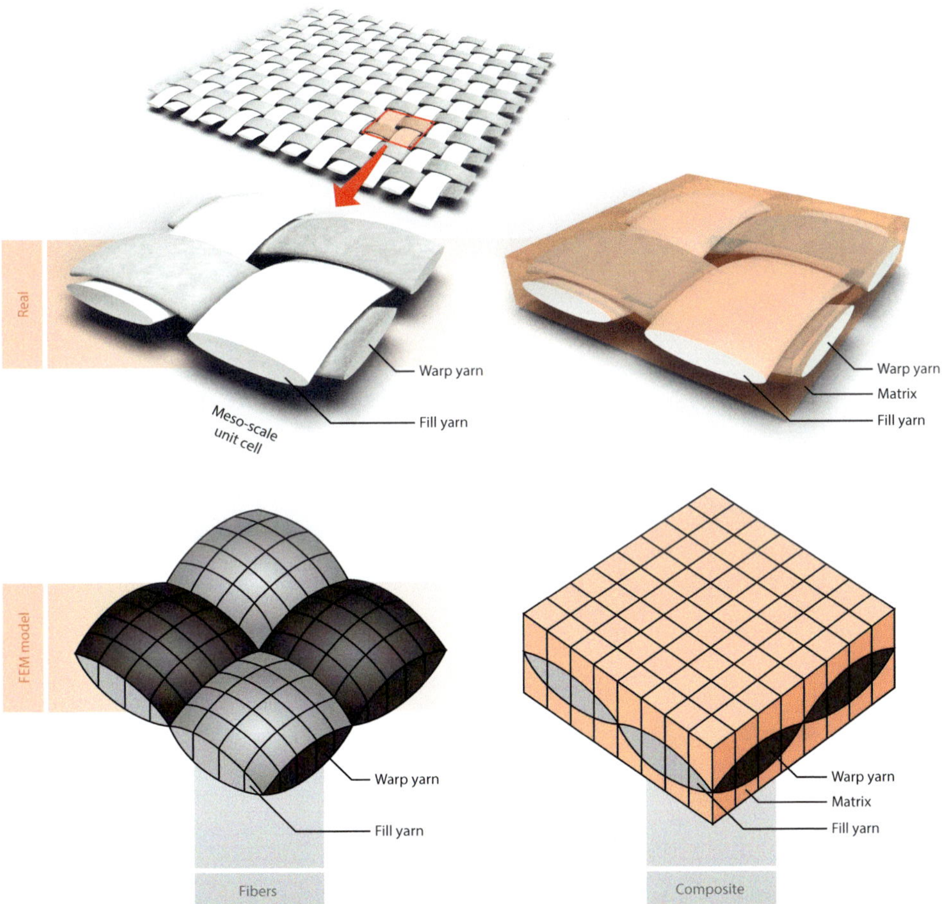

Figure 6.15 Repetitive unit cell approach to account for the microstructure details

6.3.3 Multiscale Modeling

The FE model used in the automotive applications makes use of geometry and material details to represent the components of interest. Typically, the geometry details used for the automotive or other items that involve human interaction are in the millimeter scale. This size is chosen to keep the size of the finite element model within the computable size. The structural models with approximately 3 to 4 million nodes are considered acceptable and often used in practice. Such models, depending on the purpose and loading, can be computed within hours, not days, using the current computing resources. When traditional material such as steel and aluminum, which are isotropic and homogeneous, are used, the millimeter scale (resolution) works well as geometrical details of interest such as radii, holes, etc. can be easily represented at a millimeter scale. However, when fiber-reinforced materials are used, as discussed in the previous Section 6.2, the material properties depend on the condition of the fiber, i.e., fiber length, orientation, and concentration. Therefore, it is important to account for the fiber condition to have accurate material properties. Since the glass fibers are typically 14–20 microns in diameter and carbon fibers have a 5–7 micron diameter, it becomes necessary to model the detail at micro scale.

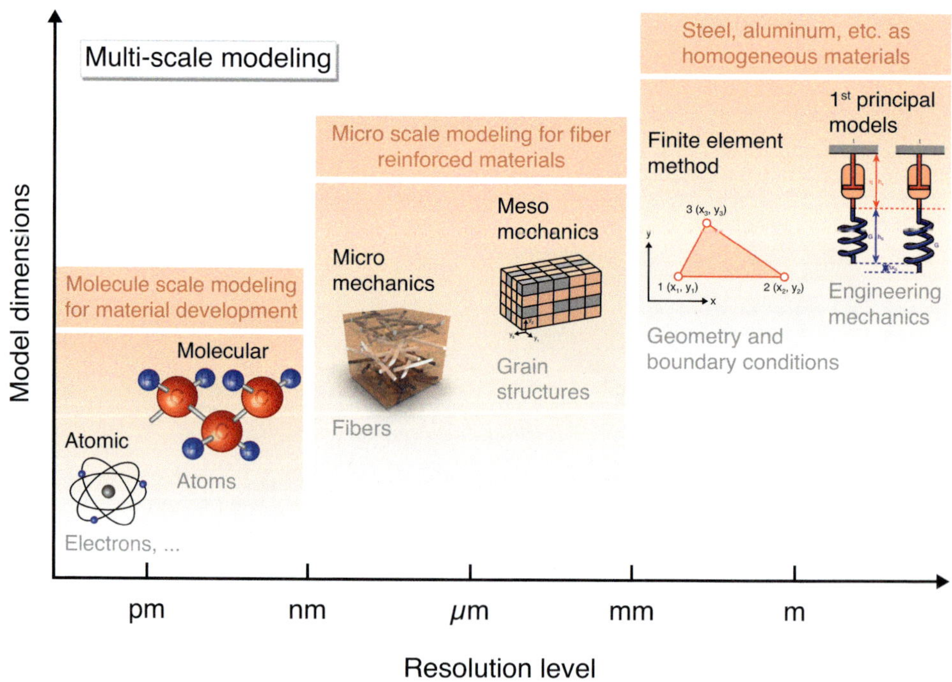

Figure 6.16 Multiscale modeling application map

Making the FE model at micro scale, with all the details of fibers, is possible but the number of elements can increase exponentially and computing such a large model can be very difficult or impossible. To address this difficulty the approach used in industry is called multiscale modeling; see Figure 6.16. The key idea is to model the material microstructure details at micro scale and then couple the material properties from the micro level to the FE model in millimeter scale. Going back and forth from micro to millimeter scale, during the FE calculation, requires many assumptions and extensive mathematical manipulations.

To implement such multiscale approach for the fiber-reinforced materials, first the microstructure details–such as length, orientation, and concentration of fibers in the finished parts of interest–need to be estimated and then, based on the microstructure condition, the material properties can be calculated. For injection or compression molded parts with discontinuous fibers, the fiber conditions depend on how the parts are processed, the geometry of the part, location of the input material gate, temperature, and pressure used to fill the mold. The resulting microstructure can vary from location to location in the finished part.

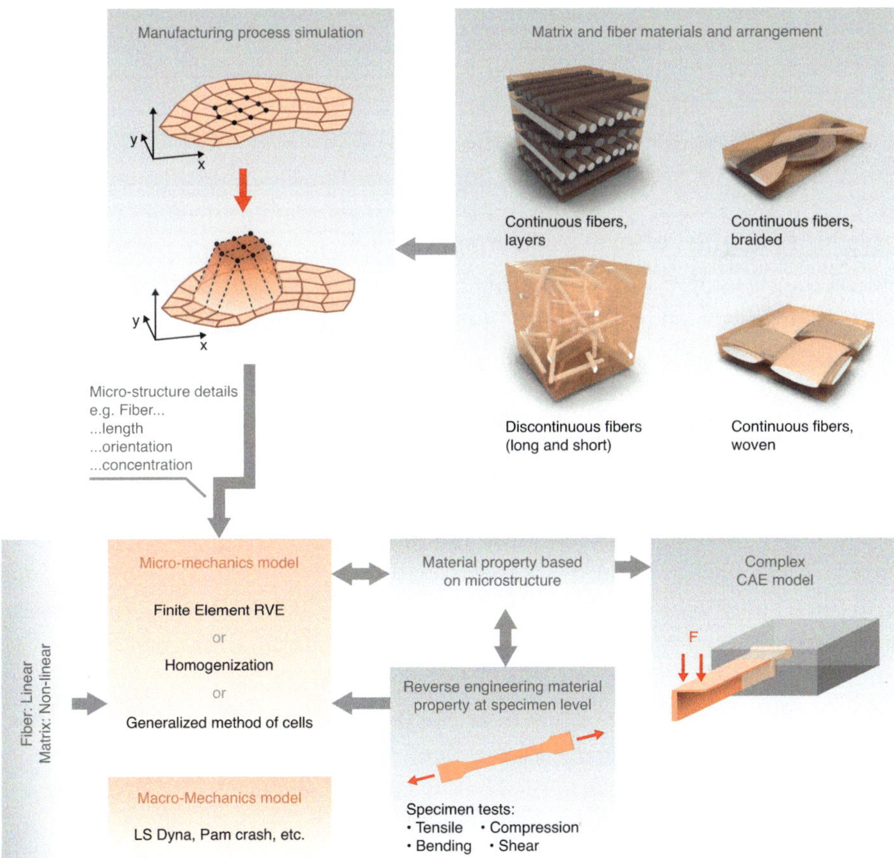

Figure 6.17 Multiscale approach key implementation steps

Measurement of microstructure based material properties can be extremely time-consuming. Therefore, the most practical approach is to use process simulation. The details of process simulation for both injection and compression molding are discussed in Chapter 7. Many fluid mechanics based formulations to model the flow of viscous melt material have been developed and software is commercially available. There are some limitations when simulating the processing of discontinuous fiber-reinforced thermoplastic materials, which are:

- The process simulation capabilities are adequate to estimate the short fiber orientation only. Simulations for materials with long fibers are in development. Fibers with aspect ratio higher than 100 are considered long.
- The capability to predict the fiber length is being introduced in the software. The confidence in the accuracy of predicting the fiber length is under verification.
- The capability to measure fiber concentration is still at a concept level; we consider that to estimate the fiber concentration distribution significant development is required.
- The good news is that orientation plays the most important role in material properties. In general, the orientation tensor can be predicted with good confidence for short fiber materials.

Once the microstructure is estimated based on the process conditions, the next task will be the development of material properties based on the predicted microstructure (i.e., fiber condition) and the material properties of the fiber and the resin. A typical multiscale process as implemented in the finite element scheme is presented in Figure 6.17. Here, the finite element program works with user-defined material subroutines. At each calculation point, i.e., integration point, the program provides the location coordinate to the user material subroutine. The user-defined material subroutine takes the location detail and identifies the fiber condition for that location from process simulation results. Then, using the fiber condition and known fiber and resin material properties, the material properties at the calculation point (integration point) are estimated. This can be done using any of the microstructure modeling schemes discussed in Section 6.3.2. Depending on the model and level of accuracy needed, one can choose a mean-field homogenization approach such as Mori-Tanaka or the generalized method of cells, or one can choose a finite element based approach, with a representative volume element (RVE). Once an estimated material property is calculated and returned by the user material subroutine, the finite element program will use it for the structural calculation at that location. Then, for the next location in the finite element model the microstructure details are again sent to the user-defined material subroutine and new material properties based on the fiber conditions at the next location are calculated. Thus, the microstructure based material properties, which vary from location to location based on the fiber condition, are implemented as "user material" in the finite element simulation. This is the key process for the multiscale analysis of the fiber-reinforced material using the finite elements method. User material subroutines, which couple the finite element software and update the material property at each integration point based on the microstructure, are the main enabler. Many different software solutions specifically developed for this purpose are commercially available; see Table 6.1.

The selection of the microstructure modeling method by the user material subroutine is one of the main differentiators in applying the multiscale approach. A relative comparison of different methods in terms of accuracy and resources is presented in Figure 6.18. The mean field homogenization by Mori-Tanaka has been one of the most popular approaches for discontinuous fiber systems. This is due to a good balance between the computing efficiency and accuracy. One of the major limitations with such an approach, as the name implies, is homogenization. The stress and strain distribution is homogenized in the selected region, which typically is an element. Whereas in the actual material each constituent such as fibers, resin, interface, and voids are a distinct phase, the stress-strain distributions in each phase are different and can form discontinuities at the interface. Therefore, the stress and strain distributions at the micro level based on the Mori-Tanaka approach are often not accurate. However, they may be acceptable at a macro level, which means that the material properties based on the Mori-Tanaka homogenization will be a good approximation for stiffness-based calculations; however, for the breakage and fatigue condition, the failure mechanism, which relies on initiation at microstructure level, may be missed. A comparison of relative accuracy and computing resources for different approaches to model the fiber-reinforced materials is presented in Figure 6.18.

The Mori-Tanaka approach discussed here is for elastic materials. Many improvements in the Mori-Tanaka basic approach to extend the result to plastic and nonlinear material behaviors have been developed [33].

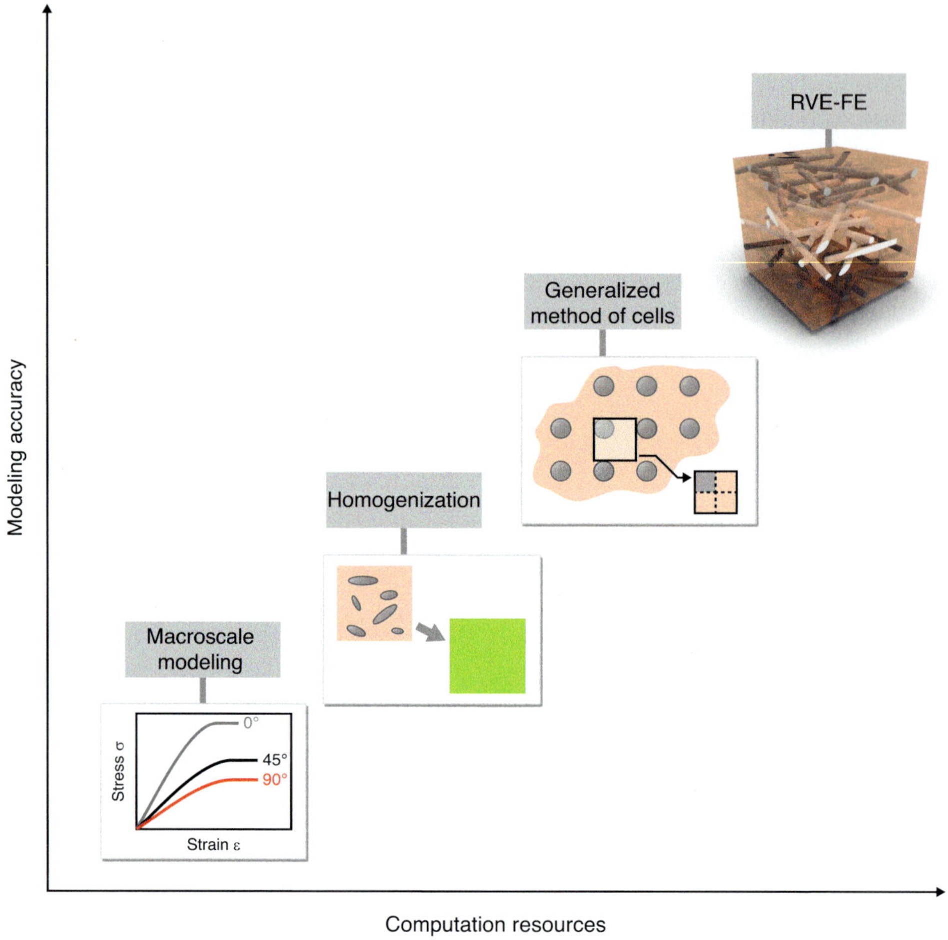

Figure 6.18 Relative comparison of different methods to model fiber-reinforced materials

6.4 Survey of Existing Options in Multiscale Modeling of Fiber-Reinforced Material

At first, multiscale modeling may seem overwhelming; however, for using such approach it is important to understand the concept. For usage in practice, many software packages are available. A brief comparison of such options, based on the available data in March of 2019, is presented in Table 6.1. This table should be used as a reference with caution as the capabilities in the software are constantly evolving. Additionally, the information presented is based on the limited resources available, so accuracy can not be assured.

Table 6.1 Survey of Multiscale Modeling Software Available for Fiber-Reinforced Materials

Software/ Provider	Application/Features (2019 Version)							
	continuous fiber composites	discontinuous fiber composites	method to include fiber orientation	homogenization or micro scale modeling of materials	interface for external solvers	matrix level nonlinearity	strain rate modeling of matrix	failure for material model
Digimat™	yes	yes	yes	homogenization	yes	yes	yes	yes
AlphaSTAR™ Products	yes	yes	yes	both	yes	yes	yes	yes
Converse™	no	yes	yes	both	yes	yes	yes	no
NASA Generalized method of cells	yes	yes	yes	micro	yes	yes	yes	yes
ESAComp™ (Altair)	yes	no	no	both	yes	yes	yes	yes
AUTODESK™	yes	yes	yes	both	yes	yes	yes	yes
Moldex3D™	yes	yes	yes	both	yes	yes	yes	no
LMS™	yes	limited	unknown	unknown	no	limited	unknown	yes
MultiMech™	yes	yes	yes	both	yes	yes	no	yes
Swiftcomp™ Micromechanics	yes	yes	yes (PreVABS)	homogenization	yes	yes	yes	unknown
HyperSizer™	yes	no	no	unknown	yes	no	no	yes

■ 6.5 Comments and Guidance

The fundamental principles of physics are often described using differential equations. While such equations can be solved for simpler shapes where boundary conditions can be mathematically expressed, it is very difficult to apply such equations to complex geometries. The finite element method is an excellent numerical method, which discretizes the complex shapes into simpler shapes to help overcome these challenges. Such numerical approach can be easily implemented for digital computers and applied for large scale applications.

Discontinuous fiber-reinforced polymer properties depend on the microstructure, i.e., fiber conditions such as fiber length, fiber orientation, and fiber concentration. In this chapter, it was also discussed how a material model that accounts for such fiber microstructure can be developed and integrated with the finite element models for structural and other calculations. Since the material microstructure is at a micron scale and finite element models typically are at millimeter scale, such integrated models are called multiscale models.

There are a number of approaches presented to represent the microstructure effect in the material properties. Homogenization is the simplest; it uses little computational resources and works well for the global properties, such as stiffness, force deflection, vibration study, etc. When it comes to local behavior, such as damage, crack propagation, and material failure, which is highly localized, the homogenization approach may not work well as the variation in the stress and strain field and material properties are not well defined. To address these limitations, microstructure details can be modeled using the finite elements approach, which works well but can add tremendous computing resources. The generalized method of cells has been developed to offer something in the middle. The microstructure is modeled using predefined cell and sub-cell structures, which allow a limited variation of properties within the representative volume element of the microstructure.

While using such models, it is important to keep in mind the well-known statement by G. Box, *"Essentially, all models are wrong, but some are useful"*. There are many assumptions involved in developing such multiscale finite element models for fiber-reinforced materials. For example, we assume there is perfect attachment between the fiber and resins. The fibers are assumed to be straight and not touching each other, etc. These assumptions are not always valid. In addition, the actual physical parts have many variations, e.g., the diameter of the fibers, the length of the fibers, material properties, all have some distributions. It is impossible to account for such variations in the model accurately all the time. Therefore, the results estimated using the models will have some limitations on the accuracy. This does not mean the models are not useful; in fact, such models can help get much deeper insights into the material behavior and can be very useful in developing designs. The point is that the models should be used with caution. While interpreting results from such models, it is very important to have a deep understanding of the physics of the event being modeled along with the limitations and assumptions involved in the models. The application of modeling and simulation on real world complex parts is presented in Chapters 8, 9, and 10. These case studies will help understand the modeling approaches further.

References

[1] Zienkiewicz, O.C., Taylor, R.L., Zienkiewicz, O.C., and Taylor, R.L., *The finite element method* (Vol. 36), McGraw-Hill, London (1977)

[2] Bathe, K.J., *Finite element procedures*, 2nd ed., Prentice Hall (2006)

[3] Reddy, J.N., *An introduction to nonlinear finite element analysis: with applications to heat transfer, fluid mechanics, and solid mechanics*, OUP Oxford (2014)

[4] Hrennikoff, A., Solution of problems of elasticity by the framework method, *J. Appl. Mech.* (1941)

[5] Turner, M.J., Stiffness and deflection analysis of complex structures, *Journal of the Aeronautical Sciences*, 23(9), 805–823 (1956)

[6] Argyris, J.H. and Kelsey, S., *Energy theorems and structural analysis* (Vol. 960), Butterworths, London (1960)

[7] Clough, R.W., The finite element method in plane stress analysis. In Proceedings of 2nd ASCE Conference on Electronic Computation, Pittsburgh Pa., Sept. 8 and 9 (1960)

[8] Belytschko, T., Gracie, R., and Ventura, G., A review of extended/generalized finite element methods for material modeling, *Modelling and Simulation in Materials Science and Engineering*, 17 (4), 043001 (2009)

[9] Osswald, P.V. and Osswald, T.A., A strength tensor based failure criterion with stress interactions, *Polymer Composites*, DOI:10.1002/pc.24275 (2017)

[10] Tsai, S.W. and Melo, J.D.D., *Composite materials design and testing*, JEC Group (2015)

[11] Verpoest, I. and Lomov, S.V., Virtual textile composites software WiseTex: Integration with micromechanical, permeability and structural analysis, *Composites Science and Technology*, 65 (15–16), 2563–2574 (2005)

[12] Harper, L.T., Qian, C., Turner, T.A., Li, S., and Warrior, N.A., Representative volume elements for discontinuous carbon fibre composites–Part 1 and 2: Boundary conditions, *Composites Science and Technology*, 72(2), 225–234; 204–210 (2012)

[13] Aboudi, J., *Mechanics of composite materials: a unified micromechanical approach* (Vol. 29), Elsevier (2013)

[14] Pierard, O., Friebel, C., and Doghri, I., Mean-field homogenization of multi-phase thermo-elastic composites: a general framework and its validation, *Composites Science and Technology*, 64 (10–11), 1587–1603 (2004)

[15] Klusemann, B. and Svendsen, B., Homogenization methods for multi-phase elastic composites: Comparisons and benchmarks, *Technische Mechanik*, 30(4), 374–386 (2010)

[16] Voigt, W., On the relation between the elasticity constants of isotropic bodies, *Ann. Phys. Chem.*, 274, 573–587 (1889)

[17] Reuss, A., Berechnung der fließgrenze von mischkristallen auf grund der plastizitätsbedingung für einkristalle, *ZAMM Journal of Applied Mathematics and Mechanics/Zeitschrift für Angewandte Mathematik und Mechanik*, 9 (1), 49–58 (1929)

[18] Hashin, Z., Analysis of composite materials–a survey, *Journal of Applied Mechanics*, 50(3), 481–505 (1983)

[19] Eshelby, J.D., The determination of the elastic field of an ellipsoidal inclusion, and related problems, *Proc. R. Soc. Lond. A*, 241 (1226), 376–396 (1957)

[20] Mori, T. and Tanaka, K., Average stress in matrix and average elastic energy of materials with misfitting inclusions, *Acta metallurgica*, 21 (5), 571–574 (1973)

[21] Benveniste, Y., A new approach to the application of Mori-Tanaka's theory in composite materials, *Mechanics of materials*, *6*(2), 147–157. Papthanasiou, T. D. and Guell, D. C. (Eds.), *Flow-induced alignment in composite materials*, Elsevier (1997)

[22] Zhao, Y. H., Tandon, G. P., and Weng, G. J., Elastic moduli for a class of porous materials, *Acta Mechanica*, *76*(1–2), 105–131 (1989)

[23] Papthanasiou, T. D. and Guell, D. C. (Eds.), *Flow-induced alignment in composite materials,* Elsevier (1997)

[24] Ortolano González, J. M., Hernández Ortega, J. A., and Oliver Olivella, X., *A comparative study on homogenization strategies for multi-scale analysis of materials,* Centre Internacional de Mètodes Numèrics en Enginyeria (CIMNE) (2013)

[25] Nguyen, V. P., Stroeven, M., and Sluys, L. J., Multiscale continuous and discontinuous modeling of heterogeneous materials: a review on recent developments, *Journal of Multiscale Modelling*, *3*(04), 229–270 (2011)

[26] Pahr, D. H. and Arnold, S. M., The applicability of the generalized method of cells for analyzing discontinuously reinforced composites, *Composites Part B: Engineering*, *33*(2), 153–170 (2002)

[27] Aboudi, J. The generalized method of cells and high-fidelity generalized method of cells micromechanical models–A review, *Mechanics of Advanced Materials and Structures*, *11*(4–5), 329–366 (2004)

[28] Bednarcyk, B. A. and Arnold, S. M., A framework for performing multiscale stochastic progressive failure analysis of composite structures (2007)

[29] Murthy, P. L., and Pineda, E. J., Tool for generation of MAC/GMC representative unit cell for CMC/PMC analysis (2016)

[30] Arnold, S.M., Bednarcyk, B. A., Wilt, T. E., and Trowbridge, D., Micromechanics analysis code with generalized method of cells (mac/gmc): User guide, version 3 (1999)

[31] Harris, B. (Ed.), *Fatigue in composites: science and technology of the fatigue response of fibre-reinforced plastics*, Woodhead Publishing (2003)

[32] MacNeal, Richard H., "The NASTRAN Theoretical Manual" (1972)

[33] Doghri, I. and Ouaar, A., Homogenization of two-phase elasto-plastic composite materials and structures: study of tangent operators, cyclic plasticity and numerical algorithms, *International Journal of Solids and structures*, *40*(7), 1681–1712 (2003)

Appendix

Eshelby Tensor Computer Implementation

Eshelby [19] developed an analytical solution for strain in the elliptical inclusion embedded in an infinite elastic body

$$\varepsilon_{\text{inclusion}} = \mathbb{A}\varepsilon_m$$

where ε_m = eigen-strain and the tensor $\mathbb{A}$:

$$\mathbb{A} = \left[\mathbb{I} - \mathbb{S}E_m^{-1}(E_f - E_m)\right]^{-1}$$

$\mathbb{I}$ is the fourth-order unit tensor and E_f and E_m are the stiffness of inclusion and surrounding elastic body, respectively. $\mathbb{S}$ is the fourth-order Eshelby tensor, whose components are defined as a function of $a = L/D$, aspect ratio of the elliptical inclusion, and ν_m, Poisson's ratio of the surrounding elastic body.

$$S_{1111} = \frac{1}{2(1-\nu_m)}\left[1-2\nu_m+\frac{3a^2-1}{a^2-1}-\left(1-2\nu_m+\frac{3a^2}{a^2-1}\right)g\right]$$

$$S_{2222} = \frac{3a^2}{8(1-\nu_m)(a^2-1)}+\frac{1}{4(1-\nu_m)}\left[1-2\nu_m-\frac{9}{4(a^2-1)}\right]g$$

$$S_{2233} = \frac{1}{4(1-\nu_m)}\left[\frac{a^2}{2(a^2-1)}-\left(1-2\nu_m+\frac{3}{4(a^2-1)}\right)g\right]$$

$$S_{2211} = -\frac{a^2}{2(1-\nu_m)(a^2-1)}+\frac{1}{4(1-\nu_m)}\left[\frac{3a^2}{(a^2-1)}-(1-2\nu_m)\right]g$$

$$S_{1122} = \frac{1}{2(1-\nu_m)}\left[1-2\nu_m+\frac{1}{a^2-1}\right]+\frac{1}{2(1-\nu_m)}\left[1-2\nu_m+\frac{3}{2(a^2-1)}\right]g$$

$$S_{2323} = \frac{1}{4(1-\nu_m)}\left[\frac{a^2}{2(a^2-1)}+\left(1-2\nu_m-\frac{3}{4(a^2-1)}\right)g\right]$$

$$S_{1212} = \frac{1}{4(1-\nu_m)}\left[1-2\nu_m-\frac{a^2+1}{a^2-1}-\frac{1}{2}\left(1-2\nu_m-\frac{3(a^2+1)}{a^2-1}\right)g\right]$$

$$S_{3333} = S_{2222}$$

$$S_{3322} = S_{2233}$$

$$S_{3311} = S_{2211}$$

$$S_{1133} = S_{1122}$$

$$S_{1313} = S_{1212}$$

$$g = \frac{a}{(a^2-1)^{3/2}}\left[a(a^2-1)^{1/2}-\cosh^{-1}a\right]$$ for a prolate shape where $a > 1$.

$$g = \frac{a}{\left(1-a^2\right)^{3/2}}\left[\cosh^{-1} a - a\left(1-a^2\right)^{1/2}\right]$$ for an oblate shape where $a < 1$.

Mori-Tanaka Computer Implementation

Mori-Tanaka [20] extended the Eshelby's tensor to two phases, i.e., fiber and matrix system.

$$\varepsilon_{\text{inclusion}} = \mathbb{T}\varepsilon_m$$

where ε_m = eigen-strain.

The tensor $\mathbb{T}$ can be expressed in terms of $\mathbb{A}$ and the volume fractions ϕ of the inclusions as

$$\mathbb{A} = \left[(1-\phi)\mathbb{I} + \phi\mathbb{T}\right]^{-1}\mathbb{T}$$

The elastic modulus for composites can be expressed as

$$E_c = E_m + \phi(E_f - E_m)\ \mathbb{T}$$

The computer implementation of composite properties (E_c), adopted from Zhao [22], is given below. For the calculations, the constants μ and λ are Lame constants defined as

$$\mu = G = \frac{E}{2(1+v)}$$

$$\lambda = (Ev)/(1+v)/(1-2\lambda = (Ev)/(1+v)/(1-2v))$$

For these equations above and below, E is the elastic modulus, G is the bulk modulus, V is the volume fraction, and v is Poisson's ratio. Then, using the subscript f for fiber as inclusion and m for the matrix, the following constants are defined:

$$D_1 = 1 + \frac{2(\mu_f - \mu_m)}{\lambda_f - \lambda_m}$$

$$D_2 = \frac{(\lambda_m + 2\mu_m)}{\lambda_f - \lambda_m}$$

$$D_3 = \frac{\lambda_m}{\lambda_f - \lambda_m}$$

$$B_1 = V_f D_1 + D_2 + V_m(D_1 S_{1111} + 2S_{2222})$$

$$B_2 = V_f + D_3 + V_m(D_1 S_{1122} + S_{2222} + S_{2233})$$

$$B_3 = V_f + D_3 + V_m((1+D_1)S_{2211} + S_{1111})$$

$$B_4 = V_f D_1 + D_2 + V_m(D_1 S_{2222} + S_{2233} + S_{1122})$$

$$B_5 = V_f + D_3 + V_m(D_1 S_{2233} + S_{2222} + S_{1122})$$

Furthermore,

$$A = 2B_3 B_2 - B_1(B_4 + B_5)$$

$$A_1 = D_1(B_4 + B_5) - 2B_2$$

$$A_2 = (1+D_1)B_2 - (B_4 + B_5)$$

$$A_3 = B_1 - D_1 B_3$$

$$A_4 = (1+D_1)B_1 - 2B_3$$

$$A_5 = (1-D_1)/\left(B_4 - B_5\right)$$

Composite properties can be calculated in terms of constants defined above as

$$\frac{E_{11}}{Em} = \frac{1}{1+\left[V_f(A_1 + 2V_m A_2)\right]\Big/A}$$

$$\frac{G_{12}}{Gm} = 1 + \frac{V_f}{\dfrac{G_m}{G_f - G_m} + 2V_m S_{1212}}$$

$$\frac{G_{23}}{Gm} = 1 + \frac{V_f}{\dfrac{G_m}{G_f - G_m} + 2V_m S_{2323}}$$

$$\frac{v_{12}}{v_m} = 1 - V_f \frac{v_m(A_1 + 2v_m A_2) + (A_3 - v_m A_4)}{v_m\left[A + V_f\left(A_1 + 2v_m A_2\right)\right]}$$

$$\frac{k}{k_m} = \frac{(1+v_m)\left(1-2v_m\right)}{1 - v_m\left(1+2v_{12}\right) + \dfrac{V_f\left[2\left(v_{12} - v_m\right)A_3 + \left[1 - v_m\left(1+2v_{12}\right)\right]A_4\right]}{A}}$$

where k_m is the plain strain bulk modulus, the bulk modulus of composites, and the S_{ijkl} are from the Eshelby tensor defined earlier.

7 Process Simulation for Discontinuous Fibers

Huan-Chang Tseng[1], Jim Hsu[1], Anthony Yang[1], Sebastian Goris, Yu-Yang Song, Umesh N. Gandhi, and Tim A. Osswald

7.1 Modeling Fiber Motion During Processing

This chapter summarizes and describes modeling approaches that aim to establish the link between processing and the flow-induced configuration of the fibers in the molded part. Furthermore, the models discussed in this chapter are commonly implemented in commercially available process simulation software packages.

7.2 Process Simulation

As described in Chapter 3, injection molding and compression molding are the most important industrial processes for producing large quantities of complex plastic parts made from discontinuous-fibers-filled thermoplastics. New polymers and the demand for high-quality electronics, consumer products, automobiles, and airplanes have forced engineers and designers to improve mold-tooling efficiency and the quality of final parts. Consequently, researchers have focused on developing numerical solutions to replace the experimental trial-and-error process and expedite the development of new products. The progress in both hardware and software over the past several decades has made computer-aided-engineering (CAE) a highly effective tool for analyzing the complicated physical phenomena inherent in the injection or compression moldings and made it an integral step in all industries. Commercial software such as Moldex3D™ (CoreTech System, Hsinchu County, Taiwan) [1] and Moldflow™ (Autodesk Inc., San Rafael, CA) [2] have been available for process simulation for many years.

[1] CoreTech System Co., Ltd., Taiwan

7.2.1 Injection Molding

During the injection molding process, the polymer melt is forced to flow, under constant high pressure, through the sprue, runner, and then fill into the empty cavity. The plastic part is ejected after the polymer melt is sufficiently cooled and hardened. All of those steps need to be modeled and simulated. Over the last several decades, numerous researchers have attempted to analyze such a complex process by distinct simplifications and approximations under the limited computational resources [3–8]. The Hele-Shaw model has been extensively adopted in the current commercial CAE packages. The Hele-Shaw model neglects the inertia and the gap-wise velocity component for polymer melt flow in thin cavities. The flow governing equations are simplified into a single Poisson equation based on these assumptions [8]. In this manner, the requirements of the computational storage and CPU time requirements can be considerably reduced. However, such a simplified model possesses some limitations. First, the shell element employed in the Hele-Shaw model needs the construction of the mid-plane, which is time-consuming. Furthermore, some significant three-dimensional flow regions, i. e., flow around corners, over bosses, ribs, or thickness-change regions, or the fountain effect in the vicinity of melt fronts, may remain in thin cavities. These regions will not only complicate the identification of the mid-plane but will also cause pressure, heat transfer, or stress prediction errors. Moreover, the dominant three-dimensional flow present in thick parts such as gears and connectors invalidates the Hele-Shaw approximation in these cases.

Therefore, for the past decade, the true three-dimensional analysis model has been developed to reduce the time spent in the construction of the mid-plane, but also to accurately predict the flow in the thick parts. A true three-dimensional analysis can give more insight into the microstructure of parts by providing detailed information about fiber orientation, residual stress, or degree of cure distributions, etc. However, the highly nonlinear interplay between transient heat transfer, non-Newtonian fluid flow, moving interface, and phase change challenges the accuracy, efficiency, and stability of the fully three-dimensional mold filling analysis.

7.2.1.1 Mathematical Models and Assumptions

Let u, v, w denote the velocity components along the x, y, and z planar coordinates in Figure 7.1. General polymer melts used in injection or compression molding are considered as viscous fluids. When such fluid is passed through the mold cavity there is resistance to the flow as well as a cooling effect, resulting in fountain flow.

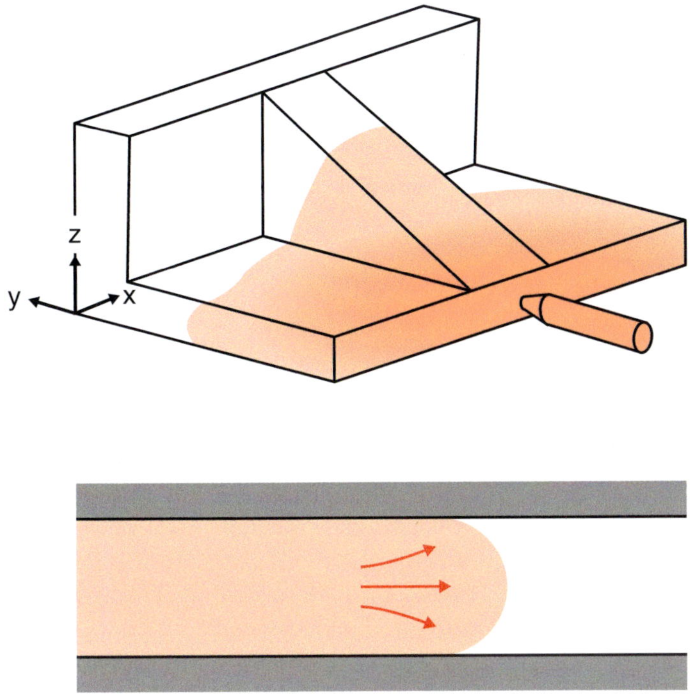

Figure 7.1 Schematic diagram of the viscous melt through the mold cavity and the fountain flow front effect

At the filling stage, the polymer melt is assumed to behave as a generalized Newtonian fluid (GNF). Hence, the non-isothermal 3D flow motion can be mathematically described by the following:

$$\frac{\partial \rho}{\partial t} + \nabla \cdot \rho \boldsymbol{u} = 0 \tag{7.1}$$

$$\frac{\partial}{\partial t}(\rho \boldsymbol{u}) + \nabla \cdot (\rho \boldsymbol{u}\boldsymbol{u} - \boldsymbol{\sigma}) = \rho \boldsymbol{g} \tag{7.2}$$

$$\boldsymbol{\sigma} = -p\boldsymbol{I} + \eta\left(\nabla \boldsymbol{u} + \nabla \boldsymbol{u}^T\right) \tag{7.3}$$

$$\rho C_P \left(\frac{\partial T}{\partial t} + \boldsymbol{u} \cdot \nabla T\right) = \nabla \cdot (k \nabla T) + \eta \dot{\gamma}^2 \tag{7.4}$$

where $\boldsymbol{u}$ is the velocity vector, T the temperature, t the time, p the pressure, $\boldsymbol{\sigma}$ the total stress tensor, ρ the density, η the viscosity, k the thermal conductivity, C_P the specific heat, and $\dot{\gamma}$ the rate of deformation tensor. To solve this problem, the polymeric feature needs to be described suitably. For example, the modified Cross model with Arrhenius temperature dependence is employed to describe the viscosity of the polymer melt:

$$\eta(T,\dot{\gamma})=\frac{\eta_o(T)}{1+\left(\eta_o\dot{\gamma}/\tau^*\right)^{1-n}} \tag{7.5}$$

$$\eta_o(T)=B\exp\left(\frac{T_b}{T}\right) \tag{7.6}$$

where n is the power law index, η_o the zero-shear viscosity, and τ^* is the parameter that describes the transition region between zero shear rate and the power law region of the viscosity curve. A volume fraction function f is introduced to track the evolution of the melt front. Here, $f = 0$ is defined as the air phase, $f = 1$ as the polymer melt phase, and then the melt front is located within cells with $0 < f < 1$. The advancement of f over time is governed by the following transport equation:

$$\frac{\partial f}{\partial t}+\nabla\cdot(\boldsymbol{u}f)=0 \tag{7.7}$$

The flow rate or injection pressure is prescribed at the mold inlet. No slip is assumed at the mold wall. Note that only an inlet boundary condition is necessary for the hyperbolic transport equation of the volume fraction function.

7.2.1.2 Numerical Method

There are several different numerical methods that can be used to discretize the above-mentioned governing equations, including the finite difference method (FDM), the finite element method (FEM), and the finite volume method (FVM). Each method has its own advantages and limitations.

The FDM is purely based on the partial differential equations (PDE) defining the problem. The FVM and FEM are based on a numerical integration of the underlying PDE. FDM reveals a high precision but is not easy to extend for the practical tests. The strength of the FEM and FVM lies on the flexibility to deal with the complex application [9] with good precessions.

Different commercial packages use different methods for their flow solvers. For example, Moldex3D™ uses FVM and Moldflow™ uses FEM. In this section, the FVM approach is discussed in more detail and it is explained how to use it to solve the mold filling for an injection molded part (Figure 7.2).

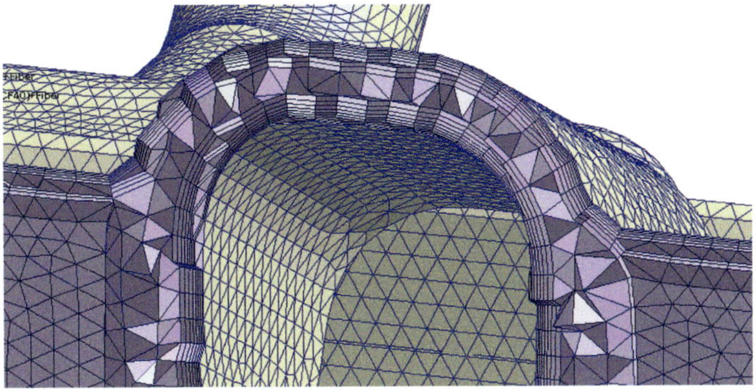

Figure 7.2 Cross section of 3D mesh of an injection molded part

The previously described governing equations can be cast into the following generic form with an appropriate choice of φ, Γ, and Q_φ:

$$\frac{\partial}{\partial t}(\rho\varphi)+\nabla\cdot(\rho\boldsymbol{u}\varphi)-\nabla\cdot(\Gamma\nabla\varphi)=Q_\varphi \tag{7.8}$$

where φ denotes the dependent variable, which can be a scalar or vector components, Γ is the diffusivity, and Q_φ is called the source term, which contains all the terms that cannot be accommodated in the convection and diffusion term. In order to discretize the equation, the whole computational domain is subdivided into a finite number of non-overlapping control volumes in the FVM. The transport variables are stored at the center of the control volumes. The transport equations are then integrated over each of the control volumes in the domain:

$$\left(\frac{\partial\rho\varphi}{\partial t}\right)_P V_P+\sum_{j=1}^{n}(\rho\varphi\boldsymbol{u})_j\cdot\boldsymbol{A}_j-\sum_{j=1}^{n}(\Gamma\nabla\varphi)_j\cdot\boldsymbol{A}_j=(Q_\varphi)_P V_P \tag{7.9}$$

Next, different schemes such as the central differencing scheme can be combined with the upwind scheme to approximate the transport variables at the cell faces. The resulting algebraic equation can be rearranged as below:

$$a_P\varphi_P+\sum_{nb}a_{nb}\varphi_{nb}=S_\varphi \tag{7.10}$$

Then, the modern matrix solver, like the CG family solver or the multigrid solver, can be used to solve the algebraic equation.

The collocated cell-centered finite volume method proposed by Chang and Yang [10] is modified. Base on a SIMPLE-like iteration procedure, the method has improved numerical stability and robust convergence. Pressure, velocity, and temperature are segregated in the solver so that the efficiency can be achieved even for 3D calculation. Based on these velocity and temperature files, the coefficients of the discretized equations are updated, and a new outer iteration is activated. The process is repeated until convergence.

In order to track the melt front advancement, the algebraic volume-tracking method based on fixed mesh can be used because of its relative topological robustness compared with the moving-mesh method. The algebraic volume-tracking method is a whole-domain approach, in which both air and liquid are considered during computation. In the algebraic volume-tracking method, the volume of the fluid is tracked by solving the volume fractional transport equation algebraically [11].

After the flow and heat transfer governing equations are solved by the FVM, the advancement of interfaces at each time step is determined by solving the volume fractional advection equation according to the obtained velocities. The material properties are calculated from the updated volume fractional function, and then the next computation of thermal flow field is initiated. This procedure is repeated until the cavity is completely filled.

7.2.2 Compression Molding

Before the compression molding step, the bulk charge is extruded using the twin screw extruder mixing fibers, resin, and additives. After mixing, the bulk material is placed in the mold cavity and compressed, as shown in Figure 7.3. In general, compression molding results in longer fibers and, hence, better mechanical properties compared to injection molding. This is because the shear flow rate through the cavity in compression molding is smaller and, hence, there is less fiber breakage. The finished part properties are highly dependent on the fiber configuration in the part, i.e., fiber length, fiber orientation, and volume fraction distribution through the parts.

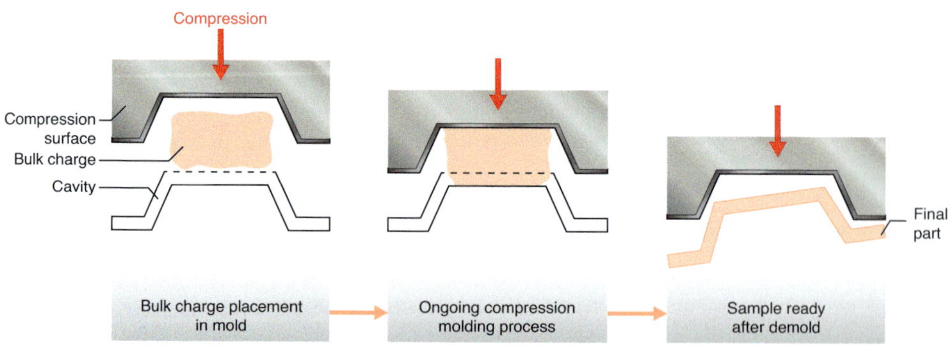

Figure 7.3 Schematic of the compression molding process

The earliest commercial software programs for compression molding were developed in the late 1980s using the control volume approach [12]; CADPRESS was marketed by The Madison Group in Wisconsin in the 1990s around the same time as EXPRESS was being marketed by MBase GmbH in Germany [13]. Both software packages were designed to predict the filling and the subsequent warpage of the part in SMC compression molding. However, the simulations were limited to the 2D shell element, which means the flow in the shell surface is calculated while the flow along the thickness of the shell is assumed to be constant, based on experience. Just like the injection molding simulation, the true 3D simulation has been successfully extended in the last decade to simulate the compression molding with the commercial package. The material models used in these software tools are based on non-Newtonian fluid flow and can consider the material consisting of high fiber volume content. By using a novel modified fiber orientation algorithm based on the Folgar-Tucker model, the simulation tool can predict the fiber flow and fiber orientation during compression molding simulation [14].

In this compression molding simulation, we only have to define the mesh with a different property to distinguish the moving mesh boundary. The compression action is modeled by a moving mesh boundary technique, where the computational domain will dynamically deform with time within filling stages as in Figure 7.4.

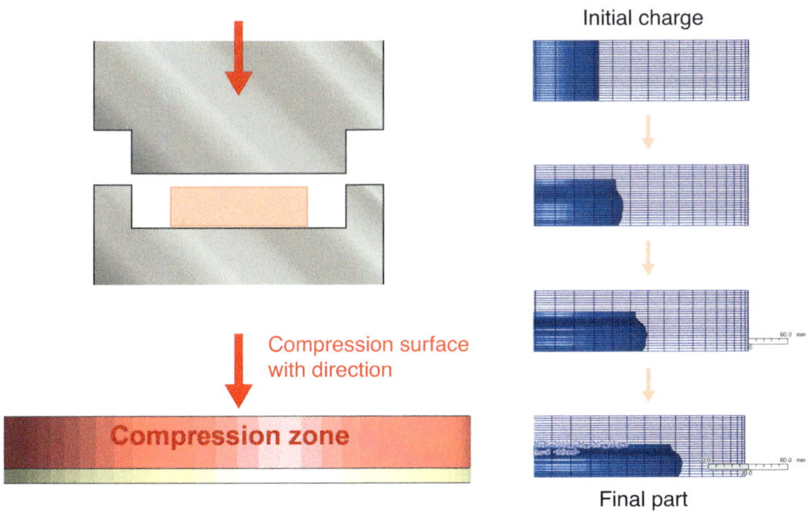

Figure 7.4 Illustration of the moving mesh boundary technique

The melt front moves forward along the compression direction. During closure of the mold, the resin moves radially toward the edges. Since the upper layer yields to the pressure first, after a few seconds of preheating and compression, the resin front moves faster than the bottom layer. Then, a stable plug-like flow is formed along the flowing direction to fill into the cavity region as in Figure 7.5.

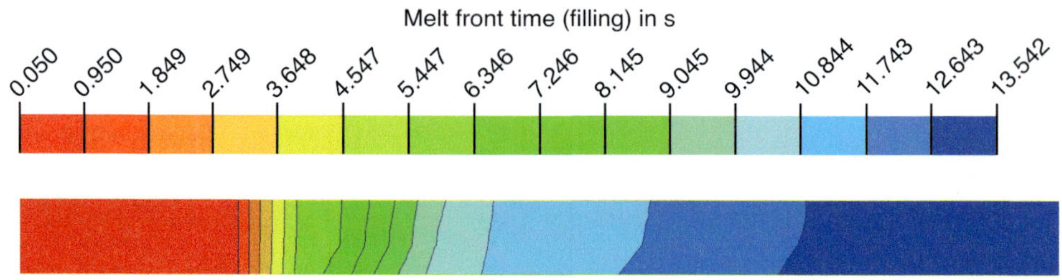

Figure 7.5 The behavior of squeezing flow during compression

Hence, the non-isothermal 3D flow motion can be mathematically described by the same governing equations described in the previous section. As in the compression molding process, the compressibility of polymer melt plays a dominant role in creating the primary and secondary penetration; the density of polymer melt is modeled through the two-domain modified Tait equation. The compressibility of the polymer is considered during the molding process. The two-domain modified Tait equation is then modeled as follows:

$$\rho(p,T) = \frac{1}{V_0(T)[1 - C\ln(1 + \frac{p}{B(T)})] + V_t(p,T)} \tag{7.11}$$

$$V_0(T) = \beta_1 + \beta_2(T - \beta_5),$$

$$B(T) = \beta_3 e^{-\beta_4(T-\beta_5)},$$

$$T_t(p) = \beta_5 + \beta_6 p,$$

$$V_t(p,T) = \begin{cases} \beta_7 e^{-\beta_8(T-\beta_5)-\beta_9 p} & ,T > T_t \\ 0 & T < T_t \end{cases}$$

where T_t denotes the transition temperature and C, β_1, β_2, β_3, β_4, β_5, β_6, β_7, β_8, and β_9 are parameters of material properties. Then, the similar discretization methodology described in the previous injection molding sections can be applied to solve the equations.

7.2.3 Discussion on Key Challenges in Simulation of Compression Molding

The viscosity is resistance to flow due to interaction between the polymer molecules and collision and friction of the suspended particles in the material. Thermoplastic polymers in general have longer polymer chains, so their viscosity is higher, and when fibers are added, the resistance to flow further increases. If the fibers are longer, the situation becomes worse. When carbon fibers that are stiffer are used, the viscosity is even higher. So, paying attention to viscosity becomes very important for any process, including compression molding.

The viscosity is defined as the ratio of shear stress to shear rate; see Figure 7.6.

$$\eta(T,\dot{\gamma}) = \tau_{xy} / \dot{\gamma}_{xy} \tag{7.12}$$

where τ_{xy} is shear stress, $\dot{\gamma}_{xy}$ is shear rate, and $\eta(T,\dot{\gamma})$ is viscosity. $\dot{\gamma}$ is the magnitude of the rate of deformation tensor. In general, for a given material the viscosity is a function of temperature and shear rate. The fluids for which the viscosity is only a function of temperature and does not change with the shear rate are called Newtonian fluids. The relation between the shear stress and shear rate for such fluids is linear and, therefore, the computations are simpler. When the viscosity changes as a function of shear rate as well as temperature, they are referred to as non-Newtonian fluids. Such fluids have a nonlinear relation between the shear stress and shear rate. As shown in Figure 7.6, some polymers show a tendency to thicken as shear rate increases and some show a tendency to become thinner and flow more easily. For fiber-reinforced polymers, as the fluid starts to flow the fibers and polymers tend to align along the flow, which results in reduced interaction between the fibers and, hence, reduced viscosity, similar to a shear thinning effect. Non-Newtonian computational models, such as the power-law model, Bird-Carreau-Yasuda model, modified Cross model, and Herschel-Bulkley model, are available to represent the relation between shear stress and shear rate [15]. Also, these models are implemented in mold filling software such as Moldex3D™—a CAE software package used for injection/compression molding of plastics—developed by CoreTech company [1]. Generalized Newtonian fluids (GNF) are a special case of non-Newtonian fluids, where the shear stress depends on the shear rate in a nonlinear fashion, but only for a particular point in time. The effect of time, such as stress relaxation, is not included in GNF models.

Earlier in Section 7.2.1.1, the theoretical background for process simulation is presented using the Generalized Newtonian Fluid (GNF) equations, which represent the nonlinear relation between the shear stress and shear rate for the fiber-reinforced thermoplastic melt material. For the injection molding process, a volume fraction function approach is used with the hyperbolic transport equation to track the flow front progress during mold filling. For the compression molding simulation, the material is expected to be compressible (Equation (7.13)) and highly viscous. Also, the flow front of the melt material moves differently compared to the injection molding, due to different boundary conditions. Therefore, a moving mesh boundary method, where the computational domain deforms and follows the filling stages, is used to track the flow in the mold cavity [16].

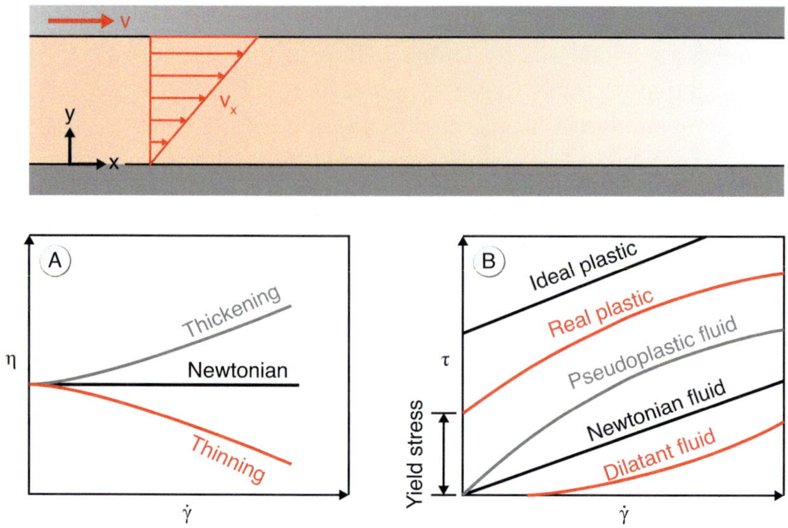

Figure 7.6 Viscosity, shear stress, and rate of deformation relationships

In addition, it is observed that for compression molding the bulk material charge, which was prepared by mixing the fibers and thermoplastic resin, is in a semi-solid state at temperatures above the melting temperature of the resin. When such bulk or sheet-shaped charge is compressed at a high pressure, the charge flows like a combination of deformable elastic continuum and highly viscous liquid, as shown in Figure 7.7. Depending on the shape of the mold cavity, the charge can undergo large deformations to fill the mold cavity. For the mold cavity shown in Figure 7.7(b) one sees that the bulk material moves more like a solid material to fully contact the cavity first, and then flows like fluid. These deformations can be mathematically expressed as a combination of solid material undergoing elastic deformation as well as viscous fluid flow. In other words, the viscoelastic behavior of the fiber-reinforced polymer parts under study must be included when simulating the deformation and stress relaxation in the compression molding process.

To capture the viscoelastic behavior, the material model is modified at two different stages: one at the liquid stage [17], the other one at the solid stage [18].

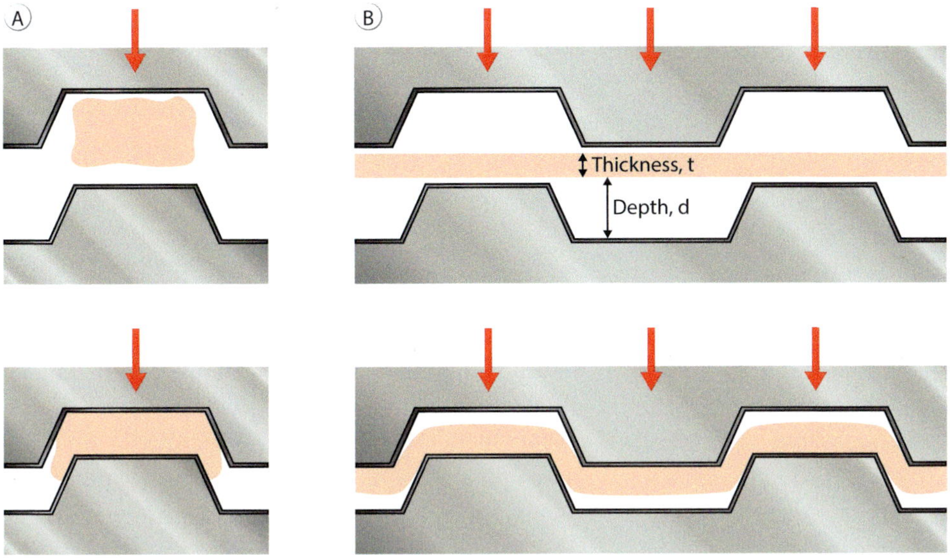

Figure 7.7 Importance of viscoelasticity in compression molding simulation. a) Bulk charge flows when compressed; GNF can describe the process. b) Bulk charge first deforms elastically and then flows; viscoelastic properties are needed to describe the process

A simple intuitive explanation of viscoelastic terms of differential form added to a liquid state is the co-rotational Maxwell model given by

$$\underline{\underline{\tau}} + \lambda_0 \underline{\underline{\tau}}_{(0)} + \lambda_1 \underline{\underline{\tau}}_{(1)} = \eta_0 \underline{\underline{\dot{\gamma}}} \tag{7.13}$$

where $\underline{\underline{\tau}}_{(0)}$ is the co-rotational or Jaumann derivative of the stress tensor defined by

$$\underline{\underline{\tau}}_{(0)} = \frac{\mathcal{D}}{\mathcal{D}t}\underline{\underline{\tau}} = \frac{D}{Dt}\underline{\underline{\tau}} + \frac{1}{2}\left\{\underline{\underline{\omega}} \cdot \underline{\underline{\tau}} - \underline{\underline{\tau}} \cdot \underline{\underline{\omega}}\right\}$$

and $D\underline{\underline{\tau}} / Dt$ is the substantial derivative and $\underline{\underline{\omega}}$ is the vorticity tensor given by

$$\underline{\underline{\omega}} = \nabla \underline{u} - \left(\nabla \underline{u}\right)^{\dagger}$$

where $\nabla \underline{u}$ is the velocity gradient

$$\nabla \underline{u} = \begin{pmatrix} \frac{\partial u_x}{\partial x} & \frac{\partial u_y}{\partial x} & \frac{\partial u_z}{\partial x} \\ \frac{\partial u_x}{\partial y} & \frac{\partial u_y}{\partial y} & \frac{\partial u_z}{\partial y} \\ \frac{\partial u_x}{\partial z} & \frac{\partial u_y}{\partial z} & \frac{\partial u_z}{\partial z} \end{pmatrix}$$

and $(\nabla \underline{u})^{\dagger}$ is the transpose of the velocity gradient.

The basic idea is to add the effect of time-dependent parameters in the calculation. Many different variations to the differential formulation in Equation (7.13), such as the Upper Convected Maxwell (UCM) model, the White-Metzner model, and the Giesekus model, to name a few, are available [19, 20]. Also, these formulations are implemented in the Moldex3D™ software as optional flow viscoelastic models [7] to help with the simulation.

For the solid stage, viscoelastic effects play a critical role in the warpage performance of the material. After the product is ejected from the mold, a free thermal shrinkage occurs during cooling. The equilibrium equation representing the stress is expressed with the divergence of the stress tensor as

$$\nabla \cdot \underline{\underline{\sigma}} = 0 \tag{7.14}$$

The relation between the stress and the strain is expressed as

$$\underline{\underline{\sigma}} = \mathrm{C}\left(\underline{\underline{\varepsilon}} - \alpha_{\mathrm{CLTE}} \underline{\underline{I}} \Delta T\right)$$

with

$$\underline{\underline{\varepsilon}} = \nabla \underline{X} + (\nabla \underline{X})^{\dagger} \tag{7.15}$$

where $\underline{\underline{\sigma}}$ is the stress, C is a fourth-order tensor and a function of the relaxation modulus $E(T_{ref}, t/a_T)$, α_{CLTE} is the coefficient of the linear thermal expansion tensor, $\underline{\underline{\varepsilon}}$ is the strain tensor, which is transferred by the volume shrinkage during molding, and $\underline{X}$ is the displacement vector, respectively. To consider the viscoelastic behavior after molding, a master function for the relaxation modulus is described by using the Generalized Maxwell model, often referred to as the Prony series, which is a function of time and temperature with the temperature shift factor a_{T} described as

$$a_T = 10^{\frac{-C_1(T-T_{ref})}{C_2+T-T_{ref}}}$$

where C_1 and C_2 are constants and T_{ref} is the reference temperature.

In the compression molding simulation, the non-Newtonian fluid models with the flow viscoelasticity terms are used to model mold cavity filling. Once the cavity is filled and the material is sufficiently cooled for demolding, then solid viscoelastic terms are used to calculate shrinkage and warpage during the cooling phase. As the material solidifies, depending on the fiber alignment and crystallization of the resin materials, the anisotropic properties start to emerge. The anisotropy, mainly due to variation in the fiber orientation, plays a very important role in the performance of the part, including the final shape due to warpage.

The formulation to include the viscoelastic properties discussed here is implemented in commercial mold filling software such as Moldex3D™. While the objective of this chapter is not to provide a full scale theoretical background, the focus is on providing insights into practical challenges in simulating compression molding and how such challenges can be addressed within current theoretical frameworks. It needs to be highlighted that the actual material properties involved are quite complex and the computational models used here are only a partial improvement to represent them correctly in computations. Further improvements are possible and active research in both industry and academia is consistently pushing the capabilities for the models.

In addition to including the viscoelastic properties, compression molding, with both the bulk and sheet charge, needs further considerations that are unique for each charge based on their geometry. These considerations are discussed in the next sections.

7.2.3.1 Key Challenges in Compression Molding Simulation Using a Bulk Charge

For the bulk charge, one of the key challenges that needs to be addressed while simulating the compression molding process is how to handle the initial fiber orientations in the charge. It is convenient to assume the fibers are distributed randomly or have a known uniform orientation, such as perfectly aligned in one direction. However, the physics of the plasticization process in the extruder leads us to conclude that a distinctive non-random periodic fiber orientation pattern exists [21]. This has been confirmed with the μCT scan study of the bulk charge; see Figure 7.8 [22, 23]. For the accurate prediction of fiber orientation in the finished part, it is important to start with the accurate description of the initial fiber orientations in the bulk charge. Therefore, in Chapter 9, Section 9.1, a method to define the initial orientation for the bulk charge based on actual measurement is developed and demonstrated with examples.

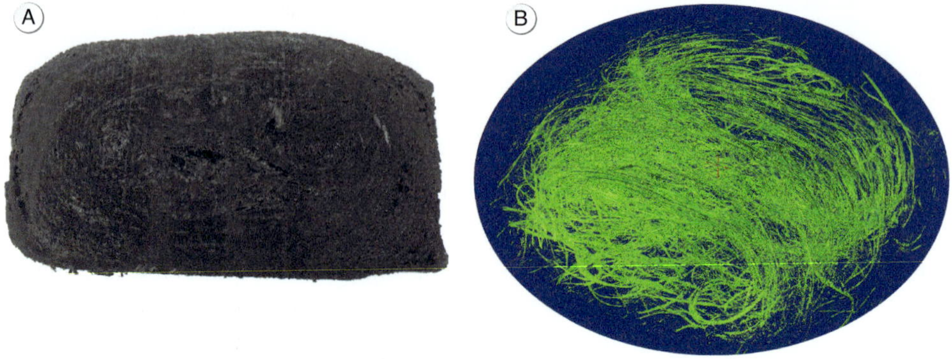

Figure 7.8 CT scan image of the charge material cross section shows fibers are not randomly distributed; A) charge cross section image, B) charge CT scan image

7.2.3.2 Key Challenges in Compression Molding Simulation Using Sheet Charge

The sheet material differs from the bulk material in the manner it is prepared and its initial shape. In most cases, it is reasonable to assume that the fibers in the sheet material are in-plane random, which means that defining the initial fiber condition is much simpler compared to the bulk material. However, the sheet material geometry is quite different, which brings challenges of its own in the simulation.

The key phases for compression molding of sheet material are presented in Figure 7.9. First the fiber-reinforced polymer sheets are heated and placed on the mold cavity. Next, the heated sheets are compressed in the mold cavity at very high pressure until solidification. Finally, the solidified finished parts are ejected from the mold and cooled. Depending on the process parameters, the ejected parts can warp during cooling due to differential shrinkage.

Since fiber-reinforced sheets are thin, when they are heated they lose much of their stiffness. As a result, when the heated sheets are placed on the mold cavity, the sheets tend to drape easily on the mold surface under small compressive force. This draping process involves significant rigid body motion of the heated sheets. The fluid mechanics-based approach used for compression molding of bulk material [24], even with the viscoelastic terms included, is not adequate to simulate the large rigid body motions occurring during the draping process.

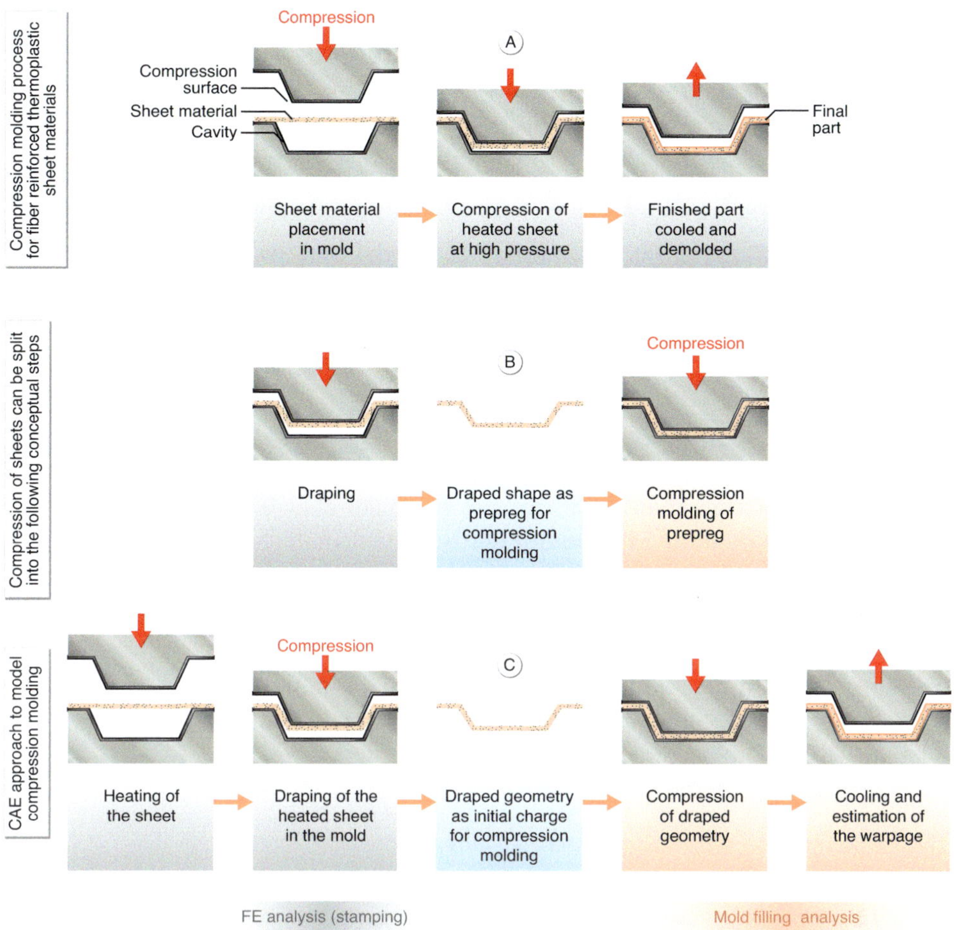

Figure 7.9 Key steps in applying CAE to the compression molding of sheet materials [9]

This challenge can be addressed in CAE by splitting the simulation of the compression of the heated sheet into three conceptual steps, as shown in Figure 7.9. First, the draping of the heated sheet is simulated using an explicit finite element approach. The explicit finite element approach, available in LS-Dyna–explicit FEA analysis software developed by Livermore Software Technology Corporation (LSTC)–or PamStamp–a software package from ESI Group, used for metal sheet stamping simulations–has been proven to be effective for large structural deformation as well as rigid body motion [25, 26]. Next, the draped shape of the fiber-reinforced thermoplastic sheet is transferred as a prepreg to the compression molding software. This step may look trivial, but it did require significant development effort, since one is transferring the geometry from one kind of software to the other. A translator that can map overall geometry as well as mapping of element type and shape is needed. Finally, the transferred prepreg is compressed at very high pressures. This is similar to the compression molding of a bulk charge, with the only difference that the shape of the bulk charge is the output from the draping analysis. The compression molding of the bulk charge is already developed and was discussed earlier.

Obtaining the material properties required for the simulation is a critical step but requires cumbersome testing. One of the major attractions in using the compression molding process is to use material with higher fiber length and the ability to retain fiber length (5–25 mm) in the finished part. The rheological properties for the material with such long fibers are not readily available. Also, the method to characterize such longer fiber length material is under development. In the work presented here an experimental approach being developed by Huang et al. [69] to estimate such properties was used.

The sheet material undergoing draping usually is heated above its melting temperature. Accurate material properties for such material that can be used for explicit finite element simulation are not readily available either. Also, since the material properties of interest are at high temperature, the measurement of structural properties for such material is quite challenging. For accurate simulations, attention to characterization of the materials to get the desired properties under operating temperature, which are high, is quite important.

The key steps used in the compression molding CAE are presented in Figure 7.9. Also, the key challenges during each step of CAE are presented in Table 7.1. In Chapter 9, Section 9.2, the approach to address these challenges is discussed, and examples that explain best practices are illustrated.

Table 7.1 Challenges for Compression Molding Simulation of Discontinuous Fiber-Reinforced Thermoplastic Sheet Materials

Steps (mechanism)	Details	Key challenges/Comments
1. Heat the sheets	Resin softens at high temperature. Fibers straighten/separate creating voids, etc.	Material properties needed for the draping of the sheets at high temperature are difficult to obtain.
2. Drape the heated sheets in the mold	The sheets drape on the mold with very little pressure. Material can fold, stretch, and wrinkle.	Explicit finite element approach can be used to model draping. Adaptive mesh enhancement may be needed to localize large deformation, e.g., edge of mold cavity.
3. Transfer the draped geometry to compression molding as initial charge	The draped shape and thickness are transferred to mold fill simulation software.	Special software for transferring the shape and material condition details from one software system to other software system is needed. The element type in each simulation system can be different.

4. Compress the draped geometry	Resin and fiber flow and fill the gaps in the mold. Prepreg shape and orientation are used as initial condition.	Viscoelastic flow is important to represent the material and fill the cavity. The material properties for the long fiber material are difficult to measure.
5. Cool and estimate warpage	The cooling can be different in different areas, which can cause warpage.	Temperature distribution is a function of time during cooling. Representative material properties including the viscoelastic effect are important.

7.3 Fiber Modeling

In processing discontinuous fiber-reinforced polymers, the three important microstructural variables are fiber orientation, fiber length, and fiber concentration. The following sections describe what predictive models are currently available to obtain the evolution of the fiber configuration during processing.

7.3.1 Fiber Orientation

Over the last three decades, theoretical researchers in the fiber suspension rheological field have made considerable effort to determine the dynamic fiber orientation states. The classic Folgar-Tucker Isotropic Rotary Diffusion (IRD) model is the pioneer work that is still widely applied to predict short fiber orientation. More recently, potential models have been developed in relation to concentrated suspension rheology with slow fiber orientation kinetics and anisotropic orientation distribution. Therefore, Tucker and co-workers further developed the RSC (Reduced Strain Closure) model [28] and the ARD (Anisotropic Rotary Diffusion) model [29] for short and long fiber composites, respectively. Tseng et al. also derived the iARD-RPR (improved ARD and Retarding Principal Rate) model [30].

7.3.1.1 Folgar-Tucker Model

Based on the Jeffery hydrodynamics of a rigid ellipsoidal particle immersed in a dilute viscous Newtonian fluid undergoing a homogenous flow, [31] used the probability density function of a fiber orientation vector, $\psi(\boldsymbol{p})$, to describe the motion for a group of fibers by the equation of continuity developed on the fiber orientation kinetic theory of fiber suspensions:

$$\frac{d\psi}{dt} = -\nabla_s \cdot (\dot{\boldsymbol{p}}\psi - \dot{\gamma} C_I \nabla_s \psi) \tag{7.16}$$

$$\dot{\boldsymbol{p}} = \boldsymbol{W} \cdot \boldsymbol{p} + \xi(\boldsymbol{D} \cdot \boldsymbol{p} - \boldsymbol{D} : \boldsymbol{ppp}) \tag{7.17}$$

where $\frac{d}{dt}$ and ∇_s are the material derivative operator and the surface spherical gradient operator, respectively; $\dot{\boldsymbol{p}}$ is the material derivative of vector **p** with respect to time. $\boldsymbol{L} = \nabla \boldsymbol{v} = \boldsymbol{W} + \boldsymbol{D}$ is the velocity gradient tensor with its component of $L_{ij} = \nabla_j v_i$, wherein v_i is the component of the velocity in the x_i direction; $\boldsymbol{W} = \frac{1}{2}(\boldsymbol{L} - \boldsymbol{L}^{\mathrm{T}})$ is the vorticity tensor; $\boldsymbol{D} = \frac{1}{2}(\boldsymbol{L} + \boldsymbol{L}^{\mathrm{T}})$ is the rate-of-deformation tensor and $\dot{\gamma} = (2\boldsymbol{D} : \boldsymbol{D})^{1/2}$ is the strain rate, namely, $\dot{\gamma} = (2\boldsymbol{D} : \boldsymbol{D})^{1/2}$; C_I is the fiber–fiber interaction parameter; $\xi = (a_r^2 - 1)/(a_r^2 + 1)$ is a shape factor of a particle; a fiber's aspect ratio a_r is the ratio of its length l to its diameter d, $a_r = l/d$. The factor ξ equals 1.0, 0.0, and −1.0, which indicates a thin rod, a spherical molecule, and a thin disc of a particle, respectively.

Using the orientation tensor **A**, Advani and Tucker simplified the Folgar-Tucker equation of continuity [see Eq. (7.16)] based on the probability density function of a fiber orientation vector, $\psi(\boldsymbol{p})$, as below:

$$\dot{\boldsymbol{A}} = \dot{\boldsymbol{A}}^{\mathrm{HD}} + \dot{\boldsymbol{A}}^{\mathrm{IRD}} \tag{7.18}$$

$$\dot{\boldsymbol{A}}^{\mathrm{HD}} = (\boldsymbol{W} \cdot \boldsymbol{A} - \boldsymbol{A} \cdot \boldsymbol{W}) + \xi(\boldsymbol{D} \cdot \boldsymbol{A} + \boldsymbol{A} \cdot \boldsymbol{D} - 2\boldsymbol{A}_4 : \boldsymbol{D}) \tag{7.19}$$

$$\dot{\boldsymbol{A}}^{\mathrm{IRD}} = 2\dot{\gamma} C_I (\boldsymbol{I} - 3\boldsymbol{A}) \tag{7.20}$$

where the time-evolution equation of the second-order orientation tensor is fixed on the material derivative, denoted as $\dot{\boldsymbol{A}}$, $\dot{\boldsymbol{A}}^{\mathrm{HD}}$ is the Jeffery hydrodynamic (HD) contributor, and $\dot{\boldsymbol{A}}^{\mathrm{IRD}}$ the isotropic rotary diffusion (IRD) contributor.

7.3.1.2 RSC Model and ARD-RSC Model

In the Folgar-Tucker model the fiber to fiber interaction is very simple; as a result, the predicted change in fiber orientation over thickness is quite faster compared to experimental observations. For this, Wang et al. [28] developed the RSC (Reduced Strain Closure) model to slow down such a response rate:

$$\begin{aligned}\dot{\boldsymbol{A}}^{\mathrm{RSC}} = {} & \boldsymbol{W} \cdot \boldsymbol{A} - \boldsymbol{A} \cdot \boldsymbol{W} + \xi\{\boldsymbol{D} \cdot \boldsymbol{A} + \boldsymbol{A} \cdot \boldsymbol{D} - 2[\boldsymbol{A}_4 + (1-\kappa)(\boldsymbol{L}_4 - \boldsymbol{M}_4 : \boldsymbol{A}_4)] : \boldsymbol{D}\} \\ & + 2\kappa \dot{\gamma} C_I (\boldsymbol{I} - 3\boldsymbol{A})\end{aligned} \tag{7.21}$$

$$\boldsymbol{L}_4 = \sum_{i=1}^{3} \lambda_i \, \boldsymbol{e}_i \boldsymbol{e}_i \boldsymbol{e}_i \boldsymbol{e}_i \tag{7.22}$$

$$M_4 = \sum_{i=1}^{3} e_i e_i e_i e_i \tag{7.23}$$

where the factor κ constrained between 0 and 1 is a slow-down factor fitted by experimental data, and generally suggested in a range of 0.05–0.2. When $\kappa = 1$, the RSC equation can recover to the standard Folgar-Tucker equation. λ_i and $\boldsymbol{e}_i$ are eigenvalues and eigenvectors of the second-order orientation tensor **A**, respectively; $\boldsymbol{L}_4$ and $\boldsymbol{M}_4$ are fourth-order orientation tensors depending on λ_i and $\boldsymbol{e}_i$.

Previously, Phan-Thien et al. [32] proposed the Folgar-Tucker isotropic rotary constant C_I extension to an anisotropic rotary (ARD) tensor $\boldsymbol{D}_r$. Thereby, Phelps and Tucker [29] constructed the ARD model coupled with $\dot{\boldsymbol{A}}^{\mathrm{HD}}$ for the time-evolution equation of the second-order orientation tensor,

$$\dot{\boldsymbol{A}} = \dot{\boldsymbol{A}}^{\mathrm{HD}} + \dot{\boldsymbol{A}}^{\mathrm{ARD}} \tag{7.24}$$

$$\dot{\boldsymbol{A}}^{\mathrm{ARD}} = \dot{\gamma}[2\boldsymbol{D}_r - 2\mathrm{tr}(\boldsymbol{D}_r)\boldsymbol{A} - 5\boldsymbol{D}_r \cdot \boldsymbol{A} - 5\boldsymbol{A} \cdot \boldsymbol{D}_r + 10\boldsymbol{A}_4 : \boldsymbol{D}_r] \tag{7.25}$$

Furthermore, the ARD model is coupled with the RSC model as ARD-RSC, as below:

$$\begin{aligned}\dot{\boldsymbol{A}}^{\mathrm{ARD\text{-}RSC}} &= \boldsymbol{W} \cdot \boldsymbol{A} - \boldsymbol{A} \cdot \boldsymbol{W} + \xi\{\boldsymbol{D} \cdot \boldsymbol{A} + \boldsymbol{A} \cdot \boldsymbol{D} - 2[\boldsymbol{A}_4 + (1-\kappa)(\boldsymbol{L}_4 - \boldsymbol{M}_4 : \boldsymbol{A}_4)] : \boldsymbol{D}\} \\ &+ \dot{\gamma}\{2[\boldsymbol{D}_r - (1-\kappa)\boldsymbol{M}_4 : \boldsymbol{D}_r] - 2\kappa\mathrm{tr}(\boldsymbol{D}_r)\boldsymbol{A} - 5(\boldsymbol{D}_r \cdot \boldsymbol{A} + \boldsymbol{A} \cdot \boldsymbol{D}_r) \\ &+ 10[\boldsymbol{A}_4 + (1-\kappa)(\boldsymbol{L}_4 - \boldsymbol{M}_4 : \boldsymbol{A}_4)] : \boldsymbol{D}_r\}\end{aligned} \tag{7.26}$$

For the anisotropic rotary tensor $\boldsymbol{D}_r$, Hand [33] used the theory of anisotropic fluids and developed a general second-order polynomial tensor-valued function form, which depends on the orientation tensor and the rate-of-deformation tensor with nine parameters $b_{1\text{-}9}$:

$$\begin{aligned}\boldsymbol{D}_r &= b_1\boldsymbol{I} + \frac{b_2}{\dot{\gamma}}\boldsymbol{A} + \frac{b_3}{\dot{\gamma}^2}\boldsymbol{A}^2 + \frac{b_4}{\dot{\gamma}}\boldsymbol{D} + \frac{b_5}{\dot{\gamma}^2}\boldsymbol{D}^2 + \frac{b_6}{\dot{\gamma}}(\boldsymbol{D} \cdot \boldsymbol{A} + \boldsymbol{A} \cdot \boldsymbol{D}) \\ &+ \frac{b_7}{\dot{\gamma}}(\boldsymbol{D} \cdot \boldsymbol{A}^2 + \boldsymbol{A}^2 \cdot \boldsymbol{D}) + \frac{b_8}{\dot{\gamma}^2}(\boldsymbol{D}^2 \cdot \boldsymbol{A} + \boldsymbol{A} \cdot \boldsymbol{D}^2) + \frac{b_9}{\dot{\gamma}^2}(\boldsymbol{D}^2 \cdot \boldsymbol{A}^2 + \boldsymbol{A}^2 \cdot \boldsymbol{D}^2)\end{aligned} \tag{7.27}$$

In simplifying, Phelps and Tucker [29] constrain the b_i coefficients to be scalar constants and set $b_{6-9} = 0$,

$$\boldsymbol{D}_r = b_1\boldsymbol{I} + b_2\boldsymbol{A} + b_3\boldsymbol{A}^2 + \frac{b_4}{\dot{\gamma}}\boldsymbol{D} + \frac{b_5}{\dot{\gamma}^2}\boldsymbol{D}^2 \tag{7.28}$$

The ARD-RSC model contains a set of six parameters (κ and b_1–b_5), which can be measured/computed. In principle, the target orientation state is chosen, in which the injection filling flow in an edge-gated plaque is representative of a steady, simple shear flow; therefore, the identification procedure uses the average fiber orientation components in the shell layers near the gate region to determine the ARD-RSC parameters. The RSC factor κ constrained between 0 and 1 is a factor for controlling the core-width or slowing down the orientation.

7.3.1.3 iARD-RPR Model

Following the previous studies of fiber suspensions, Tseng et al. [30] improved the ARD tensor $\boldsymbol{D}_r$, depending on the identity matrix **I** and the non-objective velocity gradient tensor $\boldsymbol{L}$ with two parameters:

$$\boldsymbol{D}_r = C_I(\boldsymbol{I} - C_M \frac{\boldsymbol{L}^{\mathrm{T}} \cdot \boldsymbol{L}}{\boldsymbol{L}^{\mathrm{T}} : \boldsymbol{L}}) \tag{7.29}$$

where the tensor $\boldsymbol{L}$ is non-objective in the constitutive relation that might yield different answers at different coordinates [34]. Such a tensor $\boldsymbol{D}_r$ given to $\dot{\boldsymbol{A}}^{\mathrm{ARD}}$ [see Eq. (7.25)] is called the "iARD" model, $\dot{\boldsymbol{A}}^{\mathrm{iARD}}$. When $C_M = 0$ is given, the iARD model becomes the Folgar-Tucker IRD model. However, the iARD tensor does not pass the classic rheological rule of Euclidean objectivity, namely, material frame indifference. It is hard to ignore the non-objective effect due to the fact that different coordinate systems can yield different answers.

More recently, Tseng et al. [35] have found a new objective $\boldsymbol{D}_r$ tensor, which is related to the square of the rate-of-deformation tensor **D**,

$$\boldsymbol{D}_r = C_I(\boldsymbol{I} - C_M \frac{\boldsymbol{D}^2}{\left\|\boldsymbol{D}^2\right\|}) \tag{7.30}$$

where the scalar $\left\|\boldsymbol{D}^2\right\| = \sqrt{\frac{1}{2}\boldsymbol{D}^2 : \boldsymbol{D}^2}$ is the norm of tensor $\boldsymbol{D}^2$. It has been proven that **D** is indeed objective in the classic rheological rule [36].

To slow down a response rate of fiber orientation kinetic, Tseng et al. [30] also developed the RPR (Retarding Principal Rate) model:

$$\dot{\boldsymbol{A}}^{\mathrm{RPR}} = -\boldsymbol{R} \cdot \dot{\boldsymbol{\Lambda}}^{\mathrm{IOK}} \cdot \boldsymbol{R}^{\mathrm{T}} \tag{7.31}$$

$$\dot{\Lambda}_{ii}^{\mathrm{IOK}} = \alpha \dot{\lambda}_i \quad i,j,k = 1,2,3 \tag{7.32}$$

where $\dot{\boldsymbol{\Lambda}}^{\mathrm{IOK}}$ is the material derivative of a particular diagonal tensor and its superscript indicates the intrinsic orientation kinetics (IOK) assumption, $\boldsymbol{R}$ is the rotation matrix, and $\boldsymbol{R}^{\mathrm{T}}$ is the transpose of $\boldsymbol{R}$. This rotation matrix is also an orthogonal matrix, $\boldsymbol{R} \cdot \boldsymbol{R}^{\mathrm{T}} = \boldsymbol{R}^{\mathrm{T}} \cdot \boldsymbol{R} = \boldsymbol{I}$, where **I** is the identity matrix. The superscript T is the transpose operator of a matrix throughout this section; λ_i are the eigenvalues of $\boldsymbol{A}$, $\lambda_1 \geq \lambda_2 \geq \lambda_3$; and $\boldsymbol{R} = [\boldsymbol{e}_1, \boldsymbol{e}_2, \boldsymbol{e}_3]$ is defined by eigenvector columns of $\boldsymbol{A}$. The parameter α is in dimensionless units, and the subscripts i, j, and k are indices of permutation.

Therefore, the iARD-RPR model coupled with $\dot{\boldsymbol{A}}^{\mathrm{HD}}$ for the time-evolution equation of the second-order orientation tensor, involving Eqs. (7.16), (7.22), and (7.27)-(7.29),

$$\dot{\boldsymbol{A}} = \dot{\boldsymbol{A}}^{\mathrm{HD}} + \dot{\boldsymbol{A}}^{\mathrm{iARD}}(C_I, C_M) + \dot{\boldsymbol{A}}^{\mathrm{RPR}}(\alpha) \tag{7.33}$$

where $\dot{\boldsymbol{A}}^{\mathrm{iARD}}$ has two available parameters: the fiber–fiber interaction parameter C_{I} (or Folgar-Tucker coefficient) and the fiber–matrix interaction parameter C_{M}; $\dot{\boldsymbol{A}}^{\mathrm{RPR}}$ has one parameter α, which is meant to slow down the response rate of the fiber orientation.

In summary, the iARD-RPR model has only the three physical parameters: a fiber–fiber interaction parameter C_{I}, a fiber–matrix interaction parameter C_{M}, and a slow-down parameter α. The available region of these parameters, $0 < C_{\mathrm{I}} < 0.1$, $0 < C_{\mathrm{M}} < 1$, and $0 < \alpha < 1$, is originated from the theoretical development of the iARD-RPR model [30, 35]. These authors also provided the default parameters obtained from experiments: the short fiber parameters, $C_{\mathrm{I}} = 0.005$, $C_{\mathrm{M}} = 0.1$, $\alpha = 0.7$, are based on the 30 wt % short glass fiber/PBT (polybutylene-terephthalate) [37]; for the long fiber parameters of the 40 wt % long glass fiber/PP (polypropylene) [35], $C_{\mathrm{I}} = 0.01$, $C_{\mathrm{M}} = 1.0$, $\alpha = 0.1$. In addition, the shell-orientation layer depends on the iARD parameters, while the core width is controlled by the RPR parameter. The value of fiber orientation in the shell increases with increasing C_{I} and decreasing C_{M}. The core width increases with increasing α. According to these criteria, three parameters are fine-tuned to obtain an accurate orientation prediction compared with experimental data, namely, determining an optimal parameters set.

7.3.2 Fiber Breakage

As described in Chapter 4, the fibers are exposed to substantial bending, friction, and buckling forces that cause fiber breakage. The initial fiber length of LFT pellets is as long as the pellet length and a common length is 10 to 15 mm. However, few fibers with this length survive the process and the vast majority of fibers break during any stage of the molding process. In addition, the fiber length distribution is not uniform in the finished parts and simulations suggest that the average fiber length in the core region is larger than those of the outer layers due to high shear rates in the shell layers.

As described in Chapter 4, the common descriptions of fiber length values are the number average, L_{N}, and the weight average, L_{W}, which represent the full fiber length distributions of the fiber population.

Phelps et al. [38] pioneered the fiber length attrition model, which is capable of accurately simulating a fiber breakage history during injection mold processing.

The fiber length attrition model defines the change rate of the number of fibers with length l_i:

$$\dot{N}_i = -P_i N_i + \sum_{k=1}^{N} R_{ik} N_k \tag{7.34}$$

$$R_{ik} = N_{norm}(l_i, \frac{l_k}{2}, Sl_k) \tag{7.35}$$

$$R_{ki} = R_{(i-k)i} \tag{7.36}$$

$$\sum_{k=1}^{N} R_{ji} = 2P_i \tag{7.37}$$

where a time-evolution equation $\dot{N}_i$ is fixed on the material derivative; P_i is a breakage probability for parent fibers of length l_i at time Δt; and R_{ik} is the rate of production of child fibers of length l_i by breaking parent fibers of length l_k, thus with the mass conservation law. When the number of fibers N_i is obtained, the number average length L_N and the weight average length L_W based on Eq. (7.31) and Eq. (7.32) can be calculated, respectively.

R_{ik} is given by the normal distribution function N_{PDF}:

$$N_{PDF}(x, \mu, \sigma) = \frac{1}{\sqrt{2\pi\sigma^2}} e^{-\frac{(x-\mu)^2}{2\sigma^2}} \tag{7.38}$$

where x is a variable, μ is the mean, and σ is the variance; the variable S is a dimensionless fitting parameter that controls the shape of the Gaussian breakage profile. A small value of S corresponds to a high probability of breakage occurring near the fiber center, while a large S value distributes the probable breakage points more evenly along the fiber length.

Furthermore, the probability of fiber breakage P_i can be derived by the Dinh-Armstrong hydrodynamic compression force and the Euler critical buckling force [38], as below:

$$P_i = \begin{cases} 0 & B_i < 1 \\ C_B \dot{\gamma}[1 - \exp(1 - B_i)] & B_i \geq 1 \end{cases} \tag{7.39}$$

$$B_i = \frac{\dot{\gamma}}{\dot{\gamma}_{crit,i}} \tag{7.40}$$

$$\dot{\gamma}_{crit,i} = \frac{\pi^3 E_f d_f^4}{64\zeta\eta_m l_i^4} \tag{7.41}$$

where B_i is the ratio between the actual shear rate $\dot{\gamma}$ and the critical shear rate $\dot{\gamma}_{crit,i}$; C_B is a constant used to scale the breakage rate to control the time required to reach that state. S influences the exact shape of the fiber length distribution; ζ is a dimensionless drag coefficient to control the location of the peak in the near-steady-state fiber length distribution; η_m is the matrix viscosity; d_f is the fiber diameter; and E_f is the fiber modulus. All details of the fiber length attrition model are available elsewhere [38].

7.3.3 Fiber Concentration

The fiber concentration in the composite is a main property that determines the mechanical performance of the molded part. As discussed in Chapter 4, the concentration can be heterogeneous due to the processing conditions. Fundamentally, the volume fraction φ_V and weigh fraction φ_W are given as

$$\varphi_V = \frac{V_F}{V_T} \tag{7.42}$$

$$\varphi_W = \frac{W_F}{W_T} \tag{7.43}$$

where V_F and W_F are the total of fiber volume and fiber weight; V_T and W_T are the sample volume and weight.

Short fibers easily result in rotary diffusion, and quickly align along the flow direction. Some experimental data [39] indicated that the steady-state concentration profile of fibers in the wide-gap Couette flow was relatively independent of the fiber aspect ratio, and was close to that of spheres. In addition, a clear and systematic variation in fiber concentration within their molded parts was observed: the fiber concentration in the core region is obviously higher than in the shell layers [40]. Along flow length, there are lower weight fractions near the gate and higher weight fractions at the far end of the cavity. Because of a non-uniform fiber concentration distribution, such a phenomenon is called the *matrix-fiber separation* or *fiber migration*. In practice, the flow-induced fiber migration corresponds to the gradient of the fiber concentration in the mold filling [41].

In a first approach, short fibers may be simplified as powders or particles for modeling purposes. Description of the flow in a two-phase suspension requires mass and momentum balances for the bulk suspension and one of the components.

Following previous studies, Nott and Brady [42] and Morris and Boulay [43, 44] developed the suspension balance model, which uses the balance equations of the bulk mixture and particle phase to develop the suspension balance model for the suspensions flow of rigid spherical particles in a Newtonian fluid. The particle phase is approximated as a pseudo-continuum. The dominant interaction between the particles involves hydrodynamics, viscous, and non-Brownian forces without external field, except for gravity. Therefore, the suspension balance model is introduced below.

The conservation equation of particle-phase mass balance [43] may be expressed as:

$$\frac{\partial \varphi}{\partial t}+\langle \boldsymbol{u}\rangle \cdot \nabla \varphi=-\nabla \cdot \boldsymbol{j}_{\perp} \tag{7.44}$$

$$\boldsymbol{j}_{\perp}=\varphi(U_p-\langle \boldsymbol{u}\rangle) \tag{7.45}$$

where φ is the particle volume fraction, $\langle \boldsymbol{u}\rangle$ is the suspension average velocity, $\boldsymbol{j}_{\perp}$ is the particle flux relative to the mean suspension motion and the subscript $\perp$ is used to emphasize the usual focus on the cross-stream migration, and U_p is the local average velocity of the particulate phase.

In assuming the Stoke flow, the particle flux $\boldsymbol{j}_{\perp}$ [43] may be obtained from the particle momentum balance over the particle phase, and restricting the attention to low Reynolds numbers and the neutrally buoyant condition, it is given by:

$$\boldsymbol{j}_{\perp}=\frac{2a^2}{9\eta_f} f(\varphi)\nabla \cdot \Sigma_p \tag{7.46}$$

where a is the particle radius, η_f is the viscosity of the suspending fluid, $f(\varphi)$ is termed the sedimentation function, and Σ_p is the particle stress.

The sedimentation function is described in Richardson and Zaki [45]:

$$f(\varphi)=(1-\varphi/\varphi_m)(1-\varphi)^{n-1} \tag{7.47}$$

where φ_m is the maximum volume fraction and $\varphi_m=0.63$ for the random close packing monodisperse spheres; the power index, n, is given as $n=2-4$.

The particle stress tensor Σ_p is a symmetric matrix containing both shear and normal stress portions, as suggested by Morris and Boulay [43, 44] for shear flows:

$$\boldsymbol{\Sigma}_p=\begin{bmatrix} \Sigma_p^{11} & \Sigma_p^{12} & \Sigma_p^{13} \\ \Sigma_p^{12} & \Sigma_p^{22} & \Sigma_p^{23} \\ \Sigma_p^{13} & \Sigma_p^{23} & \Sigma_p^{33} \end{bmatrix}=2\eta_f\eta_p(\varphi)\boldsymbol{E}-\eta_f\eta_n(\varphi)\dot{\gamma}\,\boldsymbol{Q} \tag{7.48}$$

where η_p and η_n are the particle stress contributors to the shear-stress viscosity and the normal-stress viscosity made dimensionless with η_f, respectively. **E** is the rate-of-strain tensor and $\dot{\gamma} = \sqrt{2\boldsymbol{E} : \boldsymbol{E}}$ is the magnitude of rate of strain; $\boldsymbol{Q}$ is an anisotropic tensor.

Both η_p and η_n are given in dimensionless volume fraction $\overline{\varphi} = \varphi / \varphi_m$,

$$\eta_p(\varphi) = 2.5\varphi_m(1-\overline{\varphi})^{-1} + K_s\overline{\varphi}^2(1-\overline{\varphi})^{-2} \tag{7.49}$$

$$\eta_n(\varphi) = \quad K_n\overline{\varphi}^2(1-\overline{\varphi})^{-2} \tag{7.50}$$

where two factors, K_s = 0.1 and K_n = 0.75, are fit to match the experimental data of the wide-gap Couette flow [43]. Note that the relative viscosity is defined as:

$$\eta_s(\varphi) = 1 + \eta_p(\varphi) \tag{7.51}$$

The anisotropic tensor $\boldsymbol{Q}$ describes the anisotropy of the normal stresses [43]:

$$\boldsymbol{Q} = \begin{bmatrix} \lambda_1 & 0 & 0 \\ 0 & \lambda_2 & 0 \\ 0 & 0 & \lambda_3 \end{bmatrix} \tag{7.52}$$

where the tensor $\boldsymbol{Q}$ is a diagonal matrix and its components: $\lambda_1 = 1$, $\lambda_2 = 0.8$, and $\lambda_3 = 0.5$ provide reasonably good agreement with concentrated suspension rheology and with observed migration behavior in viscometric flows [43]. The directions of the tensor **Q** correspond to the principal directions of a viscometric shear flow, as indexes 1, 2, and 3 denote flow, gradient, and vorticity, respectively.

Physically, the particle stress tensor of Eq. (7.47) implies that all normal stresses are negative, and the shear flow results in a positive suspension pressure. It is clear that the anisotropic particle stress tensor's three diagonal components are given in terms of this modeling by:

$$\Sigma_p^{11} = -\eta_f\eta_n(\varphi)\dot{\gamma}\lambda_1 \tag{7.53}$$

$$\Sigma_p^{22} = -\eta_f\eta_n(\varphi)\dot{\gamma}\lambda_2 \tag{7.54}$$

$$\Sigma_p^{33} = -\eta_f\eta_n(\varphi)\dot{\gamma}\lambda_3 \tag{7.55}$$

Thus, the first and second normal stress differences (N_1 and N_2) and suspension pressure (Π) are obtained:

$$N_1 = \Sigma_p^{11} - \Sigma_p^{22} = -\eta_f\eta_n(\varphi)\dot{\gamma}(\lambda_1 - \lambda_2) \tag{7.56}$$

$$N_2 = \Sigma_p^{22} - \Sigma_p^{33} = -\eta_f\eta_n(\varphi)\dot{\gamma}(\lambda_2 - \lambda_3) \tag{7.57}$$

$$\Pi = -\frac{\Sigma_p^{11} + \Sigma_p^{22} + \Sigma_p^{33}}{3} = \eta_f\eta_n(\varphi)\dot{\gamma}\left(\frac{\lambda_1 + \lambda_2 + \lambda_3}{3}\right) \tag{7.58}$$

7.4 Fiber Configuration Prediction

7.4.1 Fiber Orientation and Fiber Breakage

In this section, a case study is used to show the application of the iARD model for fiber orientation prediction and the Phelps model for fiber breakage prediction for a simple injection molded disk geometry [46]. The material of interest was a fiber-reinforced thermoplastic composite of polypropylene (PP) resin containing 40% by weight of long glass fibers (40 wt % LGF/PP). Based on Nguyen et al. [47–49] and Phelps et al. [38, 50], the center-gated disk of mold filling is illustrated in Figure 7.10. Here, using cylindrical coordinates, the radial distance r describes the flow or radial direction, the azimuth θ is the cross-flow or tangential direction, and the z-axis is the through-thickness direction. The data on the fiber orientation and fiber length through the thickness were measured at three regions: the near-entry, the center, and the near-end-of-filling along the radial direction of the disk, marked as regions A, B, and C, respectively. The disk is 177.8 mm in diameter with a thickness of $2h$ = 3 mm. Regions A, B, and C are located at 6, 34, and 64 mm in the radial direction, respectively. The operating conditions of the disk-mold filling are shown in Table 7.2.

The density of the 40 wt % LGF/PP composites was reported as 1.22 g/cm^3 with a nominal fiber weight fraction of 40%. The composite material, MTI-PP40G, was provided by Montsinger Technologies, Inc. The average glass fiber diameter is about 17 μm. The average fiber length is about 13 mm. The post-extruded average fiber length is about 3 mm. Fibers longer than 1 mm are generally considered long fibers. The useful material properties, including rheological and thermal properties, are explained elsewhere [50].

Here, Moldex3D was utilized to perform an injection molding simulation in conjunction with the iARD-RPR fiber orientation model [46]. The iARD-RPR model parameters are listed in Table 7.3. In addition, the predicted and measured orientation tensor components are compared, in dealing with long fibers flowing in an injection-mold filling process of the center-gated disk. The objective of this work was to examine the difference between the ARD-RSC model [49, 50] and the iARD-RPR model at the same level of using the 3D numerical computation technology; the results using both models were compared with the experimental data.

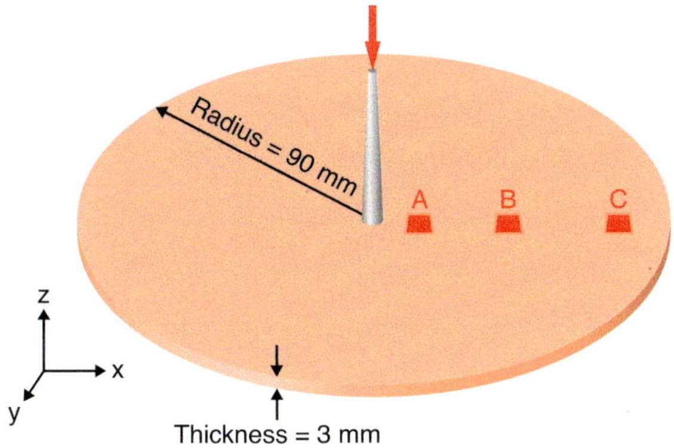

Figure 7.10 Geometry of center-gated disk with three regions, A, B, and C, of measuring fiber orientation

Table 7.2 Processing Conditions for the 40 wt % LGF/PP Composite during the Injection Mold Filling

Processing conditions	Value
Fill time (s)	1.58
Fill rate (cm^3/s)	131
Mold Temperature (°C)	85
Melt Temperature (°C)	270

Table 7.3 Optimal Parameters of the iARD-RPR Fiber Orientation Model for the 40 wt % LGF/PP Composite during the Injection Mold Filling

Model parameters	Value
C_I	0.01
C_M	1.0
α	0.1

Figure 7.11, Figure 7.12, and Figure 7.13 show these predictions of orientation distributions through the normalized thickness obtained from regions A, B, and C, respectively, marked in the disk; the available data in the figures, including the experimental points [49] (see square symbols), the iARD-RPR model curves (see solid lines), and the ARD-RSC curves (see dashed lines) are referred to in the PNNL report [49].

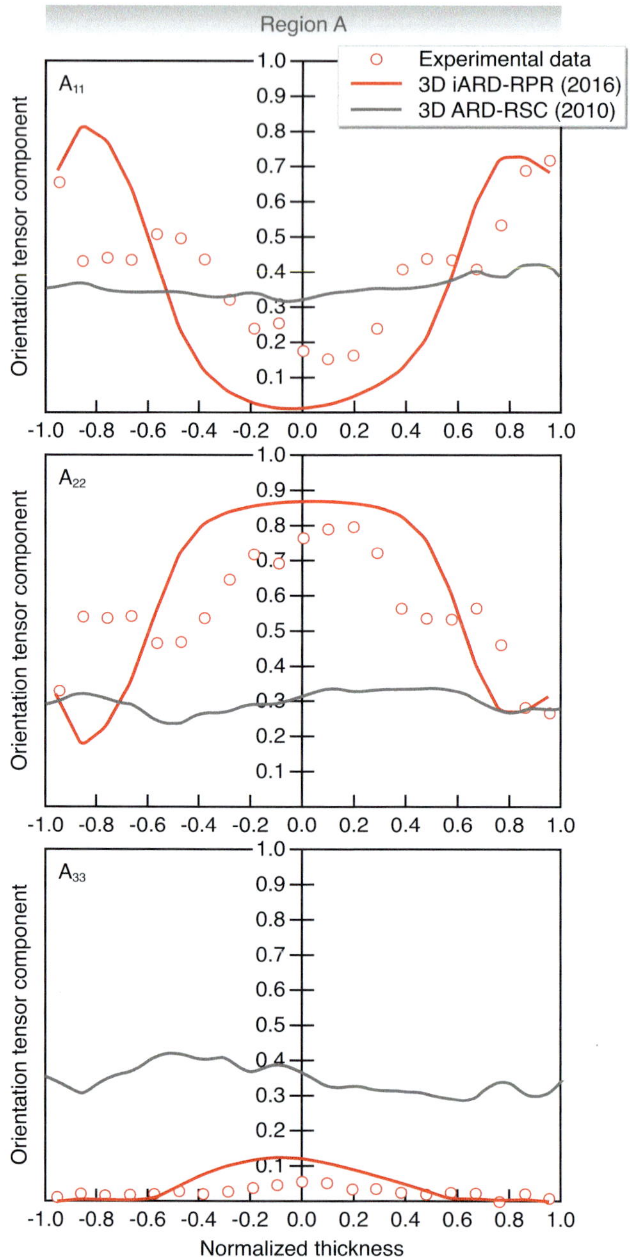

Figure 7.11 The 3D orientation predictions using the iARD-RPR model were compared with the previous 3D orientation predictions using the ARD-RSC model [49] and experimental data [49] for three diagonal components of the orientation tensor, (a) A_{11}, (b) A_{22}, and (c) A_{33}, through the normalized thickness at **Regions A** of the center-gated disk

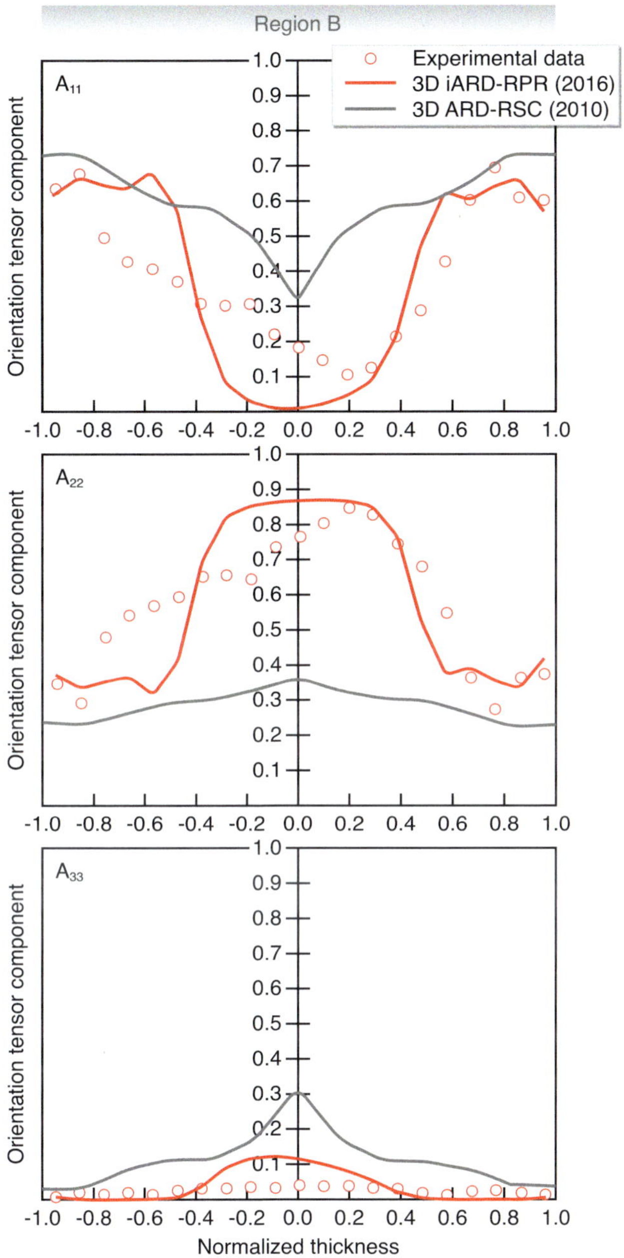

Figure 7.12 The 3D orientation predictions using the iARD-RPR model were compared with the previous 3D orientation predictions using the ARD-RSC model [49] and experimental data [49] for three diagonal components of the orientation tensor, (a) A_{11}, (b) A_{22}, and (c) A_{33}, through the normalized thickness at **Regions B** of the center-gated disk

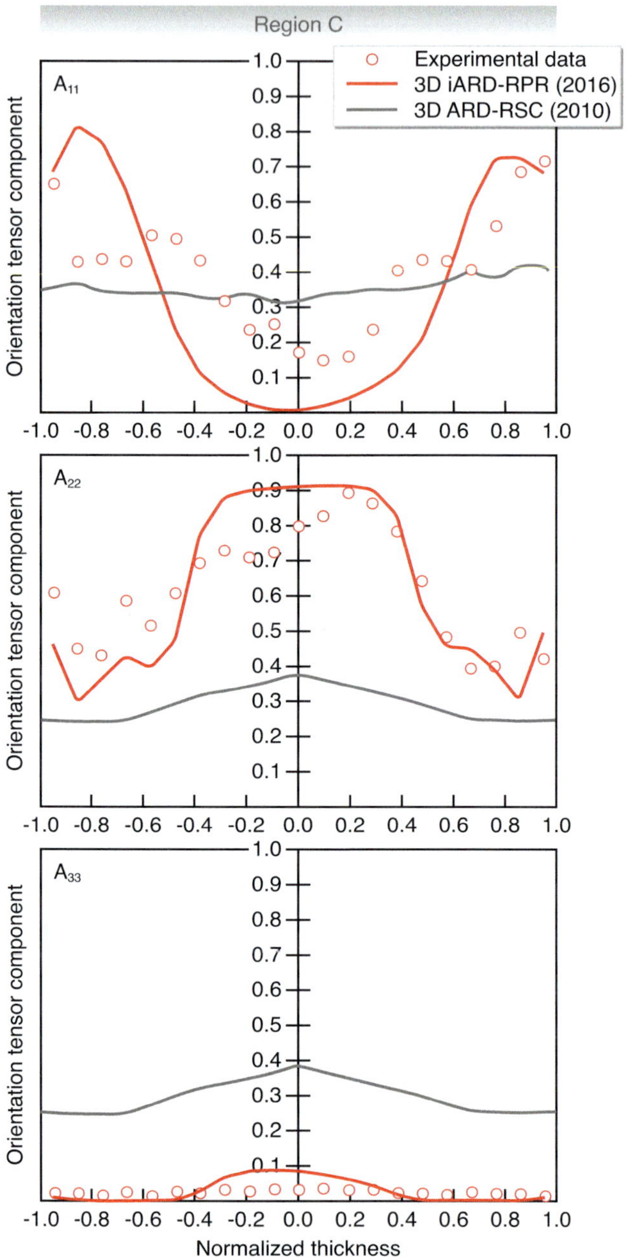

Figure 7.13 The 3D orientation predictions using the iARD-RPR model were compared with the previous 3D orientation predictions using the ARD-RSC model [49] and experimental data [49] for three diagonal components of the orientation tensor, (a) A_{11}, (b) A_{22}, and (c) A_{33}, through the normalized thickness at **Regions C** of the center-gated disk

Using the iARD-RPR model, the predicted orientation curves nearly pass the experimental points in all the studied regions. Clearly, such an orientation distribution exhibits the shell-core structure; in particular, there is a wide core in the orientation pattern. Basically, the plurality of fibers found in the shell region are mainly aligned in the flow direction due to the high shearing near the mold surface, while others at the core are transverse to the flow due to low shearing.

Furthermore, the descriptors of the orientation tensor components are essential: A_{11}, A_{22}, and A_{33} correspond to the flow, cross-flow, and thickness directions, respectively. In general, a high value of A_{11} would indicate a great deal of fibers lying in the flow direction, whereas a near-zero value of A_{33} would indicate little or no orientation in the thickness direction. As shown in Region A, the predicted orientation distributions by the ARD-RSC model present an almost random core, and do not match the transversely aligned core as observed in the experiments. With a continued comparison in other regions, both components A_{11} and A_{33} are over-predicted, whereas component A_{22} is under-predicted. It appears that the predictions do not agree with the results of the experiments. Obviously, the 3D predictions from using the iARD-RPR are superior to the previous results of using the ARD-RSC (2010). Referring to the PNNL report [37], this deviation of the predictions was attributed to the nature of the RSC model. In addition, this report explained that "the inlet condition of the orientation tensor" and "the calculation accuracy of the velocity gradient tensor" strongly influenced the fiber orientation predictions.

The attrition/breakage history of fiber length or fiber aspect ratio also strongly influences the mechanical properties of fiber-reinforced composite parts. The Phelps-Tucker fiber length attrition model [38] has been incorporated into commercial software for injection molding simulation, such as Moldex3D. Following the same injection molding system from the last section, Moldex3D was used to predict the fiber length distribution during the injection mold filling. The attrition parameters (C_D, C_B, and S) are listed in Table 7.4.

Table 7.4 Optimal Parameters of the Phelps-Tucker Fiber Breakage Model for the 40 wt % LGF/PP Composite during the Injection Mold Filling

Model parameters	Value
C_D	5
C_B	0.0088
S	1

The initial fiber length was 13 mm. Before flowing into a runner entrance, the initial fiber length distribution was given by the Gaussian distribution with the weight-average fiber length L_w = 2.89 mm, as shown in Figure 7.14. Bailey and Kraft [51] performed a number of experimental studies to examine changes in fiber length for LFRT materials. They found that significant attrition occurs before the polymer melt enters the mold: fiber breakage as a function of process parameters such as screw back pressure, injection speed, gate size, and nozzle design. In addition, they found the mean fiber lengths in the inner core to be three times those of the skin layer, a phenomenon attributed to the high shear rates in the skin during mold filling.

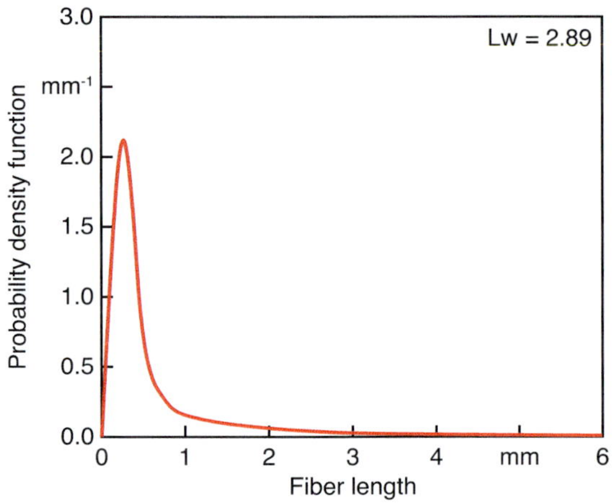

Figure 7.14 Initial condition of fiber length probability density distribution at the melt entrance

Figure 7.15 shows the weight-average fiber length L_w distribution through the normalized thickness obtained from the three regions, A, B, and C, marked in the disk. The longer fiber was clearly found in the core (the center of the thickness) due to a low-shear-rate flow field, whereas the shorter fiber was near the shell with high shear rates. This finding agrees well with the earlier study of Bailey and Kraft [51].

Moreover, Figure 7.16 shows that the change in fiber length was not dramatic with respect to different regions. The predicted L_w values were close to the experimental data [38]. Note that Phelps et al. [38] attributed the result of the fountain flow effect to an increment of the average lengths near the end of the flow path. In the following, the primary effect of fiber length on structural analysis of mechanical testing is discussed.

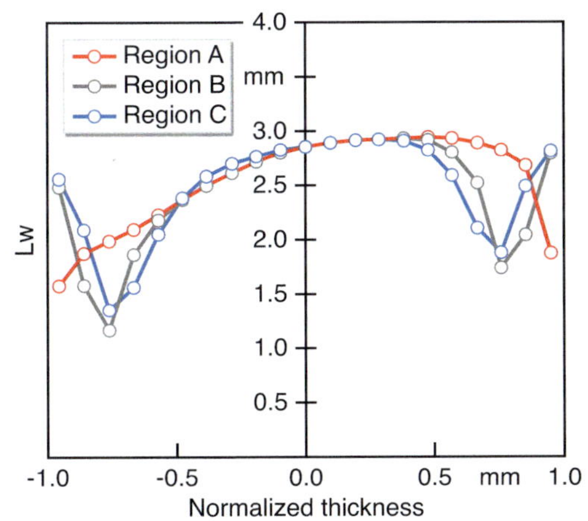

Figure 7.15 Weight-average fiber length probability density distribution

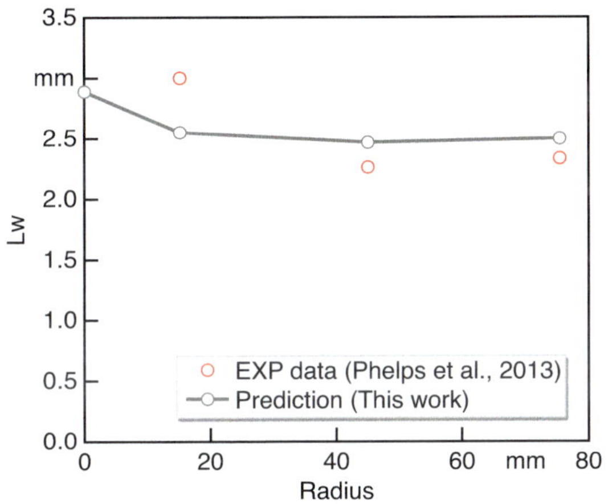

Figure 7.16 Weight-average fiber length along the radius of the disk

Because the fiber orientation can vary strongly with respect to the part's thickness, it is not meaningful to make point-by-point comparisons between the predicted and experimental data. For an overall assessment of the prediction's accuracy as defined in the PNNL-ORNL report [Pacific Northwest National Laboratory (PNNL) and the Oak Ridge National Laboratory (ORNL) in the USA], the average values of the predicted flow-orientation tensor component A_{11} at selected regions were compared to the corresponding experimental data based on the 15% relative-error criterion [52]. Consequently, the deviations of the predicted A_{11} component and the weight-average fiber length L_w are arranged in Table 7.5 and Table 7.6, respectively. The overall deviation is less than the 15% deviation of the PNNL-ORNL quality criterion. The predictions are satisfactory.

Table 7.5 The Predictive Flow Orientation Component A_{11} Using the iARD-RPR Model at Regions A, B, and C, as Compared with the Experimental Values for the 40 wt % LGF/PP Composite

	Experimental value	Predictive value	Deviation (%)
A	0.408	0.351	-14.1
B	0.390	0.389	-0.1
C	0.346	0.348	0.7

Table 7.6 The Predictive Weight-Average Fiber Length *(L_w)*, at Regions A, B, and C, as Compared with the Experimental Values for the 40 wt % LGF/PP Composite

	Experimental value	Predictive value	Deviation (%)
A	3.01	2.56	-15.0
B	2.28	2.47	8.7
C	2.34	2.49	6.6

7.4.2 Fiber Concentration

This section shows a simulation case study focusing on the process-induced change in fiber concentration in injection molding of short fiber-reinforced thermoplastics. The experimental results are obtained from Vélez-García et al., who molded a center-gated disk mold [53, 54]. A schematic of the center-gated disk is shown in Figure 7.17, wherein r describes the flow or radial direction, θ is the cross-flow or tangential direction, and z is the through-thicken direction. The disk is 152.4 mm in diameter with a thickness of h = 1.38 mm.

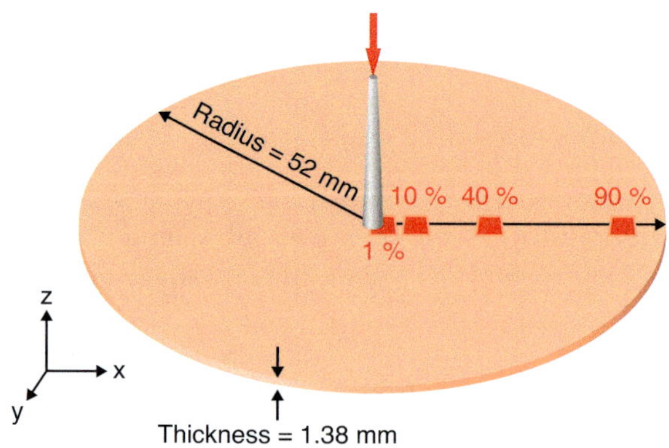

Figure 7.17 Illustration of center-gated disk of mold filling

The commercial fiber-reinforced thermoplastic material (Valox 420, Sabic Co.) is short glass fiber filled polybutylene terephthalate (PBT) with an average volume fraction of ϕ_v = 17.7 vol %. The average fiber length for 949 fibers was found to be 362.8 μm. The average diameter for 1000 fibers was determined as 12.9 μm. The wall temperature was held at approximately 90 °C, while the melting temperature was 260 °C. The filling time was about 3.0 s. Details of related experiments are available elsewhere [53, 54].

Figure 7.18 shows the fiber concentration pattern in the runner flow. Near the wall, the concentration is relatively low, whereas the central concentration is high. However, the overall concentration is close to the average volume fraction of 17.7 vol %. Thus, the concentration gradient is not obvious. In Figure 7.19, flowing through the gate, the concentration distribution of the disk is very vivid from the low wall concentration to the high central concentration. The average concentration near the wall is roughly 15 vol % and a maximum concentration of over 20 vol % occurs at the channel's central region. Furthermore, it is essential to quantitatively compare the difference between the prediction and experiment data.

Vélez-García et al. [54] estimated the volume fraction via optical micrographs through adaptations of the method of ellipses (MoE) approach (see Chapter 4 for details of the measurement technique). The measured fiber volume fraction is plotted along the normalized thickness and is shown in Figure 7.20. Note that the lubrication region is at a middle point along the radial direction. There is an obvious minimum concentration near the channel wall due to a high shear rate and a maximum concentration at the channel's central region due to a low shear rate. This shows shear-induced fiber migration causing the fiber concentration gradient in the mold filling.

Furthermore, our predicted fiber concentration is in good agreement with the experimental data [54]. This is a satisfying assessment of fiber concentration available for injection molding. Most importantly, we realize that the center region's high concentration corresponds to the low fiber orientation of the core region, whereas near the wall the low fiber concentration corresponds to the high orientation of the shell layers.

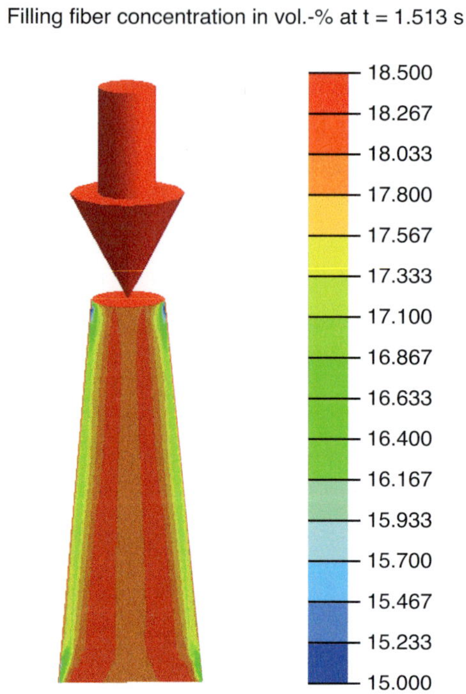

Figure 7.18 Fiber concentration pattern for the runner filling with the average concentration of 17.7 vol %

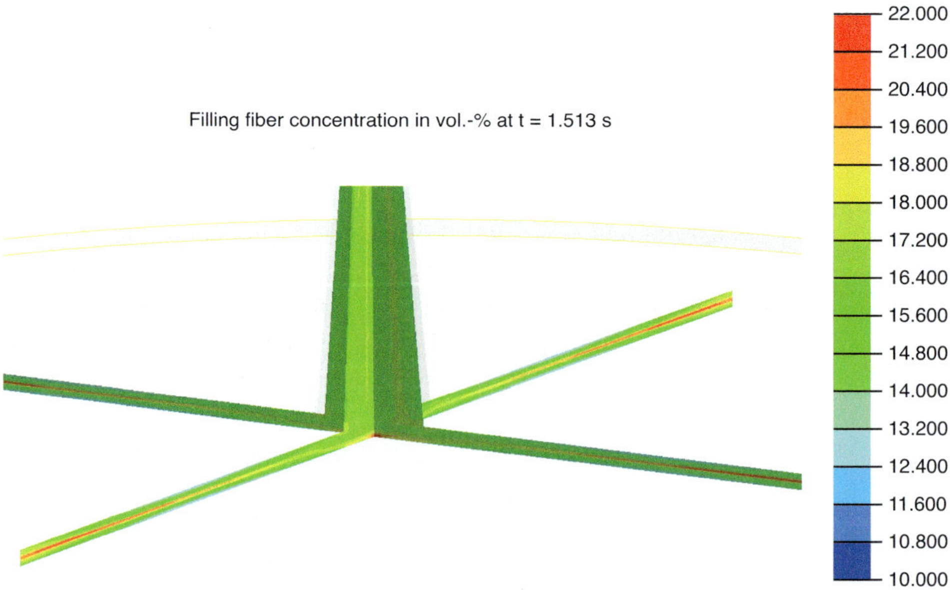

Figure 7.19 Fiber concentration pattern for the center-gated disk filling with the average concentration of 17.7 vol %

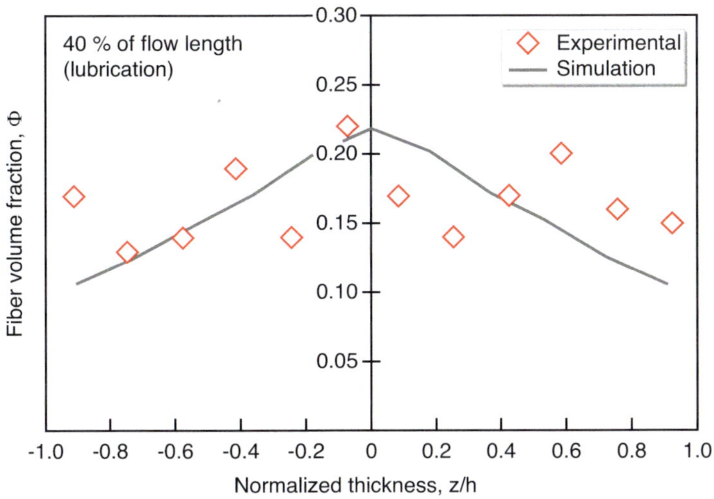

Figure 7.20 Fiber volume fraction against the normalized thickness, for the center-gated disk; simulation and measurements are compared

7.5 Direct Fiber Simulation

The previously described continuum models aim to describe the mechanisms of fiber-matrix separation, fiber attrition, and fiber alignment individually, but lack a linked theory between the three mechanisms and their independencies. A more fundamental modeling approach is a particle level simulation (PLS), which solves the forces and moments acting on individual fibers during processing to account for the complex interactions of fibers and matrix at a particle level. The state and motion of the fibers are not described as averaged and homogenized properties like in continuum-mechanical approaches, but by solving the governing equations for each fiber explicitly. Hence, PLS enables the analysis of the micro-mechanical behavior of single fibers in a suspension and the observation of fiber-fiber, fiber-matrix, and fiber-mold interactions during the processing. The limiting factor of a PLS approach is the high computational cost and that it generates a large amount of data that has to be analyzed. Solving the interactions and motion of millions of fibers for a typical injection molding or compression molding process is currently not practical. Nevertheless, particle level simulation is an invaluable tool to learn about the mechanisms present in processing discontinuous fibers and obtain a fundamental understanding of the underlying physics that cannot be captured.

A PLS was developed at the Polymer Engineering Center (PEC) at the University of Wisconsin - Madison, USA, which was based on the work of Schmid et al. [55], who focused on modeling fiber flocculation in papermaking. In this mechanistic model, each individual fiber is discretized as a chain of rigid segments, called rods, which are interconnected on their nodes (ends of the rods). The nodes elastically couple the segments of a fiber, as illustrated in Figure 7.21. The model includes effects such as hydrodynamic forces, fiber flexibility, and excluded volume forces due to fiber-fiber and fiber-wall contacts. The fiber length, the diameter of the fibers, and their stiffness can be varied to match the properties of the actual fiber while the number of elements per fiber can be modified.

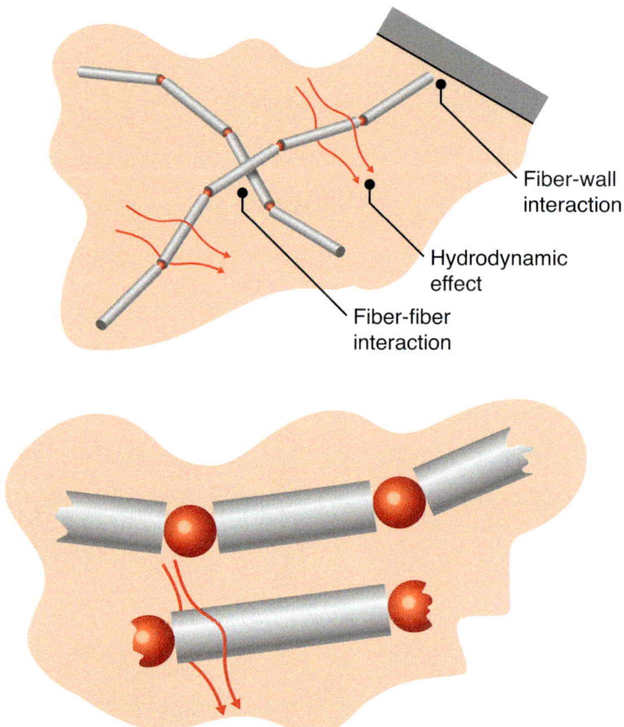

Figure 7.21 Particle level simulation: modeling single fiber and macroscale interactions

In this model, the position x_i, velocity v_i, and angular velocity ω_i are defined at each node i. The fibers are immersed in a viscous fluid that is flowing at a low Reynolds number; therefore, inertial effects can be neglected. The force and torque balances are composed for each segment to describe the motion and deformation of the fiber. The acting forces are the drag forces from the surrounding fluid, F^H, the fiber-fiber interaction forces (volume forces and friction forces excluded), F_{ij}^C, and the intra-fiber forces, X_i, exerted by internal connection loads at the nodes.

The translational equation of motion is written as

$$0 = F^H + \sum_j F_{ij}^C + X_i - X_{i+1} \tag{7.59}$$

The rotational equation of motion is analogous but includes an elastic recovery term M_b and a hydrodynamic torque T^H

$$0 = T^H - r_i \times X_{i+1} + \sum_j r_{ij} \times F_{ij}^C + M_i^b - M_{i+1}^b \tag{7.60}$$

If a fiber has more than one element, an extra constraint that enforces connectivity between the different elements is used

$$v_i + \omega_i \times (x_{i+1} - x_i) - v_{i+1} = 0 \tag{7.61}$$

where x_i is the position of node i, and x_{i+1} is the position of node i+1.

The hydrodynamic forces on the fiber are approximated by modeling each segment of the fiber as a chain of beads, as illustrated in Figure 7.22. The flow field, U∞, is known from the mold filling simulation. The hydrodynamic force F_i^H is the sum of the forces experienced by the beads F_k^H and is calculated as follows:

$$F_i^H = \sum_{k=1}^{m} F_k^H \tag{7.62}$$

In the above equation, F_k^H is represented by the well-known Stokes equation given by

$$F_k^H = 6\pi\mu a \left(U_k^\infty - u_k\right) \tag{7.63}$$

where k denotes the number of beads, a is the bead radius, μ is the fluid viscosity, U_k^∞ is the surrounding fluid velocity, and u_k is the velocity of bead k.

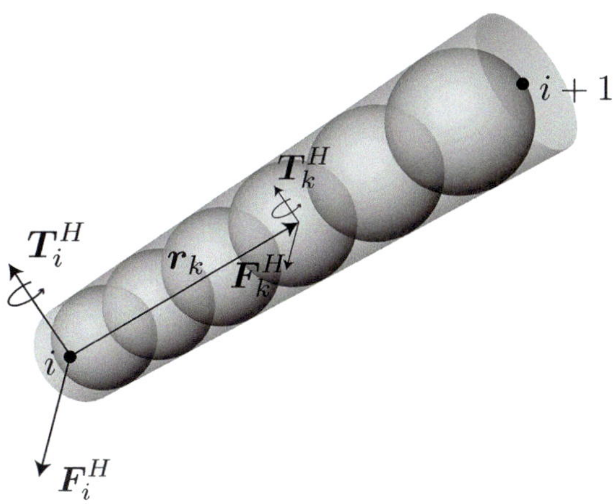

Figure 7.22 Single fiber modeled as a chain of beads to account for hydrodynamic interactions with the surrounding matrix

With the use of the second-order orientation tensor, it is now easier to develop an equation of orientation change to describe how the orientation tensor changes over time under the flow field of mold filling process. In the next several sections, we will explain how to predict the filling process of injection and compression molding, and then how to model the change of orientation tensor with flow simulation.

The fiber–fiber interaction force F_{ij}^C between two segments i and j consists of two components, a normal force F_{ij}^N acting as an excluded volume force and a tangential force F_{ij}^T, which describes the friction between the segments. The fiber–fiber interaction force F_{ij}^C acts when the distance between adjacent fibers, d_{ij}, is below a defined threshold $d_{threshold}$:

$$F_{ij}^C = F_{ij}^N + F_{ij}^T \quad \text{for } d_{ij} < d_{threshold} \tag{7.64}$$

The excluded volume force models the collision-response of the fibers and is implemented with a discrete penalty method, as illustrated in Figure 7.23.

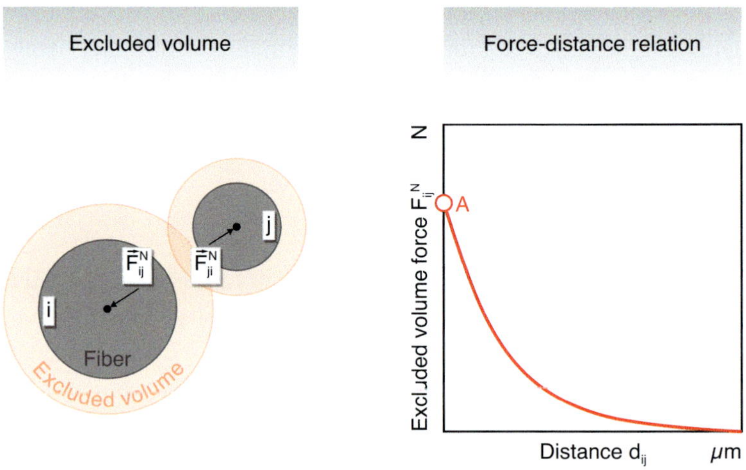

Figure 7.23 Excluded volume force between two adjacent fibers

The friction force between elements i and j is calculated as a force acting in the direction of the relative velocity of the elements, and is proportional to the magnitude of the normal interaction force, computed by

$$F_{ij}^T = \mu_f \left|F_{ij}^N\right| \frac{\Delta v_{\{ij\}}}{\left|\Delta v_{\{ij\}}\right|} \tag{7.65}$$

μ_f is the friction coefficient between fibers, $\Delta v_{\{ij\}}$ is the vector of the relative velocity between elements i and j.

The bending of a fiber is described by the theory of elastic beams as illustrated in Figure 7.24. The radius of curvature between two segments, r_{curv}, can be computed by

$$\alpha = \beta$$

$$\frac{1}{r_{curv}} = \sin\left(\frac{\beta}{2}\right) = \sin\left(\frac{\alpha}{2}\right) \approx \frac{\alpha}{2}$$

$$\frac{1}{r_{curv}} \approx \frac{\alpha}{2} \tag{7.66}$$

The bending moment is described by

$$M_i^b = \frac{\alpha}{l} E\,I\,e \tag{7.67}$$

with the normalized vector e calculated by

$$e = \frac{r_i \times r_{i-1}}{\left|r_i \times r_{i-1}\right|} \tag{7.68}$$

Fiber attrition can be modeled through a critical radius of curvature at which the fiber fails, r_{crit}, which can be derived from the inherent strength of the fiber. If the deflection of a fiber exceeds r_{crit}, the fiber is divided into two new fibers along the interconnection nodes (see Figure 7.24 right).

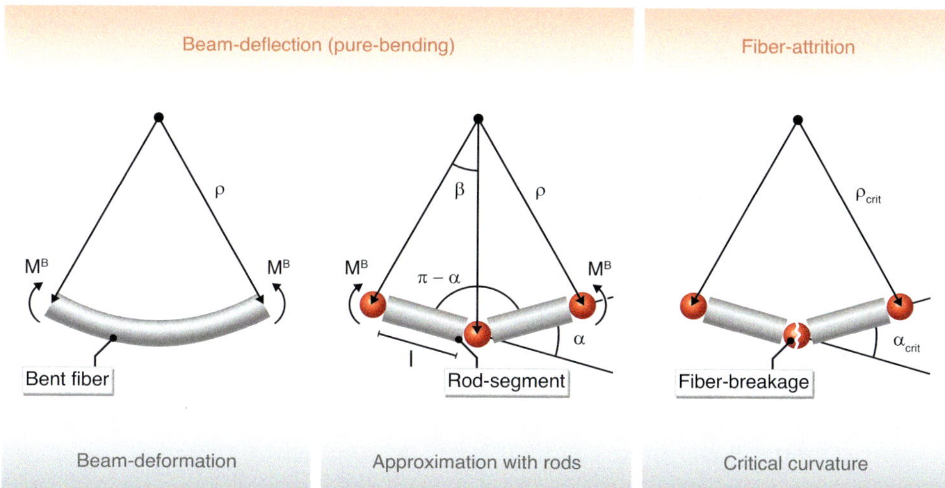

Figure 7.24 Schematic of the modeling approach for flexibility and deflection of fibers

Each time step, a force and torque balance are performed on the fiber element, which results in a system of linear equations that yields the velocities and connective forces in each fiber. These velocities are then integrated over time using an explicit Euler scheme to determine the fiber trajectory during the simulation. Figure 7.25 illustrates the full work flow of the PLS.

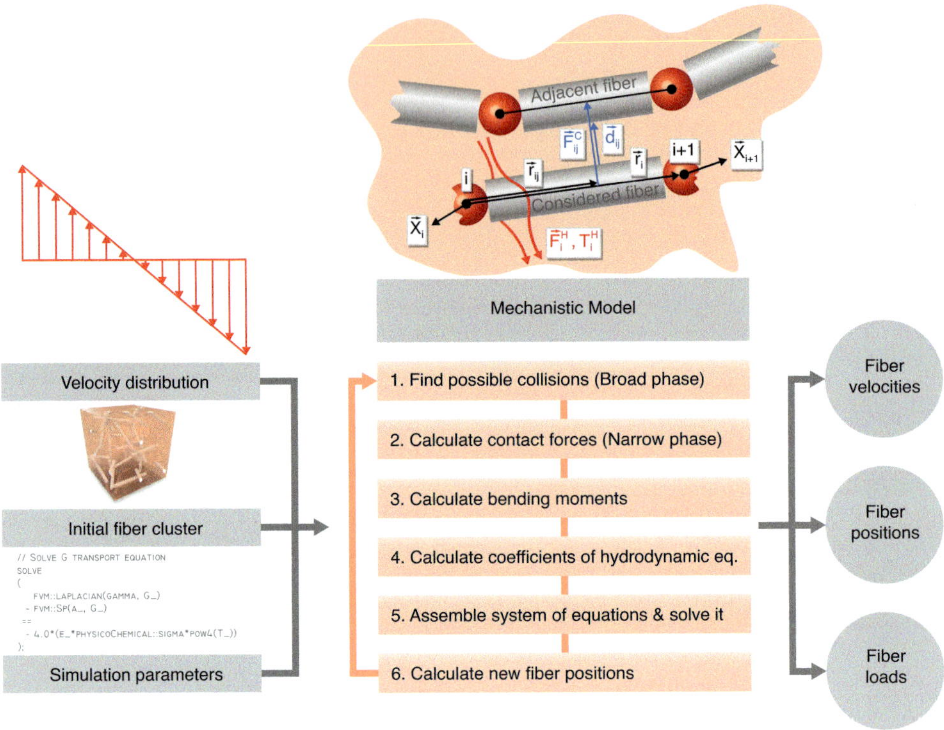

Figure 7.25 Work flow and summary of the PLS (particle level simulation)

7.5.1 Application Example: Fiber Orientation Evolution during Compression Molding

Jackson and Tucker [56] performed experiments with a sheet molding compound compression molding process. In the process, a charge produced by impregnating rovings of chopped fibers with a thermosetting resin is compressed in a hot mold. Charges were compressed at different total strains and the fiber orientation was measured in the final parts. The total strain was varied by changing the initial mold coverage, using 100%, 67%, 50%, and 33%. The 100% mold coverage was used as the initial configuration. With 33%, the largest strain was obtained. The material used was a thermoset polyester reinforced with 2.54 cm long glass fiber bundles, with a fiber content of 65 wt % [57].

The experiments were replicated numerically by simulating the rovings as single bodies, with a diameter of 0.5 mm [58] and aspect ratio of 50. 5100 fibers were used in the simulation; each fiber was discretized with 16 segments.

Due to intrinsic limitations on the fiber content that can be achieved by randomly positioning fibers in a confined volume, a two-step process was followed to generate a high-volume fraction fiber cluster. First, a low volumetric fraction cluster close to the theoretical volume fraction limit [59] was generated. The cluster was generated so it had the planar orientation measured by Jackson for the 100% mold coverage charge. In a second step the cluster was pre-compressed by imposing a fictitious flow field, solely compressing the fiber cluster in *z*-direction as illustrated in Figure 7.26. This flow field is not solenoidal, which is physically unrealistic for this type of materials, but it is only a computational tool to increase the fiber concentration beyond its theoretical limit.

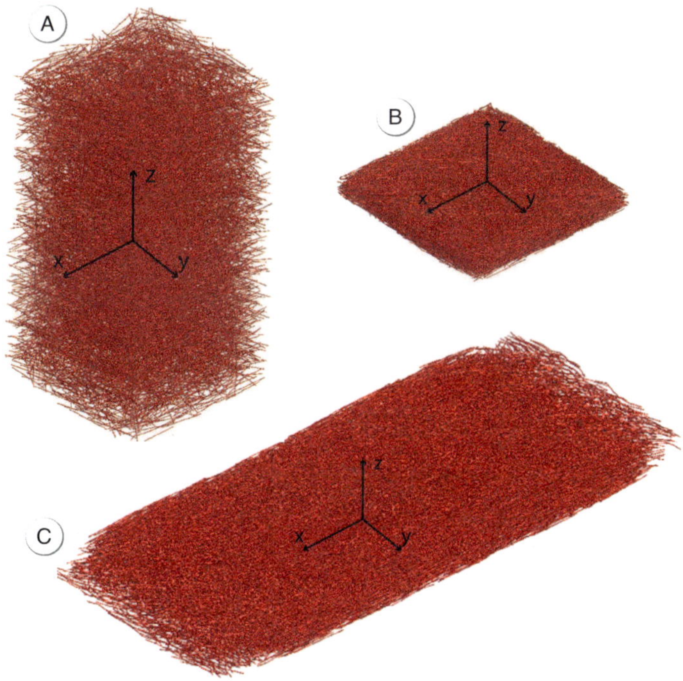

Figure 7.26 Pre-compression step: to generate high fiber content clusters, low fiber content (2.5%) clusters are generated (A) and then are compacted into a higher volume fraction (40%) cluster (B); finally, the compacted charge is compression molded (C). Compaction is only a conceptual step to help the modeling of charge

After the cluster was pre-compressed, the actual simulation of the compression molding process took place. To emulate the flow field of the experiments, we assume the material was restricted by the wall of the mold in the *y*-direction, the material was squeezed in the *z*-direction, and fills the mold in the *x*-direction. Although this flow has very simple kinematics, Barone and Caulk [60] showed that the deformation of SMC charges is extensional with a hydrodynamic friction at the mold surface, and does not exhibit fountain flow effects.

Figure 7.27(a) shows the orientation distribution for 33% mold coverage obtained after the pre-compression step, compared to the orientation measurements reported by Jackson. During the pre-compression the planar orientation changed slightly when compared to the measured orientation. However, it is still very close to the experimental distribution.

Figure 7.27 (b), (c), and (d) show the results obtained for mold coverages 50%, 67%, and 100%, respectively. For all coverages, the orientation distributions obtained with the simulation are close to the experimental results. As expected, higher deformations result in higher degrees of orientation.

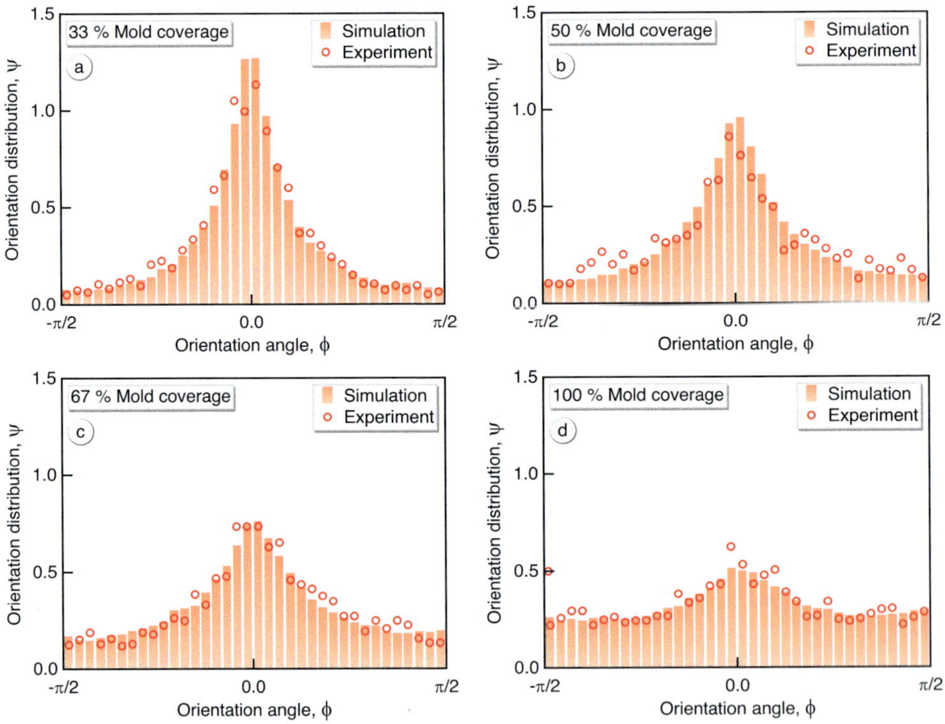

Figure 7.27 Fiber orientation results comparing simulation outcome to measurements for (a) 33% mold coverage, (b) 50% mold overage, (c) 67% mold coverage, and (d) 100% mold coverage

7.6 Conclusions

Current models to predict the process-induced fiber microstructure of discontinuous fiber-reinforced composites show great capabilities to predict the state of the fibers in the molded part. A research project administered by the Pacific Northwest National Laboratory (PNNL) focused on evaluating the performance of numerical tools for carbon fiber-reinforced thermoplastics, including process-simulation and the structural analysis. In their study, the researchers defined a relative error of 15% as the threshold for the prediction of fiber orientation. Applying the PNNL criterion for the evaluative study of the simulation tools, the current modelling approaches show that the simulation results can meet this threshold. However, it can also be concluded that continuous research and development followed by rigorous validation is needed to exploit the full potential of modeling and simulation to obtain the complex fiber configuration in real world parts.

A substantial challenge for modeling fiber evolution during processing is fiber-fiber interactions present in LFT injection molding combined with the highly dynamic flow conditions in injection or compression molding. Current trends indicate that an accurate description might only be possible through direct fiber simulation, which is currently limited due to computational resources.

Most of the current predictive tools depend on empirical fitting parameters but can lack a defined procedure on how to obtain these parameters and limitations in a constitutive relationship. However, the outcome of the simulations can depend substantially on the correct choice of the parameters. It needs to be highlighted that default values for the model parameters cannot be assumed to result in accurate simulation outcomes. Nevertheless, a truly predictive tool used in industrial product development processes requires that fitting parameters must be characterizable a priori, with little experimental input and ideally through constitutive relationships. Academic research continues to push the capabilities of the predictive tools and the importance of CAE, coupled with rigorous measurements, will only grow for LFT processing. A potential big leap forward will stem from fully coupling phenomena of fiber breakage, fiber alignment, and fiber matrix separation as they are currently treated as independent models.

7.7 Future Challenges

3D-solid simulation of injection molding, material property calculation, and structural analysis may be integrated to effectively improve the mechanical performance related to injection-molded long fiber-reinforced composite parts. Over the past few years, researchers and code developers have integrated state-of-the-art predictive engineering tools in relation to the injection simulation of long fiber-reinforced thermoplastic composites. In the past, studies sponsored by the department of energy, such as the ones by Nguyen et al. [47, 48], used the ORIENT program [61, 62], which was based on the Hele-Shaw approximation [63], to predict long fiber orientation by employing the ARD-RSC model. Furthermore, his team examined the midplane-mesh using the Autodesk Simulation Moldflow Insight (ASMI) version with ARD-RSC, and set an inlet condition around the gate area to obtain good orientation predictions for the long carbon fiber composites [64]. However, their orientation predictions using the 3D-mesh calculation model were inaccurate; in regard to this result, they explained that “the inlet condition of the orientation tensor” and “the calculation accuracy of the velocity gradient tensor” strongly influenced the fiber orientation predictions. This was addressed recently, using the so-called “3D inlet condition” in the research version of Autodesk Simulation Moldflow Insight, to enhance the accuracy of fiber orientation and fiber length calculations [65, 66]. Furthermore, van Haag et al. [67] performed an automated optimization procedure to obtain a material-specific set of the iARD-RSC parameters, thereby providing substantial help for researchers.

The objective iARD-RPR fiber orientation model and the fiber length attrition model have been demonstrated to obtain realistic results. All of the 3D numerical predictions are in good agreement with the experimental data. Such a result is satisfactory. In future work, determining how to obtain the optimal parameters of the fiber orientation and fiber length is important for real automotive products with complex geometries and requisite features, such as various changes in the direction of the flow, the inclusion of ribs, changes in thickness, and holes. In addition, measuring fiber orientation and fiber length of long-fiber composites will play an assistive role in providing reliable experimental data.

In particular, curvature and bending of long flexible fibers has been always a concern. This is a challenging issue. However, current state-of-the-art packages cannot provide such fiber curvature simulation. In injection molding of fiber-reinforced polymer melt, the flow advances faster near the edge than in the center. Additionally, for the practical compression molded glass-mat thermoplastics (GMT), it is important that the flow pattern in squeezing circular disks usually develops into an elliptical shape due to fiber-orientation-induced anisotropic flow. So far, predictive engineering tools have always provided unsatisfactory isotropic flow results.

Compression molding simulation is a new area. The key challenges are identified and the possible solutions are presented as examples in Chapter 9.

References

[1] CoreTech Inc., Manual of Moldex3D Software (2014)

[2] Autodesk Inc., Manual of Moldflow Software (2014)

[3] Karmal, M. and Kenig, S., The Injection of Molding of Thermoplastics, Part I: Theoretical Model, *Polym. Eng. Sci.,* vol. 12, pp. 294–301 (1972)

[4] Tadmor, Z., Broyer, E., and Gutfinger, C., Flow Analysis Network (Fan) – a Method for Solving Flow Problems in Polymer Processing, *Polym. Eng. Sci.,* vol. 14, pp. 660–665 (1974)

[5] Hieber, C.A. and Shen, S.F., Flow Analysis of the Non-Isothermal Two-Dimensional Filling Process in Injection Molding, *Israel Journal of Technology,* vol. 16, pp. 248–254 (1978)

[6] Stevenson, J.F., A Simplified Method for Analyzing Mold Filling Dynamics, Part I: Theory, *Polym. Eng. Sci.,* vol. 18, pp. 577–582 (1978)

[7] Chiang, H.H., Hieber, C.A., and Wang, K.K., A Unified Simulation of the Filling and Post-Filling Stages in Injection Molding, Part I: Formulation, *Polym. Eng. Sci.,* vol. 31, pp. 116–124 (1991)

[8] Kenny, P., Flow Analysis Reference Manual, MoldFlow Pty Ltd (1993)

[9] Lecheler, S., *Numerische Strömungsberechnung,* Vieweg Teubner (2009)

[10] Chang, R.-Y. and Yang, W.-H., Numerical Simulation of Mold Filling in Injection Molding Using a Three-Dimensional Finite Volume Approach, *Int. J. Numer. Methods Fluids,* vol. 37, pp. 125–148 (2001)

[11] Hsu, C.C., Huang, C.T., and Chang, R.Y., Simulation of Dynamic Gas Penetrations on Fingering Behaviors During Gas-Assisted Injection Molding, *J. Polym. Eng.,* vol. 38, pp. 93–105 (2018)

[12] Osswald, T.A. and Tucker III, C.L., Compression Mold Filling Simulation for Non-planar Parts Using the Finite Element/Control Volume Approach, *Int. Polym. Process.*, 5, 2, pp. 79–87 (1990)

[13] Rios, A., Davis, B., and Gramann, P., Computer Aided Engineering in Compression Molding, Composites Fabricators Association Annual Conference (2001)

[14] Hsu, C.C., Chiu, H.S., Huang, C.T., and Chang, R.Y., Through Fiber Orientation Investigation to Visualize Compression Molding, in The 29th International Conference of the Polymer Processing Society (2014)

[15] Owens, R.G. and Phillips, T.N., *Computational Rheology* (Vol. 14), Imperial College Press, London (2002)

[16] Barone, M.R. and Osswald, T.A., Boundary integral equations for analyzing the flow of a chopped fiber reinforced polymer compound in compression molding, *Journal of Non-Newtonian fluid mechanics*, 26(2), pp. 185–206 (1987)

[17] Williams, M.L., Landel, R.F., and Ferry, J.D., The temperature dependence of relaxation mechanisms in amorphous polymers and other glass-forming liquids, *Journal of the American Chemical Society*, 77(14), pp. 3701–3707 (1955)

[18] Jain, A., Firestone, G.C., and Yi, A.Y., Viscosity measurement by cylindrical compression for numerical modeling of precision lens molding process. *Journal of the American Ceramic Society*, 88(9), pp. 2409–2414 (2005)

[19] Cherizol, R., Sain, M., and Tjong, J., Review of Non-Newtonian Mathematical Models for Rheological Characteristics of Viscoelastic Composites, *Green and Sustainable Chemistry*, 5(01), 6 (2015)

[20] Wang, M.L., Chang, R.Y., and Hsu, C.H.D. (Eds.), *Molding Simulation: Theory and Practice*, Hanser, Munich (2018)

[21] Osswald, T.A., and Hernandez, J. P, *Polymer Processing: Modeling and Simulation*, Chapter 11 – Radial Functions Method, Hanser, Munich (2006)

[22] Perez, C., Roppers, S., Osswald, T. A., Ghandi, U., and Mandapati, R., Fiber length and orientation distribution measurement of a charge for D-LFT compression Molding. In SPE ANTEC Conference, Technical Papers (2013, April)

[23] Goris, S., Herchet, M., Perez, C., Gandhi, U., and Osswald, T. A., Impact of fiber orientation distribution within a d-LFT strand on warpage of a compression molded part. In SPE ANTEC Conference, Technical Papers (2014, April)

[24] Song, Y., Gandhi, U., Sekito, T., Vaidya, U. K., Vallury, S., Yang, A., and Osswald, T., CAE method for compression molding of carbon fiber-reinforced thermoplastic composite using bulk materials, *Composites Part A: Applied Science and Manufacturing*, 114, pp. 388–397 (2018)

[25] LS-DYNA manual, LSTC Company

[26] Pan-Form manual, ESI group

[27] Lin, G., Lai, D. L., Huang, C. T., and Wang, C. C., Numerical Simulation for the Viscoelastic Effects on the Birefringence Variation for an Injected Optical Lens, SPE ANTEC, Indianap., pp. 1231–1235 (2016)

[28] Wang, J., O'Gara, J. F., and Tucker III, C. L., An Objective Model for Slow Orientation Kinetics in Concentrated Fiber Suspensions: Theory and Rheological Evidence, *Journal of Rheology*, vol. 52, pp. 1179–1200 (2008)

[29] Phelps, J. H. and Tucker III, C. L., An Anisotropic Rotary Diffusion Model for Fiber Orientation in Short- and Long-Fiber Thermoplastics, *J. Non-Newtonian Fluid Mech.*, vol. 156, pp. 165–176 (2009)

[30] Tseng, H.-C., Chang, R.-Y., and Hsu, C.-H., Phenomenological Improvements to Predictive Models of Fiber Orientation in Concentrated Suspensions, *J. Rheol.*, vol. 57, pp. 1597–1631 (2013)

[31] Folgar, F. and Tucker III, C. L., Orientation Behavior of Fibers in Concentrated Suspensions, *Journal of Reinforced Plastics and Composites*, vol. 3, pp. 98–119 (1984)

[32] Phan-Thien, N., Fan, X.-J., Tanner, R. I., and Zheng, R., Folgar–Tucker Constant for a Fibre Suspension in a Newtonian Fluid, *J. Non-Newtonian Fluid Mech.*, vol. 103, pp. 251–260 (2002)

[33] Hand, G. L., A Theory of Anisotropic Fluids, *Journal of Fluid Mechanics*, vol. 13, pp. 33–46 (1962)

[34] Svendsen, B. and Bertram, A., On Frame-Indifference and Form-Invariance in Constitutive Theory, *Acta Mechanica*, vol. 132, pp. 195–207 (1999)

[35] Tseng, H.-C., Chang, R.-Y., and Hsu, C.-H., An Objective Tensor to Predict Anisotropic Fiber Orientation in Concentrated Suspensions, *J. Rheol.*, vol. 60, pp. 215–224 (2016)

[36] Morrison, F. A., *Understanding Rheology*, Oxford University (2001)

[37] Tseng, H.-C., Chang, R.-Y., and Hsu, C.-H., Accurate Predictions of Fiber Orientation and Tensile Modulus in Short-Fiber-Reinforced Composite with Experimental Validation, *Polym. Compos.* (2017)

[38] Phelps, J. H., Abd El-Rahman, A. I., Kunc, V., and Tucker, I. C. L., A Model for Fiber Length Attrition in Injection-Molded Long-Fiber Composites, *Composites Part A: Applied Science and Manufacturing*, vol. 51, pp. 11–21 (2013)

[39] Mondy, L. A., Brenner, H., Altobelli, S. A., Abbott, J. R., and Graham, A. L., Shear-Induced Particle Migration in Suspensions of Rods, *J. Rheol.* vol. 38, pp. 444–452 (1994)

[40] Toll, S. and Andersson, P.-O., Microstructure of Long- and Short-Fiber Reinforced Injection Molded Polyamide, *Polym. Compos.*, vol. 14, pp. 116–125 (1993)

[41] Papathanasiou, T. D. and Guell, D. C., *Flow-Induced Alignment in Composite Materials*, Cambridge, Woodhead (1997)

[42] Brady, P. R. N. a., Pressure-Driven Flow of Suspensions: Simulation and Theory, *J. Fluid. Mech.*, vol. 275, p. 157 (1994)

[43] Morris, J. F. and Boulay, F., Curvilinear Flows of Noncolloidal Suspensions: The Role of Normal Stresses, *J. Rheol.*, vol. 43, pp. 1213–1237 (1999)

[44] Miller, R.M. and Morris, J.F., Normal Stress-Driven Migration and Axial Development in Pressure-Driven Flow of Concentrated Suspensions, *J. Non-Newtonian Fluid. Mech.*, vol. 135, pp. 149-165 (2006)

[45] Richardson, J.F. and Zaki, W.N., Sedimentation and Fluidization: Part I, *Trans. Inst. Chem. Eng.*, vol. 32, pp. 35-47 (1954)

[46] Tseng, H.-C., Chang, R.-Y., and Hsu, C.-H., An Integration of Microstructure Predictions and Structural Analysis in Long-Fiber-Reinforced Composite with Experimental Validation, *Int. Polym. Process*, vol. 32, pp. 455-466 (2017)

[47] Nguyen, B.N., Bapanapalli, S.K., Holbery, J.D., Smith, M.T., Kunc, V., Frame, B.J., Phelps, J.H., and Tucker III, C.L., Fiber Length and Orientation in Long-Fiber Injection-Molded Thermoplastics—Part I: Modeling of Microstructure and Elastic Properties, *J. Compos. Mater.*, vol. 42, pp. 1003-1029 (2008)

[48] Nguyen, B.N., Bapanapalli, S.K., Kunc, V., Phelps, J.H., and Tucker III, C.L., Prediction of the Elastic-Plastic Stress/Strain Response for Injection-Molded Long-Fiber Thermoplastics, *J. Compos. Mater.*, vol. 43, pp. 217-246 (2009)

[49] Nguyen, N., Jin, X., Wang, J., Phelps, J.H., Tucker III, C.L., Kunc, V., Bapanapalli, S.K., and Smith, M.T., Implementation of New Process Models for Tailored Polymer Composite Structures into Processing Software Packages (2010)

[50] Phelps, J.H., Processing-Microstructure Models for Short- and Long-Fiber Thermoplastic Composites, University of Illinois at Urbana-Champaign (2009)

[51] Bailey, R. and Kraft, H., A Study of Fibre Attrition in the Processing of Long Fibre Reinforced Thermoplastics, *Int. Polym. Process.*, vol. 2, pp. 94-101 (1987)

[52] Kunc, V., Frame, B., Nguyen, B.N., Tucker III, C.L., and Velez-Garcia, G., Fiber Length Distribution Measurement for Long Glass and Carbon Fiber Reinforced Injection Molded Thermoplastics, in *SPE ACCE Conference, Technical Papers*, Michigan (2007)

[53] Nguyen, N., Jin, X., Wang, J., Kunc, V., and Tucker III, C.L., Validation of New Process Models for Large Injection-Molded Long-Fiber Thermoplastic Composite Structures, the US Department of Energy, Pacific Northwest National Laboratory, PNNL Report under Contract DE-AC05-76RL01830 PNNL-21165 (2012)

[54] Vélez-García, G.M., Mazahir, S.M., and Wapperom, P., Simulation of Injection Molding Using a Model with Delayed Fiber Orientation, *Int. Polym. Process.*, vol. 11, pp. 331-339 (2011)

[55] Schmid, C., Switzer, L., and Klingenberg, D., Simulations of fiber flocculation: effects of fiber properties and inter fiber friction, *Journal of Rheology*, 44, 781 (2000)

[56] Jackson, W., Advani, S., and Tucker III, C.L. Predicting the Orientation of Short Fibers in Thin Compression Moldings, *J. Compos. Mater.*, 20, 6, pp. 539-557 (1986)

[57] Chen, C. and Tucker III, C.L., Mechanical Property Predictions for Short Fiber/Brittle Matrix Composites, *J. Reinf. Plast. Compos.*, 3, 2, pp. 120-129 (1984)

[58] Dumont, P., Favier, D., Le, T., Orgeas, L., Salvo, L., and Boller, E., Characterization of the Fibrous Microstructure of SMC during Compression Moulding Using X-ray Microtomography, 16th International Conference on Composite Materials (ICCM) (2007)

[59] Evans, K., Gibson, A., Prediction of the maximum packing fraction achievable in randomly oriented short-fibre composites, *Compos. Sci. Technol.*, 25, 2, pp. 149-162 (1986)

[60] Barone, M.C. D., Kinematics of flow in sheet molding compounds, *Polym. Compos.*, 6, 2, pp. 105-109 (1985)

[61] Vélez-García, G.M., Wapperom, P., Baird, D.G., Aning, A.O., and Kunc, V., Unambiguous Orientation in Short Fiber Composites over Small Sampling Area in a Center-Gated Disk, *Compos. Pt A-Appl. Sci. Manuf.*, vol. 43, pp. 104-113 (2012)

[62] Bay, R.S. and Tucker III, C.L., Fiber Orientation in Simple Injection Moldings. Part I: Theory and Numerical Methods, *Polym. Compos.*, vol. 13, pp. 317–331 (1992)

[63] Kennedy, P. and Zheng, R., *Flow Analysis of Injection Molds*, 2nd ed., Hanser, Munich (2013)

[64] Wang, J., Nguyen, B.N., Mathur, R., Sharma, B., Sangid, M.D., Costa, F., Jin, X., Tucker III, C.L., and Fifield, L.S., Fiber Orientation in Injection Molded Long Carbon Fiber Thermoplastic Composites, in *SPE ANTEC Conference, Technical Papers*, Orlando (2015)

[65] Nguyen, B.N., Fifield, L.S., Yocum, A., and Schutte, C., Predictive Engineering Tools for Injection-Molded Long-Carbon-Fiber Thermoplastic Composites, Pacific Northwest National Laboratory (PNNL), the US Department of Energy (2014)

[66] Nguyen, B.N., Fifield, L.S., Kijewski, S.A., Sangid, M.D., Wang, J., Costa, F., Tucker III, C.L., Mathur, R.N., Gandhi, U.N., and Mori, S., Predictive Engineering Tools for Injection-Molded Long-Carbon-Fiber Thermoplastic Composites – Fy 2015 Second Quarterly Report, The US Department of Energy, Pacific Northwest National Laboratory, PNNL Report under Contract DE-AC05-76RL01830 (2015)

[67] van Haag, J., Bontenackels, C., and Hopmann, C., Fiber Orientation Prediction of Long Fiber-Reinforced Thermoplastics: Optimization of Model Parameters, in SPE ANTEC Conference, Technical Papers, Orlando (2015)

[68] Nguyen, B.N., Fifield, L.S., Yocum, A., Schu, C., Predictive Engineering Tools for Injection-Molded Long-Carbon-Fiber Thermoplastic Composites—Pacific Northwest National Laboratory (PNNL), FY 2014 Annual Progress Report-Lightweight Materials R&D, U.S. Department of Energy, pp. 192–205 (2014)

[69] Huang, C.-T. et al., Investigation of the Viscoelastic Behavior Variation of Glass Mat Thermoplastics (GMT) in Compression Molding, *Polymers* vol. 11, 2 335 (2019), doi:10.3390/polym11020335

8 Case Studies to Demonstrate Application of Multiscale Modeling for Fiber-Reinforced Polymers

Modeling fiber-reinforced materials using finite element simulation software is quite challenging because material properties depend on the fiber condition such as orientation, length, and concentration. These fiber conditions must be accounted for when estimating the material properties of the finished product. A multiscale modeling approach was presented in Chapter 6. The key point of a multiscale approach is to model the fiber conditions, based on the details at the micrometer (μm) scale, which renders a material model. Subsequently, this material model is coupled with the FE simulation of the component, which is on the millimeter scale, thus modeling in two different scales. The flow chart for multiscale modeling as applied for discontinuous fibers was discussed in Chapter 6, see Figure 6.17. While using this approach, managing the model in two different scales and sharing the data between the two scales can be quite challenging. For the injection and compression molded fiber-reinforced parts which are of interest here, the fiber conditions, i.e., orientation, length, and concentration, can vary across the part (for more details see Chapter 4). Consequently, the material properties can also vary, and as a result, such material is likely to be anisotropic and heterogeneous. A typical finite element model has thousands of elements and each element can have a number of integration points. The whole process of computing the material properties at each integration point at the microscale in the finite element model can appear quite overwhelming. It is important to understand that, in practice, once a clear flow of calculation process is established, commercially available computer programs can be easily used for the repetitive calculations.

In this chapter we will present various case studies to demonstrate how the multiscale modeling method can be applied to model a complex part. Our goal is to help understanding of the multiscale modeling process, the underlying assumptions, the limitations of the method, and applicability of the available approaches in solving real-world problems.

8.1 Study of Effect of Manufacturing Process on Flat Plaques

In Chapter 3, different manufacturing processes were discussed, such as injection molding and compression molding, which are used to manufacture parts made of discontinuous fiber-reinforced polymers. It was also observed in Chapter 4 that the microstructure, i.e., fiber orientation, length, and concentration, in the finished part depends on the manufacturing process used. The purpose of this section is to demonstrate the effect of manufacturing methods on the material properties of a simple flat demonstrator plaque. Two different materials mixes, 30% GF (glass fiber) + PP (polypropylene) and 40% GF+PP, were used to manufacture the large plaques using many different manufacturing methods. Tensile and bending tests were conducted on the coupons from the plaques to compare their structural properties at selected locations. Furthermore, the microstructure details, such as fiber length and fiber orientations in the plaques, are measured and used to relate to material properties.

Next, the simulation of the manufacturing process is carried out to estimate the microstructure, i.e., fiber orientation and fiber length. Finally, a material model based on the fiber conditions, i.e., microstructure of the plaque is estimated. The microstructure based material models are used with the multiscale finite element model to estimate the tensile and bending performance of the samples taken from the plaque. Thus, we will show the effect the manufacturing process has on the microstructure as well as structural properties of a plaque molded for a given material. In addition, we will also show how a multiscale finite element model that accounts for microstructure helps in capturing the manufacturing effects in the FE analysis. Going through these exercises will illustrate the discussion in Chapters 6 and 7. It will also help in understanding of the multiscale modeling method for the fiber-reinforced materials.

8.1.1 Plaque Manufacturing Details

A large plaque as shown in Figure 8.1 was designed and the tool was built in such a way that it can be used to build parts using different manufacturing processes. The following manufacturing processes were used to build the parts:

- Injection molding,
- Injection-compression molding, and
- Compression molding.

Furthermore, multiple materials were used for each manufacturing process. A 30% glass filled polypropylene (30% GF+PP) and a 40% glass filled polypropylene (40%

GF+PP) were prepared using in-line compounding as well as in the pellet form. The pellets were manufactured using pultrusion as well as an extrusion (wire coating) process. Based on the different combination of materials and manufacturing processes, as shown in Table 8.1, nine different types of plaques were manufactured. The main purpose behind making these plaques, produced using different manufacturing methods, was to understand the difference in mechanical properties of the finished plaque's material, based on the manufacturing process used.

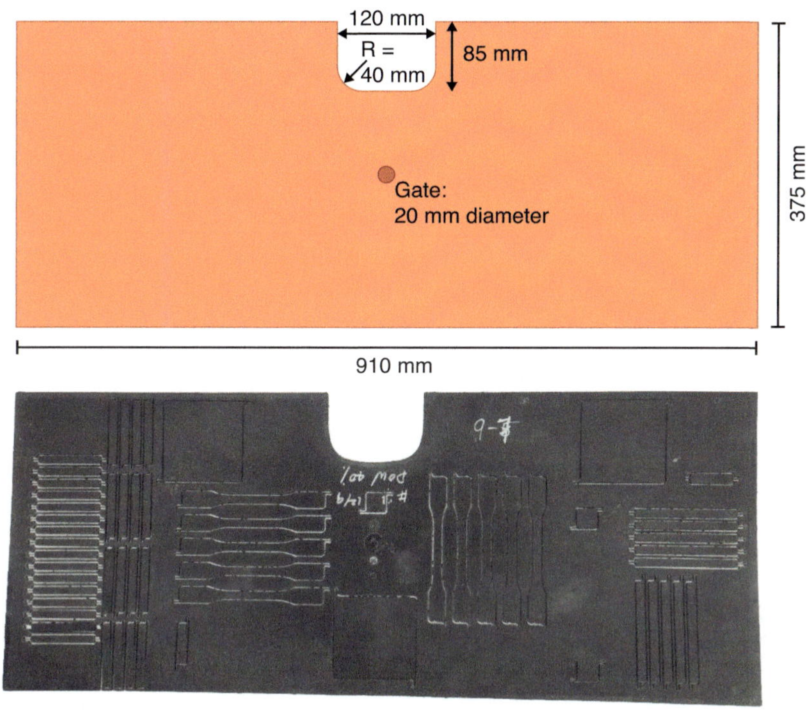

Figure 8.1 Plaque geometry and actual part used for the study

Table 8.1 Plaques Types Built Based on the Material Format and Fiber Content; the Number in Parentheses Indicates Plaque Number, Referred to in the Discussion

Manufacturing process	Material format	Glass fiber % by weight
Injection molding	In-line compounding, 15 mm long fiber	30% (2) and 40% (1)
Injection-compression	In-line compounding, 15 mm long fiber	30% (4) and 40% (3)
Compression molding	In-line compounding, 30 mm or smaller long fiber	30% (5) and 40% (6)
Injection molding	Pellet extruded	30% (7) and 40% (8)
Injection molding	Pellet pultruded	Only 40% (9)

8.1.2 Effect of Manufacturing Process on Plaque Properties

The samples for testing were cut from the plaques as shown in Figure 8.2, using a water jet cutting process, and then evaluated for tensile strength as well as bending and impact performance. Water jet cutting was used to minimize cutting related damage on the samples at the edge.

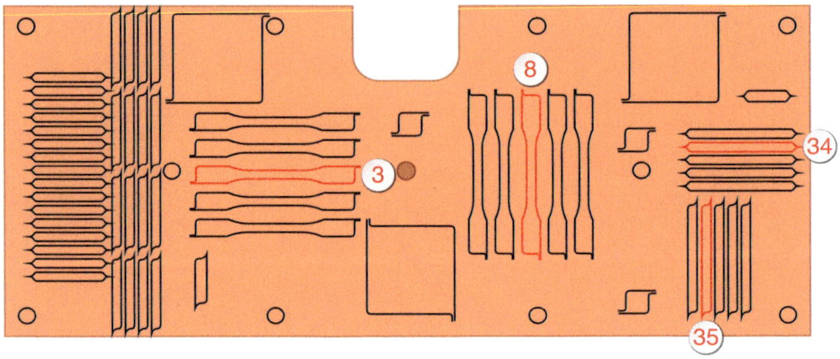

Figure 8.2 Pattern used to cut the samples for tensile and bending tests from the plaques; samples 3 and 8 are tested for tensile properties and samples 34 and 35 are tested for bending properties

The test setup used for the tensile test is presented in Figure 8.3 and that for the bending test is presented in Figure 8.4. The specimen preparation and testing are based on ISO standards for the fiber-reinforced plastics.

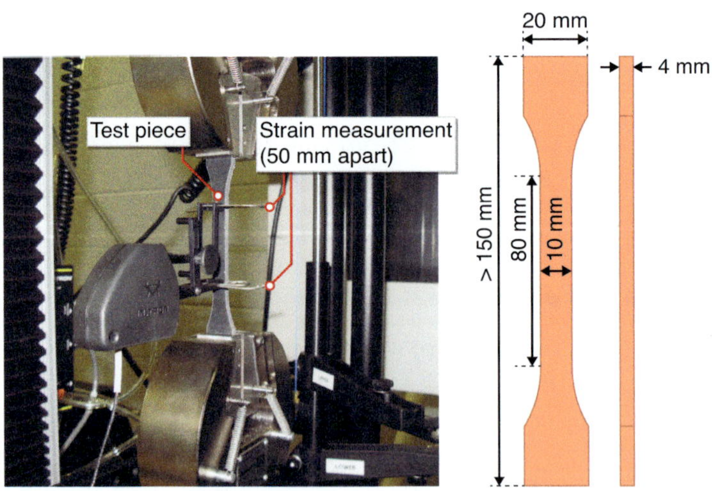

Figure 8.3 Test setup for tensile test

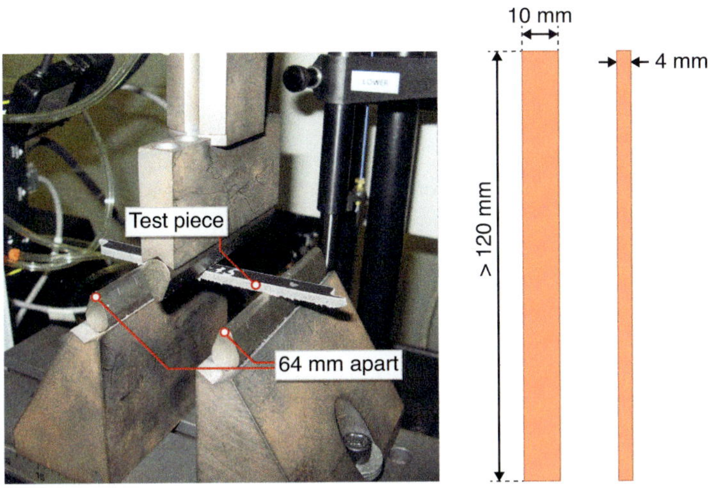

Figure 8.4 Test setup for bending test

The stress-strain curves for different plaques computed for individual test specimens are compared in Figure 8.5. Force deflection measured for the three-point bending test specimens are compared in Figure 8.6. Since plaques are likely to have variations from one to another, three samples were tested and the sample with median values of the three is used for comparison. The peak values of stress and strain at failure for the tensile test and calculated flex modulus for the three-point bending test are presented in Table 8.2. In this study, the term longitudinal is used for samples aligned along the length of longer side and transverse is used for samples aligned along width of the shorter side of the plaque, as depicted in Figure 8.5 to Figure 8.8. It is understood that the melt enters at a single point in the center and flows radially to fill the plaque cavity. However, because of the plaque geometry more material flows along the longitudinal direction. Therefore, it can be considered as *flow direction* and the corresponding transverse direction as *cross-flow direction*.

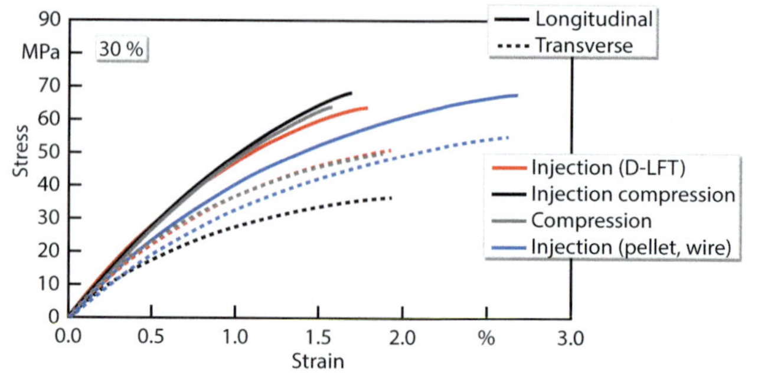

Figure 8.5 Comparison of 30% GF+PP and 40% GF+PP in the tensile test for different processes in longitudinal (L: flow) and transverse (T: cross-flow) directions; samples 3 and 8 in Figure 8.2

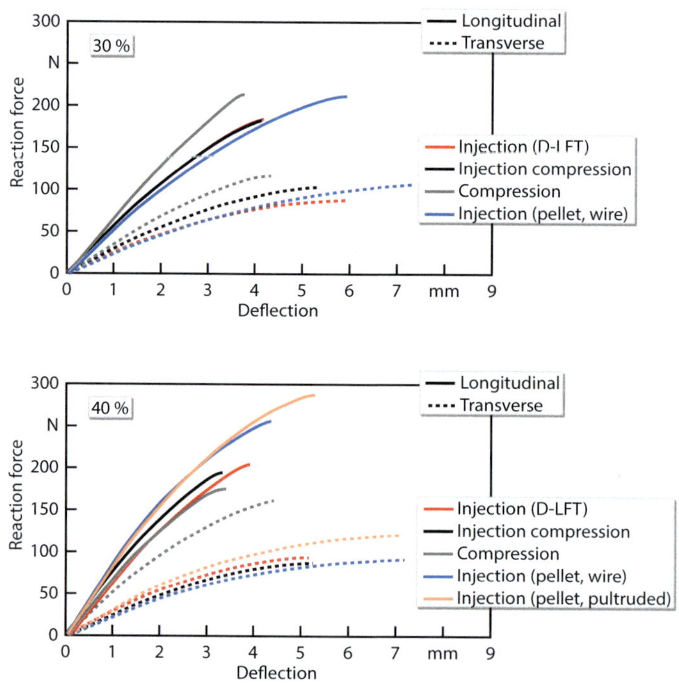

Figure 8.6 Comparison of 30% GF and 40% GF in bending test for different processes in longitudinal (L: flow) and transverse (T: cross-flow) directions; samples 34 and 35 in Figure 8.2.

Table 8.2 Comparing Results of Tensile and Bending Tests for Plaques Made from E-glass Fibers and Polypropylene

Plaque #	Plaque description method	Material preparation	Glass fiber	Measured fibers wt (%)	Tensile test (average of 3 tests)				Flexural test (average of 3 tests)				Impact test (avg. of 3 tests)	
					Stress (MPa)		Strain (%)		Flexural modulus (MPa)		Stress (MPa)		Total energy (J)	
					Longitudinal	Transverse	Longitudinal	Transverse	Longitudinal	Transverse	Longitudinal	Transverse	Far	Near
2	Injection	In-line Compounding	30%	31	64	49	1.8	1.9	6632	3172	113	60	13.3	14.7
4	Injection Compression	In-line Compounding		29	67	37	1.8	2.2	5492	3055	98	64	10.8	13.6
5	Compression	In-line Compounding		29	65	53	1.7	1.7	5385	3510	118	76	23.4	21.1
7	Injection	Pellet–Wire coated		29	67	56	2.8	2.8	5291	2434	124	66	15.0	11.3
1	Injection	In-line Compounding	40%	39	66	50	1.5	1.7	8067	3193	130	58	16.7	10.0
3	Injection Compression	In-line Compounding		39	78	39	1.6	1.8	8039	3515	117	62	13.1	18.3
6	Compression	In-line Compounding		42	60	56	1.5	1.5	5955	4536	99	93	27.4	23.9
8	Injection	Pellet–Wire coated		46	77	53	1.9	2	8617	2543	150	53	16.2	13.8
9	Injection	Pellet–Pultruded		40	84	63	2.4	2.6	8057	3082	160	68	16.4	15.8

The main purpose of comparing the mechanical performance of these samples from different plaques is to illustrate the differences in their properties. For example, different plaques made with the same base material ingredients, i.e., 40% glass fiber by weight and polypropylene, but manufactured using different manufacturing processes, show significant differences in mechanical properties. This is quite easy to notice when peak values for maximum stress in the tensile test for the longitudinal and transverse flow samples are compared, as in Figure 8.7, and the flex modulus, from the three-point bending tests, for the longitudinal and transverse samples are compared, as in Figure 8.8.

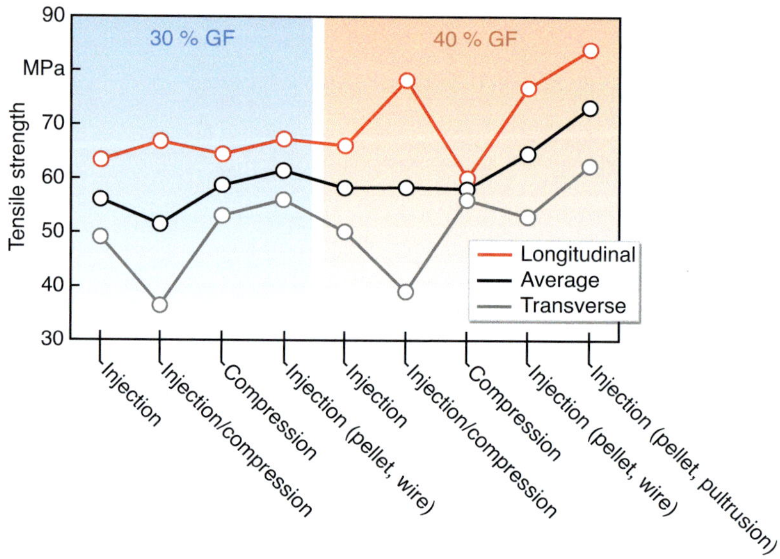

Figure 8.7 Comparison of maximum stress at failure for samples from different plaques under tensile load

The comparison shows the coupons that were along the longitudinal direction show higher strength and stiffness when compared to the transverse direction. It can be said that the difference in the strength between the longitudinal and transverse direction for the selected samples is due to the difference in the fiber orientation distributions along those directions. Hence, the difference in the strength is a measure of the anisotropy present within the plaque. We also observe that the compression molded parts, unlike the injection molded parts, show significantly lower anisotropy as the tensile strengths in the longitudinal and transverse directions are similar. The coupons from plaques with 30% fibers by weight show similar trends except that the load capacity in the longitudinal direction is slightly lower compared to similar coupons with 40% fiber content. This is consistent with the expectation that mechanical properties improve with an increase in the fiber content.

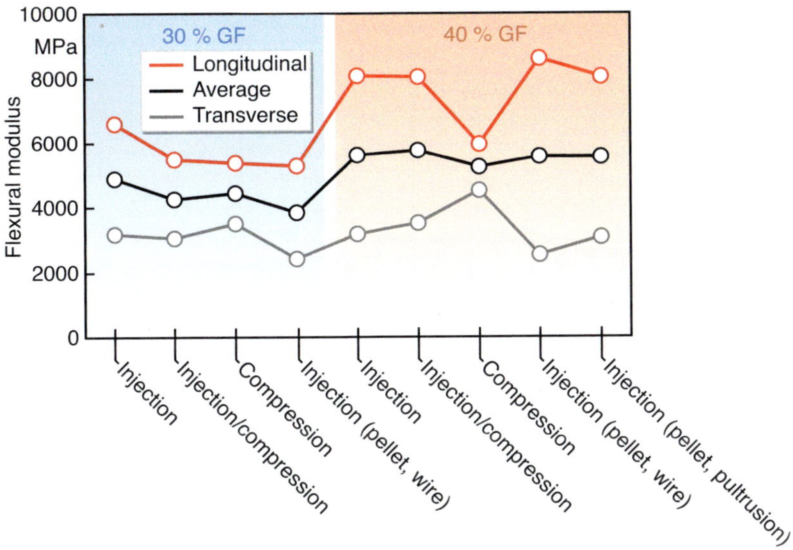

Figure 8.8 Comparison of flex modulus for samples from different plaques under bending load

The flex modulus measurements also show similar trends, with some differences. The tensile strength is proportional to the average volume of fibers aligned in the direction of loading in the sample through its cross section. If the number of fibers is higher, the tensile strength is also higher, but the distribution of aligned fibers over the thickness has a smaller effect on the tensile strength itself. However, the flexural modulus is influenced by the fiber distribution across the thickness of the sample. If the aligned fibers are in the skin regions of the sample, the fibers have higher contribution on increasing the flexural stiffness compared to the case where fibers are more random, or if the aligned fibers are in the core of the sample. This is schematically illustrated in Figure 8.9.

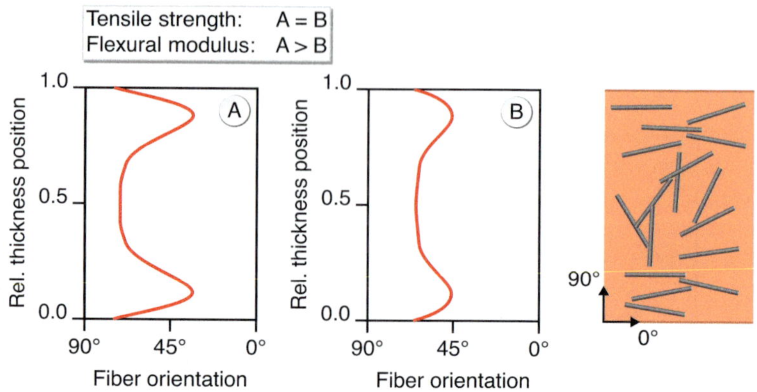

Figure 8.9 Fiber orientation distribution through thickness (1 is the direction of flow) and its effect on the tensile strength and flex modulus. Distributions A and B are hypothetical fiber orientation through the thickness

The X-ray image of the compression molded plaque and an injection molded plaque are compared in Figure 8.10. The X-ray imaging was carried out at a resolution of about 50 µm. Since the glass fibers used in this study are between 14–20 µm, it is not expected that individual fibers will be visible in the image. The X-ray image of the injection molded plaque, with pultruded pellets, does not show any fibers, indicating that the fibers are well dispersed. However, the image of compression molded plaques, with in-line compounded material shows thick and long fibers, essentially undispersed bundles of fibers. During injection molding, the fibers in the pultruded pellets are well dispersed, and then the fibers experience high rates of shearing, which helps disperse them further and break them down into individual fibers. The compression molding compounding is less severe and as are the deviatory stresses that arise during flow in a compression molding process. This results in a higher number of undispersed fiber bundles. In general, it is observed that compression molding with in-line compounded material leaves most bundles while pultruded pellets show the highest level of dispersion. Thus, X-ray images of these extreme cases are compared.

Further, to understand the fiber length resulting from the different manufacturing processes, a small piece from the finished plaque was cut and heated in the oven at approximately 450 °C to burn the resin material and to measure the fiber length distribution of the fibers using the optical methods discussed in Chapter 4. A comparison of the measured fiber length average for selected samples is presented in Table 8.3. For the compression molded plaque, the average fiber length is significantly higher compared to plaques produced by other processes; however, it can also be observed that the improvements in the mechanical properties are insignificant. This effect can be explained in that the compression molded plaques have fibers that are not well dispersed. In this study it was found that almost 55% of the fibers in the compression molded plaques were still in bundles. In general, bundles of fibers act like larger, thicker, and longer fibers and therefore their contributions to the strength and stiffness are lower when compared to the well dispersed fibers. On the other hand, the longer fibers are curved and interact with each other like a web, which improves the impact strength of the material (Table 8.2). This effect is also explained in Chapter 4, where the stress distribution and the transfer of stresses from the matrix to the fibers are shown to be due to the *L/D* ratio of the fibers, and not due to the length alone.

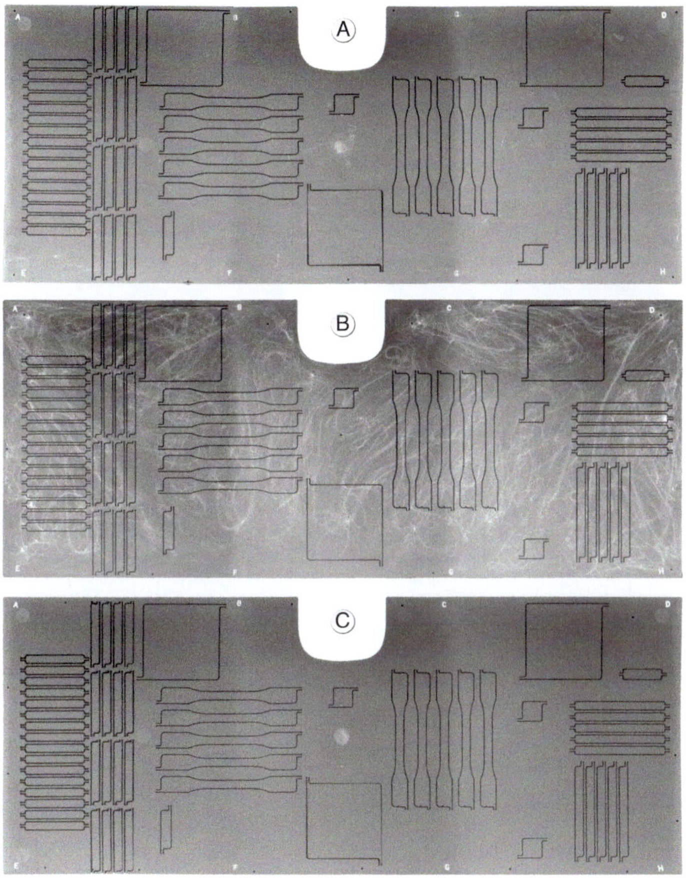

Figure 8.10 Comparison of X-ray image for plaque #1 (A) injection molded with in-line compounded material, plaque #6 (B) for the compression molded material, and plaque #9 (C) for the pultruded pellet material; all are 40% GF+PP material, but fiber bundles and dispersion level are different

The pultruded fiber pellet material usually, in this study, showed highest strength, particularly when compared with the in-line compounded material. These results are likely due to two effects. First, the pultruded fiber pellets disperse well, which is the key advantage of the pultrusion process. In the pultrusion process, the fibers are pulled through the heated bath of resin and then through a pultrusion die, which coats melted resin onto each fiber. Such a process adds cost, but also improves the wetting of fibers and hence dispersion ability of the fibers during processing. For the other processes such as direct in-line compounding, the fiber length is similar but there is some bundling of fibers, and, as shown above, bundling reduces strength. Second, during the in-line compounding, wetting is not as effective as in pultrusion, which means fiber resin bonding is poor, resulting in lower mechanical strength. Typically, the decision on what material to use is balanced based on the cost of the material, the manufacturing process being used, and the target performance. For example, in many fiber-reinforced polymer parts, a certain thickness of material is required for stiffness or appearance, such as to avoid bleed-through or readout concerns[1]. Therefore, the part will be thicker and additional strength through the improved fiber resin interface is not needed.

Table 8.3 Comparison of Fiber Length Measured for Selected Samples

	% of dispersed fibers	% of bundled fibers	Average length of dispersed fibers (mm)	Average length of bundled fibers (mm)	Total number of dispersed fibers	Average length of sample (mm)
30% injection In-line compounding Plaque #2	99.65	0.35	1.50	16.5	1278	1.50
30% injection Extruded Plaque #7	86.25	13.75	1.42	2.0	1122	1.48
40% injection pultruded pellets Plaque #9	100	0	1.43	0	655	1.43
30% injection compression Plaque #4	97.65	2.45	1.48	16.6	885	1.51
30% compression Plaque #5	44.7	55.3	6.39	19.5	132	9.25

As already mentioned in Chapter 4, fiber length and its effect on the mechanical properties such as strength, stiffness, and impact performance were systematically studied by Thomason and Vulgt [1, 2]. Their key contribution was to develop a fiber-reinforced material where fibers of known consistent length were used. The results of their study provided good reference data for trends that are expected based on mechanics of the material. The key trends are presented in Figure 8.11(A).

[1] Bleed-through or readout is when the bonding areas between shells of a panel become visible within a class A body panel side due to stresses resulting from warpage. Thin body panels are always more susceptible to this issue.

As the fiber length increases the mechanical properties such as modulus, strength, and impact resistance improve up to a certain fiber length, after which the advantage levels out. Thomason and Vulgt used short and well dispersed fibers. Since then many experiments and studies have shown that if the fibers are long and bundled, their contribution will be smaller compared to the well dispersed straight fibers. Also, they may show a decrease in mechanical properties, as illustrated in Figure 8.11(B). *Again, this effect can be explained by the L/D effect of the fibers, which is left out when the properties are plotted as a function of length alone.*

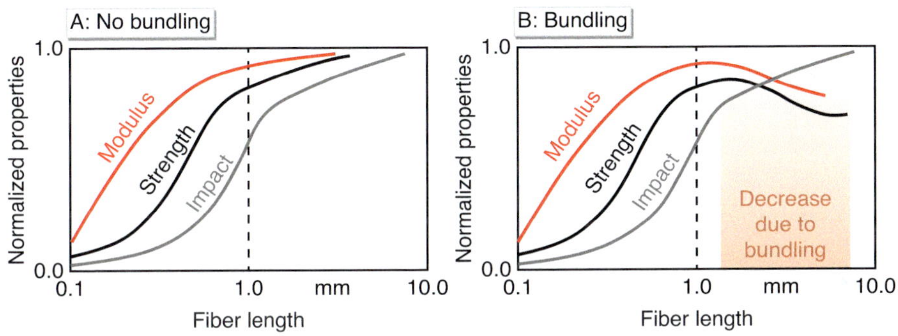

Figure 8.11 Effect of fiber length on mechanical properties and effect of bundling of the fibers

The fiber orientation for selected plaques was measured using the method of ellipses. Various methods to measure the fiber orientation have been discussed in Chapter 4. The method of ellipses was used because it has been in use in the industry for many years and is a well-established method. Recent developments in imaging technology and advances in computing tools have led to a newer µCT scan based method, which is faster and is gaining popularity. The orientation tensor is measured at the center of the longitudinal plaque's test sample. The orientation tensor distributions through the thickness are compared for 30% glass fiber-reinforced plaques in Figure 8.12 and for the 40% glass fiber-reinforced plaques in Figure 8.13. The data shows the effect of processing on the fiber orientation. For example, the pultruded fiber pellet material, which is well dispersed, shows orientation distributions that are symmetric, with higher alignment along the skin and then rapidly dropping near the center. On the other hand, the compression molded samples show a non-symmetric distribution, with lesser alignment of the fibers near the skin. There, the fibers are more random and consequently the material anisotropy is smaller. It is believed that this is a result of fiber-fiber interactions due to longer and bundled fibers within the compression molded plaque. The injection compression molded plaque shows the fiber distribution somewhere between the injection molded pellet and the compression molded material. Also, it was observed that in direct mixed long fiber material, where the fibers are chopped and added in the extruder while mixing the resin, the variation in fiber alignment through the thickness is smaller, and consequently the bending stiffness is slightly lower. This can be attributed to the longer fibers interacting with each other.

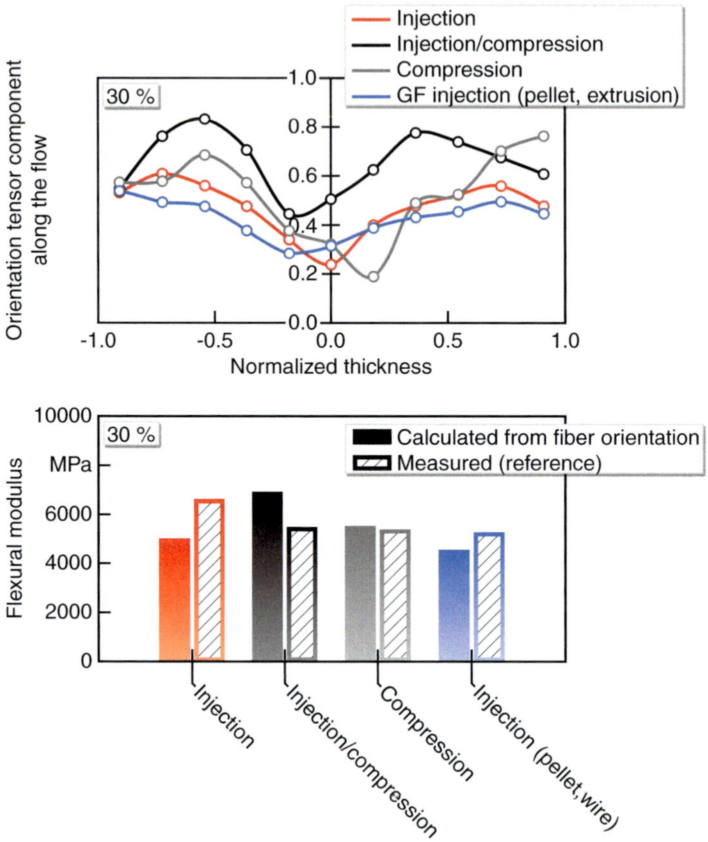

Figure 8.12 Top: 30% GF+ polypropylene, fiber orientation comparison; bottom: the flex modulus in the flow direction calculated from the measured fiber orientation, compared with the actual measured samples #34

The measured orientation through the thickness can be directly related to the flex modulus of the material with n layers. This can be achieved using a simple relation as follows:

$$E_{11} = \Sigma\, E_{11,i}\, t_k \left(i = 1 \text{ to } n\right)$$

where E_{11} is the elastic modulus of the material section with n layers and $E_{11,i}$ is the elastic modulus of the ith layer with thickness t_i. The $E_{11,i}$ for each layer can be calculated based on the fiber volume fraction, aspect ratio, and orientation tensor in that layer, using a homogenization approach such as Mori-Tanaka.

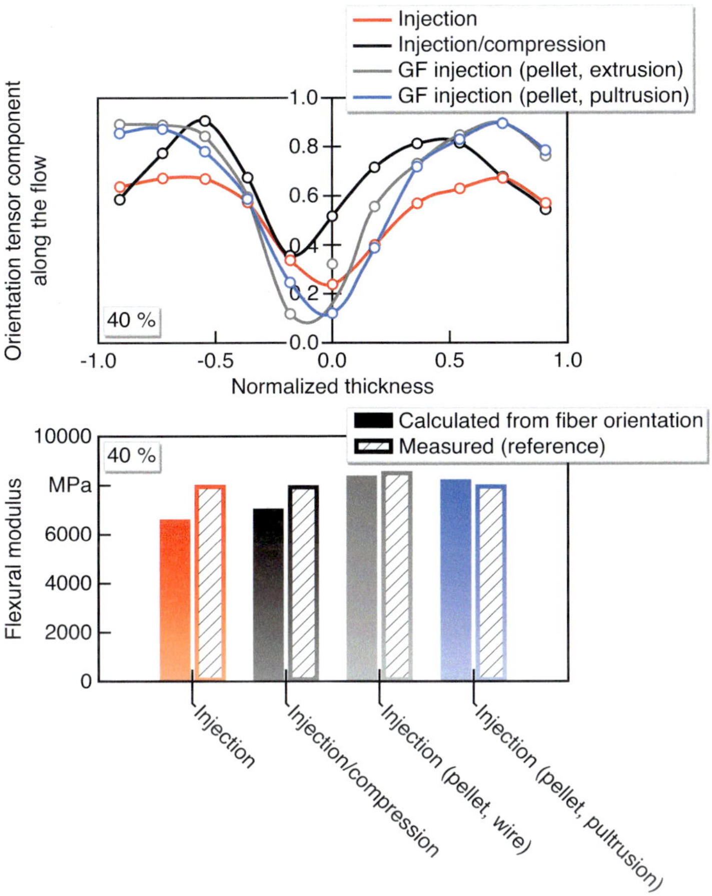

Figure 8.13 Top: 40% GF+ polypropylene, fiber orientation comparison; bottom: the flex modulus in the flow direction calculated from the measured fiber orientation, compared with the actual measured samples #34

The flex modulus can be expressed as

$D_{11} = \Sigma E_{11,i} (t_i/12 + t_i z_i^2)$;

for i from 1 to n, where z_i is distance of the ith layer from the neutral axis.

Similarly, D_{22}, D_{33}, etc., can also be calculated using the value of E_{22} and E_{33} in each layer.

The D_{11} for each layer, calculated based on the measured fiber orientation, is presented in Figure 8.12 and Figure 8.13. The measured flex modulus from the bending test (Table 8.2) is compared as reference. The difference between the two is due to two reasons. First, the orientation tensor varies through the actual part, whereas we are using measurement from only one location. Second, the orientation we are using is at the center of the sample #3, whereas test samples are at different locations. The values and trends are close as one would expect.

The fiber orientation observed in this study related well to the physics of the processing involved. Also, the fiber orientation relates well to the mechanical properties, i.e., bending stiffness and tensile strength measured. The relation is mostly based on trends, because one is only looking at fiber orientation at one location, i.e., the center of the tensile test specimen (sample #3). The actual fiber orientation throughout the part varies from one location to the other. Also, the fiber orientation for the bending specimen can be slightly different. Furthermore, other parameters such as fiber length variations based on the manufacturing process can vary and can also have an influence on the mechanical properties.

8.2 Multiscale Finite Element Simulation for the Fiber-Reinforced Flat Plaques

The study of plaques manufactured using the same material, but different processing conditions indicates the importance of processing on the final properties for the fiber-reinforced material. The plaque with same fiber and resin materials in the same proportion has different properties because the fiber condition, i.e., length, orientation or dispersion of the fibers can be different. Therefore, including the effect of manufacturing in the FE model for the fiber-reinforced material is very important. In this section, we will discuss how a multiscale modeling approach, discussed earlier in Chapter 6, can be used to model the fiber-reinforced materials. We will illustrate how to model the fiber-reinforced material for the FE model of the plaque and demonstrate the effect of different fiber microstructure (resulting from the manufacturing process) on the final mechanical properties of the plaque.

The key steps of the process are presented in the flow chart of Figure 8.14. There are two distinct simulations involved. First is the simulation of the manufacturing process and the second is the structural loading simulation. Both simulations are quite different and done separately. The results of the manufacturing simulation provide fiber microstructure in the finished part, and are used to develop the multiscale material properties used in the structural simulations. Details of the simulation methods are discussed in Chapter 6 (structure) and Chapter 7 (process).

The manufacturing process simulation, i.e., injection or compression molding, is carried out using commercial mold filling simulation software. The mold filling simulation process is validated by comparing the microstructure, such as the resulting fiber orientation distribution tensors, with actual orientations measured within the parts at selected locations. In general, measurements of microstructure from the actual parts can be quite a time-consuming process, so measurements for the actual parts are done at selected locations only. For the simulation, such detail over the whole part is easily available. This is an advantage of using simulation. The resulting data from the process simulation, such as fiber orientation and fiber length distribution, are used to develop the multiscale material model. Finally, the multiscale material model developed using the microstructure details from the process simulation is used with a structural finite element model.

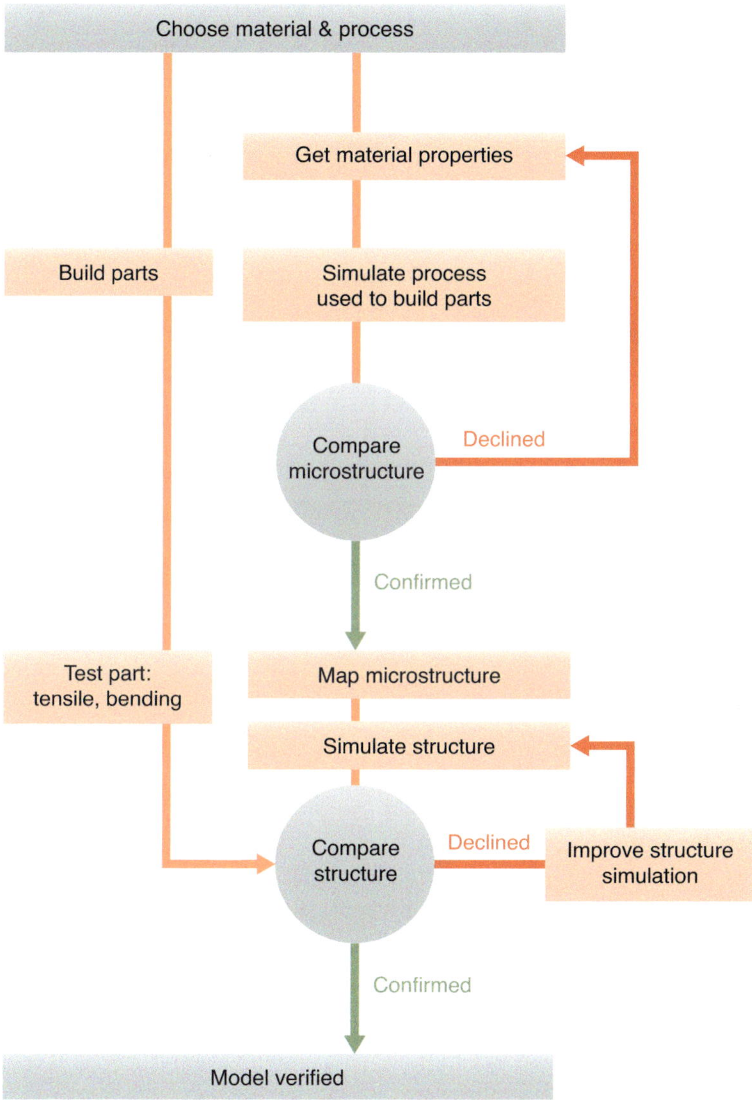

Figure 8.14 Flow chart to develop and verify the multiscale modeling for the plaque

The manufacturing process used and microstructure study for the actual parts were already discussed in Chapters 3 and 4, respectively, and key conceptual steps used for multiscale modeling were discussed in Chapter 6. The key steps needed in the implementation of the multiscale modeling using commercial software such as Digimat™ and Abaqus™ are presented in Figure 8.15. Though Digimat and Abaqus were used in this study, other software is also available to achieve the same objectives.

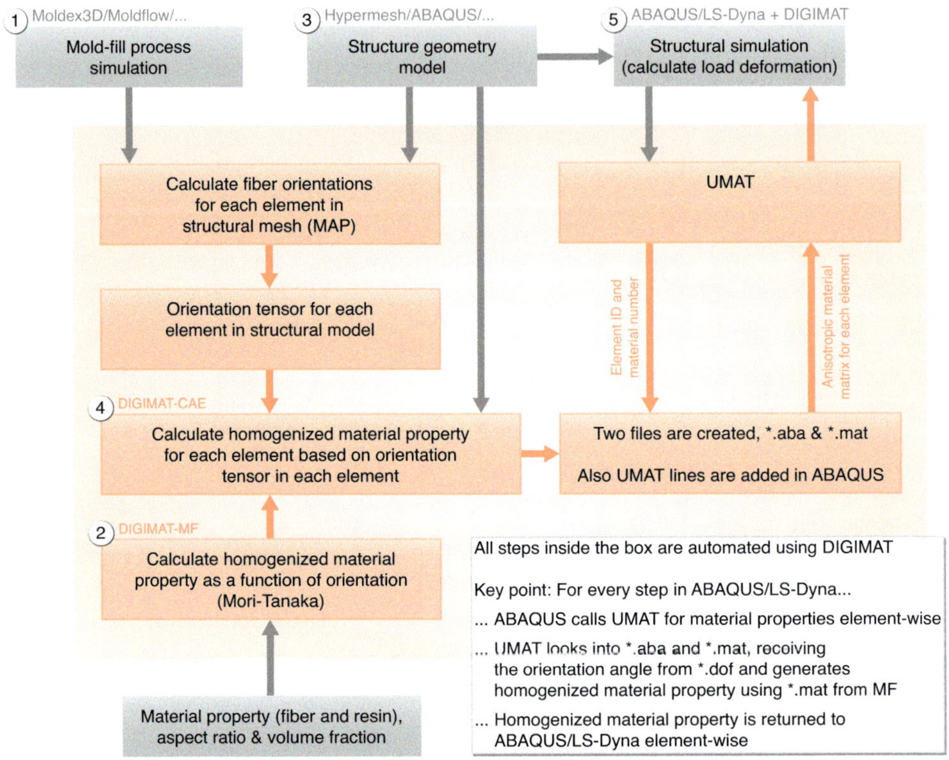

Figure 8.15 An example of practical implementation of multiscale modeling

The five key steps for the multiscale FE analysis are as follows:

Step 1. Mold filling simulation: The resin is treated as fluid with fiber suspended in it and using the flow equation and principal of conservation of momentum the mold cavity filling process is simulated. Detail of such mold filling simulation computation method is discussed in Chapter 7. This step is important as it helps create microstructure details needed for the material properties.

Step 2. Homogenization: Based on the fiber and resin properties a homogenization formulation is developed. Further detail of different homogenization is presented in the Chapter 6. Mori-Tanaka mean field homogenization is the most popular approach in use now. The homogenized properties are prepared as a numerical function of the orientation tensor.

Step 3. Structure model: A finite element mesh for the structural analysis of the component of interest is developed. Preparation of such a finite element mesh for structural purposes is quite common. The finite element mesh can be in the millimeter scale. A key difference, compared to the traditional approach, is in the next step when the material property will be assigned.

Step 4. Material property estimation: This is the key step to introduce the microstructure based material property in the finite element model for the structural analysis. The homogenized material model, which can be a numerical function of the fiber orientation tensor, fiber length, and fiber concentration, is mapped based on the fiber properties estimated through the process simulation. This mapping is done throughout the component for each element at each integration point, which means that the material property at each integration point can be different.

Also, it is important to point out that in this example the microstructure is a representative volume element (RVE), which can be a finite element or some part of a finite element, and is homogenized to get the material properties for calculation of the plaque FE model. It is possible to use other methods such as full finite element or generalized method of cells to create material properties for use in the model (see Section 6.3 for details). Generally, homogenization is cost effective and robust, and therefore it is the most popular approach.

Step 5. Finite element simulation: Finite element simulation is carried out with the "updated material" properties. The material can be updated at every step of the simulation. The finite element model is in the millimeter scale; the material properties used are based on the microstructure at micrometer scale.

In this case, Digimat™ is the key piece of software that helps in using microstructure details estimated from the process simulation to generate material properties (microstructure dependent) for use in the finite element structural simulation. Furthermore, while updating the material properties both geometric and material nonlinearities can be included.

Ideally, one would like to include fiber length, orientation, and concentration resulting from the process simulation for the structural model. However, the current process simulation software has limited capabilities. As discussed in Chapter 7, Section 7.4.1, mold filling simulation software can estimate the fiber orientation distribution quite well for short fibers. The algorithm to estimate the fiber orientation in a viscous fluid was initially developed by Folgar and Tucker [3]; the method to express fiber orientation distribution as a tensor for use in the numeral calculations was developed by Advani [4]. Both these works comprise the foundation of current fiber orientation estimation in simulations. The algorithms have improved over time to include the effect of longer fibers [4–7] and have been adopted in most mold filling simulation software.

Techniques to estimate fiber length within the molded finished parts are complex. This is because fiber attrition occurs at many locations as shown in Figure 8.16. This aspect is discussed in detail in Chapter 4.

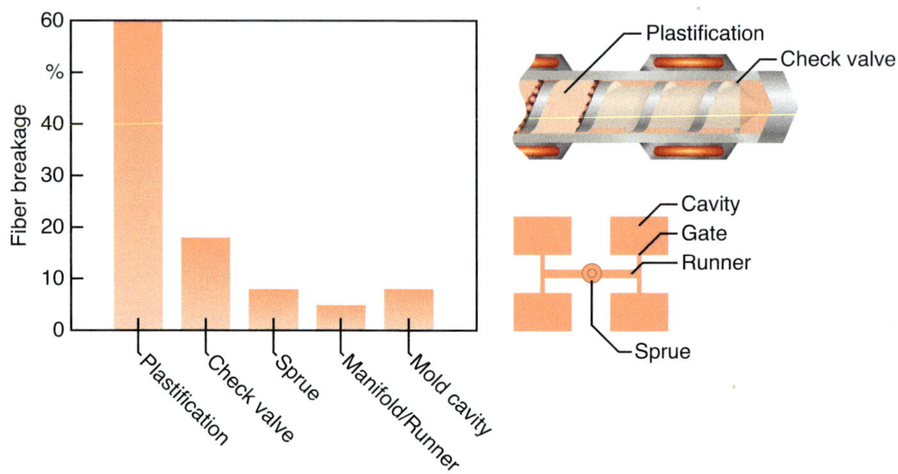

Figure 8.16 Fiber attrition in the mold filling process

Initial algorithms to estimate fiber breakage based on stresses generated during the mold filling process has been proposed by Phelps and Tucker [8] and implemented in many commercial software packages. Detailed discussion on these capabilities and theoretical approach are discussed in Chapter 7, Section 7.4.2. The accuracy and robustness of such algorithms has been demonstrated for simple shapes [9, 10].

Fiber concentration, which is usually measured as a volume fraction, varies through the part from location to location and through the thickness. This is discussed in detail in Chapter 4. Fiber variation through the thickness is due in great part to fountain flow effect. Algorithms to estimate the fiber concentration variation through the thickness are discussed in Section 7.4.2. The largest fiber concentration variations occur at edges and the areas where there is a bifurcation in the flow such as a rib, due to jamming. Furthermore, when the resin experiences a change in direction, fibers do not change direction well, leading to further fiber-matrix separation. Consequently, fiber concentration in the ribs is smaller while at the base of a rib the fiber concentration can be slightly higher. Furthermore, as Kuhn [11] has demonstrated, fibers can bridge across the thickness of a rib, creating small dams that hold fibers back. These fiber jamming effects are special situations. The currently popular fluid mechanics based approaches do not account for individual fibers and therefore cannot model such jamming. Direct-fiber or mechanistic modeling approaches, where each fiber is modeled as multiple beam elements, are useful in such situations, as demonstrated by Perez [13]. Since there are millions of fibers even in a very small part, such a direct fiber approach is computationally intensive. Often such a direct fiber model is limited to a smaller sub-volume instead of a full simulation of a part. Further discussion on such an approach in presented in Chapter 7, Section 7.5. In the case of the plaque, because of the absence of ribs, to simplify the study, it was assumed that the fiber volume fraction stays consistent throughout the part.

In summary, prediction methods to assess the effect the process has on fiber length and concentration are still under development; only the methods to estimate fiber orientations are robust and have been thoroughly proven. This can be a concern as not all microstructure details can be estimated for multiscale models. However, the good news is that the fiber orientation has a very large effect on the structural properties. Therefore, even if only the effect of fiber orientation distribution is accounted for in the models, it is still possible to have a useful prediction, especially when compared to using isotropic or orthotropic materials in the structural analysis.

8.2.1 Details of the Finite Element Model

This section presents details of how to develop a finite element model of the coupons and the assumptions involved. For the physical parts, the plaque is the molded piece, and then the coupon samples are separated using a water jet cutting process for testing. The details of the coupon locations in the plaque are presented in Figure 8.2. The coupons are tested for tensile and bending loads. The first step in the modeling process is to simulate the mold filling process of the plaques. To simulate the effect of cutting of the coupons, finite element models of the tensile and bending coupons are developed and the fiber orientation distribution tensors predicted by the mold filling simulation from the relevant locations are mapped on to the finite element models of the coupons.

8.2.1.1 Process Simulation Model

The details of the model used for mold fill simulation for the plaque is presented in Figure 8.17. The same model geometry is used for all the three processes of interest, i.e., injection molding, injection compression molding, and compression molding. The process parameters for the three processes are different; they are presented in Table 8.4. The material properties for 30% and 40% glass reinforced polypropylene used for the process simulation are taken from the standard library available in Moldex3D™. The material properties are presented in Figure 8.18 and Table 8.5. The main output of interest from the mold filling process is the fiber condition, i.e., fiber orientation throughout the finished parts in this case, which is used to develop multiscale material properties for the structural calculations. There are 875,000 elements and 466,000 nodes in the model.

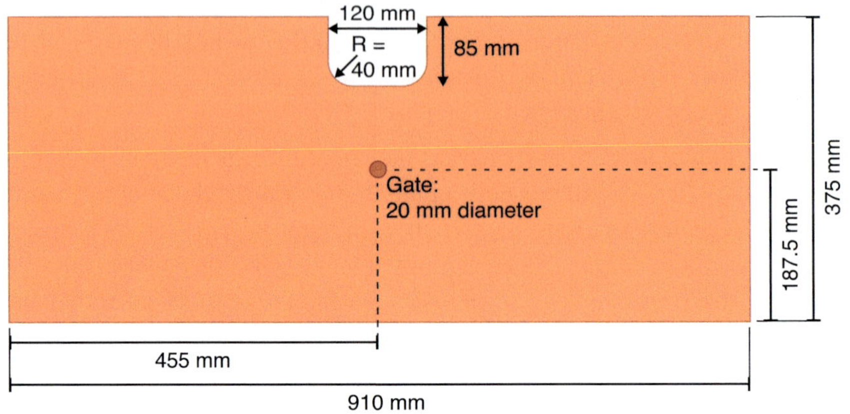

Figure 8.17 Detail of the model used for the process simulation

Table 8.4 Process Parameters Used for Injection Molding and Injection Compression of a Flat Plaque

Process parameters used	Actual process: Injection molding (reference)	Injection molding	Injection compression
Melt temperature (°F)	460	460	460
Mold temperature (°F)	190	190	190
Fill time (s)	5.3	5.34	5.8
Switch over	NA	99% of fill	NA
Gap to close (mm)	10	NA	10
Press closing speed (mm/s)	NA	NA	5
Maximum injection pressure (psi)	6000	4000	NA
Pack time (s)	6	6	6
Pack pressure (psi)	2300	2400	2500
Mold open time (s)	8	8	5
Cooling time (s)	30	30	30
C_i	NA	0.01	0.01

Process parameters for compression molding:

Temperature of charge	475 °F
Temperature of the mold	140 °F
Size of charge	200 × 335 mm
Pressure	1000 tons

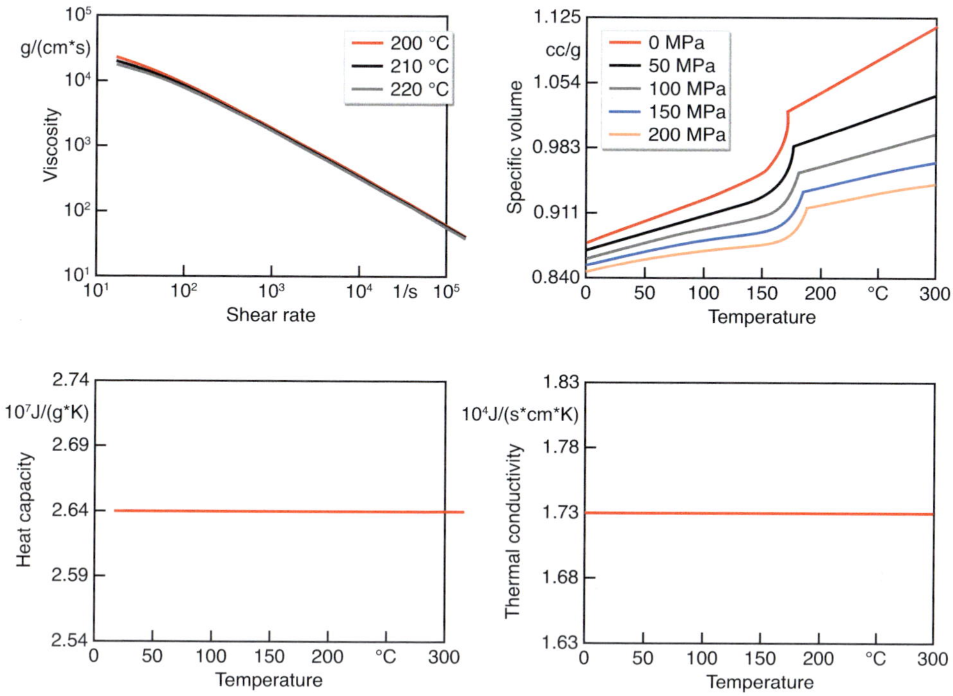

Figure 8.18 Material properties used for the mold fill simulation of the flat plaque

Table 8.5 Material Properties Used for the Process Simulation of Flat Plaque

Mechanical properties	Values	Units
Polymer density	920	(kg/m^3)
Polymer modulus E	2.0E+03	MPa
Polymer CLTE	1.0E−04	(1/K)
Fiber weight percentage	30 & 40	%
Fiber density	2.6E+03	(kg/m^3)
Fiber modulus E1	7.0E+04	MPa
Fiber modulus E2	7.0E+04	MPa
Fiber shear modulus G	3.0E+04	MPa
Fiber CLTE a1	5.0E−06	(1/K)
Fiber CLTE a2	5.0E−06	(1/K)
Fiber length/diameter	Varies (20−50)	NA
Interaction coefficient in F-T model (C_i)	0.01	NA
Polymer Poisson's ratio	0.3	NA
Fiber Poisson's ratio	0.2	NA

8.2.1.2 Structure Simulation Model

The finite element model for the coupon sample used for the tensile and bending tests are prepared using solid 3D elements. Essentially, these models are made up of a finite element mesh representing the geometry and loading condition used in the testing. The tensile coupon model is presented in Figure 8.19. The details of the three-point bending specimen are presented in Figure 8.20.

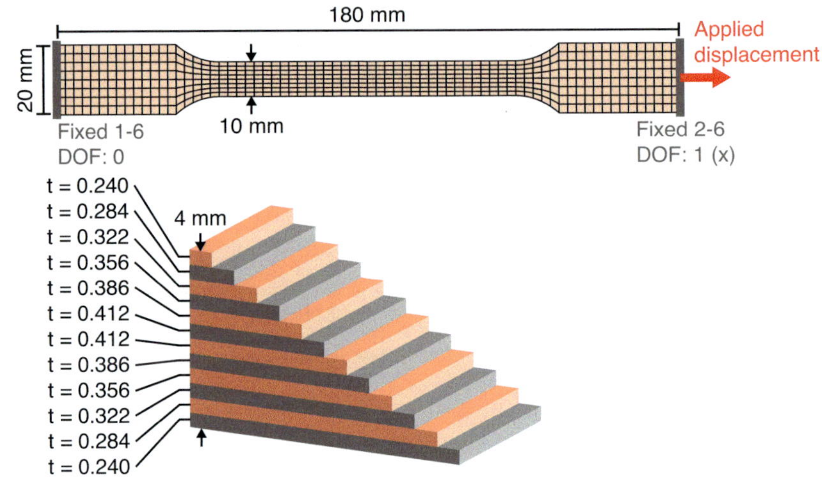

Figure 8.19 Details of the finite element model used for the coupon for the tensile test

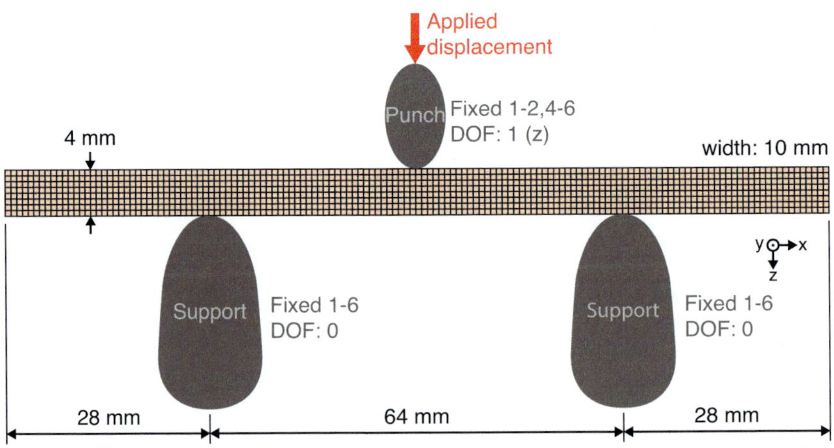

Figure 8.20 Details of the finite element model used for the coupon for bending test

Both the process and structural simulation models use finite element meshes; however, the finite element mesh used for each model can be completely different as the analysis employed is different. Usually the mold filling simulation models tend to have simple but a larger number of elements. The fiber orientation distribution tensor results are computed for each element in the process simulation model. These fiber orientations are mapped onto the structural model (which can have completely different mesh) using linear geometrical interpolation. For this case study, commercial mapping features available from Digimat™ were used to map the fiber orientation from the plaque to the structural model. Figure 8.21 shows the process. The fiber orientation from the results of process simulation is mapped on the structural finite element mesh.

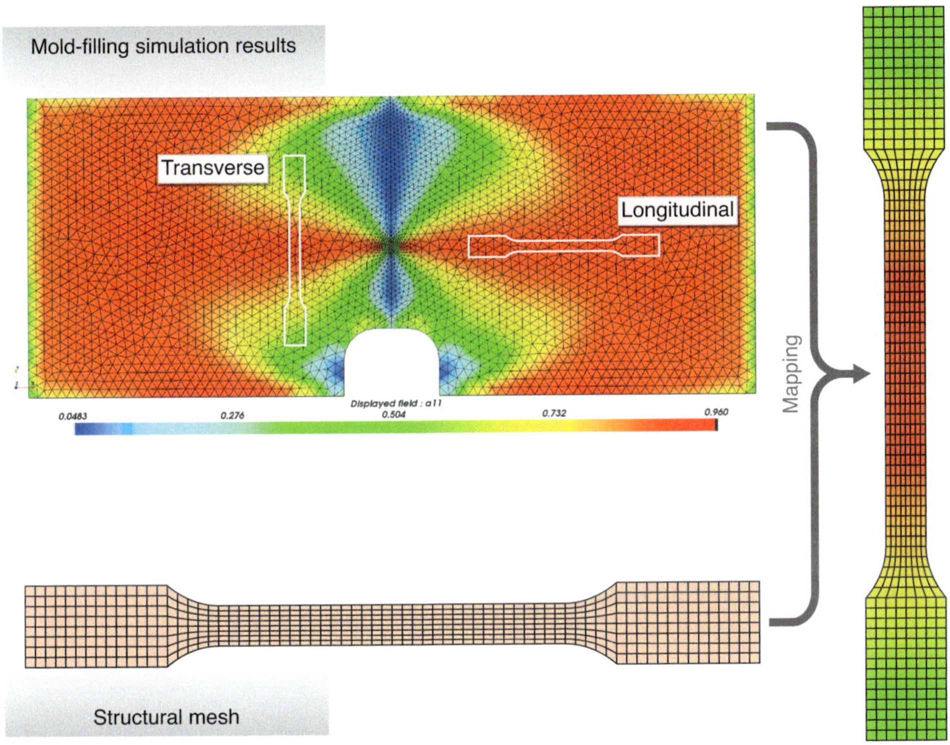

Figure 8.21 Mapping process used to transfer fiber microstructure to the structural FE model

8.2.1.3 Discussion on How to Choose the Modeling Details

While developing a mathematical model for any physical system, usually there are many assumptions and approximations involved. Therefore, discussion on the input parameters used for this case study is necessary. Parameters such as mold geometry, inlet pressure, hold time, cooling time, mold temperature, etc., are based on the parameters measured in the actual process. Usually these parameters are used based on the actual measurements from the process, or, when process parameters are not available, such input parameters are selected based on the experience and know-how gained from previous cases.

The mold filling process starts with the melted resin material pushed at high pressures from the plasticating unit of the injection molding machine into the mold cavity. As a result, the melted resin is compressed, followed by cooling in the mold cavity, and finally the material starts to solidify. For the mold filling process, the material properties, such as *PvT* (pressure, volume, temperature) curves, viscosity, thermal properties, etc., of the resin material in the molten state are required. Elastic modulus, coefficient of thermal expansion, etc., in the solid phase are also required. Measurement of such material properties is quite a time consuming and expensive process. Usually each software package maintains a library of material properties. Often the material manufacturer sponsors the evaluation of the material and the development of material property data for use with the mold filling simulation software. Within this study, generic 30% and 40% long glass fiber filled polypropylene data, as shown in Figure 8.18 and Table 8.5, were used for the calculations.

The parameters such as C_i, representing fiber–fiber interactions, and aspect ratio are selected based the process details as well as users' experience and knowledge. C_i, which is part of the fiber orientation prediction formulation discussed in Chapter 7, is often referred to as the interaction coefficient. C_i represents the relative interaction or load transfer among the fibers. C_i varies from 0 to 1, where 0 means no interaction and 1 means full interaction or highest load transfer among the fibers. Generally, users are expected to choose C_i between 0 and 1, based on the knowledge of interaction between the fibers. In practice, it is one of the input parameters that can be adjusted to improve the correlation between the simulation and the measured results. Recently, a method to estimate C_i more accurately using direct fiber simulation has been proposed [12, 13]. In this approach, each fiber is modeled as a combination of a number of beam elements. Then, the behavior of such material in pure shear load using such a model is deduced. Such a model, even if representing few millimeters of material in shear flow, can be quite large, as there will be thousands of fibers. The results from the direct fiber (or mechanistic) model are compared with the flow based mold fill simulation (e.g., Moldex3D™ or Moldflow™) for the same conditions, to identify the C_i value.

The aspect ratio, defined as ratio of fiber length to diameter of individual fibers, is another input parameter that requires additional user considerations. Aspect ratio is used in both process simulation and structural simulation models. For the process simulation model, the effect of aspect ratio is discussed in Chapter 7, Equations 7.12 to 7.14. For the Folgar-Tucker model [3] for the fiber orientation, the shape factor of a particle is defined as a function of aspect ratio a_r as $\xi = (a_r^2 - 1)/(a_r^2 + 1)$.

The relation between ξ and a_r is presented in Figure 8.22. We observe that ξ reaches almost 1.0 for an aspect ratio of 30, which indicates that the effect of aspect ratio of further higher values is not so important in process simulation with Folgar-Tucker formulation.

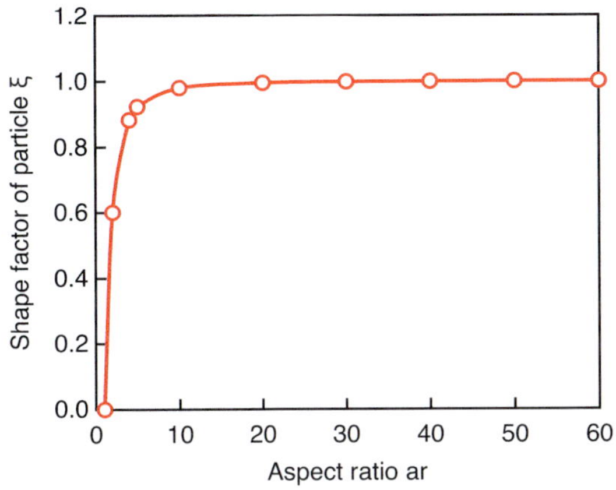

Figure 8.22 Relation between the aspect ratio (a_r) and ξ

As the aspect ratio a_r continues to increase, one of the concerns is that the Folgar-Tucker algorithm used to estimate the fiber orientation no longer captures the fiber orientation development, as the longer fibers tend to take longer to reach their final orientation. Details of the long fiber algorithms are discussed in Chapter 7. Usually the mold filling software allows different algorithms for the long fibers, which often requires knowledge of additional parameters to define the fiber to fiber interaction. Glass fibers have a typical diameter of 14–17 µm. Aspect ratios of 100 or higher, which indicate lengths around 1.4 mm in this case, are considered as long fibers. The number 100 is somewhat arbitrary, mainly based on observations where the Folgar-Tucker algorithm, intended for short fibers, shows need for improvement.

The aspect ratio is also important for the structural simulation. The fiber orientation is mapped as a tensor from the mold fill simulation results of the plaque onto the structural model. The orientation tensors are used with the fiber and resin properties to estimate the material properties using Mori-Tanaka (M-T) homogenization. The M-T homogenization assumes that the inclusions, i.e., fibers, are elliptical with two axes of dimensions. The effect of change in the aspect ratio on the strength and stiffness for the plaque material based on the multiscale simulation using M-T homogenization is presented in Figure 8.23. As expected, it was observed that the aspect ratio increased the mechanical properties, which eventually level out. This agrees with the formulation presented in Chapter 5, Section 5.3.

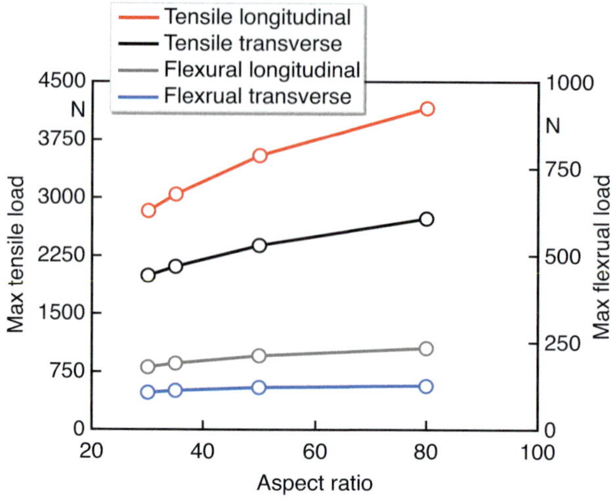

Figure 8.23 Effect of aspect ratio on the mechanical properties

The aspect ratio in the process simulation and in the structural simulation is used by completely different algorithms. The aspect ratio used should be the same for both simulations to represent the physics accurately. However, it has also been observed that the effect of aspect ratio in the process simulation when using the Folgar-Tucker model is limited when having aspect ratios that go beyond about 30 or 40. The effect of an aspect ratio higher than those values is included in the simulation using larger interaction parameters C_i, or a newer long fiber formulation for the longer fibers [8].

In practice the aspect ratio is greater than 30, and hence the value used for process simulation and structural simulation should be close to that. Since fibers are flexible and long, and consequently can be curved, the effectiveness of fiber length is reduced. Therefore, one can have a shorter "effective aspect ratio" for structural calculation. Further discussion on how to estimate "effective aspect ratio" is presented in Figure 8.32.

8.2.2 Simulation Results for the Plaques

The study of the 40% GF+PP injection molded plaque is used to illustrate the advantage of the multiscale modeling approach. In Figure 8.24 the fiber orientation distribution at the center of transverse and longitudinal samples for three different fiber interaction conditions is presented. The fiber orientation tensor at the center for both the longitudinal and transverse samples is the same, because the plaques are symmetric and the center point on the selected samples is at the same distance from the inlet gate. The fiber orientation varies through the thickness: fibers are more aligned near the skin and less aligned near the center, as expected due to fountain flow effects. Also, depending on the interaction coefficient selected, the fiber orientation tensor along the flow direction varies. We observe that lower interaction means less fiber to fiber load transfer and hence more fibers are aligned along the flow in the shell layers. The load and deformation for the tensile test are compared with the simulation in Figure 8.24. We observe that when fibers are highly aligned along the flow direction the tensile strength along the flow direction is higher than perpendicular to the flow direction, indicating highly anisotropic properties. As the fiber to fiber interaction is increased, the alignment of fibers along the flow direction is reduced, and the tensile strength along the flow direction is also reduced, while tensile strength along the cross-flow increases somewhat. As discussed earlier, C_i is the interaction coefficient, which can be varied between 0 and 1 to improve the correlation between the simulation and the test. For a specific value of the interaction coefficient C_i it is possible to match both flow and cross-flow load deformation curves for a given set of material parameters. The flow and cross-flow tensile test simulation are carried out using the same finite element model and with the same fiber and resin properties. The only difference is the resulting fiber orientation tensor. This is the key point of multiscale modeling; by varying the microstructure, fiber orientation in this example, we can get different load-elongation results from a model with the same geometry and materials.

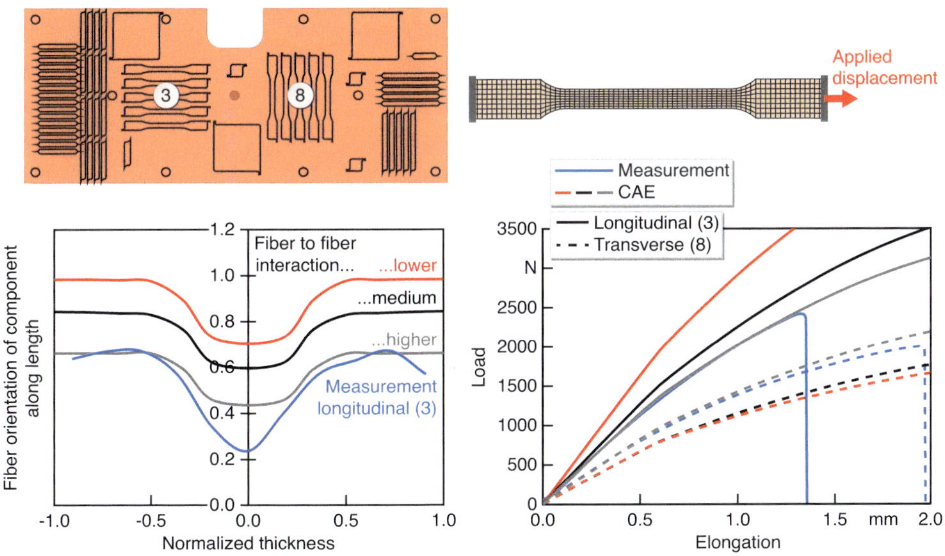

Figure 8.24 Effect of fiber orientation tensor on the load deformation in the tensile test; three different fiber orientations are considered for plaque #1: 40% GF+PP injection molded plaque. Higher fiber interaction resulting in less aligned fibers shows good correlation with the measured results

The verification is further extended for the bending test results. Two samples from the plaque positioned perpendicular to each other (Figure 8.25) were evaluated for the three-point bending test. A simulation for the bending loading is also carried out. Again, the only difference between the two samples cut at 90 degrees to each other is their position in the plaque. Hence, the finite element model is the same, but with different orientation distributions.

The comparison of the two tensile, two bending, and corresponding simulations is presented in Figure 8.24 and Figure 8.25. The material properties for the resin and fiber are same for both; however, fiber orientation distributions, depending on the location of the sample within the plaque, are different. These results show that it is possible to get a good agreement between the simulation and tests if representative fiber orientations are used in the model. This is an excellent illustration of multiscale modeling, where fiber details at the micrometer level are integrated into the material model used in the finite element model at the millimeter scale, and good simulation results are achieved.

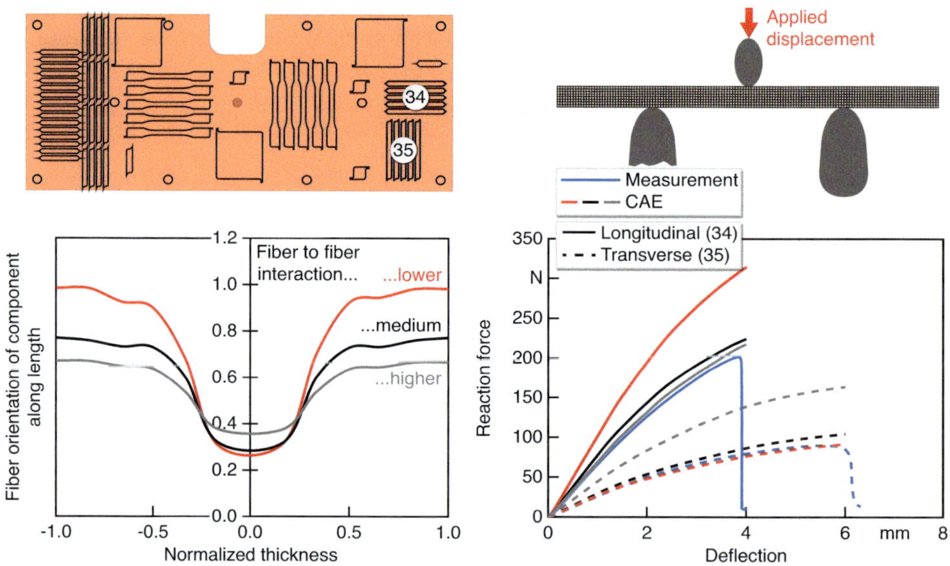

Figure 8.25 Effect of fiber orientation tensor on the load deformation in the bending test; three different fiber orientations are considered for plaque #1: 40% GF+PP injection molded plaque. Higher fiber interaction resulting in less aligned fibers shows good correlation with the measured results

Furthermore, the approach is also illustrated on tensile and bending test results for the samples from the 30% injection molded plaque, see Figure 8.26 and Figure 8.27. The base material properties are used for the resin and the fibers are the same for all calculations. The same finite element model for tensile loading is used for all the plaques as well as the same bending models. The only difference from case to case is in microstructure properties, such as the fiber orientation tensor, fiber concentration, i.e., 30%, and fiber aspect ratio for each plaque. It can also be observed that the fiber orientation has a profound impact on the material's mechanical behavior.

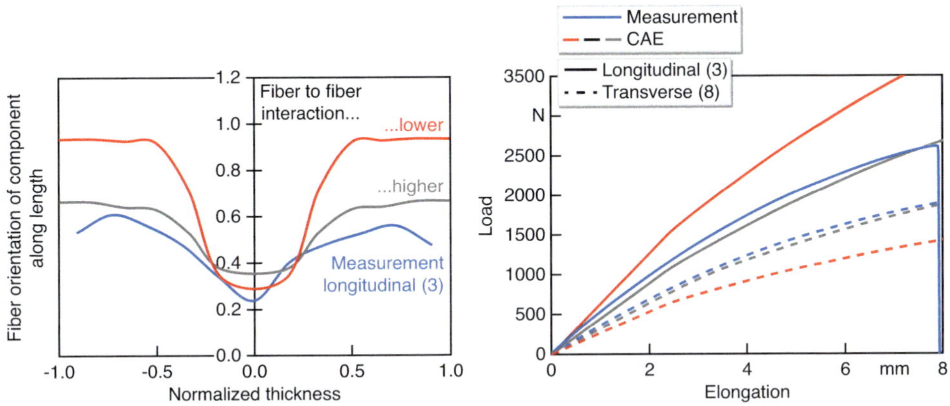

Figure 8.26 Effect of fiber orientation tensor on the load deformation in the tensile test; two different fiber orientations are considered for plaque #2: 30% GF+PP injection molded plaque

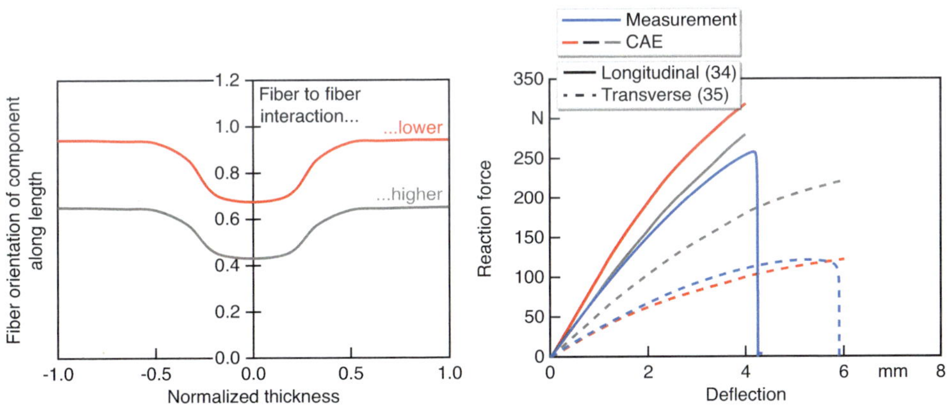

Figure 8.27 Effect of fiber orientation tensor on the load deformation in the bending test; two different fiber orientations are considered for plaque #2: 30% GF+PP injection molded plaque

8.2.3 Discussion on Applying the Multiscale Simulation

The multi-scaling approach requires updating of the material properties at each computation point, i.e., integration point for every step of the analysis. This adds significant computational burden, usually increasing computational time by a factor of 10. Therefore, one of the questions often raised is, what if the material properties are the same for all the elements, and are not updated with the deformation?

To understand the importance of multi-scaling over the whole part, this simpler material concept was evaluated for the tensile test on the longitudinal and transverse samples, i.e., coupons 3 and 8 in the injection molded plaque with 40% GF and PP, as presented in Figure 8.28. The single material property used was based on the orientation tensor at the center of the sample 3. The simulation results show that such a simplified approach does not work in this case; the force deflection is quite different compared to the measurements. Similar results have also been observed with many examples. With some special condition where fiber conditions are consistent through the parts, such an approach may work.

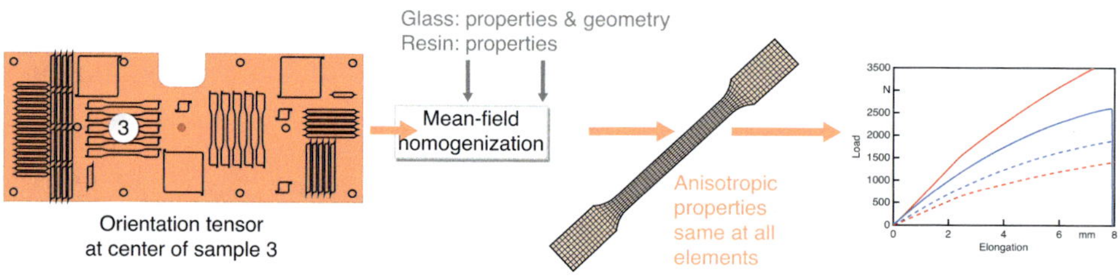

Figure 8.28 Effect assuming the same anisotropic material properties over the whole part and does not update with deformation. Solid line are longitudinal samples, broken line are transverse samples. Red color is CAE and blue is test results.

The next question is how sensitive are the material properties if there is a slight change in the position of the test sample within the plaque? To investigate such sensitivity, samples with similar orientation but slightly different locations are compared, using the multiscale simulation approach, as shown in Figure 8.29. For example, sample 3 is compared with samples 1 and 5, which are 60 mm away on either side. Similarly sample 8 is compared with samples 6 and 10. Similarly, the samples around samples 34 and 35 are also compared using multiscale simulation for bending loads. The simulation output suggests that the variation in the tensile and bending results for the selected samples, which are within 40–60 mm, oriented in the same direction, are quite small. Also, it was observed that such variations estimated in CAE are consistent with the variation observed in the actual sample tests.

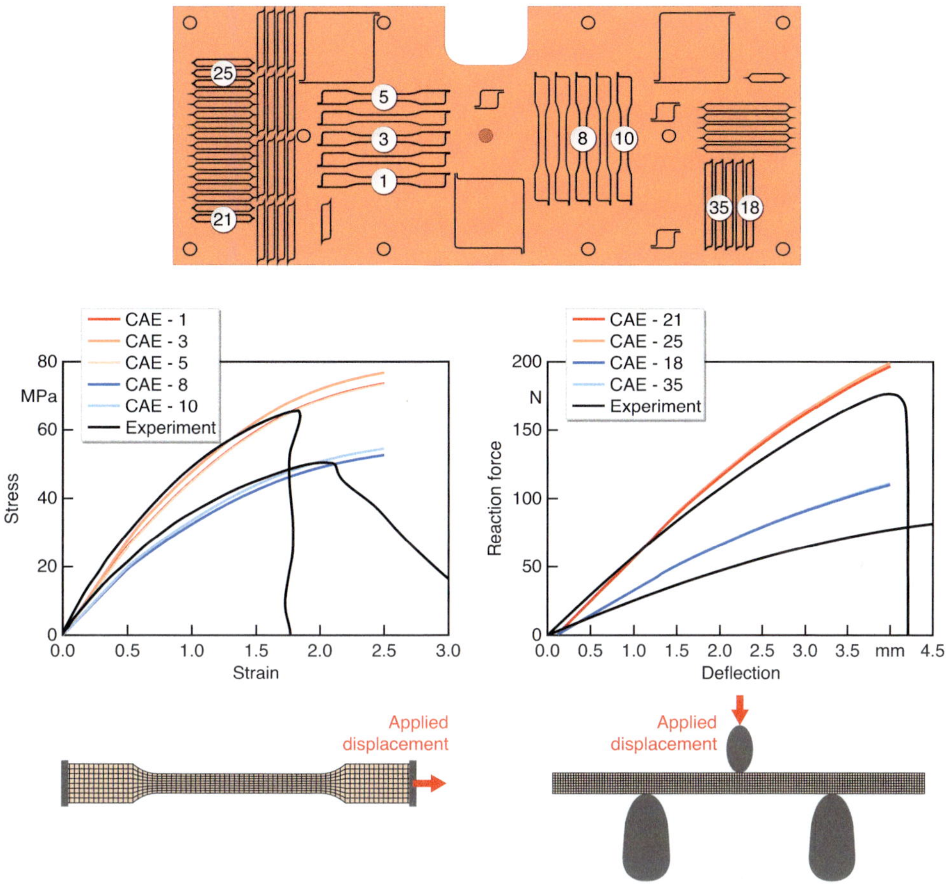

Figure 8.29 Effect of location on the final structural performance

As discussed in Chapter 7, the process simulation fundamentally starts with Jeffery's equation for an elliptically shaped particle embedded in the fluid, and extends to more complex conditions. While phenomenological terms and parameters are added to address some of the assumptions, Jeffery's algorithm assumes the fibers to be elliptical in shape. For the structural simulation, if the homogenization based on Mori-Tanaka is used to estimate the composite material properties, the fiber shape is assumed to be elliptical. The actual fibers, as shown in Figure 8.30, are cylindrical. Unless the fibers are short, they are likely to be curved and the elliptical assumption does not hold. However, it is a reasonable approximation and an essential one to develop mathematical approaches to modeling the fiber-reinforced material properties for use in the finite element simulation. Methods to address the longer and curved fibers have been under investigation for many years, and some options to address such fibers are available. In the next paragraphs, insights into addressing effects brought about by long fibers are presented.

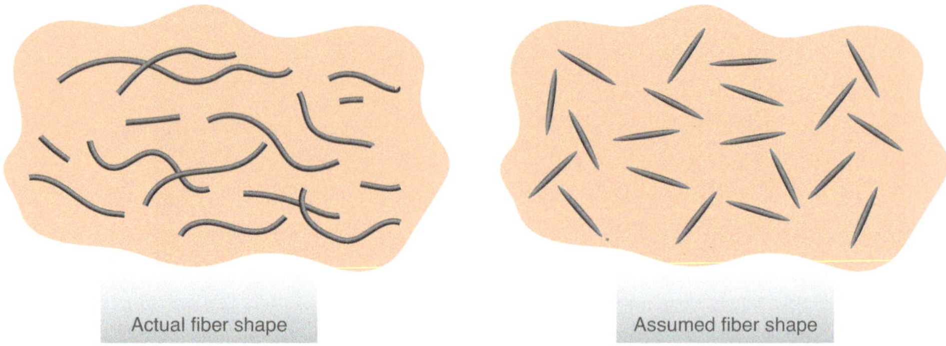

Figure 8.30 Actual (left) vs assumed (right) shape for the discontinuous fibers

The micro CT image of a compression molded plaque presented in Figure 8.31 confirms that the fibers are bundled, and are thicker and longer as well as curved. The current multiscale modeling framework discussed in Chapter 6 is not capable of simulating such fiber conditions unless finite element RVEs are used, since such RVEs are impossible for the typical parts of interest. The most practical approach has been to use the phenomenological improvements in the algorithms to account for such deviations from the assumptions.

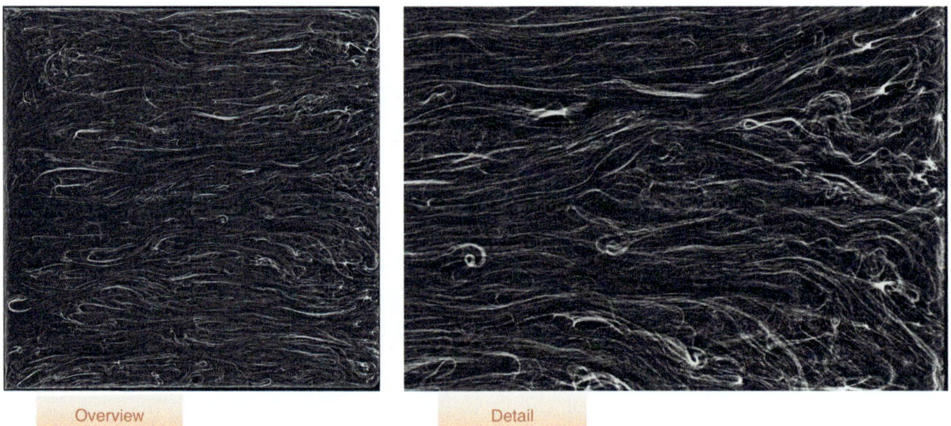

Figure 8.31 Actual condition of fibers in the compression molded plaque, with bundles and curved fibers; left side is the whole plaque, right side is zoomed-in image

To investigate this further, a detailed finite element model, where the complete geometry of the fiber and resin can be represented using solid 3D elements, was built as shown in Figure 8.32. The interface between the fiber and the resin surrounding the fibers is assumed to have perfect bonding, i.e., there is no slippage between the fiber and resin. The model is used to estimate the force required to stretch 3 mm along the length. The force required for different fiber concentrations by volume fraction as well as initial fiber curvatures are investigated using this model. It should be noted that as the curvature of the fiber is increased the load required to stretch the material reduces. This is expected as the curved fibers contribute less towards the tensile stiffness, when compared to the straight fibers. The force required to achieve the targeted stretching is highest when the fibers are straight. In the graphs, the effective force is defined as the force required for the given curvature divided by the force required for the straight fibers. The effective force for the various values of initial fiber curvature for different fiber concentrations is presented in Figure 8.32. In the graph, the curvature is defined by L'/L, where L is the length of the fiber, and L' is the distance between the ends of the fiber when curved. If the fiber curvature and concentration are known for a part, then the effective force can be estimated and used to "correct" estimates made based on the straight fiber models. For example, when a part with 20% glass fiber filled polypropylene is analyzed, a target deformation is achieved at 100 N load. However, if the fibers are known to have a curvature of 0.9, then the effective load drops to 75%, because of curvature.

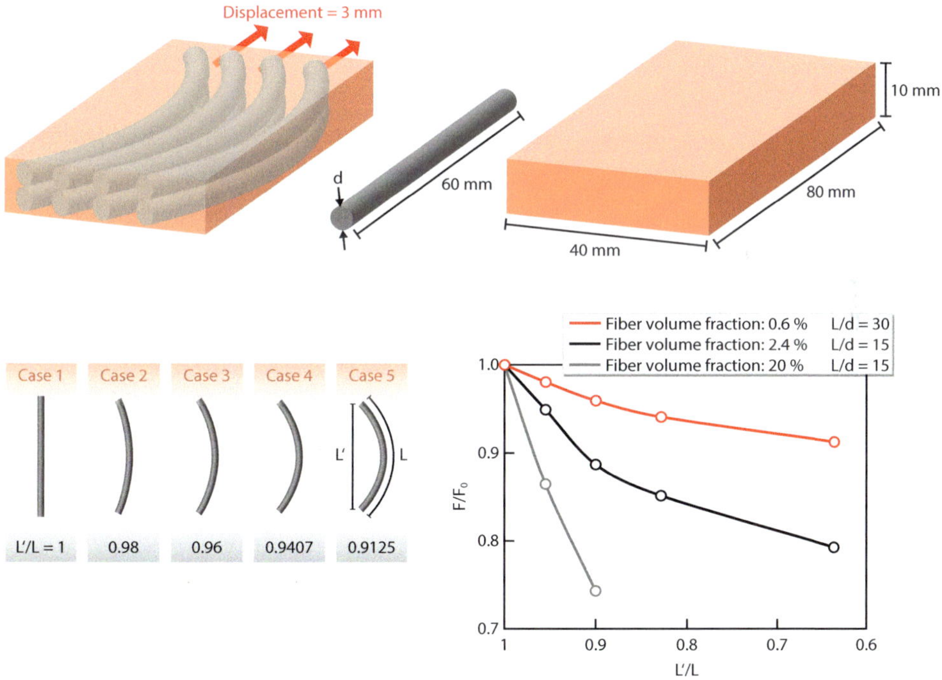

Figure 8.32 Analytical study to see effect of fiber curvature in the structural properties

Now, in practice the fiber curvature can be different for each fiber. Also, measuring such curvatures can be very challenging. However the observation that as the curvature of the fibers increases the effective load carrying capability of the fibers decreases can be empirically used. Therefore, to account for the effect of the curvature in the fibers, one of the simple but widely used approaches is to use the effective aspect ratio. Essentially, the aspect ratio is reduced by a percentage based on the curviness of the fibers. The reduction in the aspect ratio can also be based on experience and know-how. For a given fiber resin system, a finite element model of the curved fiber can also be used to estimate the reduction factors.

Another major challenge is bundling. As mentioned previously, as the fibers remain bundled their effectiveness is reduced, as shown with the compression molded plaques 5 and 6 in Table 8.2. One of the approaches to address the fiber bundles in the structural simulation is to treat the bundles as thicker fibers during the homogenization. As shown in Figure 8.33, first the bundle of fibers is homogenized into a thicker inclusion. Then, the bundles, resin, and remaining dispersed fibers are homogenized to get the final homogenized properties. Such an approach has been developed and already implemented in the commercial software Digimat™ [32]. When such bundles are used, the effective structural properties are slightly reduced compared to the situation where fibers are full dispersed. This is quite evident in Figure 8.34.

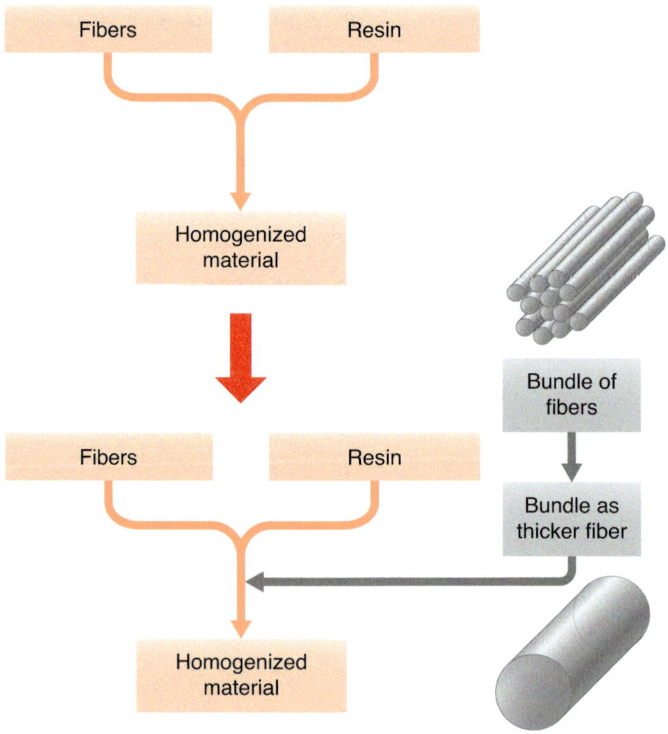

Figure 8.33 Fiber bundles and approach to model such bundles

Regarding the plaques investigated in this chapter, it was observed that in plaque #8, made with pultruded pellets, the fibers were fully dispersed and bundles were not observed. On the other hand, the compression molded plaque #6 showed that almost 50% of the fibers were in clusters and hence, as shown in Table 8.2, the tensile strength in the flow direction was significantly lower.

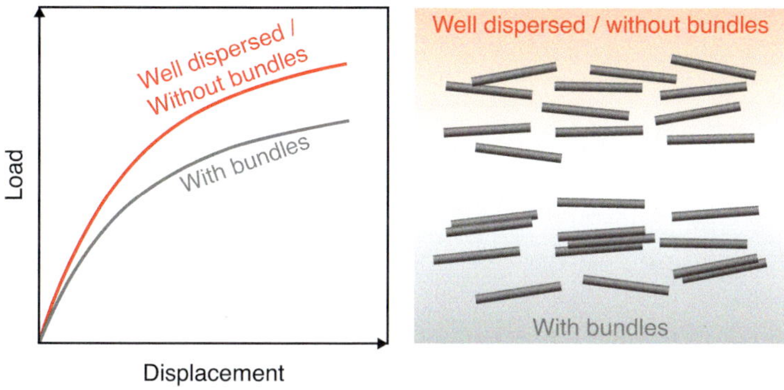

Figure 8.34 Effect of bundling on the structural performance

The number of layers of elements through the thickness of the plaque is an important consideration for both the process and structural simulation model. However, each layer requires more computing resources, so ideally one would like to see as few layers as possible. On the other hand, as discussed in Chapter 4, the microstructure details such as fiber orientation varies through the thickness. To capture these details in the structural simulation, it is important to have many layers of elements through the thickness. To understand the effect of number of layers, a finite element model for the bending sample was investigated. In this investigation, 4, 8, and 12 layers were compared. Figure 8.35 shows that the results of 8 and 12 layers are close to each other and agree with the measured results. Hence, it can be concluded that 8 layers may offer a good balance between accuracy and computing resources. The effect of number of layers on the computing resources is presented in Figure 8.36.

Using 4 layers with different thicknesses was also explored, see Figure 8.37. It can be concluded that if optimal thicknesses are selected the prediction can be improved. Using variable thickness layers is a good option; however, finding the optimum thickness for each layer for each individual case can be challenging, as it requires *a priori* knowledge of the final orientation. Also, the optimum thickness can vary from location to location due to changing fiber orientation distributions across the thickness. Therefore, eight or more layers of elements (with known thickness of layers) are recommended for a good representation of the material properties.

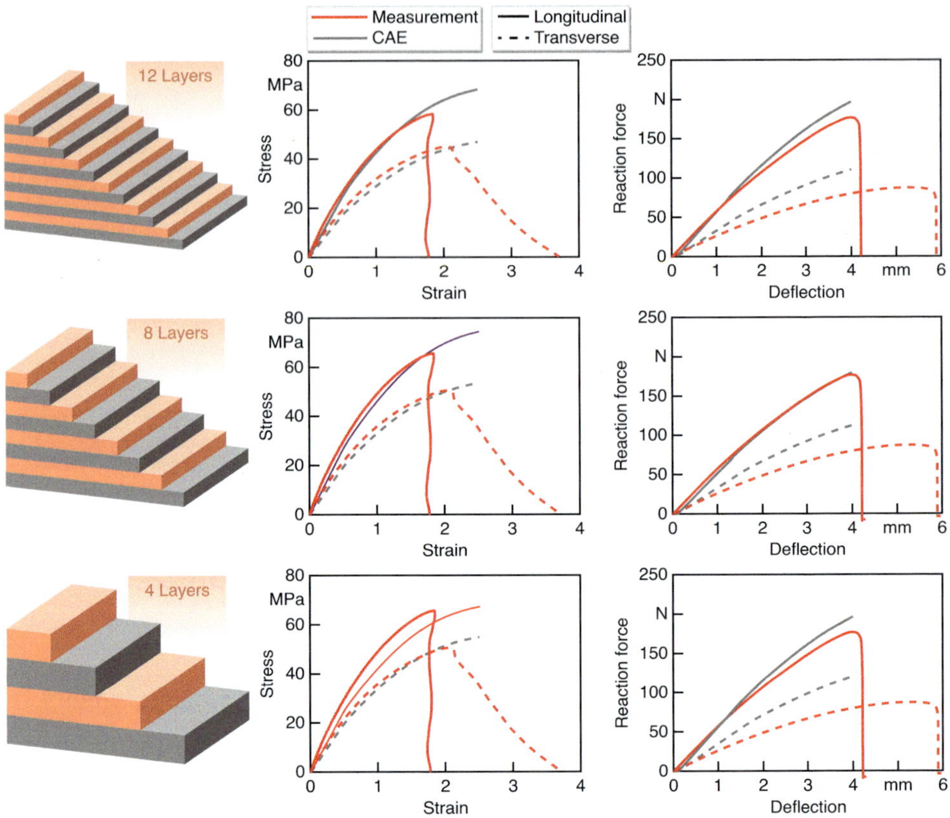

Figure 8.35 Study to decide how many layers should be used through the thickness

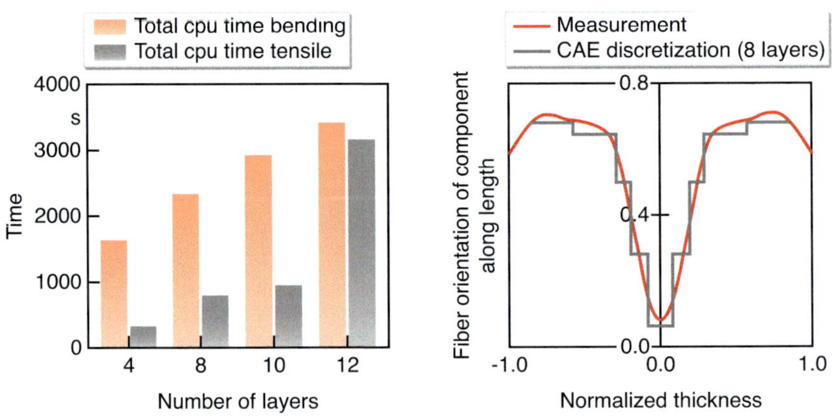

Figure 8.36 Comparison of CPU usage for the number of layers in same model for multiscale finite element analysis. We judge that eight layers provide the needed resolution for the typical fiber orientation distribution through the thickness

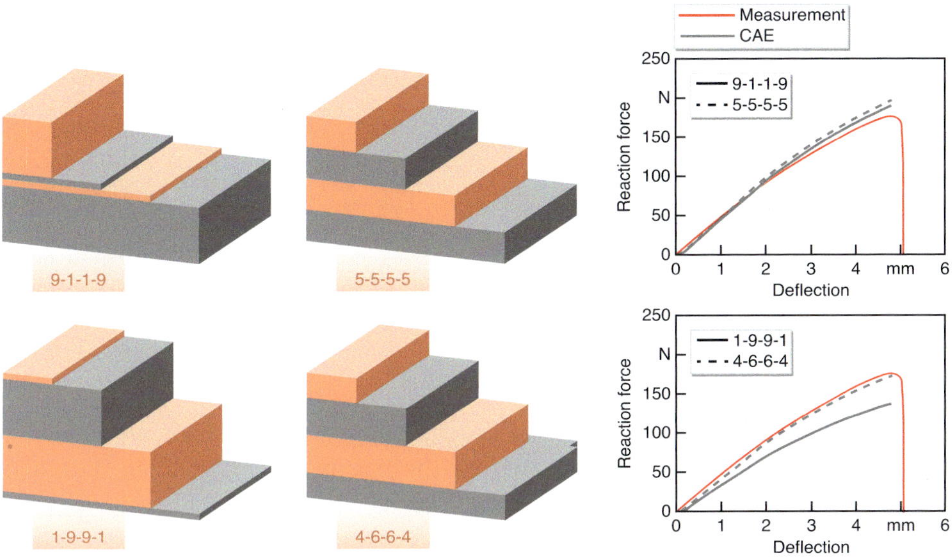

Figure 8.37 Optimizing the thickness in four layers to improve the prediction

8.2.4 Comments on Limitations of the Approach

In this case study, detailed steps of developing a multiscale model for fiber-reinforced materials were shown. The material models used are based on the Mori-Tanaka homogenization approach, and the key fiber microstructure accounted for was fiber orientation. The fiber orientation resulting during the injection molding is estimated, throughout the part, using mold filling simulation software. Most mold filling software uses the Folgar-Tucker model to estimate fiber orientation. Using the fiber orientation distribution within the part, material properties are calculated at each integration point in the structure's finite element model. The material properties at each integration point are assumed to be homogeneous and anisotropic. However, the properties can vary from one integration point to another. This approach works well for stiffness and force deflection response, but when damage is the concern it may not work well. This is mainly because the fiber-reinforced material is not a homogeneous domain; the fiber, resin, and interface are three distinct phases. Each phase has its own material properties, and therefore when such material is loaded the strain may be smooth due to material integrity but the stress distribution across the phases will have large variations. Currently, it is not possible to predict such stress variation accurately because the material is homogenized at the macro level. Therefore, prediction of damage initiation in discontinuous fiber-reinforced materials is a very challenging task. Many approaches to overcome this challenge are considered. For example, based on the knowledge of microstructure, a stress distribution function for a homogenized area can be used to estimate the stress within that homogenized area. Such an approach can help improve damage estimates depending on the function. However, the fundamental disconnect with the homogenization approach remains, and is very difficult to overcome. In general, methods to predict the damage and durability are still under development [14].

8.3 Warpage Study for a Flat Plaque

In the previous section, a multiscale modeling approach for tensile and bending behavior was illustrated and discussed for the flat plaque. For the injection and compression molded parts, where fiber-reinforced thermoplastic material fills the mold cavity and then cools to the finished shape, one of the major challenges is shrinkage and warpage. Predicting shrinkage and warpage of injection molded fiber-reinforced polymer parts is a complex issue. This is because both the structural and thermal properties depend on the microstructure conditions of the fibers in the resin, such as fiber orientation, fiber length, and fiber concentration. Furthermore, material properties of the polymers are temperature dependent. Since temperatures throughout the part vary during the process, the material properties also vary, which adds complexities to the prediction of shrinkage and warpage.

In this section, the finite element based semi-empirical approach to address these challenges and predict warpage due to cooling for a fiber-reinforced resin component in its solid phase are presented. The proposed approach is demonstrated to predict warpage of an injection molded flat plaque made of glass fiber-reinforced polypropylene, cooled from 160 °C to room temperature at 23 °C. The simulation steps include estimation of the fiber orientation using the mold filling simulation software. Measurements of the composite material properties, i.e., glass and resin combined properties, are compiled as a function of temperature. The combined material properties and calculated fiber orientations are used to estimate the "in mold" resin properties using reverse engineering. Finally, the warpage of the plaque is predicted using the estimated resin properties and fiber orientations.

Warpage predictions using such an approach compare well with the measured experimental results. This example demonstrates that accurate predictions for shrinkage and warpage of injection molded fiber-reinforced thermoplastic parts in the solid phase can be made if accurate material properties are used.

8.3.1 Discussion on Mechanism of Warpage

In the injection molding process, the molten resin and discontinuous fibers are fed into the barrel via a mixing screw. This molten mixture is then injected into the mold cavity. As this mixture of molten resin and fiber flows to fill the cavity, the fibers tend to align along the flow direction in the skin layers and perpendicular to the direction of flow in the central core of the part [15]. When the mixture cools, the resin solidifies into a finished part with fibers dispersed, and oriented based on the resin flow pattern throughout the part. A designer's goal is to design the mold cavity such that once the finished part is ejected and cooled to operating temperature, it attains near target geometry. Therefore, it is very important to accurately predict and account for the deformation, such as shrinkage and warpage, due to the manufacturing process, while designing a part [16].

The uneven shrinkage in the part generated during cooling results in residual stresses. When such a part is released from mold, the resultant stresses, if asymmetric across the thickness, lead to warpage [17–19]. If the stresses are high enough, they may lead to cracking. Ideally, it is desireable to predict the deformed shape of the part based on the process parameters and mold geometry. The warpage in a fiber-reinforced injection molded part results from various sources such as:

- Differential cooling of the mold cavity, such as uneven temperatures over the mold cavity [20, 21].
- Variation in the pressure distribution [22, 23].
- Anisotropic material properties throughout the part [24–26].

Once the mold cavity is filled under pressure the resin melt is cooled using the mold cooling channels, which, due to their lower temperature, absorb heat. As a result, the resin in the mold cavity starts to cool at the surface. This generates a temperature gradient, not only through the thickness, but also throughout the part geometry. Once the resin becomes solid, the temperature variation causes differential shrinkage, which ultimately leads to warpage when the part is released. The pressure within the molten resin has a significant influence on the shrinkage. Higher fill pressure results in higher resin density, which reduces shrinkage, whereas lower fill pressure results in lower density and higher shrinkage [23]. During the fill-and-hold stage of the injection molding process, it is likely that the pressure throughout the mold cavity will vary from one location to the other. This can also cause differential shrinkage, which can result in warpage. The thermoplastic material adds an additional factor. As the resin melt flows, the polymer molecules tend to align according to the velocity gradients experienced by the melt [16]. When the material solidifies, the polymer chain stays aligned and therefore the properties in flow and cross-flow direction could be slightly different, also contributing to part warpage.

When the polymer resin is filled with fibers, the complexities of warpage become significantly higher. This is because the composite material properties are highly dependent of the fiber conditions such as fiber length, volume fraction, and orientation. During the mold filling process, the fibers align as a result deformation and velocity gradients that result from the mold filling patterns. Consequently, as discussed in Chapters 4 and 7, the fiber conditions vary throughout the part from one location to other. Also, due to the fountain flow effect [15], the fiber alignment varies through the thickness of the part, where the fibers are aligned along the flow direction near the upper and lower surfaces, while oriented perpendicular to the flow in the central core, with randomly orientated regions that separate the various layers. Because of these process-induced variations in fiber conditions, there is considerable material property variation through the part. CLTE (coefficient of linear thermal expansion) is one such property that can contribute to warpage. The CLTE will always be lower in the direction of preferential fiber orientation and higher across that orientation direction.

For semi-crystalline resin materials, the resin properties can also add to anisotropy. This is because semi-crystalline polymers tend to change with the fiber as the polymer chains tends to align along the fiber [27, 28]. Furthermore, resin properties such elastic modulus, CLTE, and thermal conductivity vary with the temperature.

For unfilled thermoplastic resins the current approach in the industry is to use the *PvT* (pressure-volume-temperature) based approach, which accounts for shrinkage and warpage due to differential temperature and pressure of the material in the mold cavity [29]. When it comes to fiber filled thermoplastic parts, in addition to the resin, fiber anisotropy is an additional source of warpage. Generally, the fiber induced anisotropy has a larger influence on warpage compared to the resin-based effect [30]. Therefore, in the next section the effect of fibers on warpage will be explored in more depth. To predict warpage accurately, there are two related items that need close attention:

- Temperature dependent material properties: The material properties are anisotropic, nonlinear, and vary with the temperature. CLTE, thermal conductivity, and elastic modulus are temperature dependent and therefore should be treated accordingly.
- Methods to measure the temperature dependent material properties: The temperature dependent material properties are not readily available for all materials. A reverse engineering approach can be used to estimate the temperature dependent properties. First the combined fiber and resin material properties are experimentally measured and then, using reverse engineering, the resin-only properties are separated. This is important because when used with the fibers, the resin properties are slightly different compared to the resin alone, due to polymer molecule alignment. The measured resin properties are called “effective” properties. Once the resin properties are available, for given manufacturing

conditions, they can be used through the multiscale modeling approach for any combination of fiber and resin, i.e., different fiber orientation, length, and volume fractions, etc.

This approach works well if the process parameters between the two, i.e., the measured and the hypothetical case, are close to each other. In the following sections, the proposed approach will be demonstrated on a glass fiber-reinforced polypropylene plaque. For this case study a plaque was chosen because its stiffness against warpage is quite low and therefore the plaque tends to warp. In thin plaques, the warpage in the plaque is very sensitive to the material properties, and accurately predicting such warpage can be challenging.

8.3.2 Detail of the Plaques

To demonstrate warpage prediction in this case study, edge gated plaques as shown in Figure 8.38 and Figure 8.39 were used. The plaques' dimensions were 300 mm × 300 mm × 3 mm. A fan gate at the edge is used to fill the cavity. The plaques were manufactured by injecting 30% GF (by weight) polypropylene (PP) material. This material used is long fiber thermoplastic (LFT), available from Ticona™ (now Celanese™) under the trade name of CELSTRAN PP-GF30-03. The process parameters are listed in Figure 8.38.

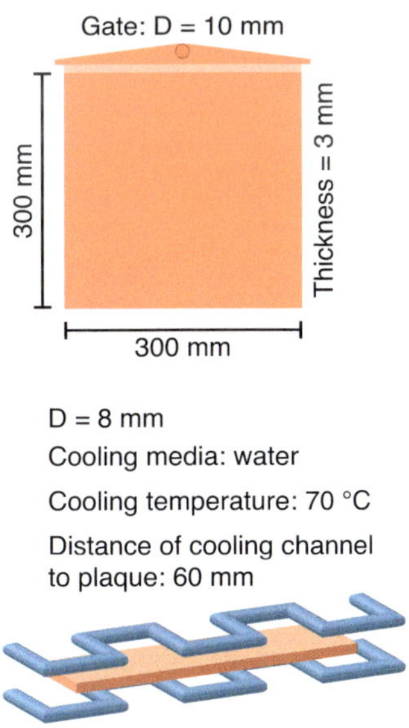

	CAE parameter	Actual process parameters
Melt Temperature	210 °C	210 °C
Mold Temperature	70 °C	70 °C
Fill Time	3.99 seconds	3.99 seconds
Cooling Time	120 seconds	N/A
Pack Time	10 seconds	N/A
Mold open Time	15 seconds	N/A
Overall Cycle Time	149 seconds	N/A
Ci	0.01	N/A
Cz	default	N/A

Figure 8.38 Detail of mold fill simulation; the material used is Ticona™ PPGF30 from the material library (PP: polypropylene, GF: glass fiber); the material properties used are from the software material library

To add additional variability in the case study, different combinations of 0.2 mm thick unidirectional tapes (70% GF+PP) were added during the injection molding. The tapes are secured in either the upper or lower surface of the mold and over molded with the fiber-reinforced material. The plaques were manufactured in three different configurations. Details of these three plaque configurations are presented in Table 8.6.

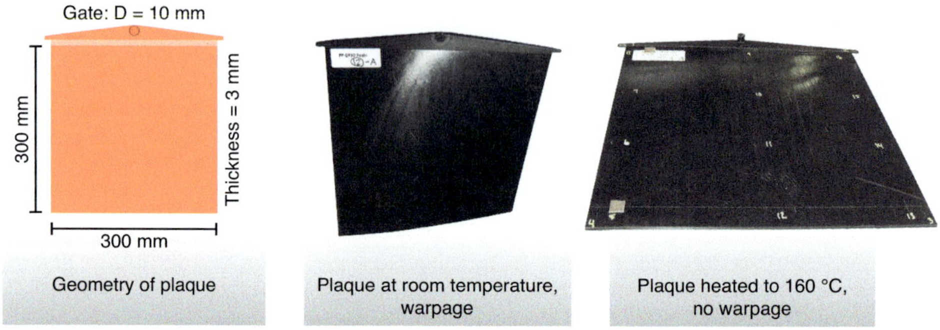

Figure 8.39 Plaque details, dimensions, and shape at room temperature and 160 °C

Table 8.6 Build Configuration and Physical Layout of Plaques from 30% Glass Fiber and Polypropylene; LFT: Long Fiber Thermoplastic

Build Configuration	Physical Layout
Injection molded LFT pellet material	
Injection molding of LFT pellet material over molded on sheet of unidirectional (UD) fibers on one side	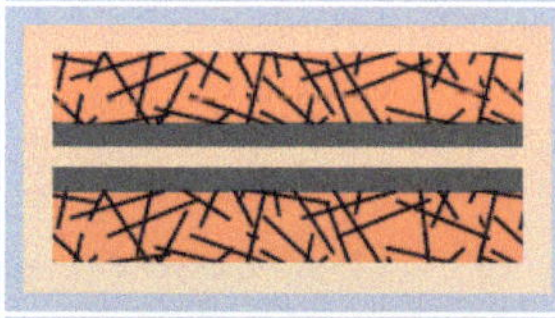
Injection molding of LFT pellet material over molded with sheet of unidirectional (UD) fibers on both sides	

It was observed that all three types of the plaques warped significantly when they were ejected from the mold and cooled to the room temperature at 23 °C (Figure 8.38). However, when re-heated and annealed at 160 °C (i.e., less than melting temperature of the resin of 180 °C, but higher than the T_g ~130 °C) for 4 hours, the plaques became flat as shown in Figure 8.39. As the plaques were allowed to cool back to the room temperature without constraint, they shrunk back to the original warped shape. This cycle was repeated three to four times to confirm the repeatability of the cycle, i.e., warpage at room temperature and no warpage at 160 °C.

Based on these observations it can be assumed that the residual stresses in the plaque at 160 °C are very low and therefore resulting in no warpage. Also, it was assumed that since the plaque is already in the solid phase, the differential shrinkage and hence warpage observed at room temperature is mainly a result of material anisotropy. In the rest of this case study the focus was directed toward developing a finite element model to predict the warpage for the three different build configurations of the plaque.

It is important to point out that in general, due to the limited bending stiffness of the plaque, the warpage prediction for a flat plate is considered quite challenging. In practice, ribs are often added to plate-like parts to add stiffness and stabilize the warpage. The primary focus of this study was to understand and model warpage of fiber-reinforced materials.

8.3.3 Prediction Method Approach

Residual stresses, which potentially lead to warpage, result from strains caused by crystallization, and strains due to differential thermal expansion.

The key steps in the entire multiscale prediction methodology are shown in Figure 8.40. First the mold filling process is simulated to compute the fiber microstructure. In step 2, the fiber microstructure is used to develop the homogenized material properties. The resin properties used during the homogenization are treated as a function of temperature.

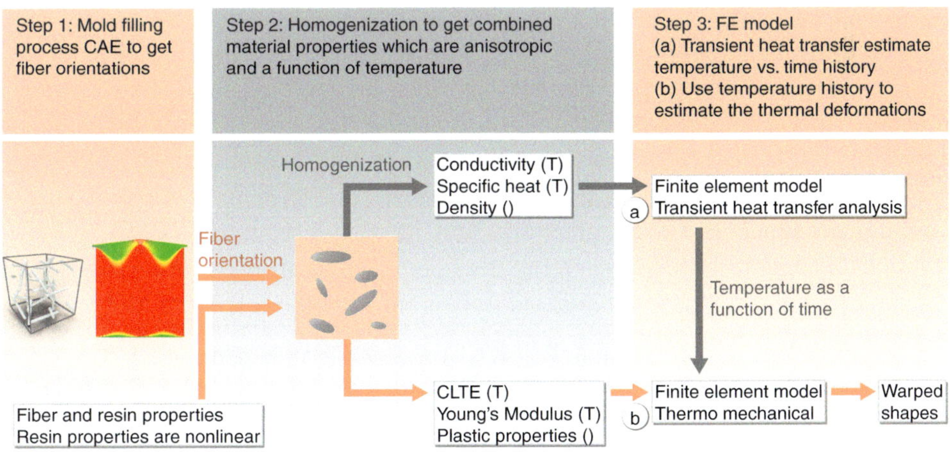

Figure 8.40 Key steps in prediction or warpage; quantities with (T) have temperature dependent properties

In step 3, the finite element model is used to estimate the warpage. The warpage computations are carried out in two phases. First, the temperature distribution throughout the plaque is predicted as a function of time using a transient heat transfer analysis. Then the temperature distribution versus time data is used to conduct the thermo-mechanical analysis to estimate the material deformation in the plaque. As shown in Figure 8.40, step 2, six different material properties are calculated for use in the finite element simulation. Out of these six thermal properties, conductivity, specific heat, and density are used in the transient heat transfer analysis, while CLTE, elastic modulus, and plastic material properties are used for the thermo-mechanical stress calculations.

These simulation steps can be performed using commercial finite element codes, provided the material properties needed for these calculations are available. The key challenges in getting the material properties are:

1. How to estimate these properties?
2. How to develop mathematical models to represent these materials for the finite element models?

These are major challenges because:

- The material properties of the fiber-reinforced resin used in the plaque depend on the fiber microstructure such as fiber length, orientation, and concentration.
- The resin properties are a function of the temperature. The elastic modulus, thermal conductivity, coefficient of thermal expansion, and specific heat of the resin vary significantly with temperature.

As with the last case study, here too fiber orientation distributions are estimated using a mold filling simulation of the plaque. The other two important microstructure features, i.e., fiber length and fiber volume, were used from actual measurements. Furthermore, it is assumed that fiber length distribution and volume fraction remain consistent throughout the plaque. The details of the mold filling process parameters are shown in Figure 8.38. Applying the multiscale modeling approach discussed in Chapter 6, the homogenized anisotropic material properties are estimated as a function of microstructure, i.e., fiber orientation, using the Mori-Tanaka homogenization approach. As the fiber orientation varies from location to location within the molded plaque, as well as through the thickness, the homogenized material properties vary accordingly.

The glass fiber properties used in the case study are assumed to be linear-elastic and constant at all the temperatures. The resin properties can be considered to be plastic. The plastic behavior of the resin is defined using the J2-plasticity model [31, 32]. In addition, it is assumed that the resin properties are also a function of temperature, between room temperature (23 °C) and 160 °C. It is known that the resin properties are influenced by the presence of fibers as polymer chains tends to align along the fiber. Therefore, as mentioned above, the resin properties that are representative of molded-in conditions are extracted through an approximate reverse engineering approach from the sample specimens. The reverse engineering is an iterative approach to estimate the resin properties, where the resin parameters in the material model are altered to match the actual measured test results. This way one can capture the process induced changes in the known material properties.

Once the resin material properties for the finished parts were estimated, the warpage predictions were carried out using the Abaqus finite element analysis software [33]. In the first step, the plaque at a uniform temperature of 160 °C is cooled to room temperature (23 °C) during which the temperature change at each location of the plaque is predicted and recorded as a function of time. In the second step, the temperature and time data predicted during first step are used to predict the mechanical stresses in the plaque. The material properties such as thermal conductivity, specific heat, coefficient of thermal expansion, Young's modulus, and nonlinear responses of the fiber-reinforced material in the plastic stage are modeled using Mori-Tanaka homogenization.

Predicting Fiber Orientation Using Mold Filling Simulation

The plaque's edge gated tool injection molding process was simulated using the commercial mold filling software Moldex3D™. First a detailed finite element model of the plaque's cavity was prepared and the simulation was carried out using the process parameters listed in Figure 8.38. A long glass fiber-reinforced polypropylene material in pellet form was used in the injection molding process. The material properties required for the simulation were extracted from the material library of the Moldex3D simulation environment. The plaque geometry was modeled using a combination of tetrahedral elements (middle areas) and prism elements (near skin areas). A total of 10 element layers were used through the 3-mm thickness of the plaque.

The calculated fiber orientation tensor components in flow and cross-flow directions at two selected locations are shown in Figure 8.41. To verify the accuracy of the prediction we measured the fiber orientation at the flow location and compared A_{11} component of the orientation tensor with the predicted values, see Figure 8.41. The difference between the predicted and experimental results can be attributed to a range typical of part to part variations. The samples around locations A and B, for which the fiber orientation is known, are cut by waterjet and used to measure material properties for further study. Details of material properties measurements and their usage in obtaining resin properties are discussed below.

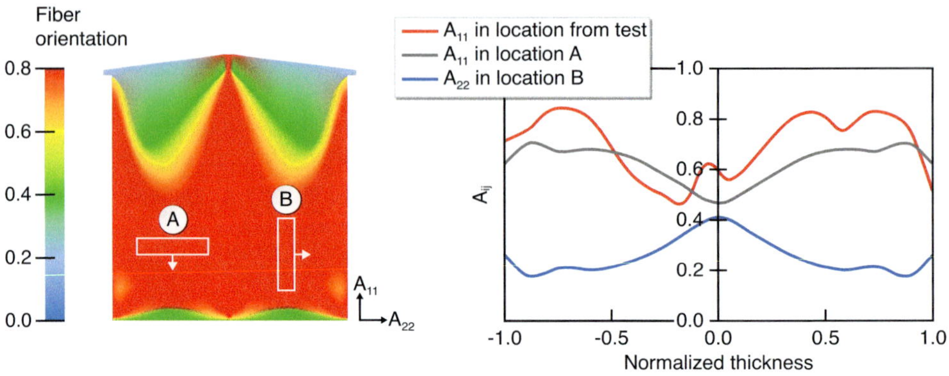

Figure 8.41 Predicted fiber orientation for samples cut in flow and cross-flow directions. The measured and predicted fiber orientation distributions are compared for the sample at location A to verify the mold fill simulation results

Fiber + Resin Homogenized Properties

The fiber and resin properties at the room temperature are presented in Table 8.7. The mechanical properties for the fiber-reinforced material in the plaque can be estimated using these fiber and resin properties and estimated fiber orientation using the mean-field homogenization approach based on a model proposed by Mori and Tanaka [28], discussed in detail in Chapter 6.

Table 8.7 Physical and Mechanical Properties of Glass Fibers and Polypropylene at Room Temperature

Property	Glass fiber	Polypropylene
Density	2.55 g/cc	0.92 g/cc
CLTE	5E–6 K^{-1}	1E–4 K^{-1}
Fiber length	0.3 mm	N/A
Fiber weight percentage	40%	N/A
Young's modulus	70 GPa	2 GPa
Poisson's Ratio	0.2	0.3

8.3.4 Mechanical Material Properties Calculations Using Reverse Engineering

This section presents the methodology to predict material properties of the composite material using the reverse engineering approach discussed earlier. Each property calculation is discussed in detail.

Young's Modulus of the Composite

The temperature dependent Young's modulus for the composite material is measured using a Q800 Dynamic Mechanical Analysis (DMA) system from Thermal Analysis (TA) Instruments. Samples cut at two locations from the plaque were evaluated in compliance with ASTM E2769-13. The sample location and measured Young's modulus for flow and cross-flow direction are shown in Figure 8.42. It is seen that both the flow and cross-flow specimens show a similar trend in the composite Young's modulus reduction with increasing temperature. Young's modulus in the flow direction is much larger than the cross-flow direction modulus, due to the higher proportion of fibers aligned in the flow direction; a result of the mold filling pattern and material deformation during mold filling.

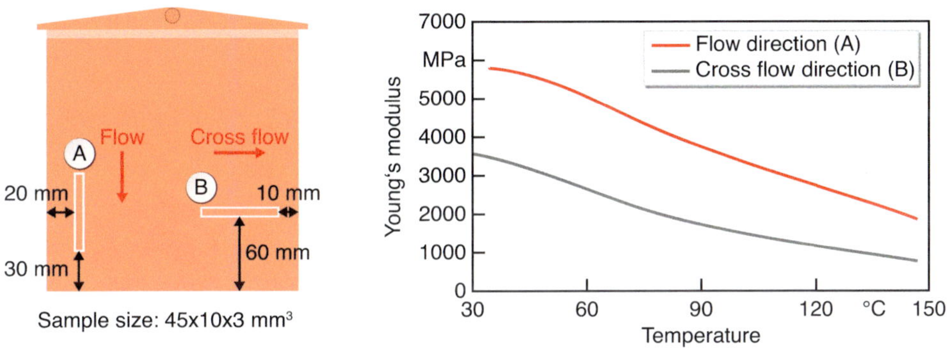

Figure 8.42 Measured Young's modulus vs. temperature for flow and cross-flow directions

Coefficient of Linear Thermal Expansion (CLTE) of the Composite

A small sample, as shown in Figure 8.43, was selected from the plaque and the CLTE was measured using Q400 series Thermal Mechanical Analysis (TMA) from TA instrument. Tests carried out for CLTE determination followed the ASTM E831-14 standard test. The influence of temperature on the coefficient of linear thermal expansion is shown in Figure 8.43. Due to the fiber orientation is this sample, the thermal expansion occurs at a lower rate in the flow direction than in the cross-flow direction. Consequently, the CLTE in the flow direction is lower when compared to the CLTE in the cross-flow direction. This is to be expected, as the thermal expansion of the fibers is much lower than the thermal expansion of the resin.

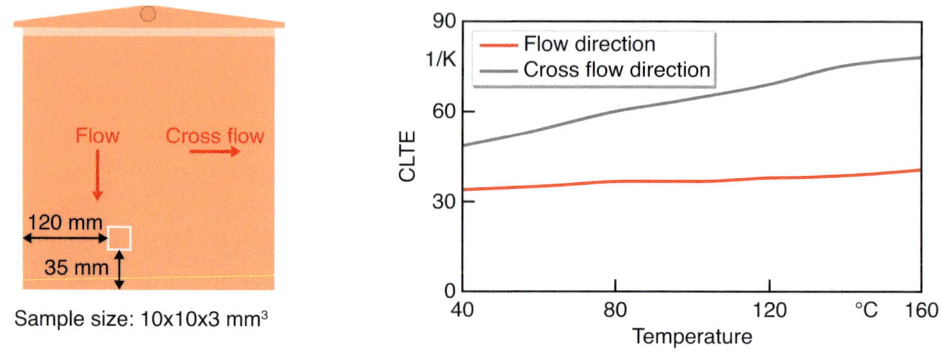

Figure 8.43 Measurement of CLTE in the flow and cross-flow directions of the plaque

Thermal Conductivity of the Composite

The thermal conductivity was measured using a modulated differential scanning calorimeter (DSC) from TA instruments. Samples from two locations were prepared and evaluated in compliance with the ASTM E1225-13 standard test. The effect of temperature on thermal conductivity is presented in Figure 8.44. The thermal conductivity in flow direction is higher, but for both flow and cross-flow samples, it increases at almost the same rate as the temperature increases.

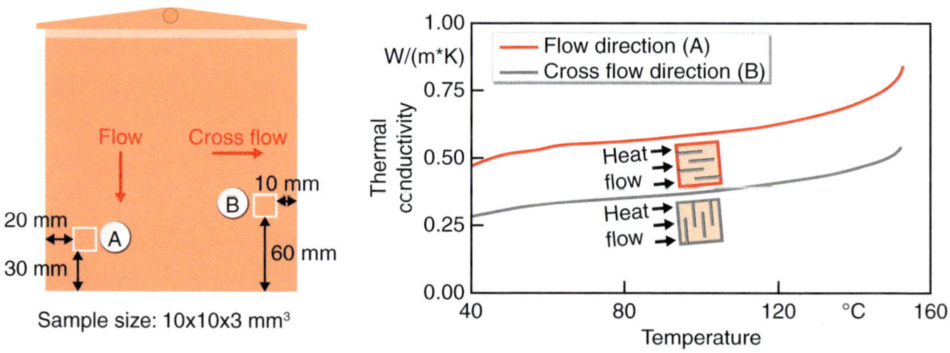

Figure 8.44 Measured thermal conductivity for the flow and cross-flow directions

Specific Heat of the Composite

Specific heat is heat required to increase the temperature per unit weight of the material. The temperature dependent specific heat is determined using a DSC Q2000 series from TA instruments and the ASTM E1269-11 standard test. The measurement results are shown in Figure 8.45. The specific heat of the composite material appears to rise steadily until 115 °C is reached, after which it increases rapidly. The specific heat is independent of flow or cross-flow directions, and only depends on the fiber volume fraction in the material.

It is assumed that the density of the material in the solid phase does not changing significantly with temperature and the small change may not have a significant impact on the structural performance of the material. Furthermore, other properties such as Poisson's ratio are assumed to be constant over the range of temperatures considered in this study.

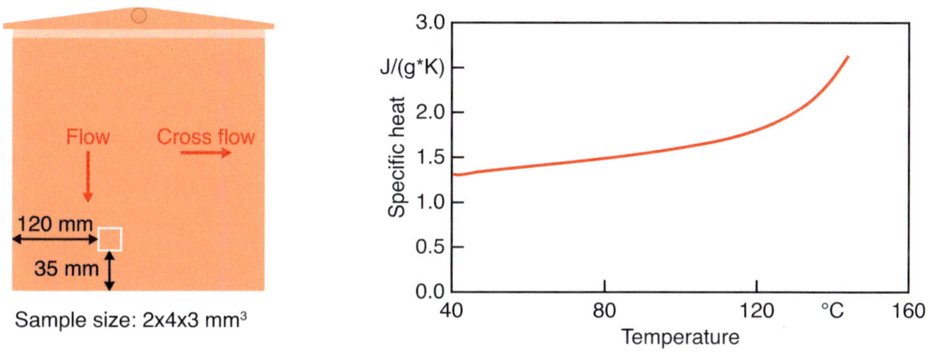

Figure 8.45 Measured specific heat of the sample, no effect of flow and cross-flow

Reverse Engineering to Extract the Resin Properties

The measured properties are for the composite material, i.e., fiber and resin, with specific fiber orientation and fiber concentration. These apply to the location at which the sample was selected for measurement. At other plaque locations, where the fiber orientation and volume fraction are different, the material properties will vary accordingly. This is a challenge in simulation, and in this section the challenge is addressed using a multiscale modeling approach.

A reverse engineering approach is used to estimate the resin properties as a function of temperature from the combined fiber-reinforced polymer properties measured at selected locations. The process flow used in the reverse engineering approach is shown in Figure 8.46. The approach used is an iterative process, where the unknown resin property parameters are estimated using the known glass fiber and composite material properties and fiber orientations.

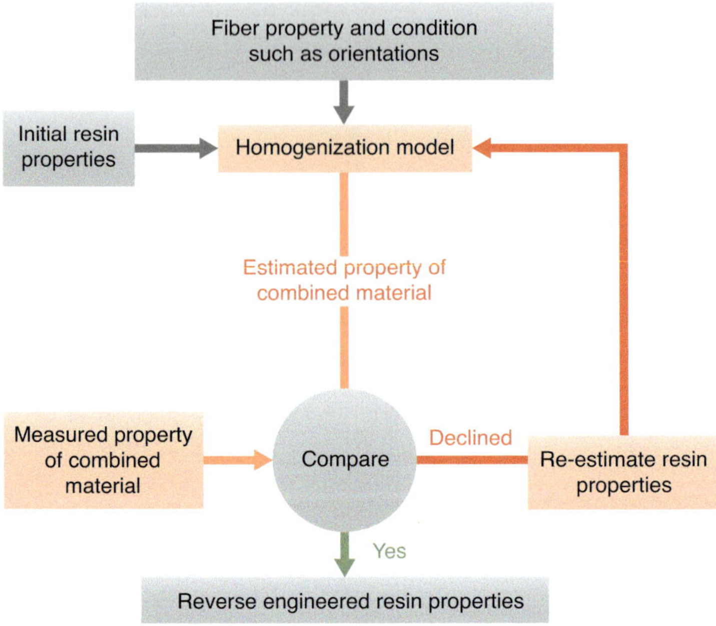

Figure 8.46 Reverse engineering process used to estimate the resin properties

The reverse engineered resin properties for Young's modulus, CLTE, and thermal conductivity are shown in Figure 8.47. Note that these are isotropic properties and are temperature dependent. These properties can now be used in conjunction with the multiscale approach to generate material properties based on the plaque's fiber microstructure predicted by the mold filling simulation software. Properties such as specific heat and density do not depend on the fiber orientation distribution, and can therefore be directly used in the finite element analysis. Separation of resin properties from the measured composite properties through reverse engineering for such properties is not necessary.

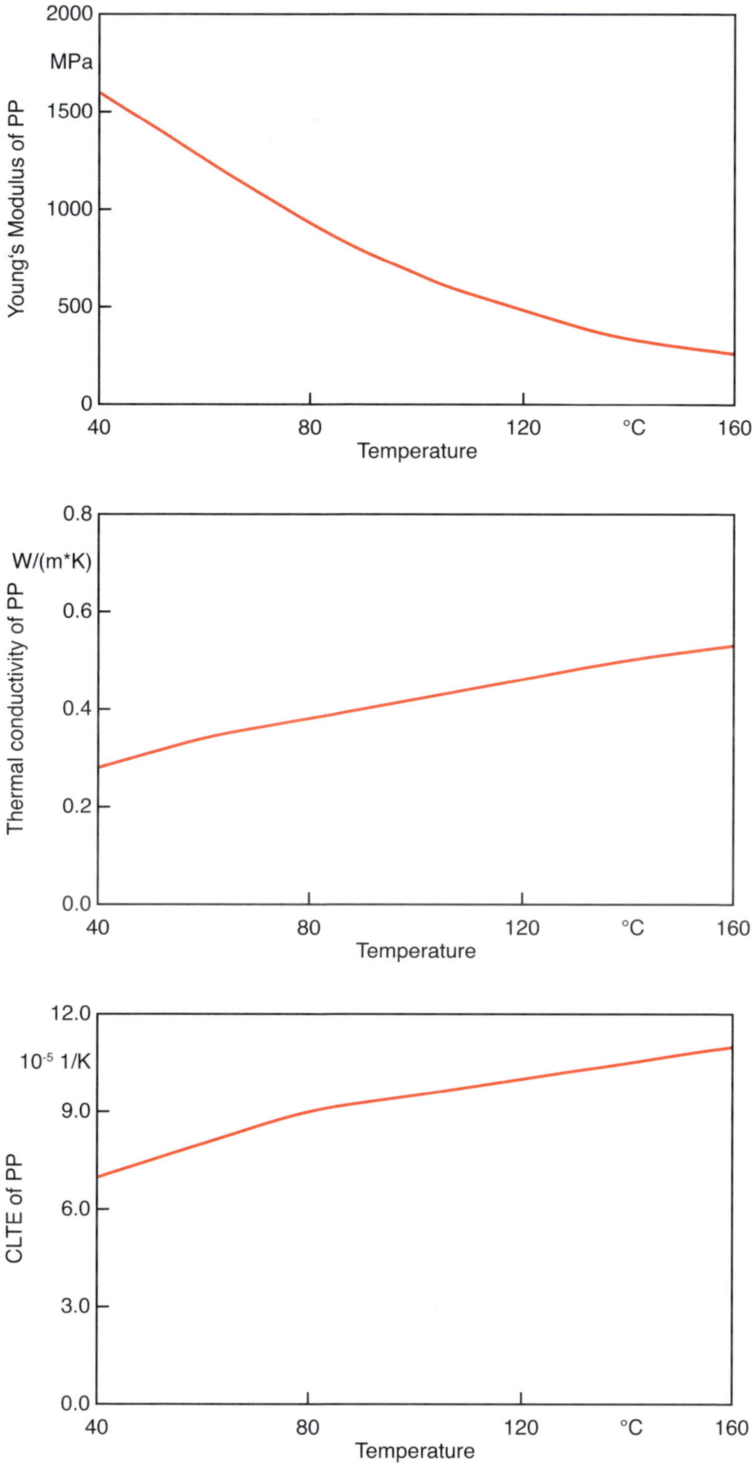

Figure 8.47 The reverse engineered properties for PP resin: (a) Young's modulus, (b) thermal conductivity, and (c) coefficient of linear thermal expansion (CLTE)

Non-Linear Plastic Material Model Properties

The constitutive model used for modeling non-linear plastic behavior of the resin is defined by the J2-plasticity model in Digimat [32]. This model is based on the Von Misses equivalent stress σ_{eq}, defined as:

$$\sigma_{eq} = \sqrt{J_2}(\sigma) = \sigma_Y + R(p) \tag{8.1}$$

where σ_Y is the yield stress, $R(p)$ is the hardening stress, and p is the accumulated plastic strain. The exponential and linear law is used for the hardening model as follows:

$$R(p) = k_p + R_\infty\left[1 - e^{-mp}\right] \tag{8.2}$$

where:

- k_p is the linear hardening modulus
- R_∞ is the hardening modulus
- m is the hardening exponent

The non-linear tensile stress stain response of the resin is shown in Figure 8.48. For simplicity it was assumed that the J2-plasticity model parameters (σ_Y, k_p, m, and R_∞) remain constant with temperature.

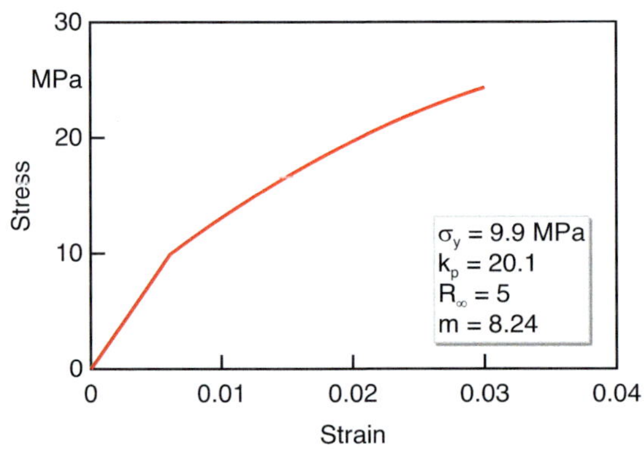

Figure 8.48 Plastic material model used in the case study

8.3.5 Finite Element Analysis to Calculate Warpage

Details of the finite element model (FEM) for the plaque analysis is shown in Figure 8.49. The focus in this case study was to demonstrate the use of a multiscale modeling approach, discussed in Chapter 6, to assess the effect of fiber microstructure on the final shrinkage and warpage of the part. An added feature in this case study was to demonstrate how temperature effects on the material properties are included.

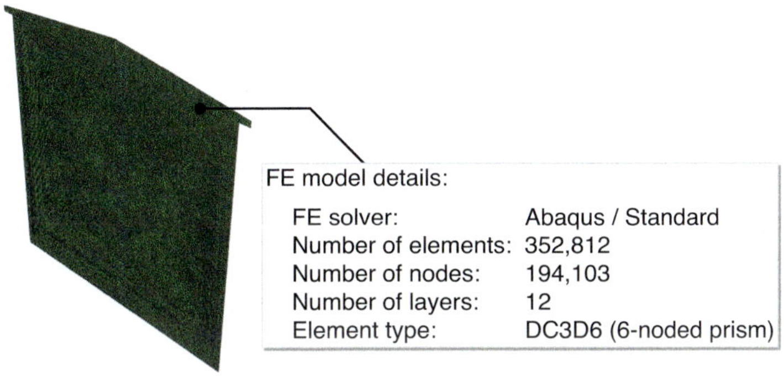

Figure 8.49 Detail of the finite element model of plaque used for the warpage study

To accomplish this, microscale fiber details are coupled with the finite element model of the plaque at the millimeter scale in Abaqus through the material user subroutine from Digimat [32]. Digimat updates the material properties for each element at every time increment based on the microstructure such as fiber orientations as well as temperature at that specific time. The updated material properties are used in the finite element calculations, as schematically depicted in Figure 8.50. The FEM analysis for the warpage calculations involves two steps: (1) calculation of temperature history at each element and (2) using the temperature history as input, the warpage of the plaque is estimated.

The plaques fabricated by over-molding LFT material on sheets of unidirectional (UD) fibers require additional considerations with the FE modeling. The added UD sheets on the outer surface are modeled as one 0.2 mm thick layer of solid element. The fiber content is assumed to be oriented along the flow direction (longitudinal) and the continuous fibers are represented by using an aspect ratio of infinity. The bonding between the tape and rest of the injection molded part is assumed to be perfect. When UD sheets are placed in the cavity, the overall thickness of the plaque remains the same; the injected material thickness is reduced by the thickness of UD tape.

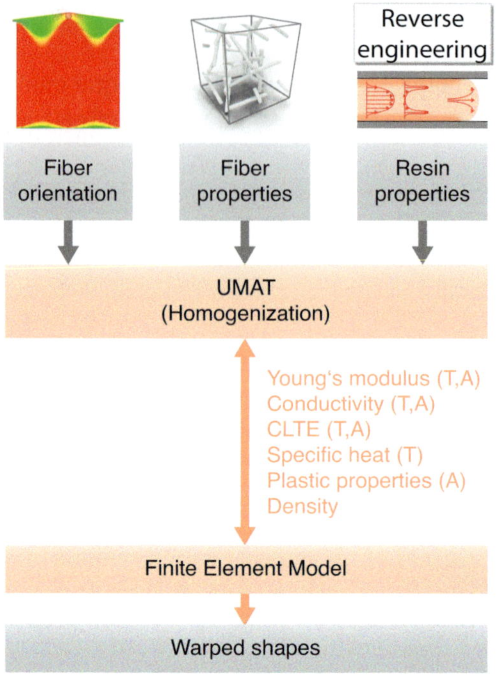

Figure 8.50 Details of the key steps used in the finite element analysis

Results

The temperature distribution in the plaque is shown in Figure 8.51(a) and temperature-time history at two selected locations is shown in Figure 8.51(b). At any given time, the temperature at different locations is different in the transient stage. Eventually the temperature becomes steady.

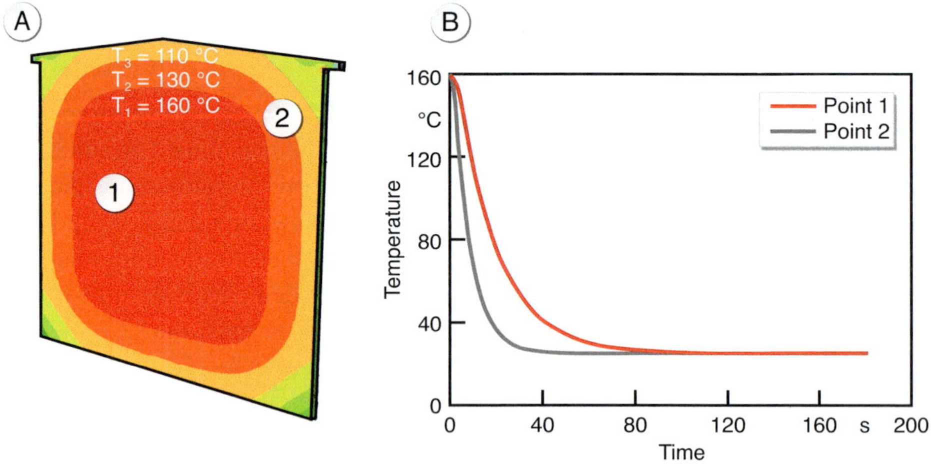

Figure 8.51 (a) Plaque temperature distribution and (b) temperature history at two locations

The comparison of the warpage predicted using the current methodology and the actual warpage measured using a coordinate measuring machine is shown in Figure 8.52. It is seen that the predicted deformation mode as well as deformation magnitude for all the three type of plaques compares well with the measured warpage.

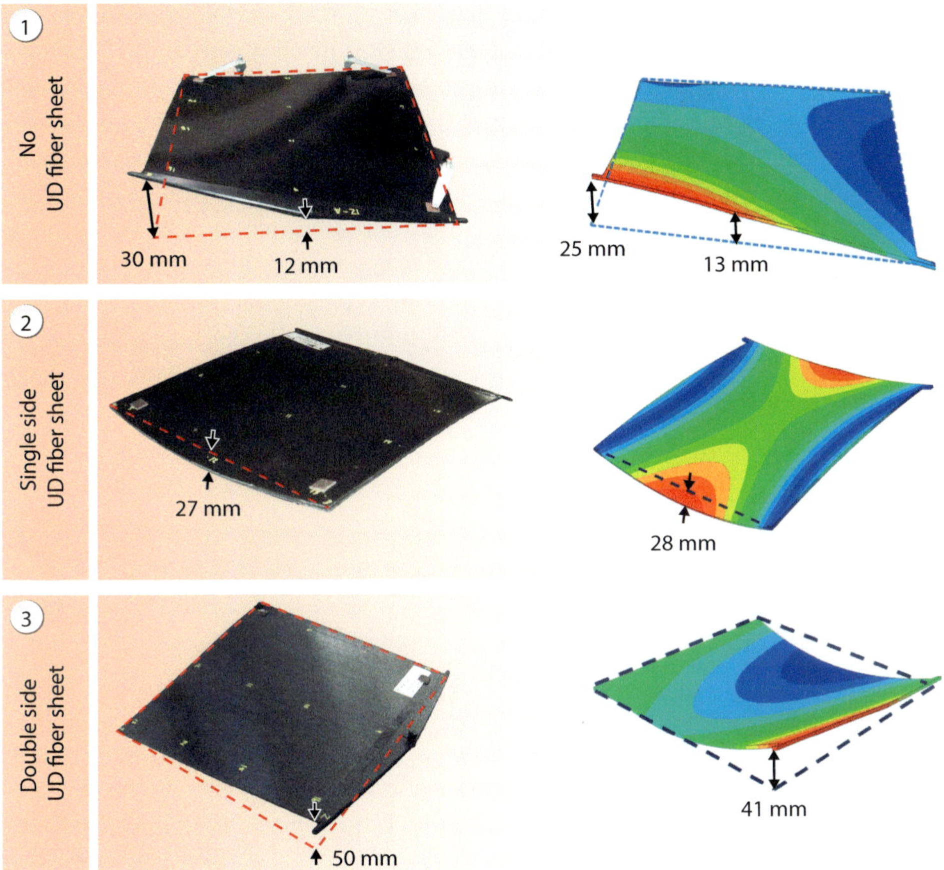

Figure 8.52 Warpage results comparison between prediction and measurement

8.3.6 Comments and Limitations

Warpage of fiber-reinforced parts has been and still remains a challenge for two reasons: (1) the anisotropy induced shrinkage due to fiber orientation variations throughout the part cannot be avoided, and (2) fiber-reinforced parts are stiffer, which means traditional approaches such as adding stiffening beads or modifying attachment schemes to force the parts to take on the desired shape are difficult.

Therefore, in order to design warp-free parts from fiber-reinforced materials, it is important to have modeling capabilities to allow warpage predictions of such parts.

In this case study, the multiscale finite element modeling approach presented in Chapter 6 used to predict the warpage for a fiber-reinforced plaque was demonstrated. The plaque was cooled from 160 °C to room temperature of 23 °C, in the solid phase. Such an approach neglects phase change effects, and assumes that residual stresses form after the whole part has reached the "solid" phase. This assumption is consistent with the literature, which states that only a small fraction of the residual stresses is caused during phase change [34]. Both the molecular crystallization shrinkage and the thermal shrinkage are lumped together as temperature dependent material properties. The predicted and measured warpage show good correlation. It was observed that the key to a successful prediction is using of accurate material conditions and their representation in the finite element models. Furthermore, it was observed that the temperature dependent resin properties are not always readily available; therefore, experimental measurement may be needed. The resin properties are strongly influenced by the fibers, as well as process conditions. A reverse engineering approach, where material properties for the resin can be estimated, becomes useful.

■ References

[1] Thomason, J.L. and Vlug, M.A., Influence of fiber length and concentration on the properties of glass fiber-reinforced polypropylene: 1. Tensile and flexural modulus, *Composites Part A: Applied science and manufacturing*, 27(6), pp. 477–484 (1996)

[2] Thomason, J.L., Vlug, M.A., Schipper, G., and Krikor, H.G.L.T., Influence of fibre length and concentration on the properties of glass fibre-reinforced polypropylene: Part 3. Strength and strain at failure, *Composites Part A: Applied Science and Manufacturing*, 27(11), pp. 1075–1084 (1996)

[3] Folgar, F. and Tucker III, C.L., Orientation behavior of fibers in concentrated suspensions, *Journal of reinforced plastics and composites*, 3(2), pp. 98–119 (1984)

[4] Advani, S.G. and Tucker III, C.L., The use of tensors to describe and predict fiber orientation in short fiber composites, *Journal of rheology*, 31(8), pp. 751–784 (1987)

[5] Phelps, J.H. and Tucker III, C.L., An anisotropic rotary diffusion model for fiber orientation in short-and long-fiber thermoplastics, *Journal of Non-Newtonian Fluid Mechanics*, 156(3), pp. 165–176 (2009)

[6] Nguyen, B.N., Bapanapalli, S.K., Holbery, J.D., Smith, M.T., Kunc, V., Frame, B.J., and Tucker III, C.L., Fiber length and orientation in long-fiber injection-molded thermoplastics–Part I: Modeling of microstructure and elastic properties, *Journal of composite materials*, 42(10), pp. 1003–1029 (2008)

[7] Hsu, C.C., Hsieh, D.D., Chiu, H.S., and Yamabe, M., Investigation of fiber orientation in filling and packing phases. In 66th Annual technical conference of the society of plastics engineers, ANTEC 2008 (Vol. 1) (2008)

[8] Phelps, J. H., El-Rahman, A. I. A., Kunc, V., and Tucker III, C. L., A model for fiber length attrition in injection-molded long-fiber composites, *Composites Part A: Applied Science and Manufacturing*, 51, pp. 11–21 (2013)

[9] Tseng, H. C., Chang, R. Y., and Hsu, C. H., An integration of microstructure predictions and structural analysis in long-fiber-reinforced composite with experimental validation, *International Polymer Processing*, 32(4), pp. 455–466 (2017)

[10] Huang, C. T. and Tseng, H. C., Simulation prediction of the fiber breakage history in regular and barrier structure screws in injection molding, *Polymer Engineering and Science*, 58(4), pp. 452–459 (2018)

[11] Kuhn, C., Walter, I., Taeger, O., and Osswald, T. A., Experimental and numerical analysis of fiber matrix separation during compression molding of long fiber reinforced thermoplastics, *Journal of Composites Science*, 1(1), 2 (2017)

[12] Walter, I., Goris, S., Teuwsen, J., Tapia, A., Perez, C., Osswald, T. A., and Madison-Madison, W. I., A direct particle level simulation coupled with the Folgar-Tucker RSC Model to predict fiber orientation in injection molding of long glass fiber reinforced thermoplastics. In Proceedings of the ANTEC (2017)

[13] Pérez, C., The use of a direct particle simulation to predict fiber motion in polymer processing. The University of Wisconsin-Madison (2016)

[14] Osswald, T. A. and Hernández, J. P., *Polymer Processing: Modeling and Simulation*, Hanser, Munich (2006)

[15] Osswald, T. A., Experimental Investigation Into the Effects of Fountain Flow on Fiber-Matrix, presented at the 44th CIRP International Conference on Manufacturing Systems (2011)

[16] Fischer, J. M., *Handbook of molded part shrinkage and warpage*, 2nd ed., Elsevier/William Andrew, Amsterdam (2013)

[17] Kamal, M. R., Lai-Fook, R. A., and Hernandez-Aguilar, J. R., Residual thermal stresses in injection moldings of thermoplastics: A theoretical and experimental study, *Polym. Eng. Sci.*, vol. 42, no. 5, pp. 1098–1114 (2002)

[18] Zoetelief, W. F., Douven, L. F. A., and Housz, A. J. I., Residual thermal stresses in injection molded products, *Polym. Eng. Sci.*, vol. 36, no. 14, pp. 1886–1896 (1996)

[19] Shen, C. and Li, H., Numerical Simulation for Effects of Injection Mold Cooling on Warpage and Residual Stresses, *Polym.-Plast. Technol. Eng.*, vol. 42, no. 5, pp. 971–982 (2003)

[20] Lee, Y. B. and Kwon, T. H., Modeling and numerical simulation of residual stresses and birefringence in injection molded center-gated disks, *J. Mater. Process. Technol.*, vol. 111, no. 1–3, pp. 214–218 (2001)

[21] Beaumont, J. P., *Successful injection molding: process, design, and simulation*, Hanser, Munich (2002)

[22] Zheng, R., Kennedy, P., Phan-Thien, N., and Fan, X.-J., Thermoviscoelastic simulation of thermally and pressure-induced stresses in injection moulding for the prediction of shrinkage and warpage for fibre-reinforced thermoplastics, *J. Non-Newton. Fluid Mech.*, vol. 84, no. 2–3, pp. 159–190 (1999)

[23] Pontes, A. J., Oliveira, M. J., and Pouzada, A. S., Studies on the influence of the holding pressure on the orientation and shrinkage of injection molded parts, presented at the ANTEC (2002)

[24] Kikuchi, H. and Koyama, K., The relation between thickness and warpage in a disk injection molded from fiber reinforced PA66, *Polym. Eng. Sci.*, vol. 36, no. 10, pp. 1317–1325 (1996)

[25] Kikuchi, H. and Koyama, K., Warpage, anisotropy, and part thickness, *Polym. Eng. Sci.*, vol. 36, no. 10, pp. 1326–1335 (1996)

[26] Kikuchi, H. and Koyama, K., Material anisotropy and warpage of nylon 66 composites, *Polym. Eng. Sci.*, vol. 34, no. 18, pp. 1411–1418 (1994)

[27] Tadmor, Z., Molecular orientation in injection molding, *J. Appl. Polym. Sci.*, vol. 18, no. 6, pp. 1753–1772 (1974)

[28] Mori, T. and Tanaka, K., Average stress in matrix and average elastic energy of materials with misfitting inclusions, *Acta Metall.*, vol. 21, no. 5, pp. 571–574 (1973)

[29] Wang, J., PVT Properties of Polymers for Injection Molding, in *Some Critical Issues for Injection Molding*, Wang, J. (Ed.), InTech (2012)

[30] Hine, P.J. and Duckett, R.A., Fiber orientation structures and mechanical properties of injection molded short glass fiber reinforced ribbed plates, *Polym. Compos.*, vol. 25, no. 3, pp. 237–254 (2004)

[31] Foss, P.H., Coupling of flow simulation and structural analysis for glass-filled thermoplastics, *Polym. Compos.*, vol. 25, no. 4, pp. 343–354 (2004)

[32] Digimat 5.0.1 Users' Manual., e-Xstream Engineering, Nov-2013

[33] Abaqus Users' Manual, Version 13, Dec-2014

[34] Osswald, T.A., *Material Science of Polymers for Engineers*, 3rd ed., Hanser, Munich (2012)

9 Special Topic: Compression Molding of Discontinuous Fiber Material

As discussed in Chapter 2 and 3, fiber-reinforced materials are possible in many forms, i.e., fibers can be continuous or discontinuous and polymer resins can be thermoset or thermoplastic. Glass and carbon fibers are the most popular choice to reinforce fiber-reinforced polymer materials. Carbon fibers are lighter and stronger compared to the glass fibers, but their cost is higher and they show a tendency for brittle failure. For typical automotive usage, large volume and low cost are extremely important. Discontinuous fibers with a thermoplastic matrix offer a short cycle time, which is important to achieve a high production rate. Injection molding, compression molding, or the resin transfer molding process are potential methods to process such materials. Among these processes, compression molding is a desired option, as depicted in Figure 9.1. This is because compression molding offers a higher possibility to maintain longer fibers and, hence, better mechanical properties. The ability to simulate the compression molding process and predict the performance of the finished part for the fiber-reinforced thermoplastic material can be an important enabler in using such materials in designs [1–4]. The basic approach to compression molding process simulation was introduced in Chapter 7. In this chapter we will demonstrate how such simulation methods can be applied to three-dimensional parts.

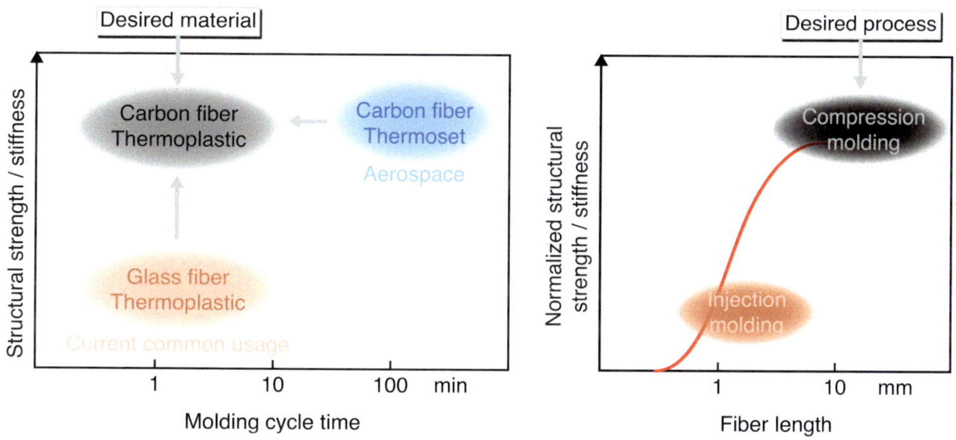

Figure 9.1 Comparison of different processes for fiber-reinforced materials

The resin used for the compression molding can be either thermoset or thermoplastic. The use of fiber-reinforced thermoset materials, such as sheet molding compound (SMC), has been quite popular in the industry for many years; however, the use of fiber-reinforced thermoplastic materials is limited [5, 6]. A short discussion on thermoset versus thermoplastic resin may be helpful to understand this.

Typically, a thermoset material, such as SMC, has a low viscosity, so it flows well and fills the cavity easily, which means a good even coverage of the mold cavity. Thermoset materials solidify through a cross-linking process, which can be time-consuming. As a result, the cycle time to manufacture the thermoset part can be longer. On the other hand, thermoplastic resins are heated to melt, and then the melt is forced to flow and fill the cavity before the resin cools and solidifies. Since thermoplastic resins do not require curing, they solidify when cooled, which is fast, leading to cycle times that are significantly shorter when compared to thermoset materials. However, the viscosity for the thermoplastic materials is higher (even higher when used with carbon fiber), bringing challenges when trying to fill the cavity. Higher pressure can be used to force the melt to fill the cavity; however, the higher pressure also means a higher shear rate, which means higher breakage of the fibers. As a result, the fiber length in the final part can be significantly smaller. For industrial applications, higher volume and shorter cycle time are very important; therefore, thermoplastic materials are preferred. Furthermore, to enhance the performance of such parts, advanced composite options, often called hybrid approaches, can also be used (see Chapter 10 for details). In a hybrid approach, the fiber-reinforced resin bulk charge is compression molded with strategically placed continuous fibers, such as unidirectional tapes or woven fabrics patches, to locally reinforce a part. Because of these challenges, a careful design of the mold cavity and selection of process parameters is very important in designing the mold cavity for the fiber-reinforced thermoplastic parts. Mold filling simulation programs, discussed in Chapter 7, are used to simulate the material flow in the mold cavity, which is helpful in designing a part and developing the process. Furthermore, as discussed earlier, it is also important to estimate the fiber microstructure, such as orientation, length, and concentration in the finished part to develop multiscale material models and, hence, predict the performance of the finished part.

Compression molding of fiber-reinforced thermoplastic polymers involves heating a mixture of fiber and polymers, often called charge, above its melting temperature and then compressing the charge in the heated mold cavity at high pressure. There are typically two different initial formats for the fiber-reinforced charge used in the compression molding process, i.e., (1) bulk charge, as seen in Figure 9.2, and (2) sheet charge, as seen in Figure 9.3. Usually the bulk charge is made using a single/twin-screw low shear plasticator. The reinforcing fibers and polymer resin are mixed in desired proportion to create a charge. The targeted size of the heated bulk coming out from the plasticator is collected and placed in the heated mold cavity for compression.

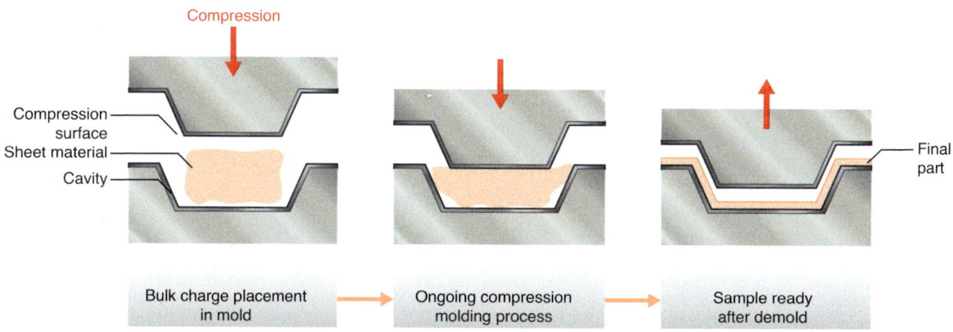

Figure 9.2 Typical compression molding process using a bulk-shaped charge

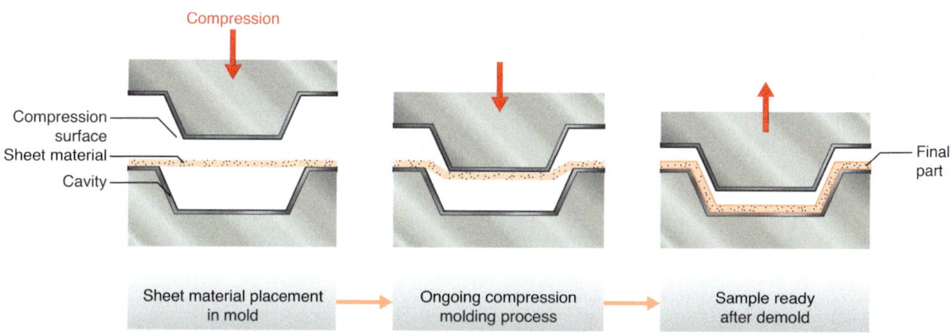

Figure 9.3 Typical compression molding process using a sheet- or mat-shaped charge

A sheet charge, which is also called pre-form, prepreg, or mat, is another format for the charge used in the compression molding process. Manufacturing of the sheet material is more complex. Reinforcing fibers such as glass or carbon fibers are chopped and dispersed in water along with the polymer resin in the discontinuous fiber form. Then the co-mingled mass of reinforcing and resin polymer fibers is extracted to form a mat using a water slurry process, like a paper-making process. Such mats consist of loosely held co-mingled reinforcing fiber and polymer fibers and tend to be fluffy. Usually they are heated and compressed until they are partially consolidated, so they can stay in mat form during handling for further processing. Chemicals such as bonding agents and stabilizers are often used to keep the mats stable, both physically as well as chemically. Such partially consolidated mats are heated and used as a charge. There are other methods used to prepare the sheets. For example, the reinforcing fibers are chopped and spread on the polymer film first. These fiber-rich polymer sheets are then heated under pressure to form a consolidated plate, which can be used for compression molding. The main advantage of using a sheet charge is the possibility of using longer fibers; fibers up to 50 mm in length are used in the prepreg stage. Depending on the mold design, there is a possibility to maintain the fiber length in the finished parts when using compression molding. One or more sheets can be tailored and spread to cover the mold cavity, which means the flow length for the melted mixture required to fill the cavity can be shortened, hence, reducing fiber attrition.

In this chapter we are presenting a method to simulate compression molding of discontinuous fiber in sheet as well as bulk form. Section 9.1 and 9.2 deal with discontinuous carbon fibers in bulk and sheet form, respectively. Section 9.3 deals with a glass mat thermoplastic material, which is a mixture of both discontinuous and continuous thermoplastic materials.

9.1 Compression Molding of Bulk Materials

In this section, the CAE method for compression molding using bulk charge material is illustrated with two examples. The theory for compression molding is presented in Section 7.2 in Chapter 7; challenges in applying the theoretical formulation to practice are also discussed in Section 7.2.3. The key challenges faced are non-linear viscosity of the material at both the liquid and solid state, non-random orientation of fibers in the initial material, etc. These challenges make the application of fundamental theory to the compression molding quite difficult. This section presents insights into how finite element simulation can be applied to compression molding of actual parts. The goal is to present the approach and demonstrate how practical challenges can be addressed. Modeling all the details of a complex process can be quite difficult and often impossible; however, if one can develop a model that is sufficiently useful in addressing the needs, then such model will be very helpful for a wide range of engineers and scientists. We strongly recommend the reader to review Section 7.2.3 as it will help understand the challenges we are addressing here.

In the first example, how the initial fiber orientation in the bulk charge can be included in the CAE is presented. The effect of fiber orientation in the bulk charge on the fiber orientation in the finished parts is also demonstrated. In the second example, a slightly more complex shape is chosen, to demonstrate the effect of visco-elastic properties along with the fiber orientation in the charge on the warpage in the finished part.

Both examples make use of three-dimensional parts with reasonable complexities. The CAE predictions are compared with physical parts for gross performance at macro level as well as fiber conditions at micro level.

9.1.1 Example 1: Single-Cavity Glass Fiber-Reinforced Polymer (GFRP) Part

In this example, a hat section part as shown in Figure 9.4 is built using 40% glass fiber-reinforced polypropylene from a bulk charge for the study.

Figure 9.4 Hat section part made using compression molding process, thickness = 2 mm

9.1.1.1 Actual Part Manufacture

First the bulk charge is prepared, as shown in Figure 9.5. A twin-screw extruder is used to mix the chopped fibers and resin melt at high temperatures to form and extrude the bulk charge for the compression molding. The process parameters used for the extruder are presented in Table 9.1.

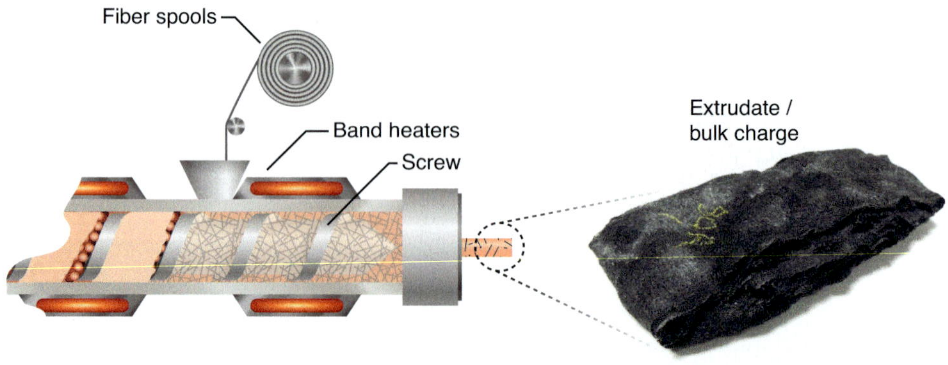

Figure 9.5 Twin-screw extruder used and bulk charge prepared for compression molding

Table 9.1 Processing Conditions Used for Compression Molding

Process conditions	Actual values
Melt temperature	210 °C
Mold temperature	70 °C
Compression time	2 s
Compression pressure	2000 kN
Charge weight	170 g

Figure 9.6 shows the computerized tomography (CT) scanned image of the bulk charge. The CT scan image is made with a 50-micron resolution. Since the diameter of individual glass fibers is around 15 microns, it is not possible to see the fibers in the CT scan image; however, the undispersed fibers in bundles are visible in the image, which gives a good representation of the orientation of the individual fibers [9, 10]. The lower resolution was to allow a CT scan of the whole part. CT scans at higher resolution (e.g., 3 microns) for such a large part were not possible.

In this section, a few observations made based on a charge material, prepared using the direct fiber extrusion process, are shared. First, generally about 50% or more fibers are in a bundle state, which are visible in the CT scan image, and the remaining fibers can be well dispersed and are not visible. Second, the fiber bundles are quite long, in the CT scan they can be 70–100 mm long. In most cases, individual fibers are shorter than they appear in the bundle form. Third, their orientation is non-random. In fact, the orientation is result of movement within the twin-screw extruder system. There is a periodicity in the fiber orientation of the charge that mirrors the screw design. Finally, it has been shown [8] that the fiber orientation measured using the bundles is a good representative of the orientation if the individual fibers are also measured and orientation is calculated.

Figure 9.6 CT scan image of initial charge used

Figure 9.7 shows the 3D CAD conceptual images of the charge and the cavity used for the compression molding process. The charge used here has similar geometry shape and size with the actual part.

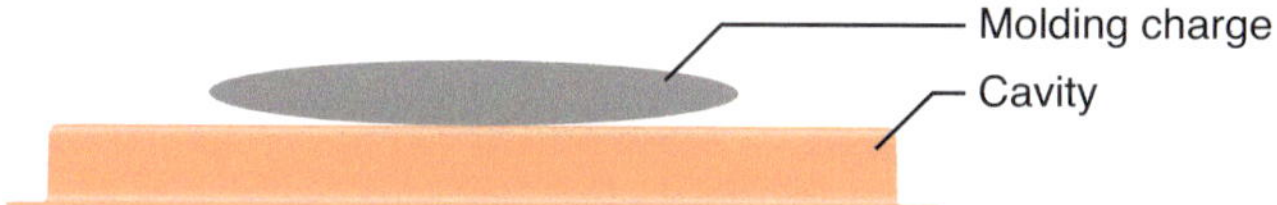

Figure 9.7 Design image of the compression molding charge and cavity used for CAE

Since the interest was in applying a CAE method to model the compression molding process, a finite element model of geometry that includes the charge and the cavity for the finished part were developed for the simulation as shown in Figure 9.10. The next task was to define the condition of the fibers within the charge.

9.1.1.2 Fiber Orientation Mapping Approach

The key steps for a method to measure the fiber orientation from the CT scan image and create an initial fiber orientation distribution for the bulk charge are presented in Figure 9.8. First, a representative section of the part is cut from the charge, then it is scanned using a CT scanner. Typically, an industrial scale CT scanner is used in order to get high quality and good images of the carbon fiber and resin materials in the charge. Next, the image is analyzed to get the fiber orientation distribution in the compression charge. There are many options for such purpose, and a detailed discussion is given in Chapter 4. Software called VG StudioMAX™ was used for analysis purposes [9, 11]. In Volume Graphics, the scanned image volume is divided into a number of small segments called voxels, and the fiber orientation tensor is estimated for each voxel, based on the fiber condition in that segment. Volume Graphics uses the amount of grey area and its distribution in each segment to quickly estimate the orientation tensor. Normally, the voxel size is an approximate 1 mm × 1 mm × 1 mm cube [11].

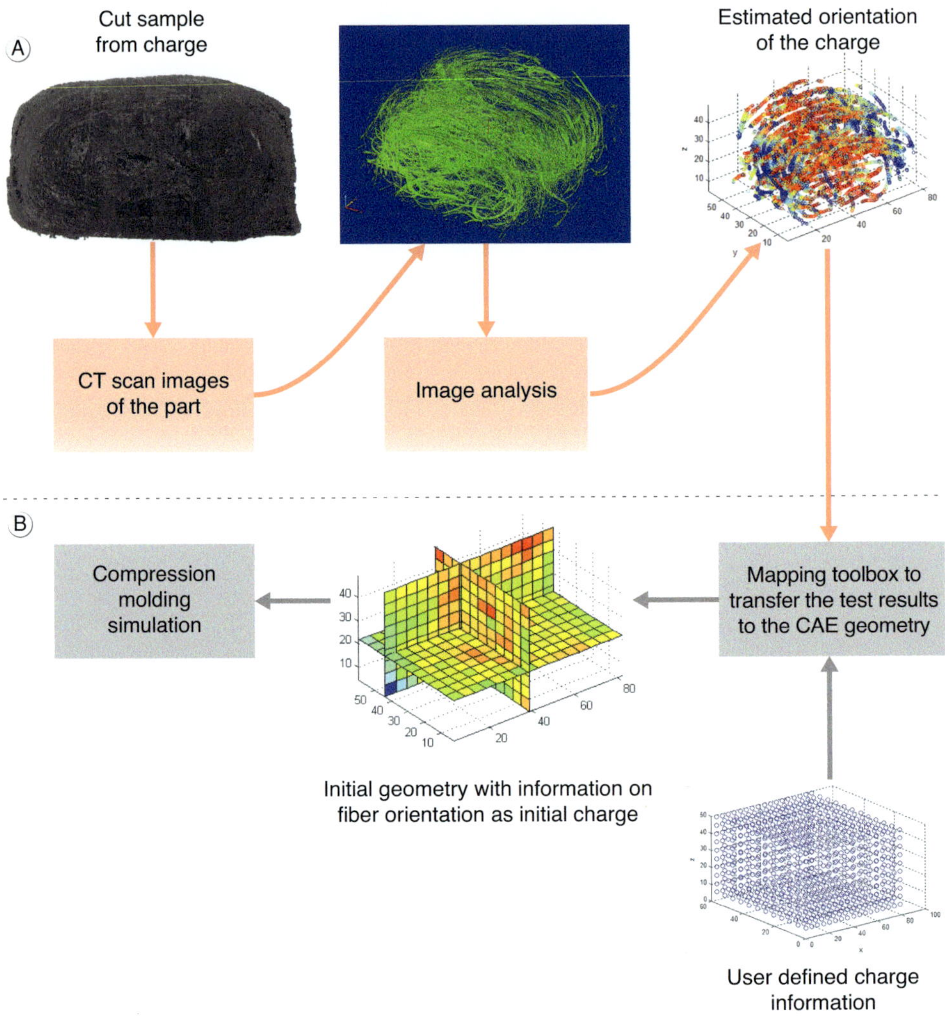

Figure 9.8 Development process for fiber orientation measurement: (a) CT scanning of the charge materials; (b) mapping of the fiber orientation from CT scan image to finite element charge used in the simulation

The measured orientations are mapped on the finite element model of the charge. The actual charge shape is not regular, and the shape can vary from one charge to other charge. Therefore, the charge shape used in the CAE for the finite element model can be slightly different than the actual CT scan data. This means that some extrapolation or interpolation is needed to map the measured fiber orientation to the geometry of the CAE model of the charge. To help this mapping process, an algorithm based on radial basis functions was developed by Pérez et al. [8, 12]. The algorithm, implemented into the interactive software, maps the measured fiber orientation from the CT scan of the actual charge to the finite element model for the charge, element by element. The mapped fiber orientation in the bulk charge model is used as an initial fiber condition in the compression molding model and simulation used by mold filling software such as Moldex3D™.

9.1.1.3 Mapping Algorithm

In the following paragraphs, the detailed mapping algorithm, developed at Polymer Engineering Center, University of Wisconsin, will be explained. In the first step, the fiber orientation distributions obtained by the μCT scans are normalized as depicted in Figure 9.9.

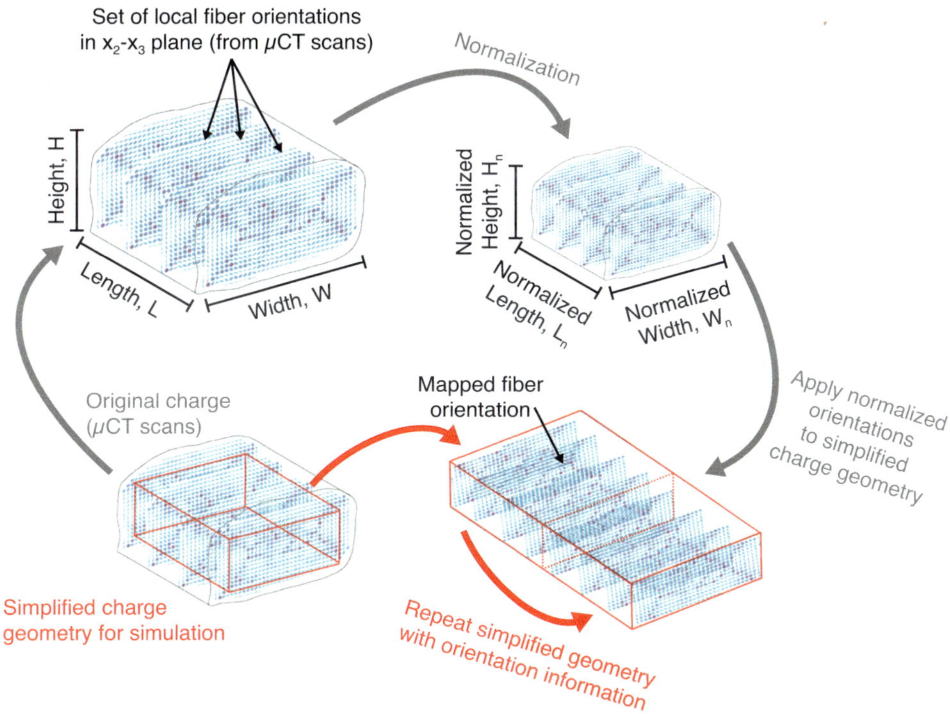

Figure 9.9 Mapping of μCT data set, normalization, and removal of overlapping nodes

Based on the mesh of the charge defined in Moldex3D™, the normalized μCT data set is rescaled to the dimensions of the mesh. The rescaling continues until the μCT data set comprises the entire mesh of the charge. This approach leads to few points lying outside of the mesh. In the end, all points outside the mesh are removed from the data set. The dimensions illustrated here are just used for the explanations of the coding algorithm.

To have a local orientation for every point of the Moldex3D mesh of the charge, an interpolation algorithm is applied to the orientation tensor at each node of the mesh. The radial basis function (RBF) [13–15] is a mesh-less interpolation technique and has been shown to be a powerful method for representing sparse, scattered data. The RBF interpolation represents interpolated data as a linear combination of the non-linear basis function that is defined at each control point and has only a radial component.

$$s(x) = \sum_{i=1}^{N} \lambda_i \phi\left(\left|x - x_i\right|\right) = \sum_{i=1}^{N} \lambda_i \phi\left(r_i\right) \tag{9.1}$$

where x represents the location of the point where interpolation is performed, x_i represents the position of the control point in a specified region of interest, ϕ is the basis function, λ_i is the weight factor to be determined, and N is the number of data points. The distance r_i is defined between the point where the function will be interpolated to and the points where the function is known from the µCT data set. For the basis function, ϕ, the thin plate spline (TPS) [16] was used because it gives accurate results for scattered data approximations

$$\phi\left(\left|x - x_i\right|\right) = \phi\left(r_i\right) = r_i^2 \log\left(r_i\right) \tag{9.2}$$

The first step in the interpolation is solving for the weight factors λ_i. A system of equations is calculated, denoted as

$$\begin{pmatrix} \phi_{11} & \cdots & \phi_{1n} \\ \vdots & \ddots & \vdots \\ \phi_{n1} & \cdots & \phi_{nn} \end{pmatrix} \begin{pmatrix} \lambda_1 \\ \cdots \\ \lambda_n \end{pmatrix} = \begin{pmatrix} s_1 \\ \cdots \\ s_n \end{pmatrix} \tag{9.3}$$

where ϕ_{ij} describes the distances between all possible combinations of points within the µCT data set. The s vector on the right-hand side contains the target values and, in the case of the fiber orientation distribution, s is one component of the orientation tensor A_{ij} with

$$\begin{pmatrix} \phi_{11} & \cdots & \phi_{1n} \\ \vdots & \ddots & \vdots \\ \phi_{n1} & \cdots & \phi_{nn} \end{pmatrix} \begin{pmatrix} \lambda_1 \\ \cdots \\ \lambda_n \end{pmatrix} = \begin{pmatrix} A_{ij1} \\ \cdots \\ A_{ijn} \end{pmatrix} \tag{9.4}$$

From the µCT data set, all ϕ and the corresponding A_{ij} are known. Thus, the system can be solved for the weight factors. Solving

$$A_{ij}(r) = \sum_{i=1}^{N} \lambda_i \phi(r_i) \tag{9.5}$$

for an arbitrary point of the Moldex3D mesh, the orientation tensor A_{ij} of that point can be interpolated. This is done for all nodes of the Moldex3D mesh and components of the orientation tensor A_{ij}. After the interpolation, the mapped fiber orientation and its corresponding coordinates are imported into Moldex3D. The mapping tool was programmed using Matlab [17].

9.1.1.4 Demonstration of Effect of Initial Charge Orientation in the CAE Model

Figure 9.10 shows the compression molding model used in Moldex3D™. The charge is represented in red and is the region where the initial fiber orientation is mapped upon, and the cavity is located below the charge. The geometry and model details are also included in Figure 9.10. The total cavity volume is 167 cm^3, and the charge volume is 193 cm^3 (about 115% of the cavity size, which accounts for the compression of the material). The detailed material properties used for the simulation are shown in Table 9.2. The processing conditions used for the simulation, such as compression time and melt temperature, reflect the actual manufacturing process. Fiber orientation tensors are extracted after the filling process at locations A, B, and C.

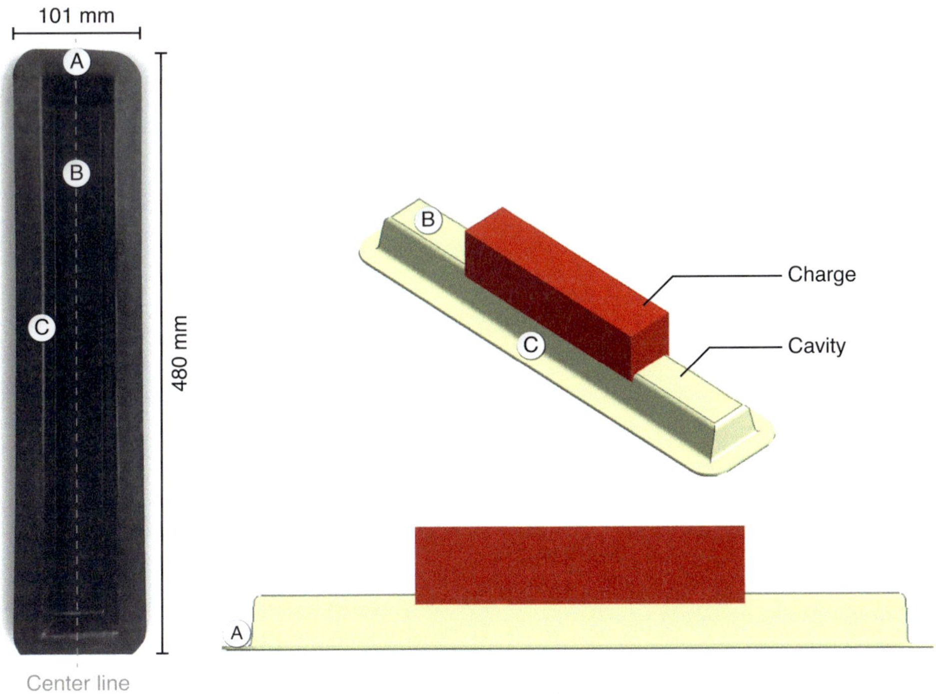

Figure 9.10 Compression molding model used and location at which fiber orientation is compared between CAE and measurement

Table 9.2 Material Properties Used for the Simulation

	Glass fiber	Polypropylene
Density	2.55 (g/cc)	0.92 (g/cc)
Fiber weight percentage	40%	N/A
Young's modulus E	70,000 MPa	2,000 MPa
Poisson's Ratio	0.2	0.3

The actual scaling process used to map the estimated orientation from the measurements to the actual finite element model used for this example's initial charge is presented in Figure 9.11. Performing a CT scan of an actual charge and creating a map of the orientation is a complex and time-consuming process, due to the large size of the data (~80 GB). To address this challenge, a smaller charge from a similar twin-screw extruder was CT scanned, and a map of orientation distribution was created. Next, the orientation from the smaller charge was mapped onto the larger charge using the radial basis function (RBF) approach. It should be pointed out that this is an approximation, which is practical and offers improvement over current methods that assume that the fibers are completely random or aligned in only one direction. It is believed that a CT scan image for a twin screw once prepared and analyzed can be reused for charge representations from the same twin-screw system but that has a different size and shape.

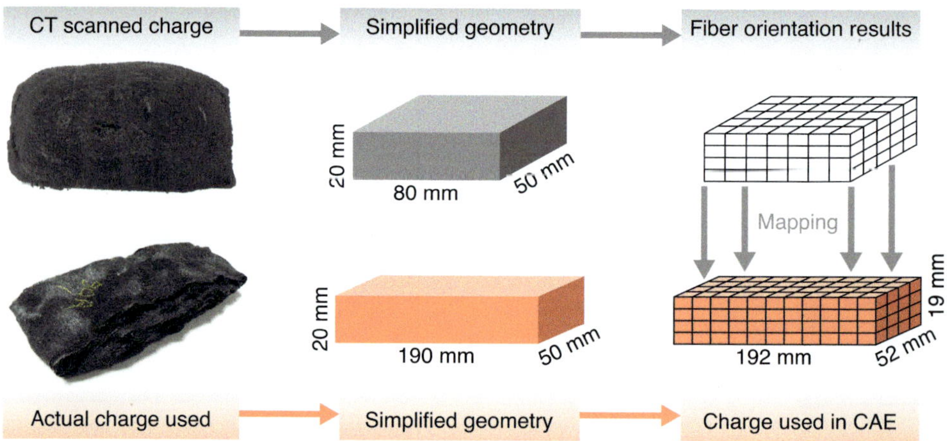

Figure 9.11 Initial fiber orientation mapping from the real charge to simulation. The dimensions used are in mm

To capture the accuracy of the compression molding simulation, the detailed microstructure information is needed. In this example, the fiber orientation tensor is used as an index to check the microstructure. Fiber orientation from points A, B, and C, shown in Figure 9.10, are compared for simulation and experiments.

Figure 9.12 shows the comparison of the A_{11} component of the fiber orientation tensor between CAE and measurement for locations A, B, and C. Two initial fiber conditions are used in the simulation:

- Initial random: the orientation in bulk charge is assumed to be random
- Initial mapped: the orientation is mapped on the bulk charge model from CT scan data of the actual charge

The comparison in Figure 9.12 shows the importance of using a mapped fiber orientation in the initial charge in compression molding. For all three locations the fiber orientation estimation in the finished part is significantly closer to the measured values in the actual part, when known fiber orientation is used in the initial charge model. It was also observed that, since the material flow lengths are smaller for compression molding, the time for the fiber to align along the flow is shorter; consequently, if one starts with random fibers, the fibers in the finished part remain close to random.

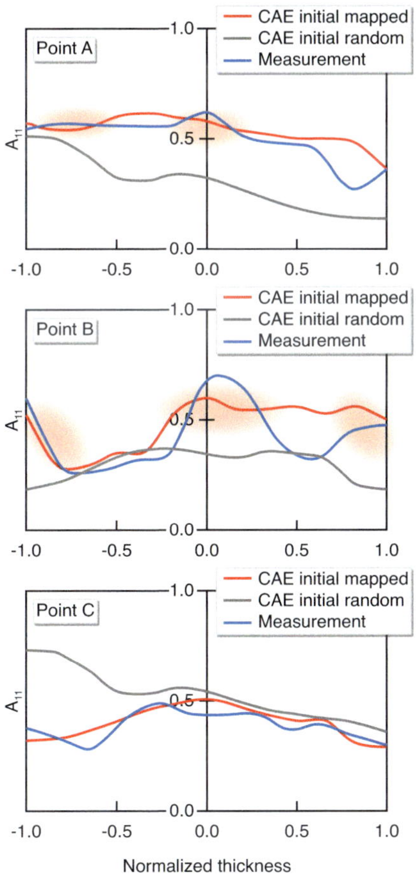

Figure 9.12 Comparison of the fiber orientation using mapping and random option with actual measurement

Here, only the A_{11} component of the orientation tensor is presented for clarity and making a point. Components A_{22} and A_{33} are available in Song [18]. Furthermore, the effect of different levels of the fiber to fiber interaction parameter is also compared in the same paper.

One of the key questions while comparing two fiber orientations (measured and predicted, in this case) is how to compare two orientations that are distributed over the thickness. Since the final goal of the simulation is to estimate the performance of the final design, it makes most sense if one compares the effect of fiber orientation distribution on the structural performance. To achieve this goal, the elastic modulus for the material based on the fiber orientation distribution was estimated. All the parameters such as fiber volume fraction, length, material properties, and geometry are assumed to remain constant. The only variable introduced was fiber orientation through the thickness. The elastic modulus calculated using different fiber orientation distributions (from Figure 9.12) are compared in Figure 9.13 and Table 9.3. The results show that, using a measured orientation of the fibers in the initial charge, the prediction using CAE improves the accuracy compared to using random fiber orientation.

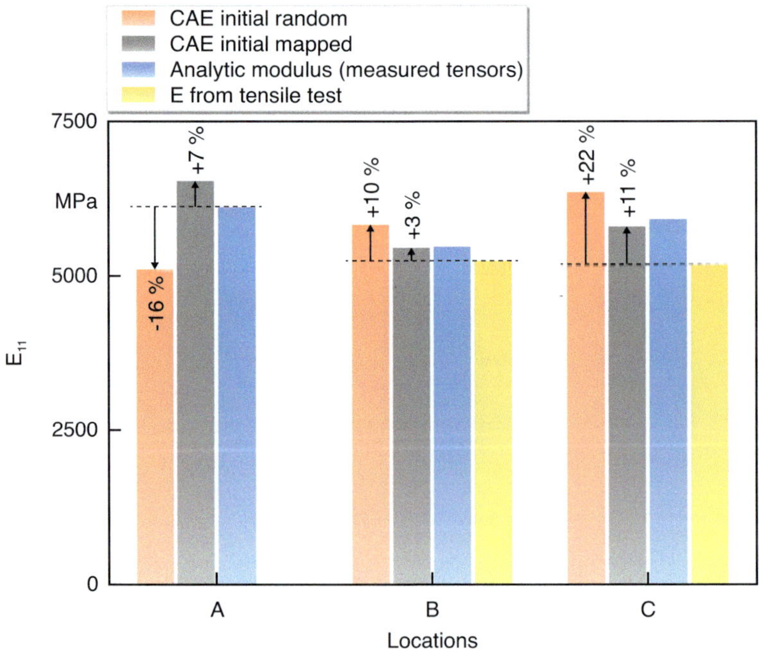

Figure 9.13 Comparison of the calculated E_{11} and measurement for these three locations and the difference between them

Table 9.3 Tensile Modulus Calculated at Three Locations for Different Fiber Orientations Assigned (in MPa)

E_{11} modulus (MPa) locations	CAE Initial random	CAE Initial mapped	Analytic modulus calculated from measured tensors	Modulus measured using MTS machine (for reference)
A	5108	6544	6100	N/A
B	5832	5460	5474	5259
C	6364	5807	5905	5202

9.1.1.5 Comments and Conclusions

This example shows the importance of defining the initial orientation in the charge. As the fiber resin melt is compressed and starts to flow, the fibers tend to align along the flow and if the flow continues for a certain length, the fibers align along the flow, erasing the effect of initial conditions. However, in compression molding, the charge flows for a shorter length, which means the time for fibers to align along the flow is shorter; as a result, the fiber orientation in the finished parts for compression molding is highly dependent of the initial fiber orientations in the charge. Therefore, accounting for the initial fiber orientation is extremely important.

9.1.2 Example 2: Three-Cavity Carbon Fiber-Reinforced Plastic (CFRP) Sample

In this example, a three-cavity tool as shown in Figure 9.14 is used to manufacture a part by compression molding of a carbon fiber-reinforced PA6 material. The main objective here is to demonstrate the effect of the viscoelastic material property terms (discussed in Section 7.2.3 in Chapter 7) in the warpage calculations. In general, carbon fiber is stiffer; consequently, the fiber-reinforced thermoplastic melts with carbon fibers show higher viscosity as well as higher elastic behavior, effects that are important for such materials. Also because of the three cavities in the tool, the filling process becomes more complex as shown in Figure 7.7(b). The melt flow of the heated charge cannot be modeled using a simple Generalized Newtonian Fluid (GNF) formulation, and a non-Newtonian formulation where time dependence of viscosity is included is required.

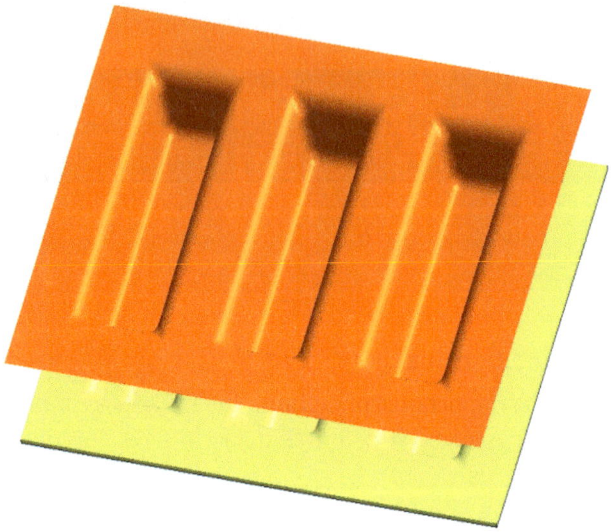

Figure 9.14 Three-cavity tool model used in Example 2

9.1.2.1 Actual Part Manufacture

The charge is prepared using a single-screw extruder from carbon fiber (35% by weight) and PA6 resin. The material used is in pellet form, with fibers in the lengthwise direction of the 12 mm long pellet. The charge used and the finished parts are shown in Figure 9.15. The details of the processing condition are presented in Table 9.4. The charge used was 120% of the cavity volume; the larger volume is usually used for the charge to account for volume shrinkage and leakage at the boundaries. The charge was placed on the center of the mold cavity for compression. The finished part was cooled, and warpage was measured. In this case, warpage is the out-of-plane displacement of one corner, where the other three corners are used to define the plane. The fiber orientation at selected locations was also measured to compare the microstructures.

Table 9.4 Process Parameters

Process Conditions	Actual Values
Melt temperature	270 °C
Mold temperature	70 °C
Compression time	60 s
Compression pressure	2000 kN
Charge weight	150 g

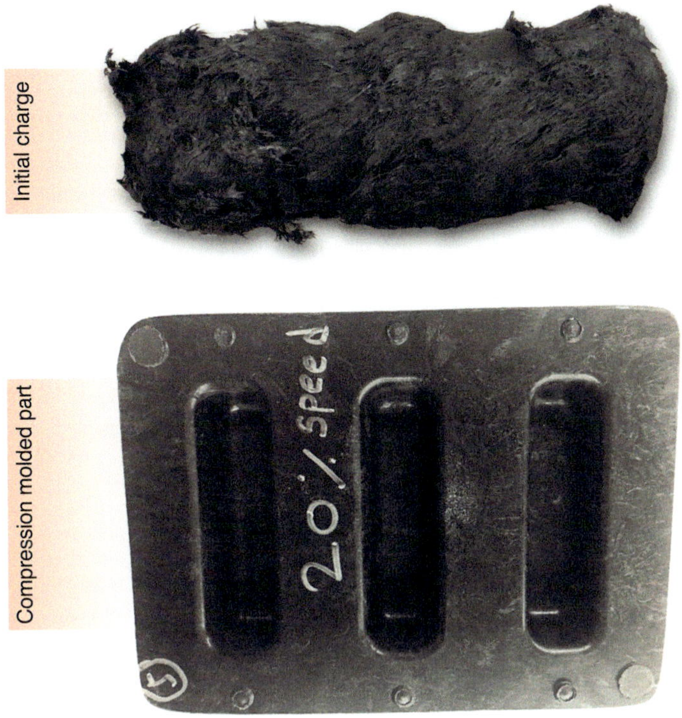

Figure 9.15 Charge used and the final part made after compression molding process

9.1.2.2 Material Properties Measurements

For the mold filling simulation, material properties such as pressure-volume-temperature *(PvT)* behavior, viscosity, and thermal properties are required. Once the charge was made using the single-screw extruder, it was cut and sent to the material test laboratory for property measurements. The measurements are done using the ASTM standards. For example, the *PvT* curve is measured using PVT-6000 based on the ISO-17744 standard [19], the viscosity is measured using rotation and oscillation MCR 502 based on ASTM-D4440 [20], the heat capacity is measured using RG-25 based on ASTM-D5930 [21], and the thermal conductivity is measured based on ASTM-E1269 using a DSC-8500 [22]. The measured material properties resulting from these tests were converted to material format for mold filling analysis as shown in Figure 9.16. The two-domain modified Tait equation was used to fit the *PvT* model. The compressibility of polymers is considered both in the filling and packing phase during the molding process [24]. The viscoelastic component of the material properties was measured using the experimental procedures developed by the Moldex3D™ laboratory, using techniques discussed by Chien [23]. These unique properties are important for the compression molding simulation of the heated fiber-reinforced thermoplastic material in the bulk form.

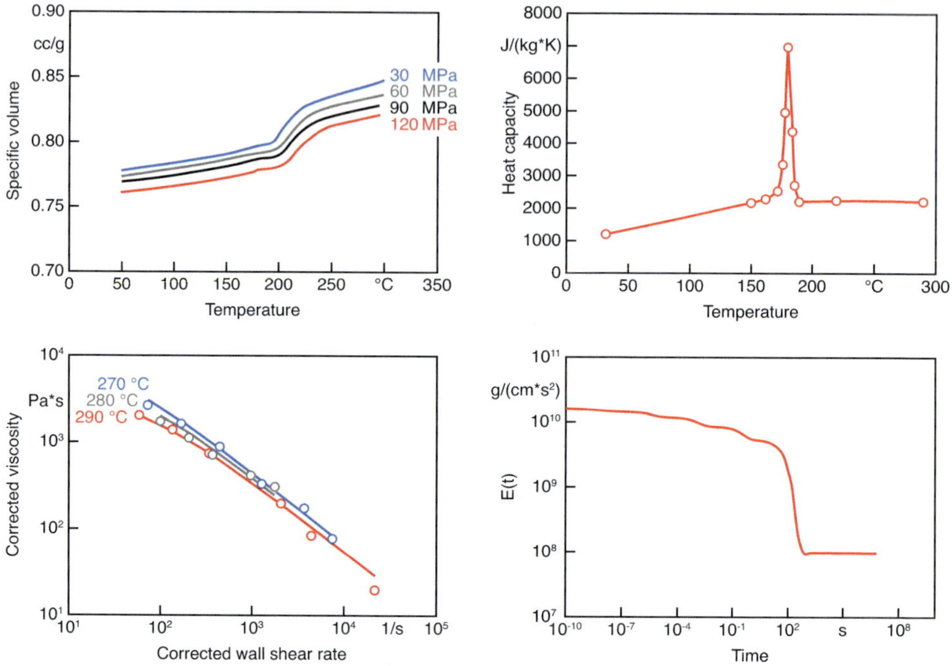

Figure 9.16 Measured charge composite material (35% CF+PA6) properties used for mold filling analysis: (a) *PvT* behavior; (b) heat capacity; (c) viscosity properties; and (d) stress relaxation modulus

9.1.2.3 Development of Model for CAE

The fiber orientation in the bulk charge is mapped using the approach discussed in detail in the previous example (Section 9.1.1.3). The bulk charge was CT scanned and the fiber orientation distribution through the volume was estimated by analyzing the images. Finally, the measured orientations were mapped on the finite element model of the charge. The key steps are presented in Figure 9.17. For this charge, based on the CT scan data, it was determined that the orientation pattern repeats itself every 10 mm, so only a 10 mm slice of charge was measured and then the same data was repeated during the mapping process to define the initial orientation for the whole charge.

Figure 9.18 shows the detailed dimensions of the final part, charge size, and charge footprint information used for the compression molding in this example. The elements used were tetrahedron and hexahedron mesh elements, with a total number of elements of about 6 million and a total number of nodes of 3 million. There was a total of 11 elements through the thickness of the cavity. The material properties used in the simulation are listed in Table 9.5.

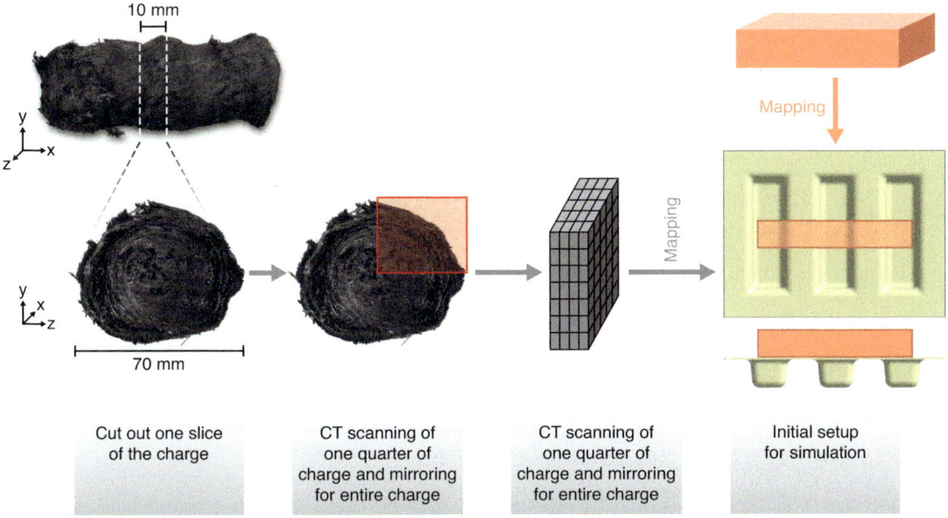

Figure 9.17 Initial fiber orientation mapping process for three-cavity tools

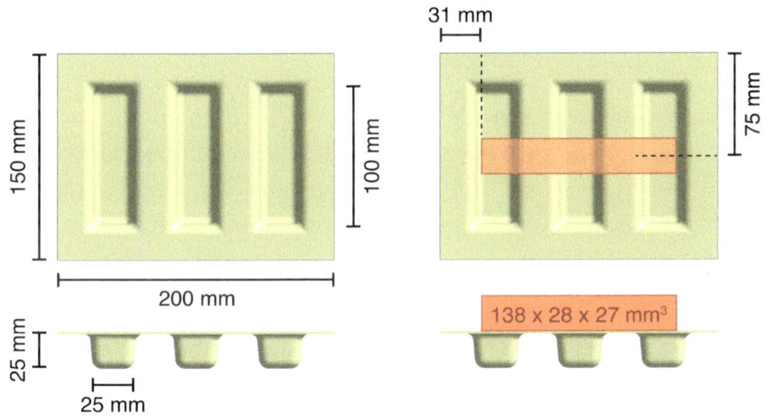

Figure 9.18 Detailed geometry used in the model and charge footprint information for compression molding (all dimensions are in mm)

Table 9.5 Material Properties Used for the Compression Molding Simulations

Items	Carbon Fiber	PA6
Density (g/cc)	1.78	1.13
Fiber weight percentage	35%	N/A
Young's modulus E (MPa)	230,000	2400
Poisson's ratio	0.26	0.42
Fiber aspect ratio	400	N/A
Fiber CLTE at fiber direction (1/K)	1e-6	N/A
Fiber CLTE at transverse direction (1/K)	1e-5	N/A
Polymer CLTE (1/K)	N/A	8.3e-5

The goal in this example is to demonstrate the effect of

- Viscoelastic terms in material properties and
- Defining fiber orientation in the initial charge

Four cases using different combinations of the above two effects, as shown in Table 9.6, were considered. In case 1, the viscoelastic (VE) model was used in the simulation and initial fiber orientation was mapped (Map) on the charge; in case 2, the VE model was not applied, but the initial fiber orientation was mapped; in case 3, the VE model was applied, but there was no initial fiber orientation mapping; and in case 4, there was no VE, nor initial fiber mapping in the simulation. When the initial fiber orientation was not mapped, the compression molding was done by using default orientation values, which is a random initial fiber orientation. The mapping was performed using the radial bases functions approach, discussed in detail in Section 9.1.1.3.

The software for compression molding process simulations used was Moldex3D™, version R14 [7]. The iARD-RPR (improved anisotropic rotary diffusion model combined with the retarding principal rate model) was used in the simulation with the C_i (fiber–fiber interaction) = 0.01. The other parameters used in the iARD-RPR model were C_m (matrix-dependent parameter) = 0.005, and the fiber matrix interaction alpha factor was 0.7. These were the default values used for the compression molding simulation within the software, the IBOF (modified orthotropic closure approximation model) 4th orientation tensor closure approximation was also used as the default [25, 26]. The processing conditions used for the simulation were the same as the one used to make the physical part and are presented in Table 9.5. The material properties used are based on actual measurements, presented in Figure 9.16.

Table 9.6 Four Cases Investigated for Comparison

Cases	Structural Viscoelastic Model (VE)	Initial Fiber Mapping (Map)
1	Y	Y
2	N	Y
3	Y	N
4	N	N

9.1.2.4 Comparison between Simulation and Experiments

The warpage predictions for the four cases are presented in Table 9.7 and are compared with the measurements from the actual parts in Figure 9.19. The displacement contour for the four cases shows significantly different results. For case 2 and case 4 we observe that the warpage prediction was anomalous in terms of the direction compared to the experiment, while the predictions in case 1 and case 3 showed the correct direction of warpage. The magnitude of the warpage predicted is slightly different between case 1 and case 3. As expected, the warpage prediction when using both VE terms in the material behavior and mapping the measured fiber orientation in the initial charge results in the best match between the simulation results and actual measurements, confirming the importance of using the VE terms and mapping the initial fiber orientation of the initial charge.

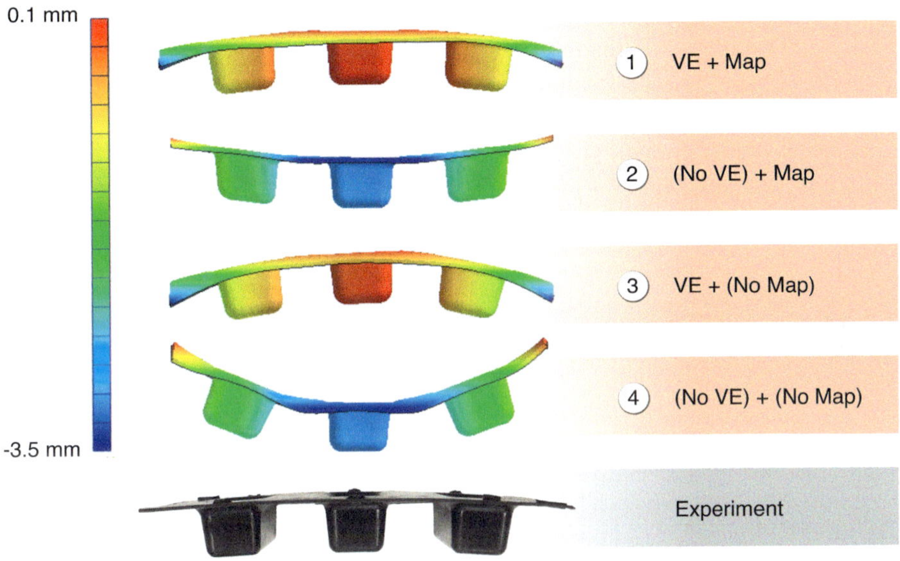

Figure 9.19 Warpage comparison for test and simulation. Four input conditions for the simulation as presented in Table 9.7 are used, VE + MAP is closest to experiment, VE is viscoelastic material properties, MAP is mapping the initial condition of fibers in the charge

For case 1 and case 3, a more detailed comparison is conducted for the points shown in Figure 9.20. These points are picked as representative locations for the whole surface warpage measurement. Since the part is symmetric, only a quarter of the part is used for analysis and these eight points are used as representative points for the comparison. Table 9.7 shows the comparison of results between these points, and the average of the errors shows that case 1 (viscoelastic material and mapped fiber orientation) had closer results than case 3 (viscoelastic material but initial random fiber orientation). The absolute value is used for the comparison.

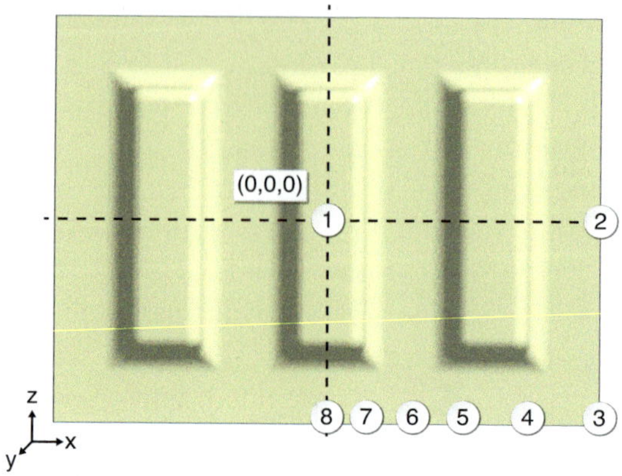

Figure 9.20 Position of out-of-plane warpage results for cases 1 and 3

Table 9.7 Comparison of Out-of-Plane Displacement for Selected Locations on the Finished Part*

Details	1	2	3	4	5	6	7	8	Average displacement	Difference
Physical part	0	3.1	3.5	2.4	1.9	1.4	0.4	0.2	1.61	NA
Case 1 (VE + Map)	0	3.1	3.3	2.2	1.8	1.2	0.3	0.1	1.50	0.11 mm
Case 3 (VE + No Map)	0	2.8	2.8	2.1	1.6	1.1	0.3	0.1	1.35	0.26 mm

* The plane orientation is defined based on the three corner points, and the location of the plane is aligned at the center point of the part.

In order to test the repeatability of the experimental results, for the case where the charge is placed in the center of the cavity, four samples were made using the same condition. The warpage measurement result for each sample is presented in Figure 9.21.

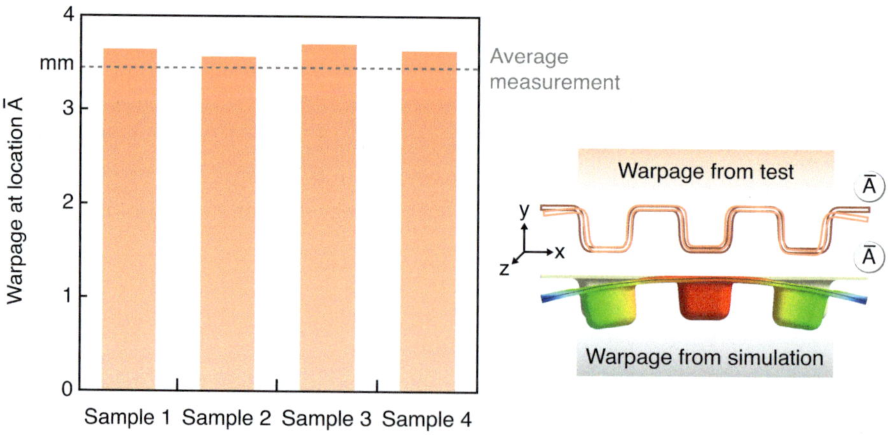

Figure 9.21 Warpage comparison for four samples to check repeatability of experiment

To verify the robustness of the simulation, the position of the compression charge configuration and location was changed, as shown in Figure 9.22:

- The charge location was moved to the side of the cavity, and
- The charge was divided into three smaller charges and placed in the cavity

The simulation and the actual measurement of warpage at point Ā for different charge footprints are also presented in Figure 9.22. The comparison shows that the simulation with viscoelastic materials and initial charge orientation mapping matches well with the measurement performed on the actual finished parts. This confirms the benefit of using the viscoelastic material and the actual orientation in the initial charge.

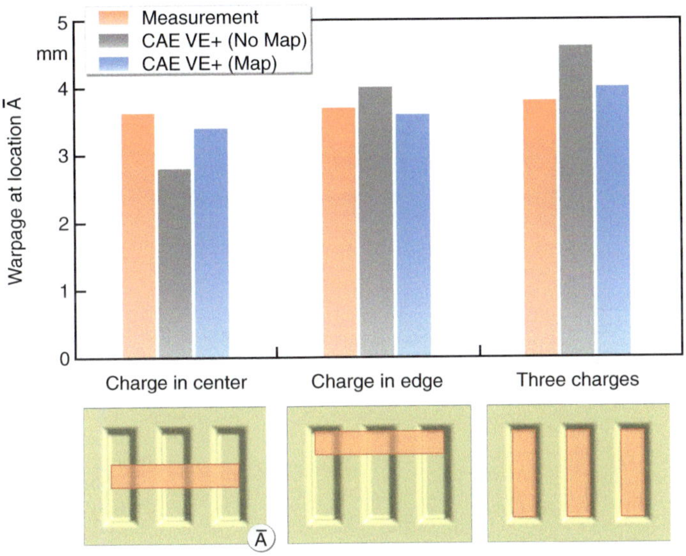

Figure 9.22 Comparison of the warpage prediction for different charge footprints (at location Ā)

It can be concluded that the key to a successful simulation is the ability to simulate the actual physics of the event. For a fiber-reinforced part, fiber length and orientation play an important role in the structural properties. Therefore, both microstructure features are included when simulating the case where the charge is in the center of the part in addition to using VE properties and mapped fiber orientation within the charge.

The fiber length of the part was measured at five selected locations using the FASEP method [27] (see Chapter 4 for more details). In the compression molding process simulation, the fiber breakage calculation was carried out using the approach presented in Section 7.4. The resulting fiber length distribution is compared with the actual measurements, for locations 1–5, in Figure 9.23. The comparison shows that the fiber length comparison of the final part was within an acceptable range.

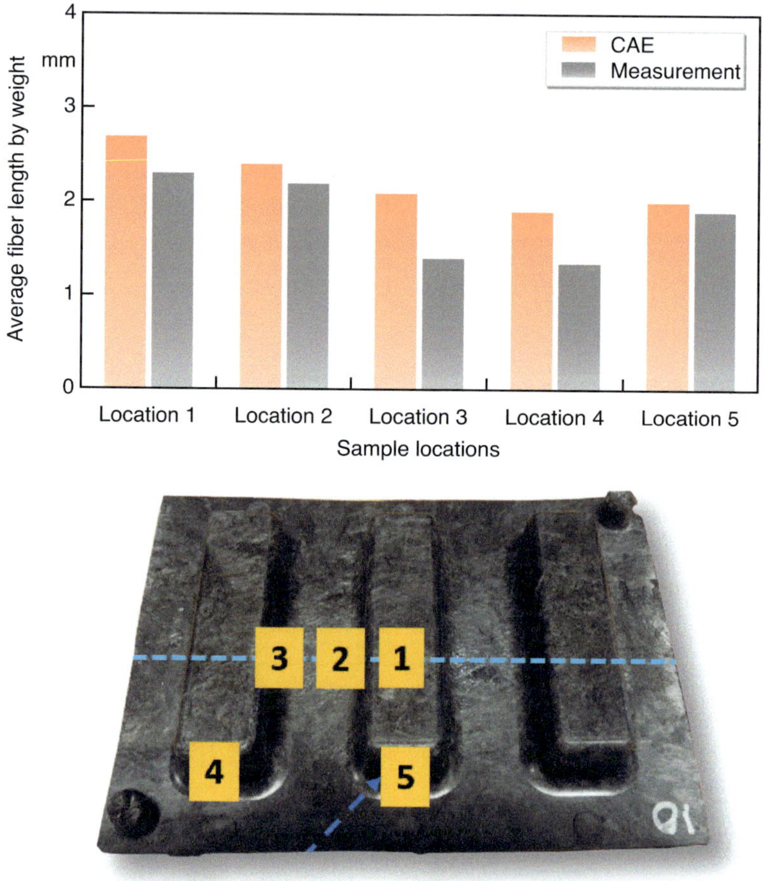

Figure 9.23 Fiber length comparison for the actual part and simulation at the selected locations

Fiber orientation is one of the major contributors to anisotropy, which causes differential shrinkage and hence warpage within the part. Therefore, we compared the fiber orientations at three selected locations in the finished part with the predicted values (see Figure 9.24).

A_{11} is the second order fiber orientation tensor component and represents the orientation distribution along the X direction. A_{22} and A_{33} are defined as corresponding fiber orientation tensor components in the global Y and Z directions. Since $A_{11} + A_{22} + A_{33} = 1$, only A_{11} and A_{22} are presented in Figure 9.24. A comparison of the orientation tensor visually showed the clear benefit of mapping the measured initial fiber orientation in the charge.

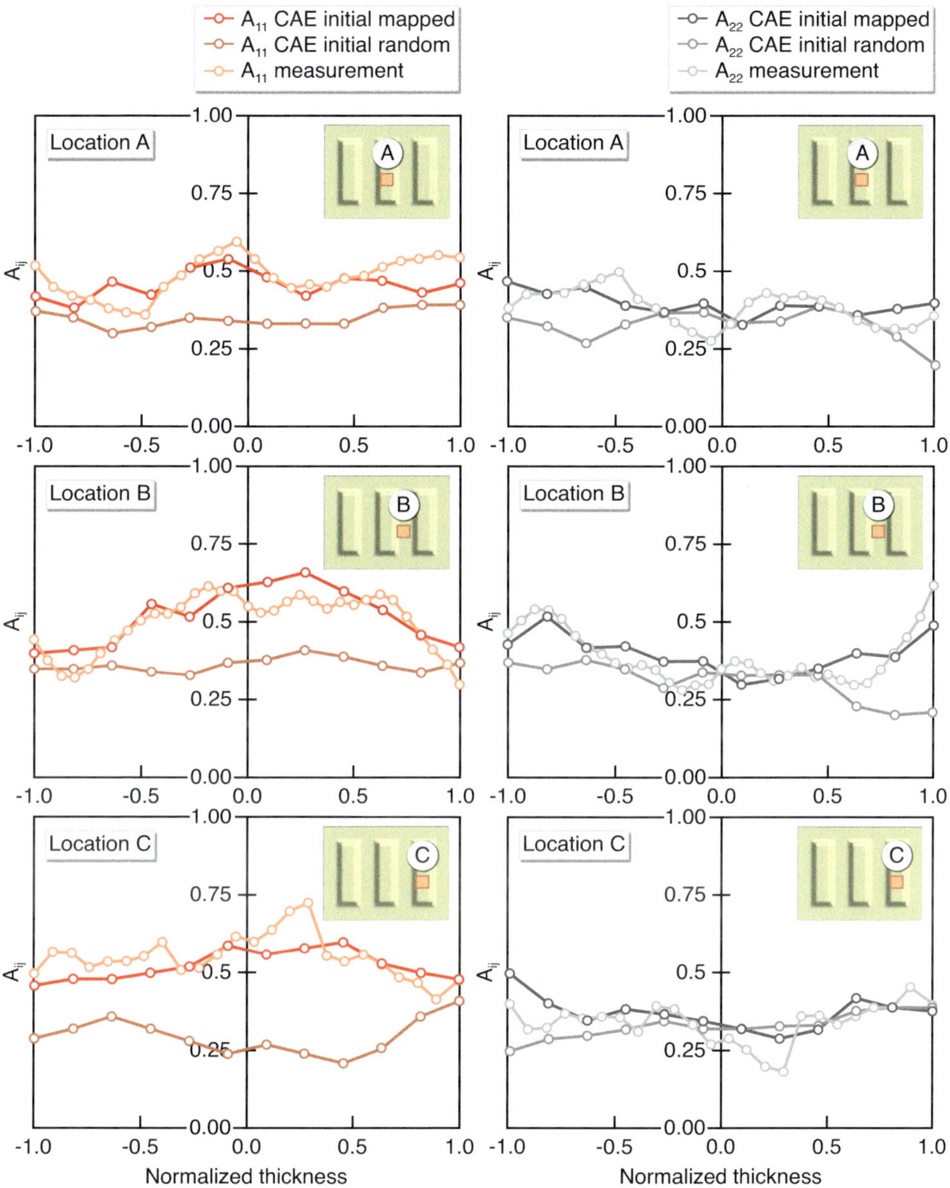

Figure 9.24 Comparison of fiber orientation tensor at locations A, B, and C using different initial fiber conditions

However, comparing the A_{11} component of the orientation tensor is a subjective comparison, as we are only visually comparing two fiber orientation distributions. To make a more objective or measurable comparison, Young's modulus (E) of the material was calculated using a Mori-Tanaka mean field homogenization approach [28, 29]. The calculation uses the fiber and resin elastic properties and available orientation tensor. The fiber volume was assumed to be 26% (35% by weight) with a fiber aspect ratio of 366. The only variation in the calculations was the fiber orientation tensor. Since ultimately the main interest with the fiber orientation is to understand its effect on material stiffness, it is believed that this is a meaningful method to compare two fiber orientation distributions. Table 9.8 compares the computed Young's modulus E_{11}, using the fiber orientation from the CAE calculations, and the measured orientation values from locations A, B, and C. The agreement in the Young's modulus was within 8%, which can be considered acceptable. The detail of computing elastic modulus from the fiber orientation tensor is available in Section 8.1.2.

Table 9.8 Computed Young's Modulus Based on Predicted and Measured Fiber Orientations at Locations A, B, and C

Locations/ Modulus	E_{11} (MPa)			E_{22} (MPa)		
	Using Predicted Fiber Orientation (CAE)	Using Measured Fiber Orientation	Agreement	Using Predicted Fiber Orientation (CAE)	Using Measured Fiber Orientation	Agreement
A	24,990	26,686	6.4%	19,089	20,661	7.6%
B	28,504	27,900	2.2%	22,084	22,041	0.2%
C	29,046	30,809	5.7%	21,054	19,616	7.3%

9.1.2.5 Comments and Conclusions

Through this example, the simulation of the compression molding of a chopped fiber filled thermoplastic bulk material was demonstrated. The importance of using viscoelastic material properties was demonstrated, when accurate predictions of warpage within a complex final part are required. Furthermore, the importance of using the correct fiber orientation within the bulk charge was also demonstrated. The inclusion of the initial fiber condition in the simulation was proven by test validation, which will improve the warpage prediction. It was also shown how the initial footprint of the bulk charge affects the final warpage of the part.

■ 9.2 Compression Molding of Sheet Materials

In the previous examples, it was demonstrated that compression molding of discontinuous fiber-reinforced thermoplastic materials in the bulk form should be simulated by treating the melt flow as a continuum with viscoelastic properties. However, the compression molding of such materials in sheet form using a similar approach does not work well. This is because the sheet deforms as a rigid body first and then the deformed shape for the sheet, when further compressed, flows to fill the cavity. As discussed in Section 7.2.3, the assumption of melt flow as a continuum does not hold for the large rigid body deformation of the sheets in the mold cavity. In this example, it will be shown how to address this challenge. A new two-step simulation approach, as discussed in Section 7.2.3.2, is used. First, the draping of the sheet in the mold cavity is simulated as a rigid body motion with structural deformation using the explicit finite element approach. Next, the draped shape is compressed as viscoelastic melt (fluid) using a compression molding formulation (Section 7.2.2). The proposed CAE method is demonstrated using the three-cavity tool discussed in the previous example. The predicted microstructure, such as predicted fiber orientation and length, as well as the predicted structural performance, such as warpage, are compared to physical parts.

9.2.1 Actual Part Manufacturing

The sheet material used for this example is prepared using a water slurry process shown in Figure 9.25. The chopped carbon fiber (~50 mm long, 35% by weight) and thermoplastic material such as PA6 or PA66 in fiber form are dispersed in the clean water. The mixture of fibers suspended in the water is passed over a sieve, which separates the comingled mesh of fibers on its surface. The loosely held mesh of comingled fibers is slightly heated, impregnated with the binder, and compressed to create fluffy mats, such that they can be handled easily for further processing. Depending on the final part geometry and process design, the material can be used in mat form or can be further heated and compressed into thinner sheets, which is called pre-consolidation. The mats used in this example were prepared from recycled carbon fibers by Carbon Conversions, a carbon fiber recycling company located in South Carolina. The compression molding was done on the fluffy mats.

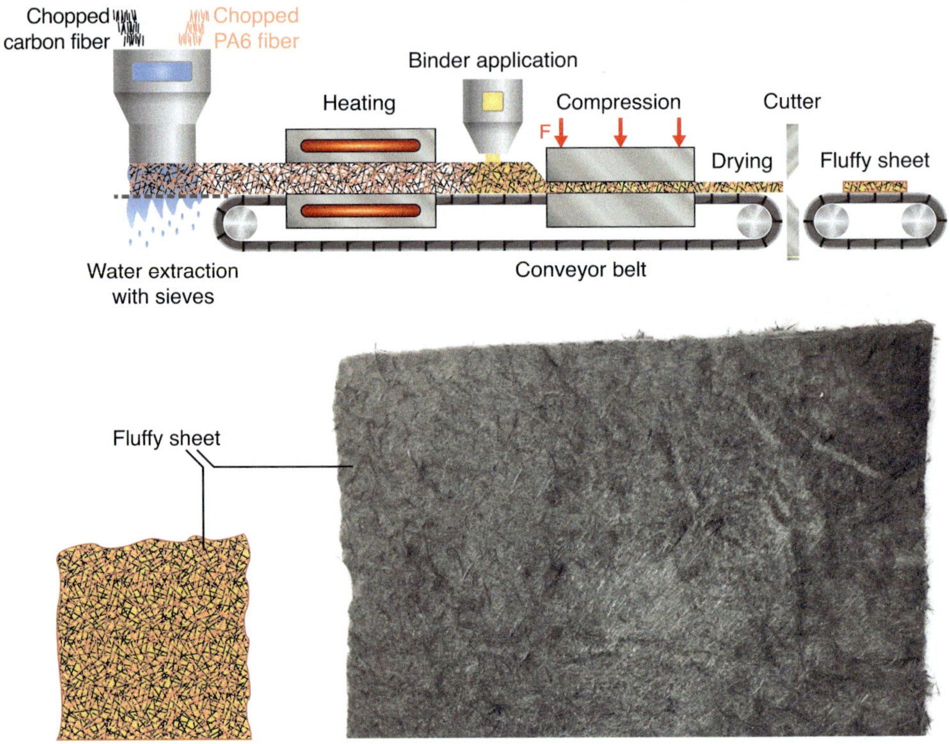

Figure 9.25 Sheet material manufacturing process using water slurry method

The three-cavity tool used in the previous example was used for this study as well. The PA6 mats with 35% carbon fiber by weight were heated to 270 °C and quickly transferred to the compression press. One of the challenges that were faced in using the mats was that they tended to cool quickly due to the large exposed surface area compared to the bulk materials. Also, the tools that are cooler than the charge tend to remove the heat out of the mats and cool them quickly. All this makes the compression of sheet materials more difficult. This can be addressed by preheating the mold and heating the mats to temperatures as high as possible, often just below the temperature at which the material starts to degrade and loose its properties. Such temperatures for each material can depend on the resin in the mat and geometry of the tools, which means that an experimental study is required to establish optimum temperatures for each situation. The key steps used to manufacture the parts are presented in Figure 9.26.

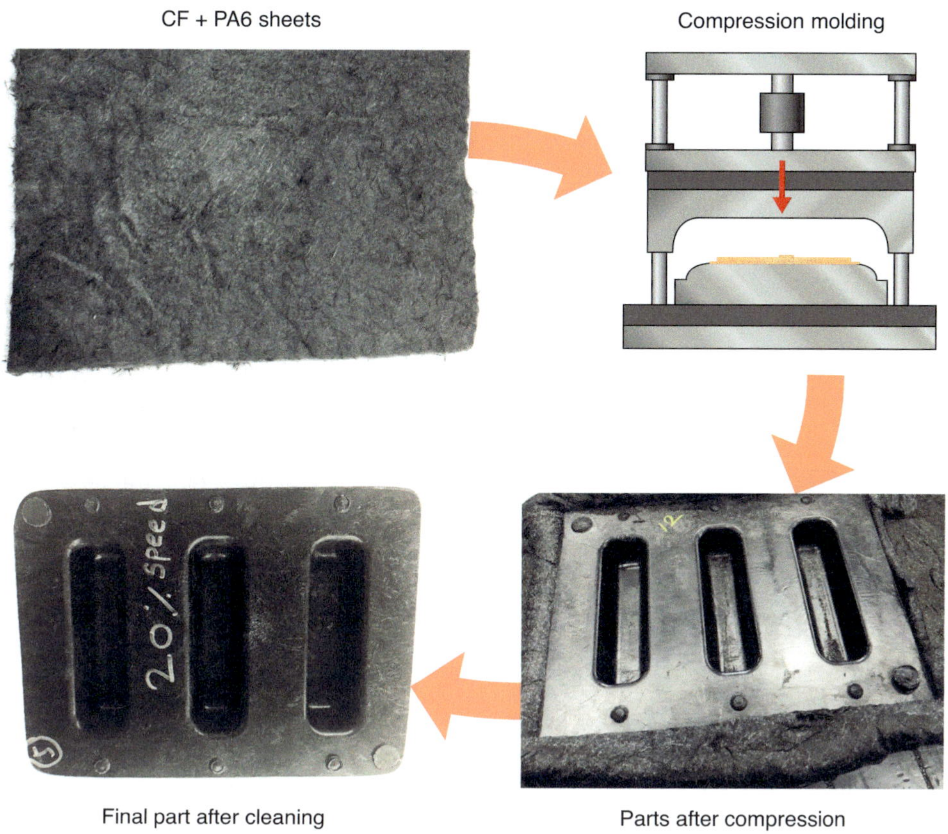

Figure 9.26 Key steps for manufacturing the parts from mats

The process parameters used in this study are presented in Table 9.9.

Table 9.9 Compression Molding Parameters Used

Process Conditions	Actual Values
Melt temperature	270 °C
Mold temperature	70 °C
Compression time	60 s
Compression pressure	2000 kN
Charge weight	150 g

The finished parts were demolded and allowed to cool to room temperature. The warpage of the finished part was measured using a scanning device attached to Faro-Arm, a technology developed by FARO for precise dimensional analysis and quality control of components. Furthermore, a 3D geometry of the surface of the finished parts was developed for comparison with the simulation, as shown in Figure 9.27.

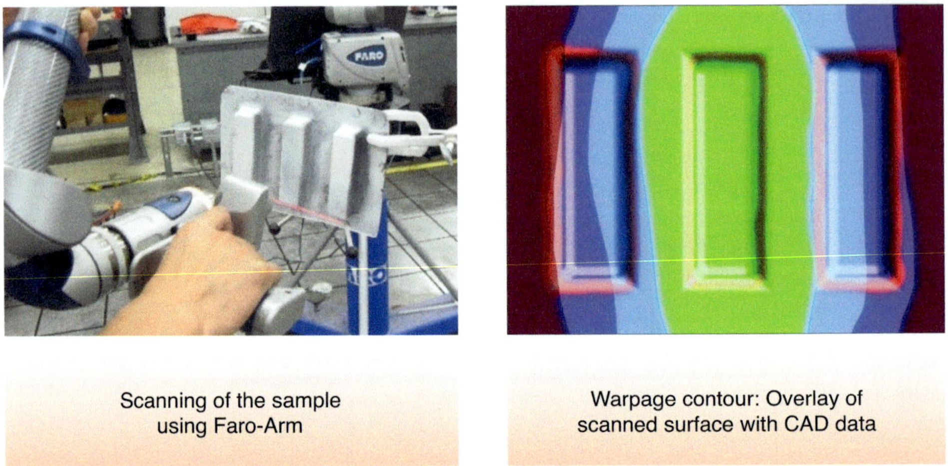

Figure 9.27 Scanning and measurement of the surface data of the finished part for the warpage study

Computed tomography (CT) scanning and Volume Graphics (VGStudio MAX^{TM}, Volume Graphics GmbH, Heidelberg, Germany) analysis were used for measuring the fiber orientation. For the fiber orientation tensor analysis, samples were cut from the molded part at locations A, B, C, etc., as shown in Figure 9.28. The sample size used was 10 mm × 10 mm × 2 mm, and the resolution used for the CT scan was 1 µm. Low resolution was required to capture the carbon fibers that are about 5 microns in diameter.

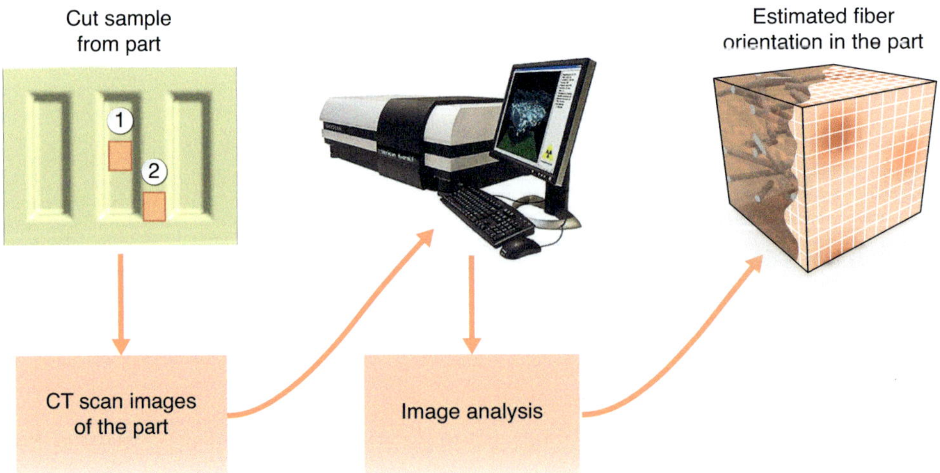

Figure 9.28 Fiber orientation measurement using CT scan and volume graphic method

9.2.2 Measurement of Material Properties of the Sheet Material

The material properties are key for a successful CAE. The two key properties needed are:

- Structural properties of the mat material at high temperature for the finite element analysis
- Flow (rheological) properties of the material for the mold fill analysis

Measurement of such properties is a new and challenging task. For the material used in this example, analysis was done at the Moldex3D™ laboratory in Taiwan to measure the material properties accurately and prepare the material data cards for use in the CAE analysis.

9.2.2.1 Material Structural Properties

For accurate draping analysis, temperature-dependent material properties such as elastic modulus and Poisson's ratio are needed for the sheets. Figure 9.29 shows the procedures used to measure the mechanical properties of the sheet/mat material. As shown, two layers of fluffy mat material are stacked together under a hot press and consolidated to form the composite plate. The tensile bar samples were cut from the consolidated plate. Tests were performed following the ASTMD638-02 standard [30], and the samples were tested in a temperature chamber to obtain the mechanical properties at different temperatures. The modulus of the material was obtained from the tensile test. Poisson's ratio was also measured as per the ASTMD638 standard, using the optical imaging system. Table 9.10 shows the measured mechanical properties. The increase of the Poisson's ratio at the end is due to the polymeric phase change at around 180 °C, due to crystallization. The modulus and Poisson's ratio results are the average of three separate tests.

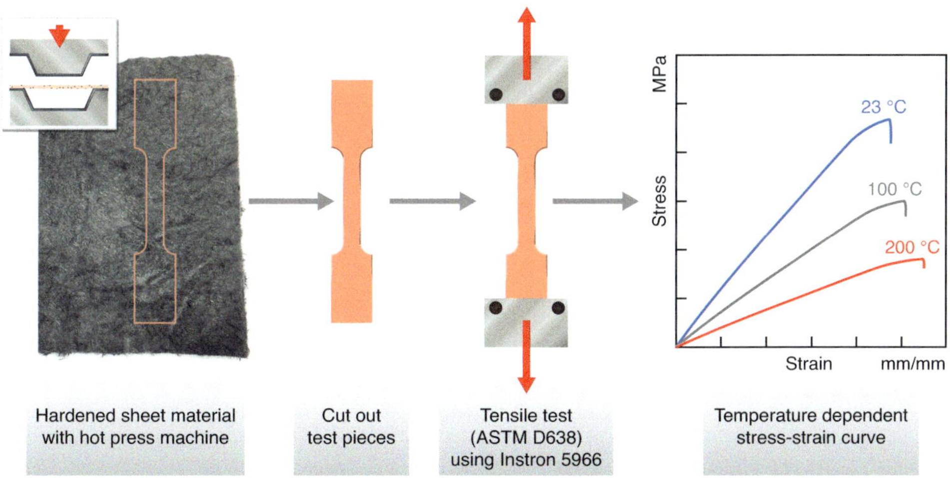

Figure 9.29 Temperature dependent Young's modulus and Poisson's ratio measurement procedures

Table 9.10 Temperature-Dependent Stress-Strain Curve for the Sheet Material

Temperature (°C)	Modulus (MPa)	Poisson's Ratio
23	14,091	0.339
80	10,444	0.239
130	10,118	0.237
180	8728	0.179
200	7380	0.403

9.2.2.2 Sheet/Mat Properties for the Mold Filling Analysis

To accurately simulate the compression molding process, actual material rheological properties such as pressure-volume-temperature *(PvT)* behavior, viscosity, and thermal properties are also needed for this process [31]. The sheet/mat material properties were measured by following the corresponding ASTM standards [32–34]. After the physical measurement, all the measurement results were converted to the Moldex3D material format. These measured composite properties were then imported into the compression molding software Moldex3D's material database and were applied to the process simulation for the part, as shown in Figure 9.30.

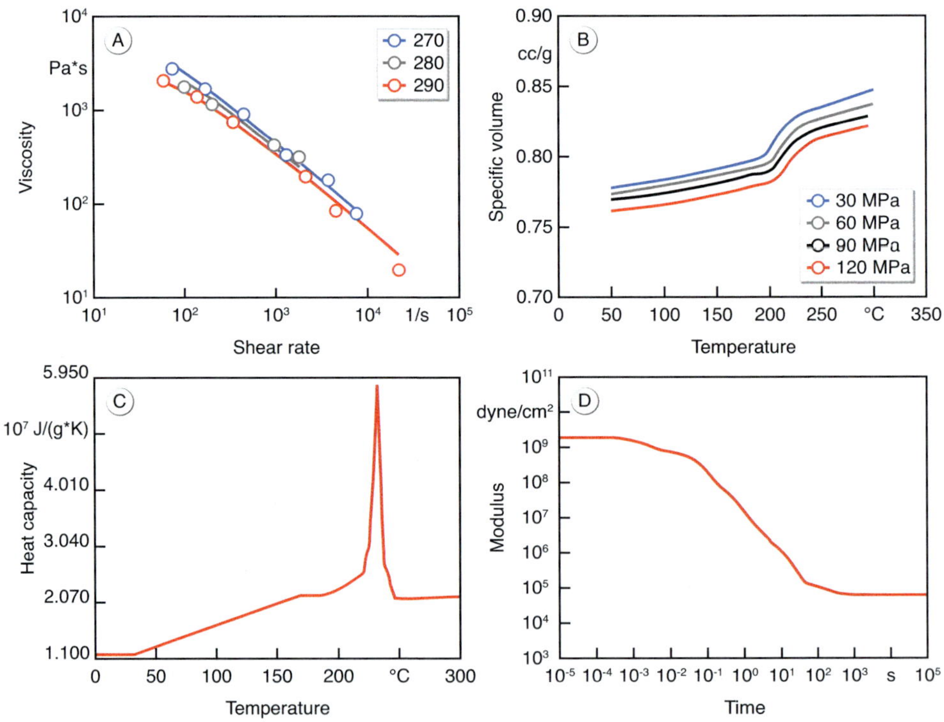

Figure 9.30 Measured sheet/mat material properties converted to the Moldex3D format: (a) viscosity curve, (b) *PvT* curve, (c) heat capacity curve, and (d) viscoelasticity curve

9.2.3 CAE Simulation for the Compression Molding of Mats

The three-cavity tool used for the study was the same from Example 2 (Section 9.1.2), with the geometry details presented in Figure 9.18. The integrated CAE approach combining LS-DYNA and Moldex3D discussed for sheet/mat material in Section 7.2.3.2 is employed here. Figure 9.31 shows the overall CAE simulation workflow for the part. The first step was the draping analysis in LS-DYNA. After the draping analysis, the draped part was transferred to Moldex3D™ for the compression molding process analysis. The details of each simulation process step will be discussed separately in the following sections.

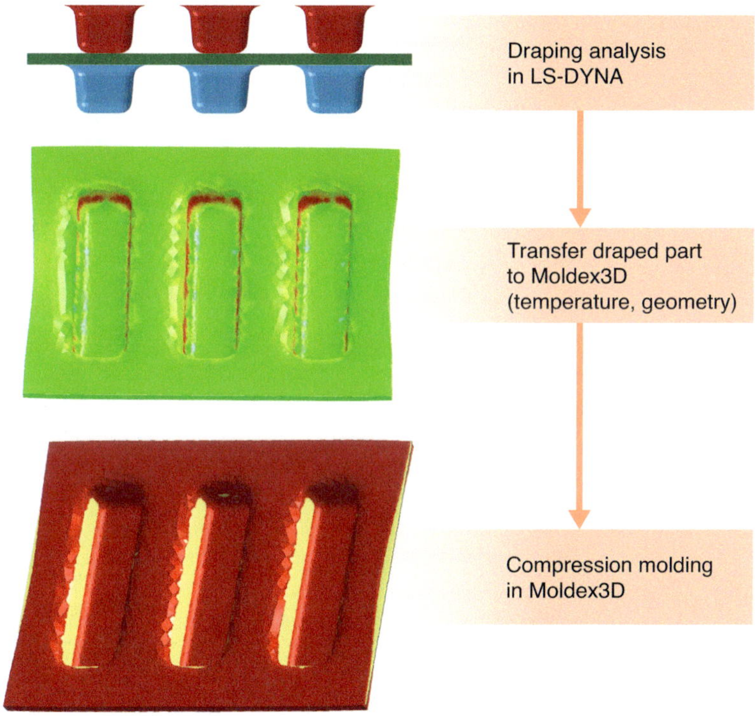

Figure 9.31 CAE steps for the compression molding of sheet materials

9.2.3.1 Draping Analysis Using LS-DYNA

For the first step of the simulation, LS-DYNA was used to predict the draping behavior of the sheet materials. A thermoforming module from LSTC was used in the simulation. The initial sheet/mat material was modeled using solid elements, as is shown in Figure 9.32. In the model, the total number of solid elements was 15,000, the total number of shell elements was about 26,000, and the total number of nodes was about 50,000.

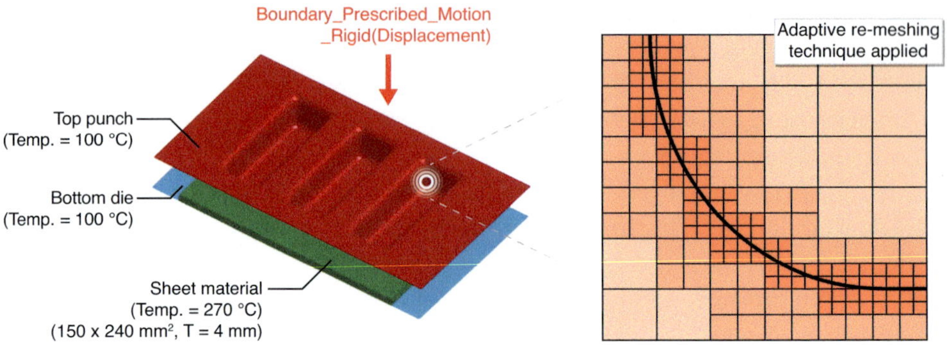

Figure 9.32 LS-DYNA model details for the draping analysis

During the draping of the sheet/mat material, at the edge of the tool, the material stretches, and large elastic deformation of the elements may occur (see Figure 9.32). Excessive stretching can result in poor element quality, which can result in convergence issues [35, 36]. In LS-DYNA, this challenge can be addressed by using a highly refined initial mesh or an adaptive re-meshing technique, where the mesh is refined as needed, as shown in Figure 9.32.

Figure 9.33 shows the displacement contour of the draped part. The geometry of the draped part was exported to Moldex3D™ for the next simulation step, i. e., the compression molding simulation of the draped part.

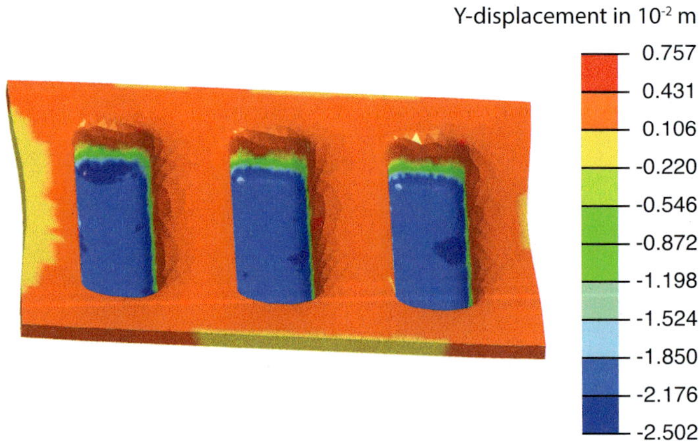

Figure 9.33 Displacement contours of the draped part (unit: m)

9.2.3.2 Compression Molding Analysis with Moldex3D

Based on the cavity design from Figure 9.18, a compression molding 3D model was built in Moldex3D™, as shown in Figure 9.34. The element types used in the mold cavity model were tetrahedron and hexahedron elements and the total number of elements was approximately 6 million, and the total number of nodes was 3 million. There were 11 elements through the thickness. The orange surface on the top is the compression surface, the pink area in the middle is the compression zone, and the bottom is the cavity.

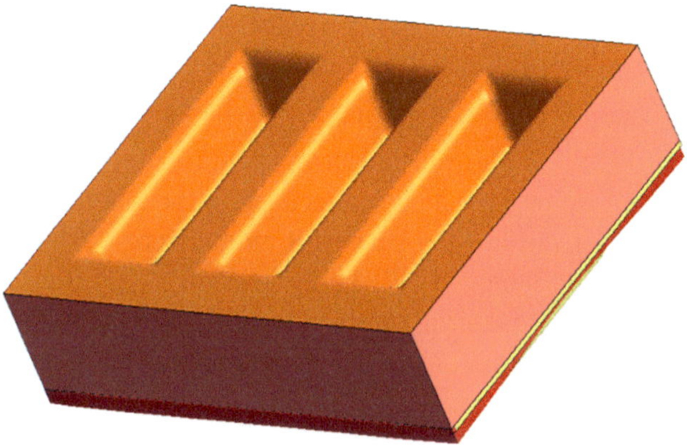

Figure 9.34 Compression molding model built in Moldex3D

Figure 9.35 shows the draped part geometry used as the compression charge. The draped geometry is re-meshed with the solid elements that are compatible for use with Moldex3D™ to create the finite element model of the charge, which is placed in the mold cavity for compression molding. Custom software is developed and used to help through this process. Wrinkling of the charge as well as temperature distribution in the draped part are also mapped while transferring the geometry from the draped shape to Moldex3D™ as an initial charge.

The fiber orientation in the initial sheets/mat is assumed to be in-plane random. During the draping process there is very little material flow; it is mostly rigid body or elastic motion. Therefore, it was assumed that the fibers stay in-plane random. During compression molding the material flows, changing the fiber orientation. However, the flow of the material is quite small compared to injection molding and, therefore, the fibers are expected to stay predominantly random throughout the part.

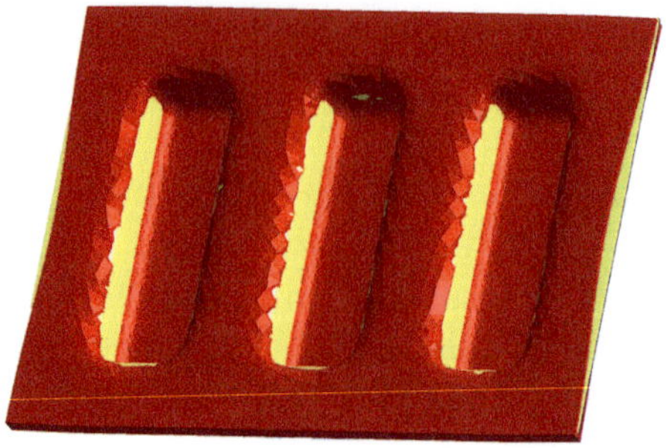

Figure 9.35 The draped part shown as charge geometry, which is mapped as a prepreg for compression molding

The compression molding process simulations were done using Moldex3D, version R14 [7]. The iARD-RPR (improved anisotropic rotary diffusion model combined with the retarding principal rate model) was used in the simulation with $C_i = 0.01$. The other parameters used in the iARD-RPR model were $C_m = 0.005$ and a fiber matrix interaction alpha factor of 0.7. The processing conditions in CAE are the same as the ones used to make the physical part. The material properties used in the simulation are the same as the ones used for the bulk charge listed in Table 9.6 and Figure 9.30. The actual material properties can be slightly different, as the material used for the bulk charge was prepared using an extruder, while the sheets are prepared using the wet-sieve process. Since the fiber conditions, which have a major influence, are accounted for, the slight difference in the resin properties is considered acceptable for the study.

9.2.3.3 Comparison between Simulation and Experiments

Figure 9.36 shows a comparison of warpage from the simulation with the actual parts. The warpage is measured as out-of-plane displacement at selected corner nodes, where other three corners are used to define the plane. The scanned surface of the actual part, obtained using the Faro-Arm with laser scanning, is compared with the simulation results. The results are compared for two different mold temperatures to verify the robustness of the simulation. With an increase in the mold temperature, the warpage increased in proportion. The overall difference between the simulation and measurement was noted to be around 6% – 8%.

For a meaningful simulation it is important to have the physics of the event represented in the simulation. Since the material properties are highly dependent on the microstructure, microstructure details such as fiber orientation and fiber length were also compared.

The fiber orientation at three selected locations (A, B, and C) are compared between the actual part and CAE in Figure 9.37. The fiber orientation tensor component in A_{11} is the major component. The overall comparison from these three locations shows a similar trend and a relatively close match.

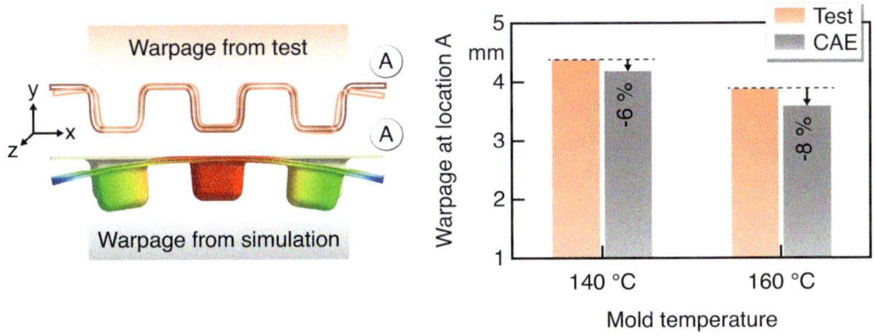

Figure 9.36 Effect of mold temperature on warpage: (a) warpage measurement locations; (b) warpage comparisons

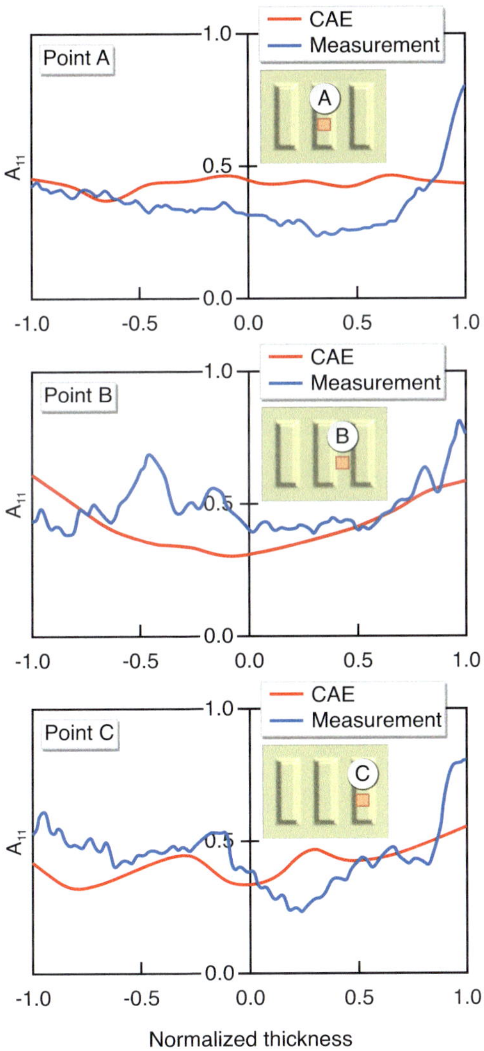

Figure 9.37 Comparison of fiber orientation tensor component A_{11} from measurement and simulation (CAE) at location (A), (B), and (C) on the finished part

The fiber lengths at five selected locations are compared in Figure 9.38. It was observed that the fiber lengths for simulation and measurement also match well. The good correlation between the actual part and the simulation for the microstructure provides good evidence that there can be reasonable confidence in the simulation approach presented here.

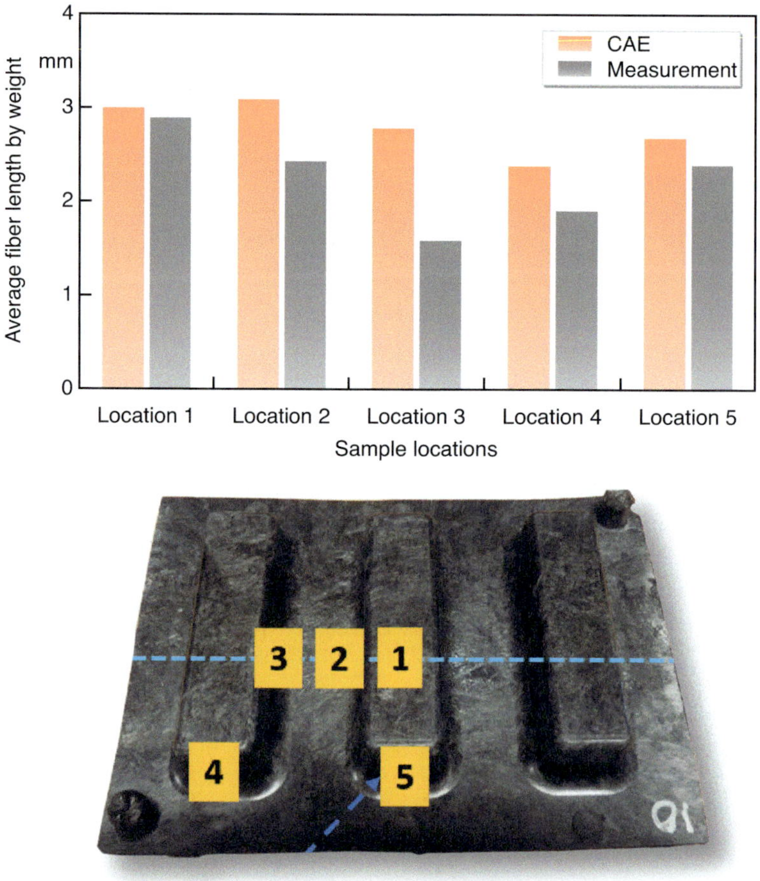

Figure 9.38 Fiber length distribution comparison. Bottom image: sample locations; top image: fiber length distribution comparison

The last question one needs to address for the compression molding simulation of sheet materials using the proposed approach is: When should the draping stop in the simulation and compression begin, and how does a partially draped part affect the warpage?

In production, the compression molding process is continuous at the established cycle times, typically 1–2 min depending on the part size. The breaking up of the compression of a sheet/mat material into three steps, as presented in Figure 7.9(b), is only conceptual, considered to help the simulation of the process. Therefore, it is important to understand how much stroke should be used in the finite element draping process and when one should switch to the compression molding.

To address this question, the LS-DYNA analysis was conducted considering three different boundary conditions. As shown in Figure 9.39, three different draping displacements, i.e., 60%, 80%, and 100% of the gap between the top punch and the bottom die, are considered. The draped parts with different draping displacement were then transferred to Moldex3D™ for the compression molding simulation.

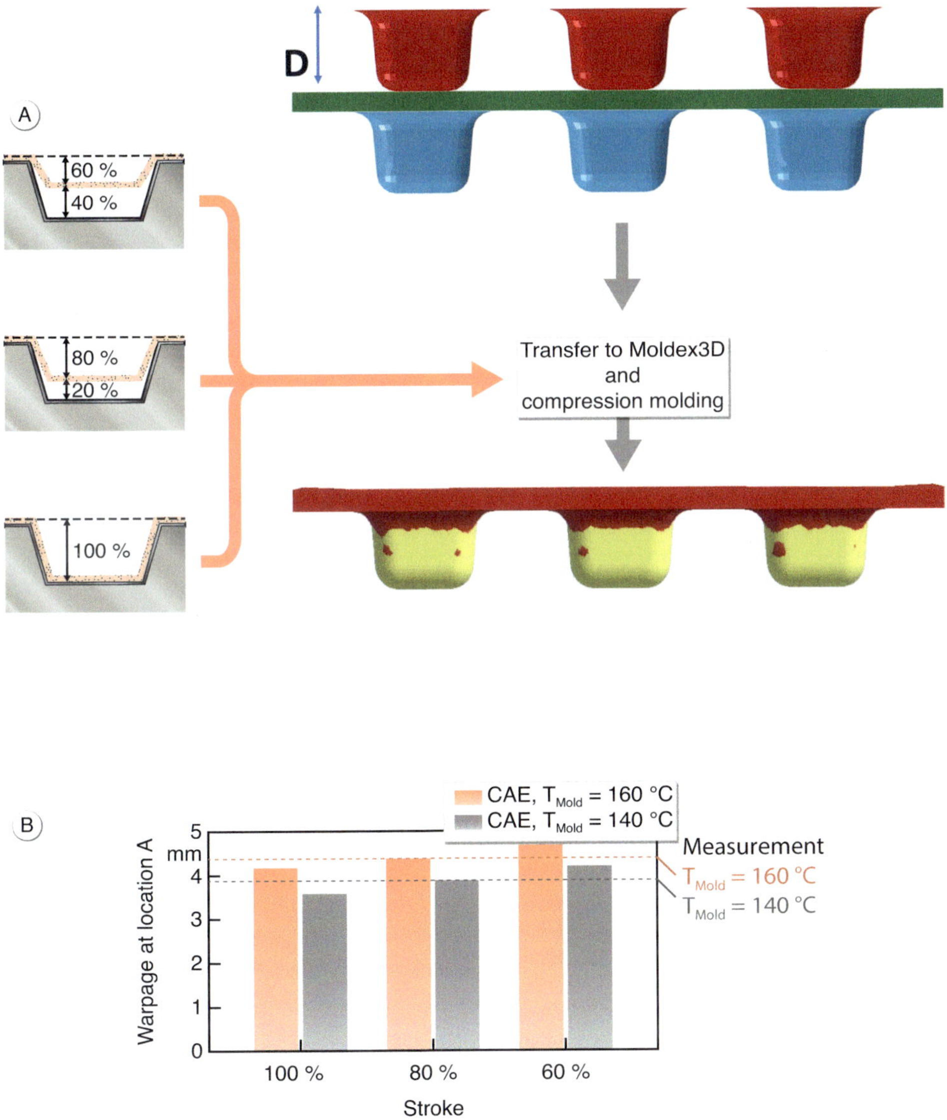

Figure 9.39 Effect of the stroke on the warpage: (A) details about stroke; (B) warpage comparison

It was observed that the draping displacement has a clear effect on the prepreg shape, therefore resulting in different warpage for each case. From the comparison of warpage measurement results, it was shown that when the draping distance is 80% of the gap, the predicted warpage is closest to the measurement. This observation is found to be consistent for two cases, i.e., molded at 140 °C as well as molded at 160 °C. Caution should be exercised here in that the draping stroke at 80% of the gap of the cavity worked well for this example. For other situations with different geometries, the meaningful draping stroke can be different. The general approach presented here should work for any case.

9.2.4 Comments and Conclusions

Compression molding of discontinuous fiber-reinforced materials is getting increasing attention, as there is a growing interest in low-cost lightweight materials for high volume production. The ability to simulate the compression molding with high confidence can help estimate the warpage and structural properties of the finished part. Such capability can be of great help in design and development of discontinuous fiber-reinforced compression molded parts.

While the basic compression molding formulation was presented in Chapter 7, its application to the practical problems needs additional considerations, which were presented in this chapter. The key challenges for successful compression molding simulation are:

- Accounting for initial fiber orientation within the compression charge
- Viscoelastic material properties of the material must be used, including the time dependency
- Draping analysis of the charge material must be performed for sheet charge format
- Measurement of material flow properties for material with long fiber materials must be performed
- Measurement of structural properties at high temperature for draping must be performed

An approach to address these challenges and successful simulation of compression molding for a complex 3D shape is demonstrated in this chapter.

■ 9.3 Compression Molding of GMT Material

Glass-mat reinforced thermoplastic (GMT) materials are widely used in the automotive industry for components such as underbody shields, seat structures, front/rear bumper, and front-end modulus. Due to the higher residual length of the glass strands, GMT materials offer better mechanical properties than injection molded fiber-reinforced thermoplastics. The most common method to manufacture GMT parts is compression molding of pre-impregnated fiber-reinforced resin mats. There are two types of GMT mats, one with discontinuous random fibers and other with unidirectional continuous fibers. A stack of different mats of various combinations is used in compression molding to tailor the mechanical properties of the final part. During compression molding, the fibers in the mat flow along with the resin, changing the fiber orientation. Due to the complex nature of the initial fiber conditions and the orientation that occurs during processing, the finished GMT parts have very unique microstructures. In this case study, an approach to model such microstructure for use with the multiscale modeling presented in Chapter 6 is demonstrated.

One of the main challenges is accounting for the initial fiber conditions, such as fiber length, orientation, and concentration in the mat, as well as the process conditions used, to predict the microstructure for the multiscale models. It has been observed in the past that the microstructure of the fiber-reinforced materials depends on all three distributions: fiber orientation, fiber length, and fiber concentration. The case studies discussed so far were mainly based on injection molding of discontinuous fibers. In such cases it was assumed that the fiber volume fraction is constant throughout the part. Also it was assumed that the fiber length is constant, and a value based on measurements or based on experience with similar parts is used. Both these approximations work well because the variation of fiber length and fiber concentration in the injection molded parts is quite small. However, with GMTs this is not the case. As multiple sheets of different format of GMT sheets are stacked for compression molding, the fiber concentration and fiber length through the finished part will vary significantly depending on the stacking sequence and type of GMT sheets used. Therefore, it is necessary to account for fiber concentration and fiber length. In this case study, an empirical measurement-based approach to address this challenge will be presented.

The fiber orientation in the finished part is estimated using compression molding simulation of the stack of mats. The fiber concentration and the fiber length distribution are estimated based on the actual measurements. The cross section of the finished part is investigated under optical microscope, and the fiber length and concentration are estimated based on the microstructure and initial stacking of mats. The microstructure properties are used for the multiscale modeling approach. The multiscale models are verified by comparing the predicted performance with the actual tensile and bending test results.

9.3.1 Discussion on GMT and Approach

GMT has been commercially available since the 1970s [37–39]. The short molding time and higher mechanical properties make the process attractive for use in the automotive industry [40, 41]. Glass fibers in a polypropylene matrix (GMT-PP) are the most common GMT material in use today.

GMT sheets are available in two formats: discontinuous random glass fibers or unidirectional continuous fibers. During the molding process, a combination of such preheated GMT sheets are stacked and squeezed until they fill the mold cavity. The structural properties of the finished parts depend on the combination of GMT sheets used in the initial stack, as well as the process conditions. Since parts are often critical structural components, it is necessary to predict the material properties for such compression molded GMT components.

In this case study the Moldex3D™ compression molding module [7] was used to simulate the compression molding of GMT sheets. The knowledge of stacking and the detailed arrangement of fibers in each layer of the GMT sheets are used to define known initial conditions for the mold fill process simulation. The estimation of fiber orientation in the finished part is obtained from the simulation results. Also, the estimated fiber orientations are verified with the actual measurements in the finished parts at selected locations.

Fiber length and concentrations in the finished part are estimated using an empirical approach. The finished part is studied under an optical microscope to identify the fiber conditions such as orientation and fiber length through the thickness. It was observed that the fibers are still bundled as layers defining specific zones, i.e., area of similar fiber length and concentration, which is typical for GMT sheets. These observations along with knowledge of initial condition in the GMT sheets are used to define the fiber length and concentration through the thickness in the final part.

This case study demonstrated two unique points:

- The effect of different initial stacking of GMT sheets on the final properties of the finished parts. Five different samples using different stacking order of the GMT sheets are built and tested for bending and tensile loading conditions. This helped understand the relation between the stacking order of GMT sheets and the mechanical properties.
- Developing the material model for a compression molded part from GMT sheets for multiscale simulation. The empirical method presented here accounts for fi-

ber length and fiber concentration in the material model. The resin is assumed to be a non-linear elastic-plastic material.

Finally, the predicted material properties are used in the finite element model for tensile and bending loads and the results are compared with the measured results to verify the accuracy of the predicted material models.

9.3.2 Actual Part Manufacturing

Two types of resin prepreg mats made by Hanwah™ Advanced Materials were used to assemble the compression stack. Figure 9.40 shows the configuration of the stacks. Each sheet had five alternating layers of polypropylene and 50% by weight glass fiber. The difference between the two mats is based on the fiber length and fiber orientation. The mats where the fibers in the two layers are random and 50 mm long will be referred to as RD mats. The mats where 50% of the fibers are continuous and unidirectional will be referred to as UD mats [43–45].

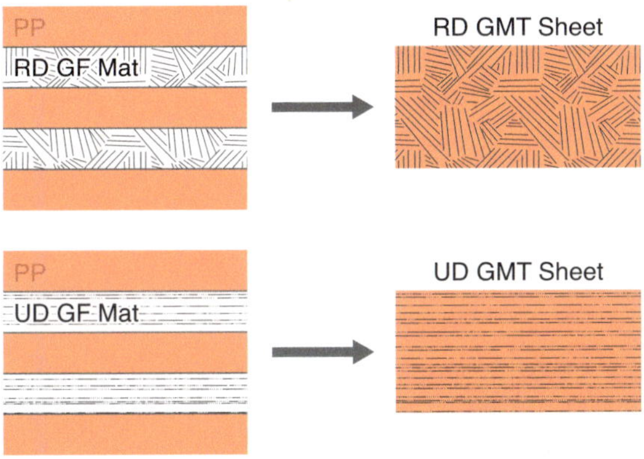

Figure 9.40 Two types of initial GMT material configuration (RD and UD GMT)

To study the effect of initial fiber alignment on the material properties of the finished part a tub-shaped part was used as a demonstrator example. Figure 9.41 shows the stack method and compression molding processing of these three GMT sheets. Five different initial lay-up configurations were considered, consisting of different combinations of three sheets made up of the UD and RD mats or sheets. Details of the configurations are presented in Table 9.11. The stacks consisting of three sheets were preheated over the glass transition temperature in an oven and then compressed in the tub-shaped mold cavity using a force of 150 ton for a period of 40 seconds. While manufacturing tubs using compression molding, the initial alignment of fibers in the UD sheets was reflected within the finished parts.

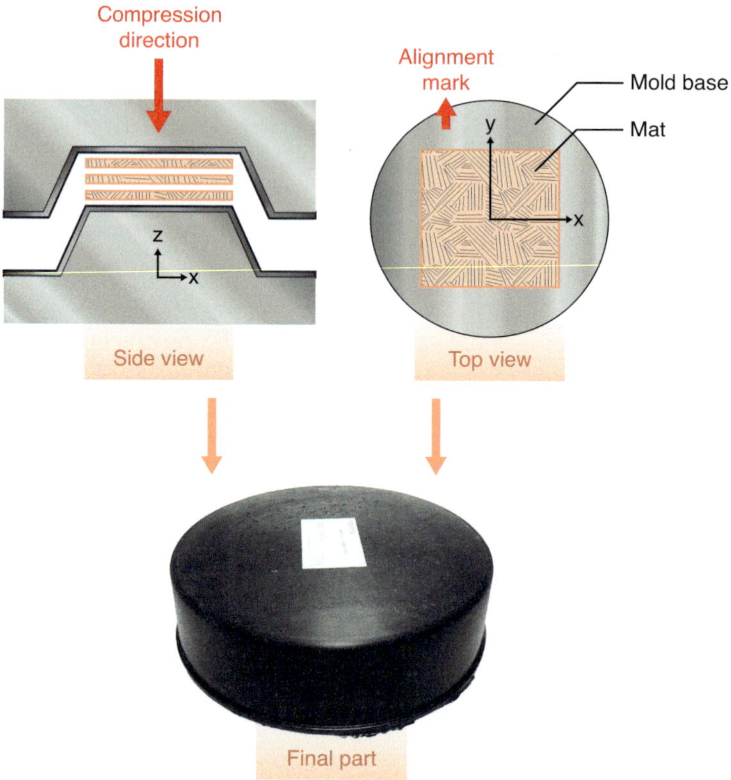

Figure 9.41 GMT material manufacturing process and the shape of the final part

Table 9.11 Sample Preparation Scenario (Five Cases) for Different GMT Sheets Initial Lay-Up Configurations

	Name	Sheet 1	Sheet 2	Sheet 3
Case 1	RRR40%	Random 40%	Random 40%	Random 40%
Case 2	R0R	Random 40%	UD45%GF (0)	Random 40%
Case 3	0R90	UD45%GF (0)	Random 40%	UD45%GF (0)
Case 4	All0	UD45%GF (0)	UD45%GF (0)	UD45%GF (0)
Case 5	RRR30%	Random 30%	Random 30%	Random 30%

9.3.3 Finished Parts Properties Measurements

Tensile and bending samples were prepared from the tub-shaped part using water-jet cutting. The detailed location of the samples is presented in Figure 9.42. The goal is to understand the effect the initial alignment of the fibers in the GMT sheet stacks has on the mechanical properties of the final part. Therefore, when the test samples are cut from the finished tub, appropriate markings are made on the test samples. The test samples aligned along the unidirectional fibers in the initial GMT sheets are called longitudinal and the test samples cut perpendicular to the unidirectional fibers in the initial GMT sheets are called transverse. In the cases where all the three GMT sheets were random (cases 1 and 5), the samples are still called longitudinal and transverse for consistency with the other tests. For those cases it is understood that the material properties are similar in both directions and choice of longitudinal or transverse is arbitrary.

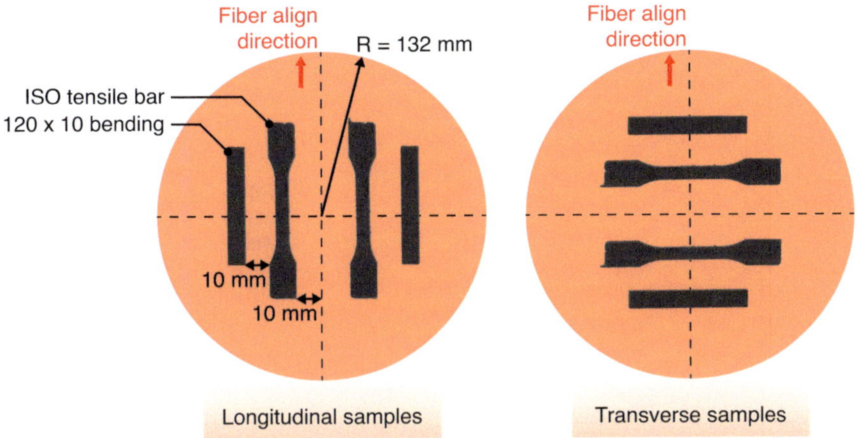

Figure 9.42 Tensile/bending specimen test sample at longitude direction in the final part

Strength and stiffness evaluations for the tensile and bending bar samples are done using an Instron machine (series 5900). Five different stacking conditions were investigated; results of the tensile tests are presented in Figure 9.43 and Figure 9.44, and detail of the bending tests results are presented in Figure 9.45 and Figure 9.46.

Tensile strength and flexural modulus show similar trends. The sample from the initial sheets where most fibers were aligned in one direction (case 4) shows highest anisotropy and the highest load capacity along the fibers. Also, when comparing case 1 and case 5, it is clear that the specimens are isotropic due to the random fiber arrangement. It is interesting to observe that in case 3 good load capability was observed in both directions. Essentially, the fibers in the finished part retain the directional trends that existed in the initial stack of the sheets before compression. It is observed that the material flow from the initial GMT sheet to the finished part is short due to part geometry. Consequently, fiber orientation from the original stack tends to change only a little, i.e., the orientation trends are maintained through the thickness while the thickness of each layer might be altered due to diffusion.

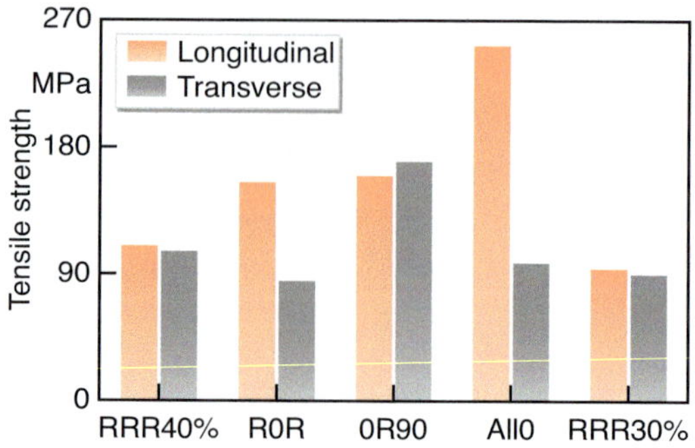

Figure 9.43 Tensile stress results for longitudinal and transverse samples for cases 1 to 5

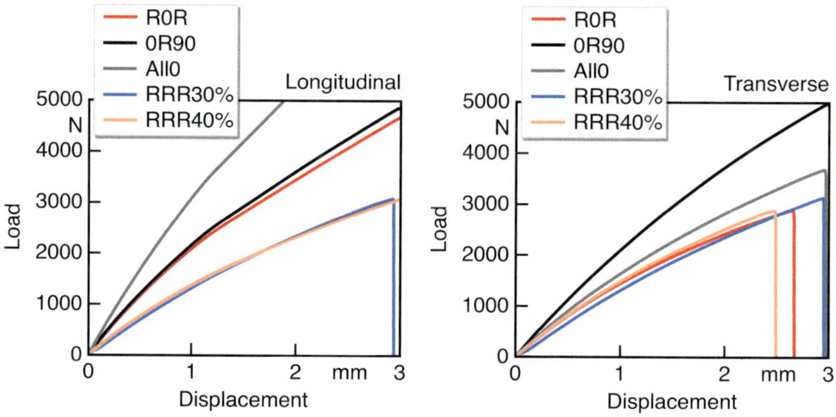

Figure 9.44 Load-displacement curve from tensile test for longitudinal and transverse samples for cases 1 to 5

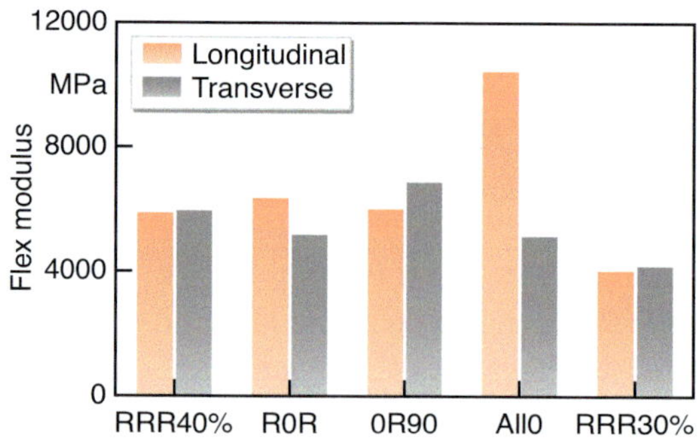

Figure 9.45 Bending results for longitudinal and transverse samples for cases 1 to 5

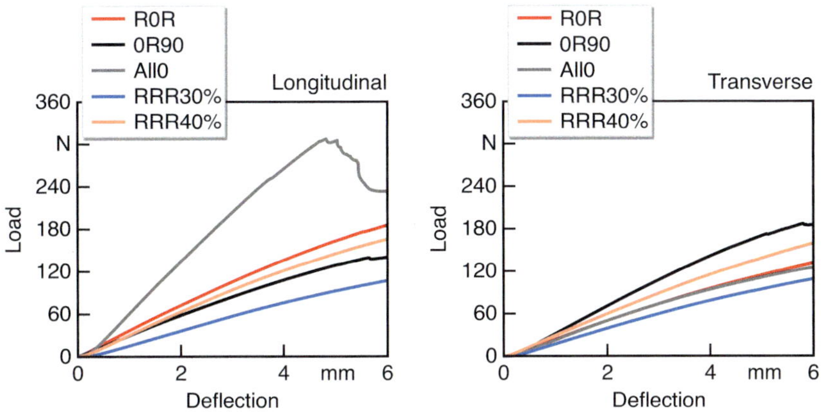

Figure 9.46 Load-deflection curve from bending test for longitudinal and transverse samples for cases 1 to 5. The loading span was 80 mm

9.3.4 FEA Analysis

This section presents the mold filling simulation fiber orientation predictions and their implementation into an FEA structural analysis program. The resulting stress-strain responses of the finite element analysis are compared to actual tests.

9.3.4.1 Compression Molding Process Simulation

The compression molding of the GMT sheet stack is simulated using Moldex3D™ R14. The details of the key theory and assumptions involved in such simulations are discussed in Chapter 7.

Since the fibers are expected to be long, the iARD-RPR algorithm in Moldex3D™ is used [46–48] to predict the fiber orientation change over time during the filling and compression phase. The RPR model is indicative of Retarding Principal Rate to slow down the fast-transient orientation rate. The iARD-RPR model uses three parameters to define fiber orientations:

c_i : Fiber–fiber interaction parameter

c_m : Fiber–matrix interaction parameter

α : Slows down the fiber orientation response

The other challenge presented in such a simulation is the inclusion of the initial fiber orientation within the charge.

In the experiment, different GMT sheets stack configurations were used (Figure 9.47) to get different final properties. It was learned that the initial condition can have a significant impact on the final fiber orientations. Therefore, it is important to have the capability to define the fiber orientation in the initial charge. This can be done based on the type of GMT sheets and their relative orientations in the stack.

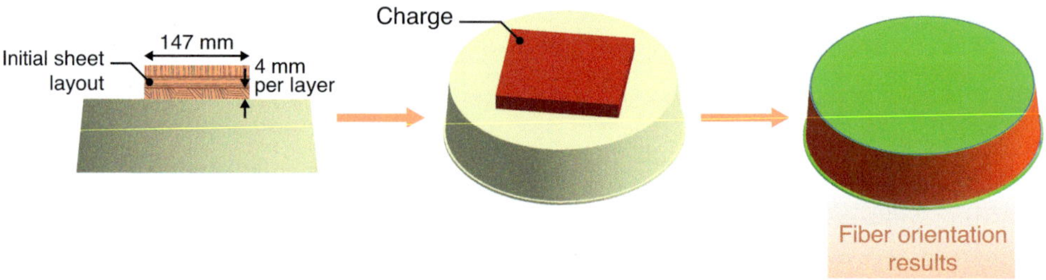

Figure 9.47 Compression molding process in Moldex3D

The stack arrangement defined in Table 9.12 can be interpreted in terms of fiber orientation as follows. The sheet called UD represents fiber orientation (1, 0, 0) and the sheet representing RD can represent the fiber orientation by (0.33, 0.33, 0.34). Since the UD sheets have some random fibers, the description is not accurate but, in general, helps in the understanding of the performance and behavior of the material at a concept level.

Table 9.12 Initial Fiber Orientation Tensor (A_{11}, A_{22}, A_{33}) Definition for the Five Cases (See Table 9.11 for More Details for Each Case)

	Case 1	Case 2	Case 3	Case 4	Case 5
Sheet 1	(0.3, 0.3, 0.34)	(0.3, 0.3, 0.34)	(1, 0, 0)	(1, 0, 0)	(0.3, 0.3, 0.34)
Sheet 2	(0.3, 0.3, 0.34)	(1, 0, 0)	(0.3, 0.3, 0.34)	(1, 0, 0)	(0.3, 0.3, 0.34)
Sheet 3	(0.3, 0.3, 0.34)	(0.3, 0.3, 0.34)	(0, 1, 0)	(1, 0, 0)	(0.3, 0.3, 0.34)

The material properties used for the compression molding simulation are shown in Table 9.13. The basic compression molding simulation parameters are shown in Table 9.14.

Table 9.13 Material Properties Used for the Simulation

	Glass fiber	Polypropylene
Density	2.55 (g/cc)	0.92 (g/cc)
Fiber length	0.3 mm	N/A
Fiber weight percentage	40%	N/A
Young's modulus E1	70,000 MPa	2,000 MPa
Young's modulus E2	70,000 MPa	2,000 MPa
Poisson's Ratio	0.2	0.3

Table 9.14 Compression Molding Parameters Used for the Simulation

	CAE parameter	Actual test
Melt temperature	210 °C	210 °C
Mold temperature	70 °C	70 °C
Compression time	40 s	40 s
C_i	1e-2	N/A
D_z	Default	N/A
Eject temperature	140 °C	140 °C

To validate the predicted fiber orientation distributions, the fiber orientations were measured at two locations labelled as A and B, and compared with the predictions. Detail locations of sample A (bottom of the tub) and sample B (vertical wall of the tub) are shown in Figure 9.48.

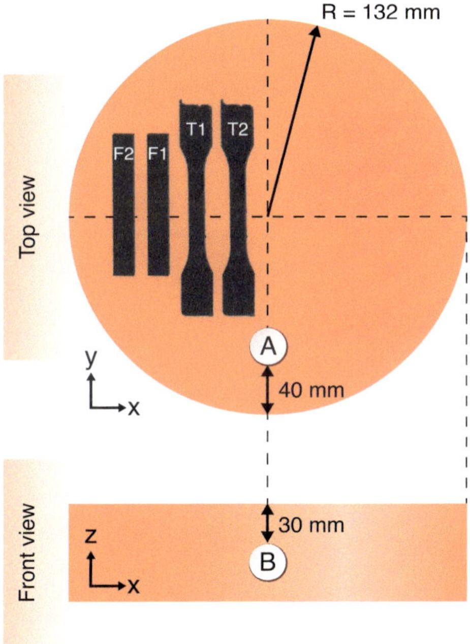

Figure 9.48 Locations of fiber orientation tensor measurement at A and B; layers with unidirectional fibers are placed such that fibers align along the y axis

In order to measure the fiber orientation distributions at the two locations, the 8 × 8 × 3 mm size samples were cut at locations A and B from the tub and 3D images with 3 μm resolution were created using CT-Scan (Skyscan μCT1172). The 3D fiber structure in the selected sample was created using Materialise software. The goal was to analyze the 3D image of the fiber geometry, identify each fiber, and then measure its orientations. A previously developed approach (Chapter 4) was used, where the scanned geometry is broken into smaller units and then in each smaller unit the fiber's geometry is reduced to a line. Based on the end points of the centerline of a fiber, the orientation for that fiber is computed. For detail of this approach the reader should consult [49].

The measured fiber orientation for cases 1, 2, 3, and 5 are presented in Figure 9.49, Figure 9.50, Figure 9.51, and Figure 9.52, respectively. The orientation for case 4 was not measured, as all the fibers are oriented in one direction. The results show that the measured and computed fiber orientation A_{11} at both locations A and B have similar trends. Also, it was observed that as the material flows the initial orientation of the fiber changes and moves such that more and more fibers are aligned in the flow direction. Since the material at location B has a longer flow history, the orientation at that location aligns more with the flow direction. This is consistent with previous studies [50]. Furthermore, it was observed that the fibers near the skin regions show only small tendencies to change the initial orientation. This is due to the fact that the flow and deformation near the mold surface (skin) is reduced due to immediate cooling, leading to less alignment of the fibers in the flow direction.

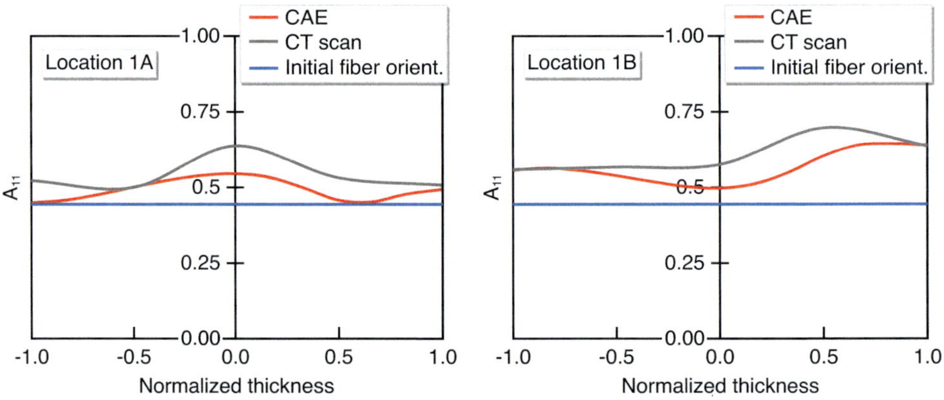

Figure 9.49 Comparison of the fiber orientation tensor at Y direction for location A and Z direction for location B, between simulation and test for case 1 (RRR40%)

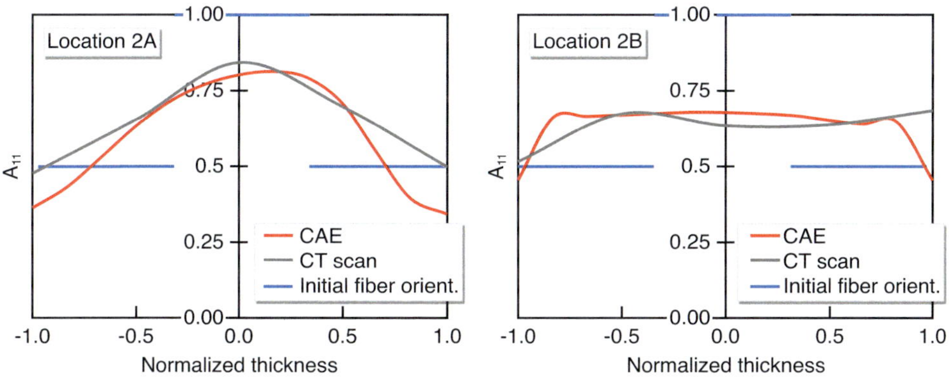

Figure 9.50 Comparison of the fiber orientation tensor at Y direction for location A and Z direction for location B, between simulation and test for case 2 (ROR)

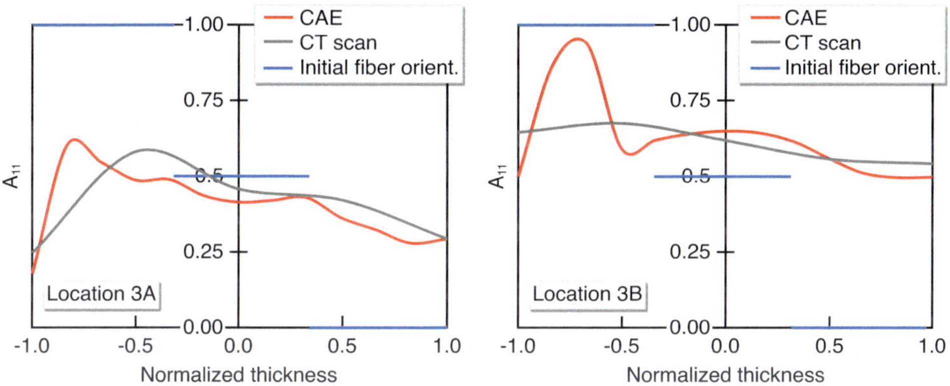

Figure 9.51 Comparison of the fiber orientation tensor at Y direction for location A and Z direction for location B, between simulation and test for case 3 (OR90)

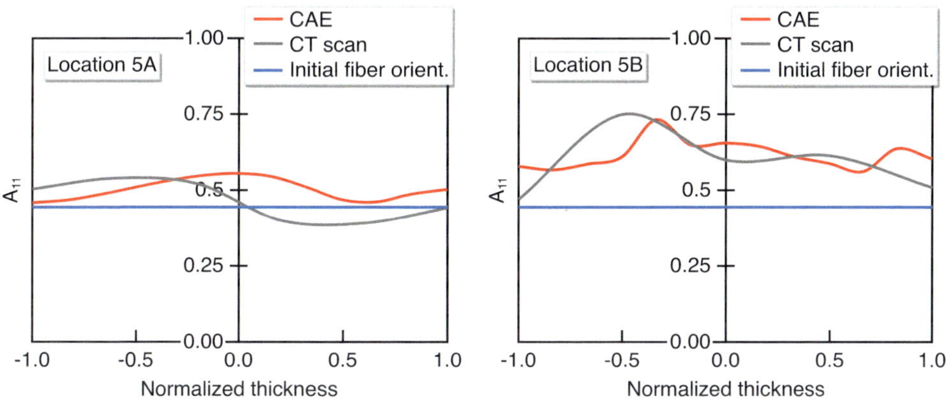

Figure 9.52 Comparison of the fiber orientation tensor at Y direction for location A and Z direction for location B, between simulation and test for case 5 (RRR30%)

9.3.4.2 Non-linear Structural FEA Simulation for the Tensile and Bending Test

The multiscale structural finite element models for the tensile and bending test samples were built and the results of the simulation are compared with the actual test to validate the modeling approach. The material properties of the finished part depend on the fiber length, fiber orientation, and fiber concentration. The traditional approach, discussed earlier, is to estimate the fiber orientation in the part at each location and develop material model using a homogenization approach to extract the material properties for use in the structural finite element analysis. The fiber orientation within the finished part was estimated through the compression mold filling simulation results presented in the previous section. The methods to predict the fiber length and fiber concentration as discussed in Chapter 7 are still evolving. Since each GMT sheet has a unique structure, alternate layers of polypropylene, and fiber mats, the fiber concentration varies significantly over the thickness. Also, the fiber length can be continuous, which needs added considerations.

In this study, this unique challenge was addressed using a semi-empirical approach to estimate the fiber length and concentration. The cross sections of finished parts for each type of sample (case 1 to 5 in Table 9.12) are examined under an optical microscope to study the microstructure. The sample location is shown in Figure 9.53, and the cross section of the selected sample under microscope is presented in Figure 9.54. The magnified cross section shows many distinct layers. It is also observed that the layers in the finished part can be loosely correlated with the layers of fibers in the initial stack of GMT sheets used to make the part. Since the condition of the layer in the initial stack of GMT sheets is known, it was postulated that the fiber length and concentration in each layer in the finished part can be judged to be the same.

Figure 9.55 shows the basic ABAQUS [51] finite element model for the tensile and bending test specimen simulations. There are 18 layers of brick elements through the thickness for both tensile and bending samples. The main purpose of using such a dense mesh is to capture the microstructure detail varying through the thickness in the material. Figure 9.54 (left) presents the optical image of the microstructure for case 4. The micrograph has detailed information about the fiber alignment and the surrounding resin. The thickness of each layer in the finished part is estimated from the microstructure image and then the empirically estimated fiber length and concentration are mapped onto the FE model on a layer by layer basis. For example, for case 4, presented in Figure 9.54 (mid and right side), the first layer is PP40GF, with a thickness of 0.778 mm, and is mapped on the first four layers of elements in the FE model. The fiber concentration for this layer is assumed to be 40%, the fiber orientation is mapped from the compression molding simulation, and fiber lengths are used based on experience of such materials. The next layer is considered to be made of UD tape, with a thickness of 0.389 mm and it is mapped on the two consecutive layers of elements in the FE model. The fibers are assumed to be continuous (aspect ratio = 999) aligned in one direction and with fiber concentration of 45%. Similarly, each distinct layer from the image of the cross section is mapped onto the FE model.

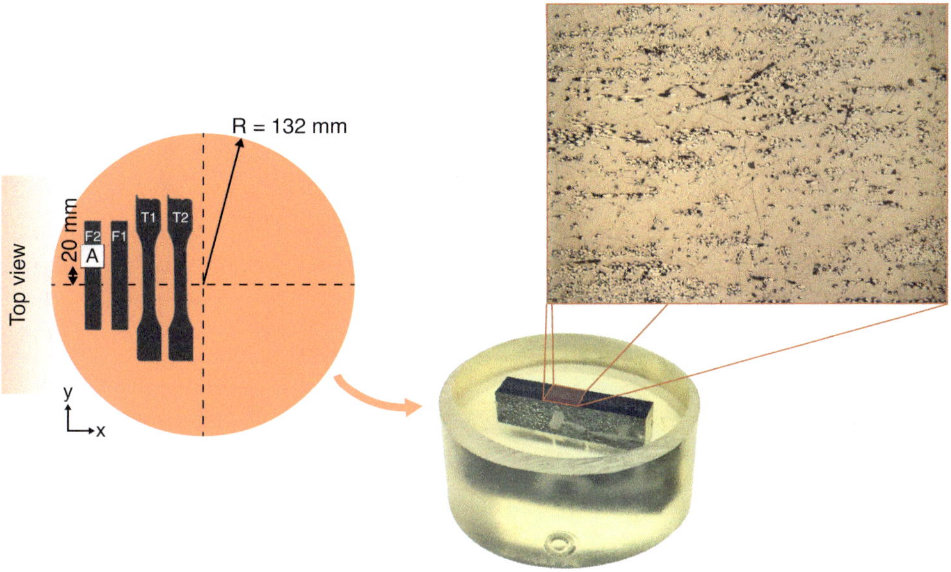

Figure 9.53 Location of the sample for optical analysis and its typical optical images

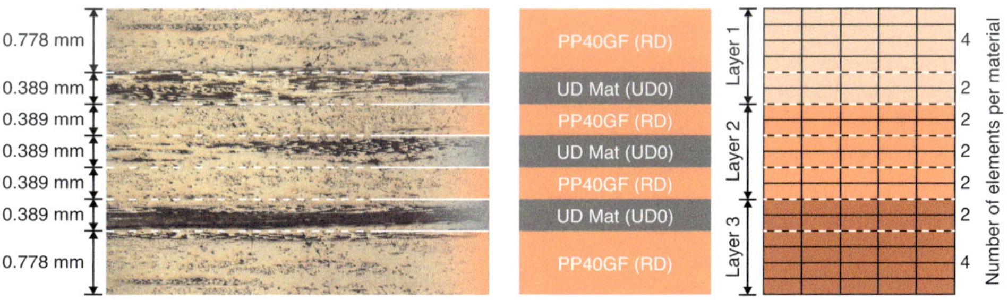

Figure 9.54 The corresponding FEA model details (cross-section view) to model the part in case 4

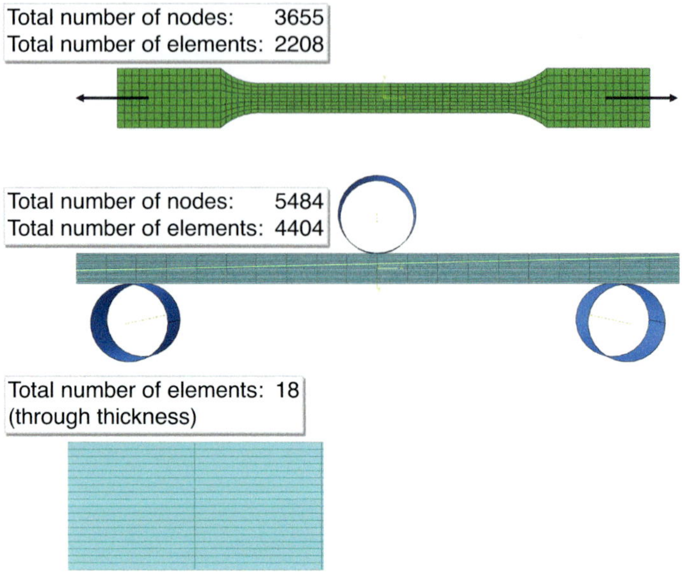

Figure 9.55 FEA model for the tensile and bending and the cross-section view through thickness for the sample

Figure 9.56–Figure 9.59 show similar optical microscopic analysis for cross sections of the part for all other cases (case 1, 2, 3, and 5). Based on the optical image analysis, the FEA model is developed for each of the five cases.

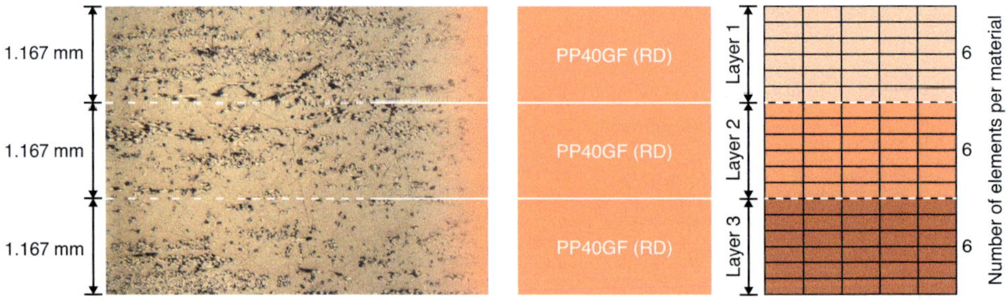

Figure 9.56 Optical microscopic image and corresponding FEA model information for case 1 (RRR40%)

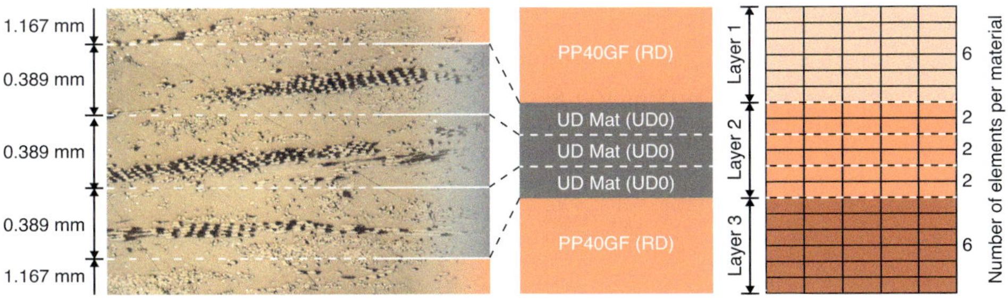

Figure 9.57 Optical microscopic image and corresponding FEA model information for case 2 (R0R)

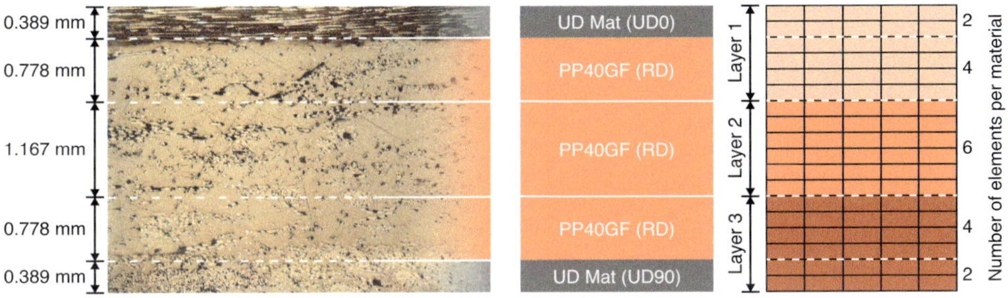

Figure 9.58 Optical microscopic image and corresponding FEA model information for case 3 (0R90)

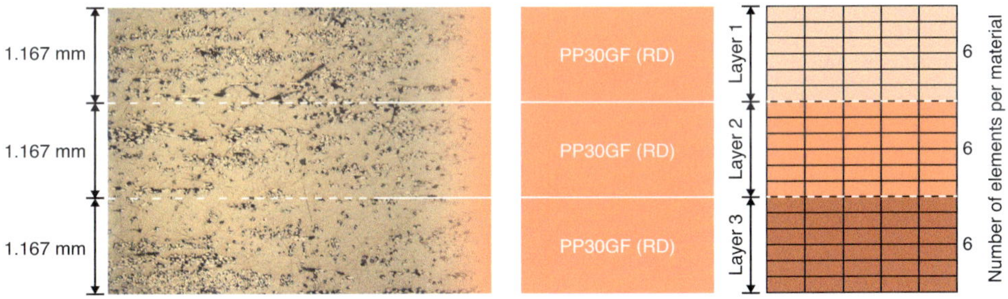

Figure 9.59 Optical microscopic image and corresponding FEA model information for case 5 (RRR30%)

Once the fiber conditions in the finished part have been estimated, the material properties in the finished part can be calculated using the Mori-Tanaka mean field homogenization approach [52]. As mentioned earlier, the purpose of the mean field homogenization approach is to compute approximate estimates of the volume averages of the stress and strain fields, both at the representative volume element (RVE) level (macro stresses and strains) and in each phase (fiber and matrix). The length and concentration are assumed to be constant in each layer. The fiber orientation is assumed to be varying and is mapped from the simulation results of the compression molded part onto the test sample's FE models for each element. When using the commercial software Moldex3D™, the fiber orientation can be exported using its FEA interface. Figure 9.60 shows the basic procedure for ABAQUS-Digimat coupled analysis for the tensile and bending bars [29, 53, 54].

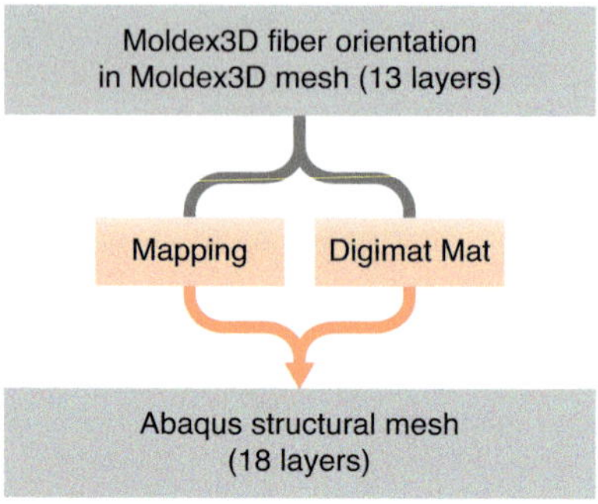

Figure 9.60 Steps for Abaqus-Digimat coupled analysis

Based on the proposed modeling approach, the material model was developed and the simulation was carried out using ABAQUS and Digimat. Figure 9.61 shows the comparison of the mechanical properties between experimental tests and the CAE results for case 4. The comparison shows a close match between the test results and the FEA analysis. The von Mises stress from Figure 9.62 shows the detailed stress contour for tensile and bending sample in case 4. The results illustrate the effect of fiber orientation on structural properties. Using a similar approach, the FEA analysis can be performed in the other four cases. Figure 9.63 through Figure 9.66 show the comparison of the load and displacement curves for the remaining four cases. As with case 4, the remaining cases also show a relatively good correlation between experimental tests and FEA analysis results.

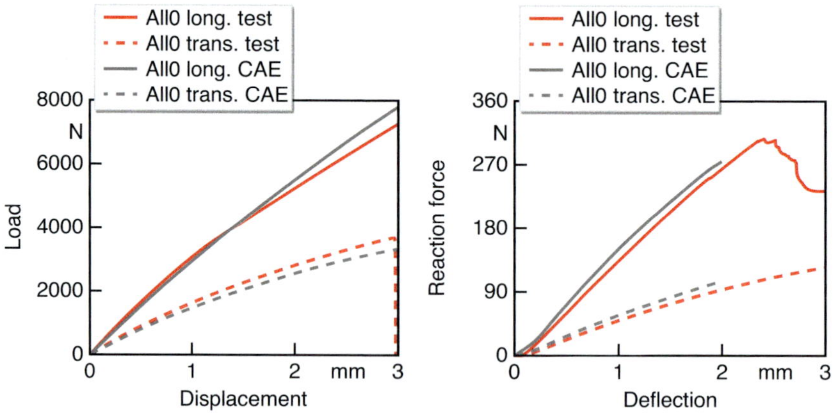

Figure 9.61 Comparison of the mechanical properties between test and CAE for case 4 (see Table 9.12); left: tensile test, right: bending test results

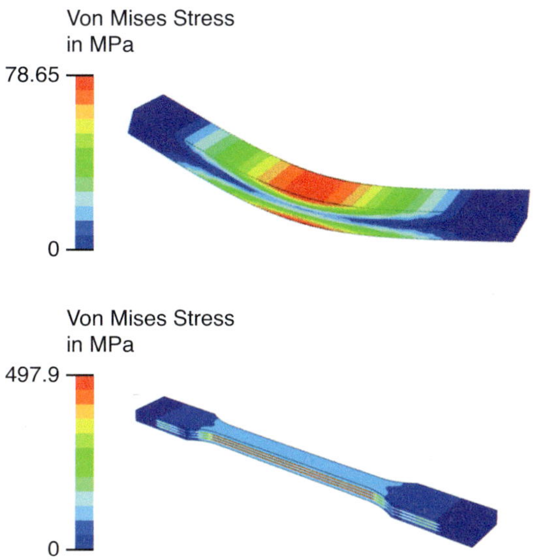

Figure 9.62 Von Mises stresses (MPa) contour for the bending and tensile test for case 4 in longitude direction

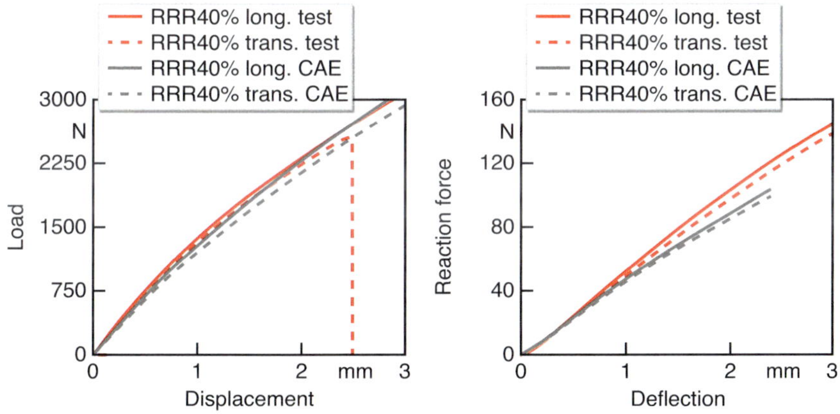

Figure 9.63 Comparison of the mechanical properties between test and CAE for case 1 (see Table 9.12); left: tensile test, right: bending test results

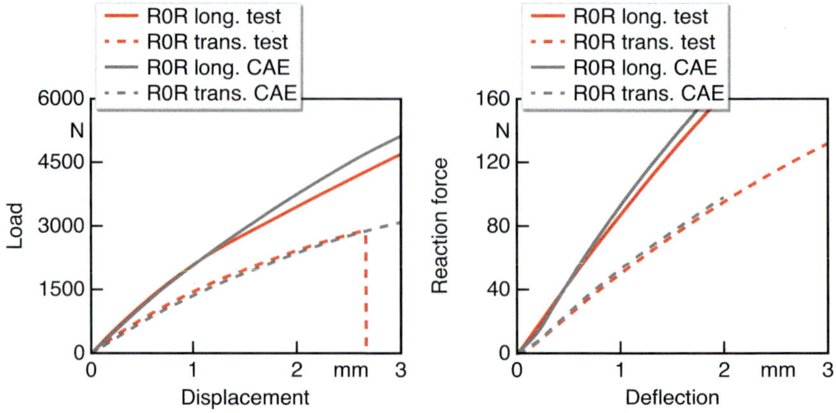

Figure 9.64 Comparison of the mechanical properties between test and CAE for case 2 (see Table 9.12); left: tensile test, right: bending test results

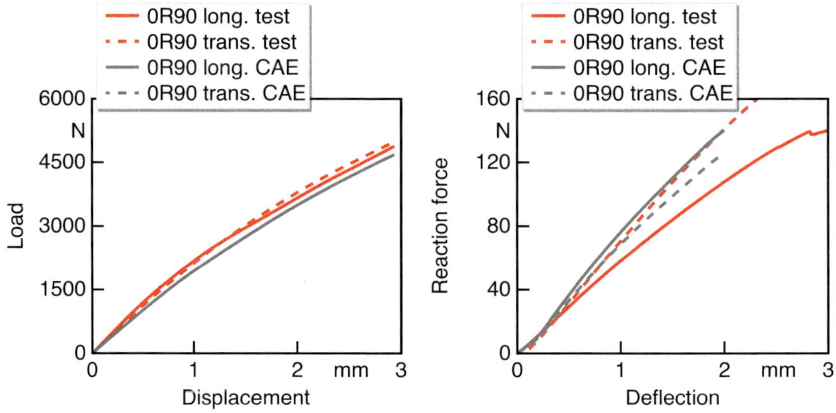

Figure 9.65 Comparison of the mechanical properties between test and CAE for case 3 (see Table 9.12); left: tensile test, right: bending test results

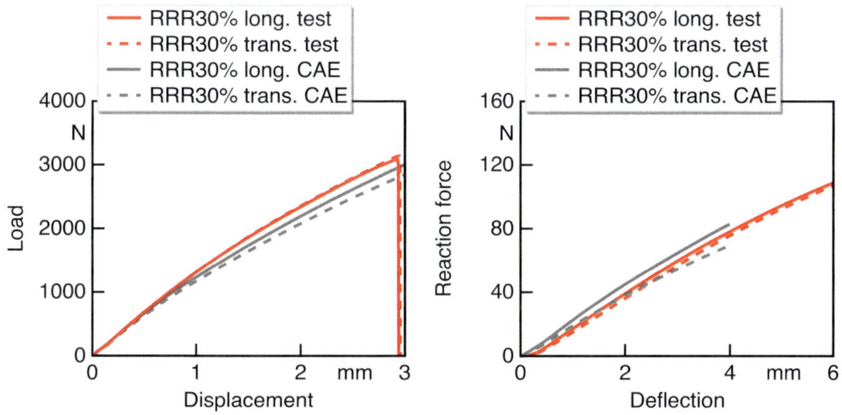

Figure 9.66 Comparison of the mechanical properties between test and CAE for case 5 (see Table 9.12); left: tensile test, right: bending test results

9.3.5 Comments and Conclusions

GMT material is quite important in developing fiber-reinforced parts. The effects of different stacking and initial orientation of GMT sheets on the structural properties of a finished three-dimensional tub-shaped part show that the initial alignment of the fibers in the GMT sheets plays a significant role in the structural properties of the finished part.

The compression molding of the GMT sheets can be simulated using Moldex3D™ to estimate the fiber orientation. It was demonstrated that the predicted fiber conditions correlate with the measured values. Finally, a unique empirical approach to account for fiber conditions such as length and concentration for the compression molded part simulation is demonstrated. The simulation results were successfully validated through physical tests of tensile and bending specimens.

This is an empirical approach because the fibers condition for some of the fibers was not completely predicted but measured based on the microstructure study and then included in the structural simulation. The key point we would like to emphasize here is that capability to predict the fibers condition can be very complex and often impossible, due to the number of fibers and the many possible variations that may exist. Therefore, it is important to consider an empirical approach based on insights and knowledge of physics of the process as well as actual materials, which can help in making useful predictions. In this case we used the microstructure of a simple part and developed knowledge of fiber orientation and fiber length distribution to develop a material model for multiscale finite element study. The demonstration was for a smaller tub; we believe such modeling approach can be extended to CAE of larger and complex shapes.

References

[1] Isayev, A., *Injection and Compression Molding Fundamentals*, CRC Press, Boca Raton, FL, USA (1987)

[2] Biron, M., *Thermoplastics and Thermoplastic Composites*, William Andrew, Norwich, NY, USA (2012)

[3] Long, A. C. (Ed.), *Composites Forming Technologies*, Woodhead Publishing, Cambridge, UK (2014)

[4] Dumont, P., Orgéas, L., Favier, D., Pizette, P., and Venet, C., Compression Moulding of SMC: In Situ Experiments, Modelling and Simulation, *Compos. Part A Appl. Sci. Manuf.*, 38, pp. 353–368 (2007)

[5] Miura, M., Hayashi, K., Yoshimoto, K., and Katahira, N., Development of Thermoplastic CFRP for Stack Frame. In SAE Technical Paper, SAE International, Detroit, MI, USA (2016)

[6] Bruce, A. D., Oswald, T. A. *Compression Molding*, Hanser, Cincinnati, OH, USA (2003)

[7] Moldex3D, User Manual and Material Database (2015)

[8] Perez, C., Roppers, S., Osswald, T. A., Ghandi, U., and Mandapati, R., Fiber Length and Orientation Distribution Measurement of a Charge for D-LFT Compression Molding. In SPE ANTEC Conference, Technical Papers (2013, April)

[9] Le, T.-H., Dumont, P. J. J., Orgéas, L., Favier, D., Salvo, L., and Boller, E., X-ray phase contrast microtomography for the analysis of the fibrous microstructure of SMC composites, *Composites Part A*, 39(1), pp. 91–103 (2008)

[10] Tseng, H.-C., Chang, Y.-J., and Hsu, C.-H., Prediction of Fiber Microstructure for Injection Molding: Orientation, Degradation and Concentration, SPE-ACCE Conference, Novi, Mi. (2014)

[11] Manual of Volume Graphics software, Volume Graphics GmbH (2014)

[12] Perez, C., Osswald, T. A., and Ropers, S., Fiber Length and Orientation Distribution Measurement of a Charge for D-LFT Compression Molding, ANTEC, Cincinnati, OH (2013)

[13] Park, J. and Sandberg, I. W., Universal approximation using radial-basis-function networks. *Neural computation*, 3(2), pp. 246–257 (1991)

[14] Lopez-Gomez, I., Estrada, O., and Osswald, T. A., Modeling and Simulation of Polymer Processing Using the Radial Functions Method, *Journal of Plastics Technology*, (3), 2 (2007)

[15] Ramírez, D., López, I., Estrada, O., and Osswald, T. A., Simulation of the Fountain Flow Effect by Means of the Radial Functions Method (RFM), SPE ANTEC, Orlando, Florida (2012)

[16] Stayton, C. T., Application of thin-plate spline transformations to finite element models. *Evolution*, 63(5), pp. 1348–1355 (2009)

[17] Manual of Matlab software, Mathworks Inc., New York (2014)

[18] Song, Y., Gandhi, U., Koziel, A., Vallury, S., and Yang, A. Effect of the initial fiber alignment on the mechanical properties for GMT composite materials. *J. Thermoplast. Compos. Mater.*, vol. 31, 1, pp. 91–109 (2017)

[19] ISO 17 744:2004, Plastics—Determination of Specific Volume as a Function of Temperature and Pressure (pvT—Diagram—Piston Apparatus Method; International Organization for Standardization, Geneva, Switzerland (2004)

[20] ASTM D4440-15. Standard Test Method for Plastics: Dynamic Mechanical Properties Melt Rheology; ASTM International, West Conshohocken, PA, USA (2015)

[21] ASTM D5930-16. Standard Test Method for Thermal Conductivity of Plastics by Means of a Transient Line-Source Technique; ASTM International, West Conshohocken, PA, USA (2016)

[22] ASTM E1269-11. Standard Test Method for Determining Specific Heat Capacity by Differential Scanning Calorimetry; ASTM International, West Conshohocken, PA, USA (2011)

[23] Chien, T., Huang, C. T., Lin, G., Sun, S., Wang, C., and Chang, R., Investigation on the Viscosity Characterization of the Glass Mat Thermoplastics (GMT) in Compression Molding System, in SPE ANTEC 2018 Conference, Technical Papers (2018)

[24] Chau, S. W., Three-dimensional simulation of primary and secondary penetration in a clip-shaped square tube during a gas-assisted injection molding process, *Polymer Engineering Science*, 48, pp. 1801–1814 (2008)

[25] Tseng, H. C., Chang, R. Y., and Hsu, C. H., Numerical prediction of fiber orientation and mechanical performance for short/long glass and carbon fiber-reinforced composites, *Composites Science and Technology*, 144, pp. 51–56 (2017)

[26] Jack, D. A. and Smith, D. E., Assessing the use of tensor closure methods with orientation distribution reconstruction functions. *Journal of composite materials*, 38(21), pp. 1851–1871 (2004)

[27] http://www.fasep.biz/

[28] Mori, T. and Tanaka, K., Average stress in the matrix and average elastic energy of materials with misfitting inclusions. *Acta Metall. Mater.*, 21, pp. 571–574 (1973)

[29] Digimat User Manual, MSC Software corporation, Newport Beach, USA (2015)

[30] ASTM D638-02a, Standard Test Method for Tensile Properties of Plastics, West Conshohocken, PA, USA (2002). Available online: https://www.astm.org/DATABASE.CART/HISTORICAL/D638-02A.htm

[31] International Organization for Standardization. ISO 17744:2004, Plastics—Determination of Specific Volume as a Function of Temperature and Pressure; International Organization for Standardization, Geneva, Switzerland (2004)

[32] ASTM D4440-15, Standard Test Method for Plastics: Dynamic Mechanical Properties Melt Rheology, West Conshohocken, PA, USA (2015). Available online: https://www.astm.org/Standards/D4440.htm

[33] ASTM D5930-16, Standard Test Method for Thermal Conductivity of Plastics by Means of a Transient Line-Source Technique, West Conshohocken, PA, USA (2016). Available online: https://www.astm.org/DATABASE.CART/HISTORICAL/D5930-16.htm

[34] ASTM E1269-11, Standard Test Method for Determining Specific Heat Capacity Differential Scanning Calorimetry, West Conshohocken, PA, USA, 2011. Available online: https://www.astm.org/Standards/E1269.htm

[35] Plewa, T., Linde, T., and Weirs, V. G., Adaptive mesh refinement-theory and applications, *Lect. Notes Comput. Sci. Eng.*, 41, pp. 3–5 (2005)

[36] Dong, Y., Lin, R. J. T., and Bhattacharyya, D., Finite element simulation on thermoforming acrylic sheets using dynamic explicit method, *Polym. Polym. Compos.*, 14, pp. 307–328 (2006)

[37] Blumentritt, B. F., Vu, B. T., and Cooper, S. L., The mechanical properties of oriented discontinuous fiber-reinforced thermoplastics. I. Unidirectional fiber orientation, *Polymer Engineering & Science* 14.9, pp. 633–640 (1974)

[38] Beardmore, P. et al., Fiber-reinforced composites- Engineered structural materials, *Science* 208.4446, pp. 833–40 (1980)

[39] Segal, L. and Steinberg, A. H., Stampable nylon sheets: Reinforcing with long glass fibers., *Polymer Engineering & Science* 15.8, pp. 615–622 (1975)

[40] Biron, M., *Thermoplastics and thermoplastic composites*, William Andrew (2012)

[41] Berglund, L. A. and Ericson, M. L., *Glass mat reinforced polypropylene: structure, blends and composites*, Chapman and Hall, London (1995)

[42] Mann, D., *Automotive plastics and composites: worldwide markets and trends to 2007*, Elsevier (1999)

[43] Shin, D. W., Kim, N. H., Lee, S. S., and Hong, S. B., Development of Seating System with GMT for ECE 17.07 (Luggage Retention) Regulation (No. 2002-01-1041), SAE Technical Paper (2002)

[44] Hosseinzadeh, R., Shokrieh, M. M., and Lessard, L. B., Parametric study of automotive composite bumper beams subjected to low-velocity impacts, *Composite Structures* 68.4, pp. 419–427 (2005)

[45] Cantwell, W. J. and Kausch, H. H., A characterization of glass mat thermoplastic composites for automotive applications, ECF9, Varna 1992 (2013)

[46] Predictive Engineering Tools for Injection-Molded Long-Carbon-Fiber Thermoplastic Composites, Pacific Northwest National Laboratory (2013)

[47] Advani, S. G. and Sozer, E. M., *Process modeling in composites manufacturing*, CRC Press (2012)

[48] Foss, P. H. et al., Prediction of fiber orientation distribution in injection molded parts using Moldex3D simulation, *Polymer Composites* 35.4, pp. 671–680 (2014)

[49] Gandhi, U., Kunc, V., and Song, Y., Method to measure orientation of discontinuous fiber embedded in the polymer matrix from computerized tomography scan data. *Journal of Thermoplastic Composite Materials*, 0892705715584411 (2015)

[50] Ericson, M. L. and Berglund, L. A., Processing and mechanical properties of orientated preformed glass-mat-reinforced thermoplastics, *Composites Science and Technology* 49.2, pp. 121–130 (1993)

[51] Abaqus/Standard 6.12.1 User Manual, Dassault System

[52] Mori, T. and Tanaka, K., Average stress in the matrix and average elastic energy of materials with mis-fitting inclusions, *Acta Metall. Mater.*, 21, pp. 571–574 (1973)

[53] Londoño-Hurtado, A., Osswald, T. A., and Hernandez-Ortíz, J. P., Modeling the behavior of fiber suspensions in the molding of polymer composites, *Journal of Reinforced Plastics and Composites*, 0731684411400227 (2011)

[54] Phelps, J. H., El-Rahman, A. I. A., Kunc, V., and Tucker, C. L., A model for fiber length attrition in injection-molded long-fiber composites. *Composites Part A: Applied Science and Manufacturing*, 51, pp. 11–21 (2013)

10 Special Topics in CAE Modeling of Composites

A key focus in this book has been on how to model discountinuous fiber-reinforced materials. The multiscale modeling approach to account for fiber orientation, length, and concentration has been discussed and case studies to demonstrate how such an approach can be used in practice have also been presented. While designing the discontinuous fiber-reinforced parts, in addition to the capability of including fiber conditions, additional complexities such as joining of such materials and dealing with mixed or hybrid material (e.g., a combination of discontinuous and continuous fiber systems), etc. are also commonly encountered. In this chapter we will demonstrate how such complexities can be modeled.

10.1 Mixed or Hybrid Fiber-Reinforced Material Modeling

Fiber-reinforced materials come in two general forms: as continuous fiber composites and discontinuous fiber composites. There are many variations within each type of fiber systems, as discussed in Chapter 2. Each type of fiber system comes with some advantages and some disadvantages. To optimize the performance, there is often interest in using a combination of two types of fibers. One of the most desired options is to selectively use continuous fibers in tape or other form, combined with discontinuous fibers in injection or compression molded parts. This is equivalent to adding reinforcement patches in stamped metal parts. A continuous fiber patch can reinforce and add strength and stiffness in selected areas and direction of a part, while maintaining the low-cost, high-volume advantage of injection or compression molding processes [1, 2]. The typical process for such hybrid materials is over molding, which essentially involves laying the continuous fiber in tape or patch form on the mold surface and filling the mold cavity with a discontinuous fiber-reinforced polymer melt. There are many challenges in executing such a process, particularly for continuous high-volume applications, where the continuous fiber patch tends to move due to interaction with the melt flowing at high speeds during the filling process. Efforts are in progress to address these challenges from a manufacturing perspective. In this section, a method to model such mixed or hybrid material systems using mold filling simulation and finite element structural analysis software is presented.

10.1.1 Study Detail

A plaque as shown in Figure 10.1 was developed with Ticona™ to study hybridization and the modeling approach for hybrid structures. The plaque was manufactured by injection molding. Three different configurations of continuous fiber tapes are considered for the plaques:

- No tape,
- Tape added on both sides of plaque, and
- Tape added on only one side of the plaque.

During the injection molding process, the molten fiber filled resin travels at high speeds, generating high stresses on the tapes, making it difficult to hold the tapes in the place. Therefore, during manufacture of the plaques, care was taken to assure that the tapes stayed secured during mold filling. Bücheler [3] addressed this issue by adding magnetic particles to the tapes, which helped to fix the tape on the mold surface and to maintain tapes in place when over molded by compression molding with SMC.

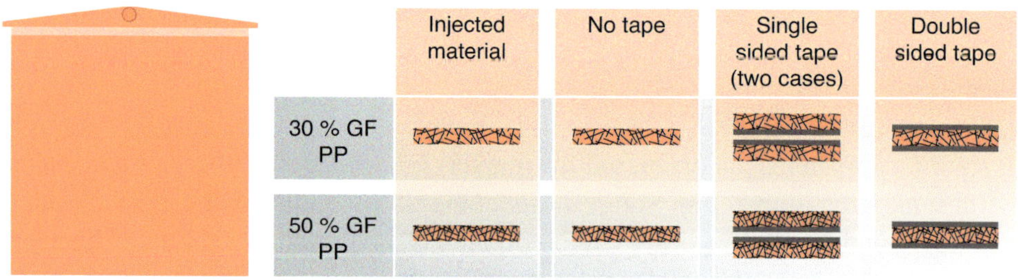

Figure 10.1 Details of plaques built for the study: the plaque material is 30% and 50% discontinuous glass fibers injection molded from pellets; the tapes used are 0.2 mm thick and have 70% continuous glass fibers by weight aligned in the flow direction; the resin used is polypropylene for both

Once the plaques were manufactured, the samples for bending and tensile test were cut using a waterjet. The cutting pattern is presented in Figure 10.2.

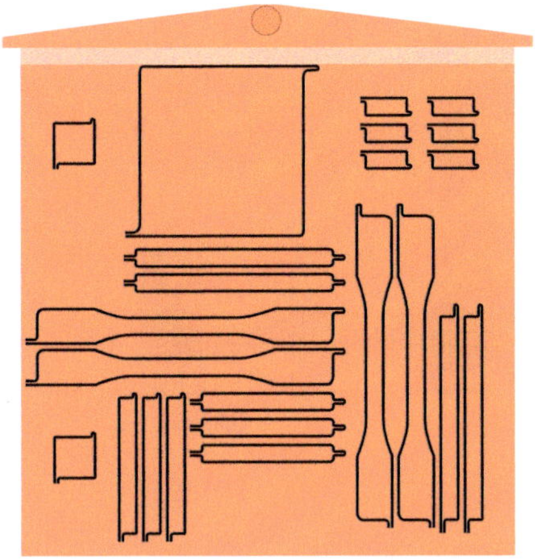

Figure 10.2 Cutting pattern used for tensile and bending samples from the plaque

10.1.2 Tensile and Bending Test Comparison

The tensile and bending tests were conducted for all the three types of plaques and compared with each other. The total thickness of all samples was maintained the same, i.e., the thickness of the tape reduced the thickness of over molded materials.

The results are presented in Figure 10.3 and Figure 10.4. It can be observed that, when loaded along the direction of fibers, there is a significant increase in the tensile strength and bending stiffness due to tapes of unidirectional fibers in the direction of fibers, whereas perpendicular to the fibers the effect is minimal. In general such tape or patches can be used to improve the properties in the direction of interest. In this study, to avoid bias due to flow for the single tape case, the continuous fibers are considered for both the surfaces, i.e., upper and lower side of the plaques.

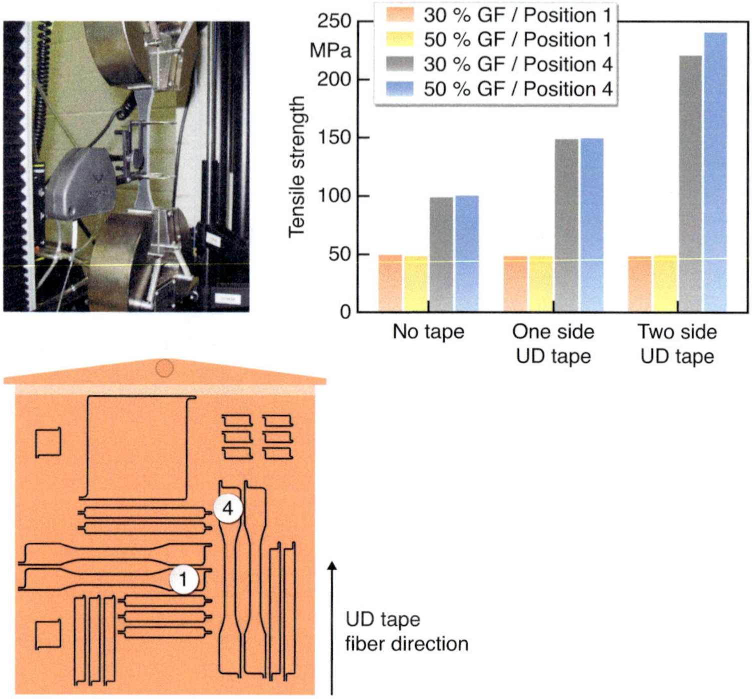

Figure 10.3 Tensile test setup and failure load comparison

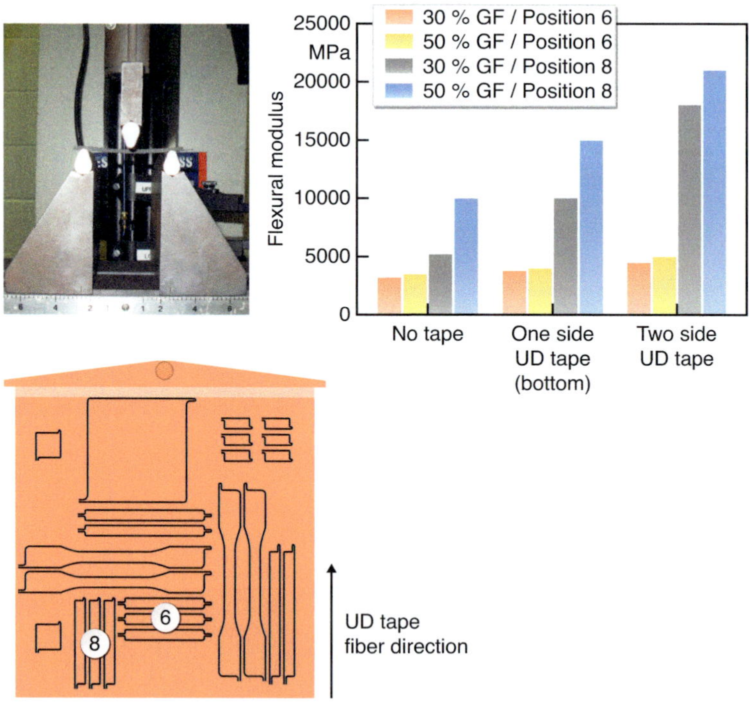

Figure 10.4 Bending test setup and flexural load comparison

10.1.3 Finite Element Modeling for the Hybrid Material

The multiscale modeling approach discussed earlier in Chapter 6 (Figure 6.17) is extended for the hybrid material here. Typically for injection molded material the process simulation is used to estimate the fiber orientation and then calculate the material properties based on the microstructure estimates [4, 5].

The microstructure of interest is a combination of two, i.e., discontinuous fibers and continuous fibers. A microscopic view of the material is shown in Figure 10.5. It was observed that there is a typical fiber distribution based on the fountain flow and then there is unidirectional tape at the outer skin. Also, there is no gap between the unidirectional tape layer and the injection molded materials. This indicates that there is good adherence between the unidirectional tape and molded material.

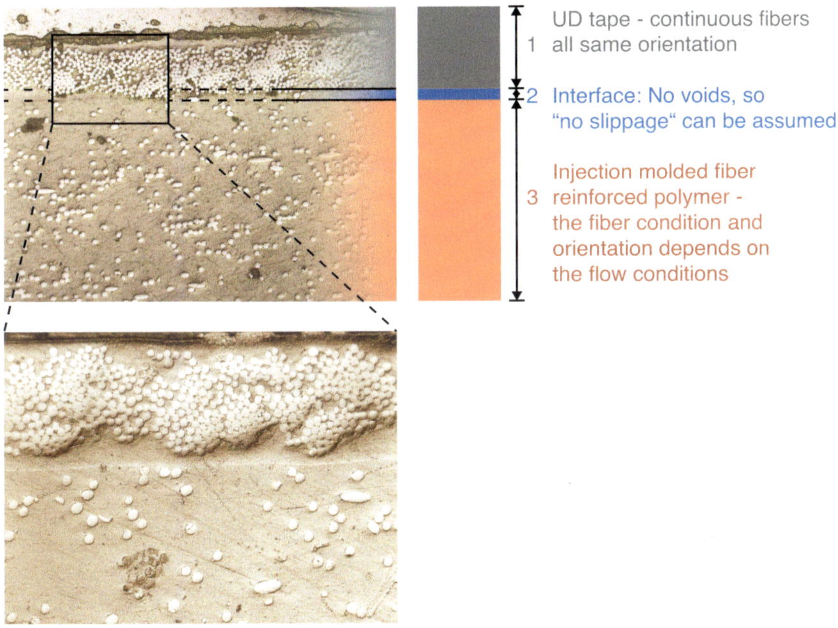

Figure 10.5 Microstructure of the hybrid material

The approach to model such hybrid microstructure with the finite element model is presented in Figure 10.6. Essentially, the approach used for the injection molding process is modified. The thickness of the part is divided into a number of layers of finite elements. One of the layers of finite elements is used to represent the unidirectional tape through the thickness. The fiber orientation in the that layer for all the fibers is assumed to be aligned in one direction, i.e., the direction of the fibers in the continuous tape attached. Also, the aspect ratio for the fibers is assumed to be infinity to represent the continuous fibers. The rest of the layers are treated as injection molded material, where the fiber orientation is mapped from the molding simulation and fiber aspect ratio is based on the measurement or simulation. It is believed that this approach is simple and yet helps achieve reasonable microstructure details for the material of interest. The flow of key steps used in the simulation and modeling is presented in Figure 10.7 [6–8].

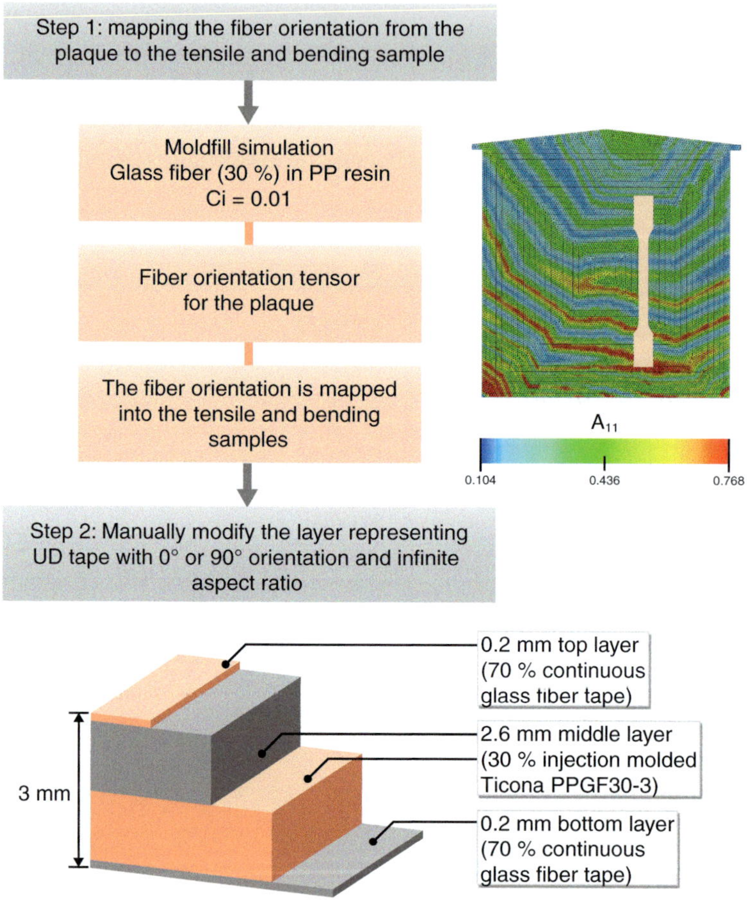

Figure 10.6 Approach to model the hybrid material

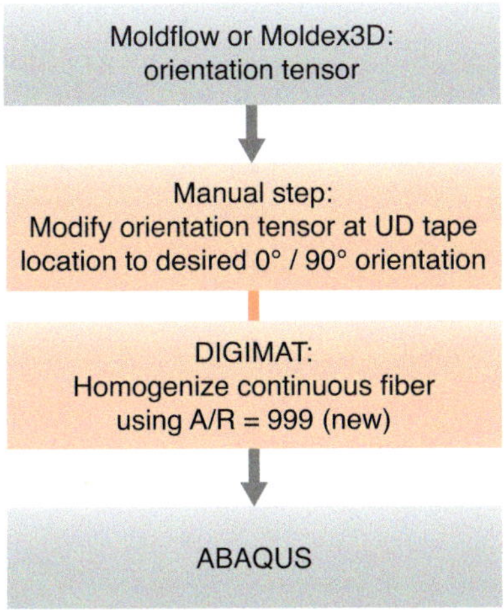

Figure 10.7 Flow of steps for implementation of continuous fiber layer in injection molded hybrid parts

10.1.4 Comparison of Simulation Result with the Measurement

To verify the accuracy of the mold filling simulation of the plaque using Moldex-3D™ [9], the fiber orientation tensor predicted is compared with the actual measurements using the CT scan-based methods discussed in Chapter 4, at a selected location, for a plaque without tapes and 30% glass fiber with polypropylene. The comparison is presented in Figure 10.8. The results are considered close enough to accept the mold fill simulation as representative of the actual part.

Further, the multiscale material model representing the fiber orientation for the injection molded layers as well as continuous fiber layer is developed using Digimat™ [10] and used as a user material subroutine in the finite element simulation using Abaqus™ [11]. The stress and strain estimates from the simulation for the selected tensile specimen are compared with the actual test results in Figure 10.9 for tensile tests and in Figure 10.10 for the bending test. Samples with no tape, tape on one side, and tape on both the sides are compared.

The comparison shows that the force displacement from the test and CAE simulation are quite comparable. The simulation was used to get the load displacement curve, but the failure point was not estimated, as CAE used in this case does not include damage or failure estimations.

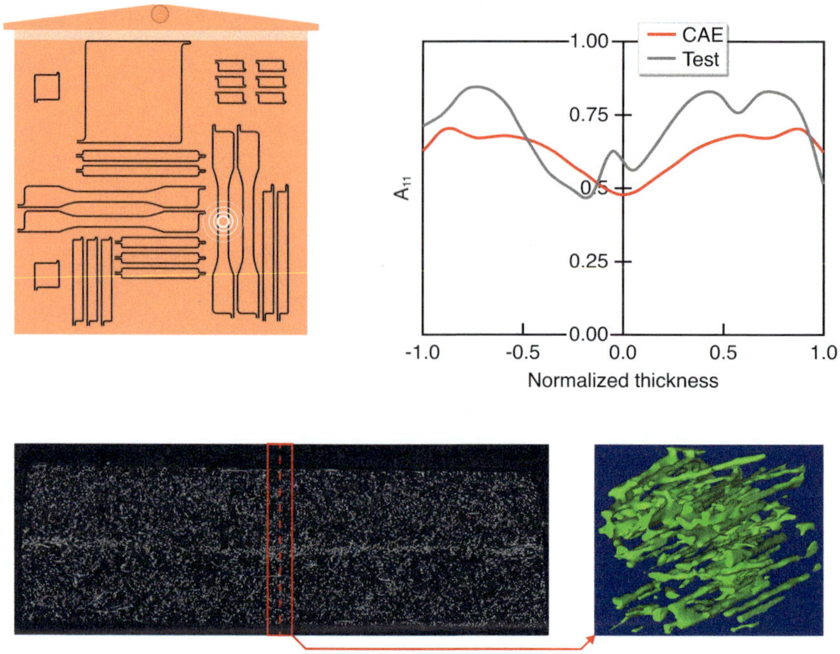

Figure 10.8 Comparison of measured and predicted fiber orientation at selected location, for the plaque with 30% glass fiber

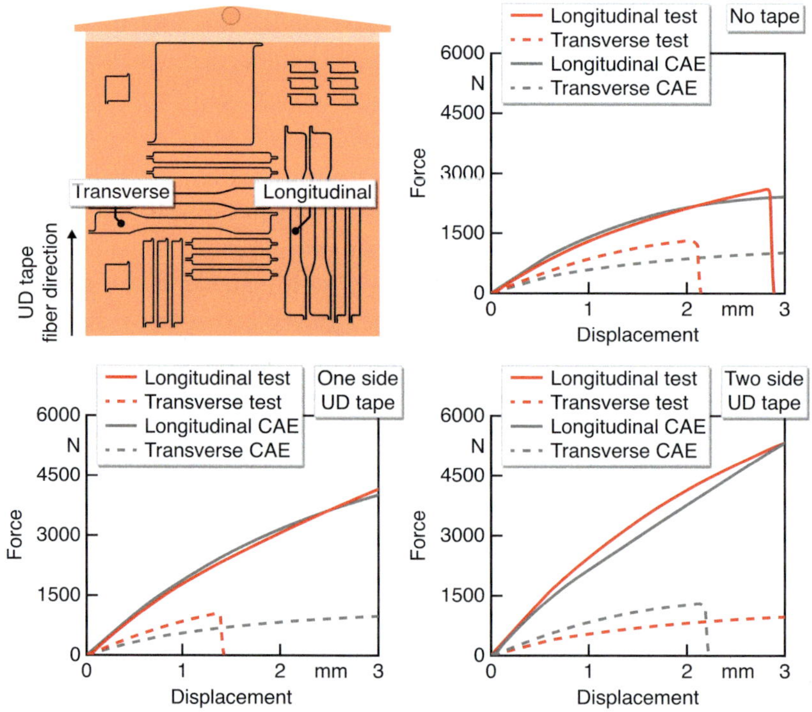

Figure 10.9 Comparison of tensile test and finite element model using multiscale material models

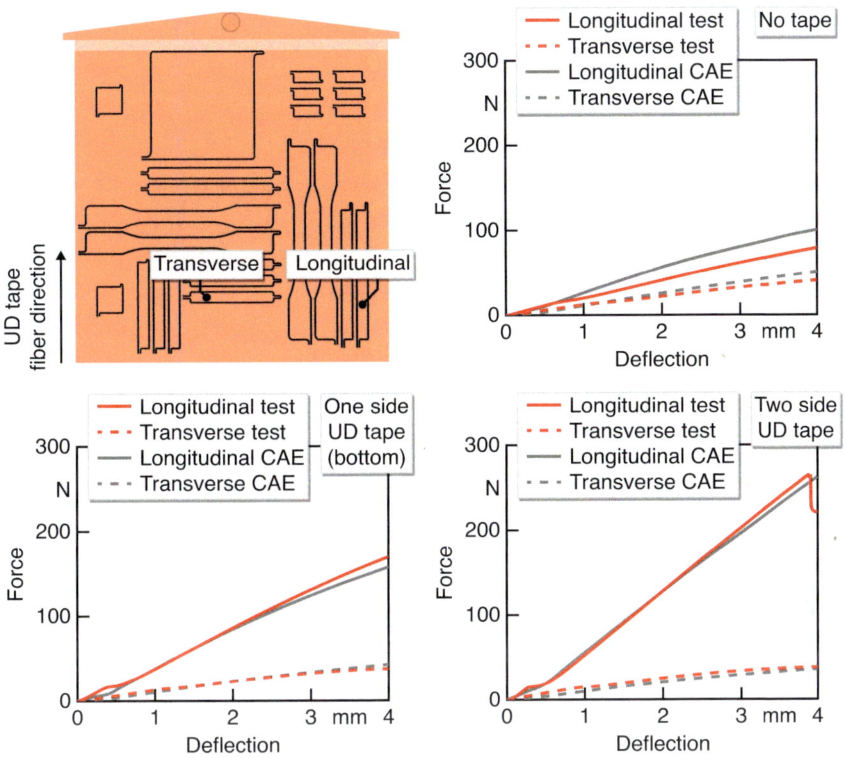

Figure 10.10 Comparison of bending test and finite element model using multiscale material models. For single sided tape both sides were tested independently to check bias through the thickness

10.1.5 Conclusion for Hybrid Material Modeling

It was observed that the uni-directional tapes added to the injection molded plaque, aligned along the loading direction, can increase strength and stiffness of the part significantly.

The multiscale material modeling approach developed in Chapter 6 can be modified to include the effect of continuous fiber, provided the thickness of the continuous fiber layer is known. It was observed that for both tensile and bending load the simulation and test results matched quite well. For the model with continuous fiber tape the correlation between the test and CAE is even better. This is because the fiber length and orientation for continuous fiber tape are accurately known compared to the discontinuous fibers.

The hybrid material can be modeled using the proposed simple approach, which can be a very useful tool in future designs with fiber-reinforced materials.

10.2 Adhesive Joining Modeling

Composite materials can be processed and developed into structural components using many of the processes discussed in Chapter 3. For a complex shape such as those which may exist within automobiles it is often necessary to join a number of components to get the final desired shape. Also, there is increasing interest in using multi-materials to optimize the weight and performance of the vehicle. This means that the capability to join different materials such as a steel to composite or aluminum to composites or polymers is very important.

There are a number of different technologies available to join different materials. Welding, fasteners, rivets, friction stir welding, adhesive bonding, etc. are some of the most commonly known options. The joining process used for a specific design will depend on the materials used, geometry of the components, manufacturing constraints, and intended usage. For example, resistance spot welding (RSW) has been one the most popular methods to join steel components in the automotive industry. This is mainly due to simplicity of design, quicker processing time, and low cost. When one deals with composite materials, resistance spot welding is not possible and therefore other methods are desired. Adhesive bonding, mechanical fasteners, and other specialized welding methods are some of the popular options available. Welding essentially refers to methods of fusing polymers or base materials to develop a proximity layer between the joining surfaces and hence forming a strong bond. Many other methods, such as friction stir, ultrasonic, impulse heating, etc. have been developed for melting and pressurizing the joining materials to form such a bond [12].

Mechanical fasteners require holes or similar modifications of the parts that need to be bonded, and they add cost, weight, and process time. For automobile structures it is highly desired to maintain the structural integrity while reducing the weight. Therefore, adhesives are preferred for joining, when possible. Adhesive bonding is a material joining process in which an adhesive, placed between the two adherent surfaces, solidifies to produce an adhesive bond. Adhesives are chemical materials whose properties can be tuned to bond with the base materials. Therefore, many variations as well as customizations of the adhesive materials based on the base materials are possible. This may seem like an added step, but for large-volume applications, it can help optimize the weight and performance. Nowadays, the use of a structural adhesive is increasing, as a wide range of off-the-shelf adhesives from various chemical companies are on the market [13–18]. One of the major concerns while using adhesives is the ability to have confidence in the quality of adhesive joint and the expected performance. Mechanical fasteners show inherent advantages in this area. Various approaches such as nondestructive testing to evaluate the quality of the adhesive bond, as well as strict process and material control that can result in desired adhesive performance for a given bond, are some of the possible approaches used to address these concerns. For traditional adhesive joining, there are typically lap-shear testing and peeling testing to characterize the joining strength of the adhesives. Debabrata Ghosh [19] systematically compared different adhesive properties in adhesive joining for automotive industry application from experiment either for metal or for composite substrates.

This section will focus on how adhesive joints can be modeled and simulated. Adhesive joints can be represented with finite element models using a cohesive zone approach. Conceptually the cohesive zone can be an area defined between two continuum materials as shown in Figure 10.11.

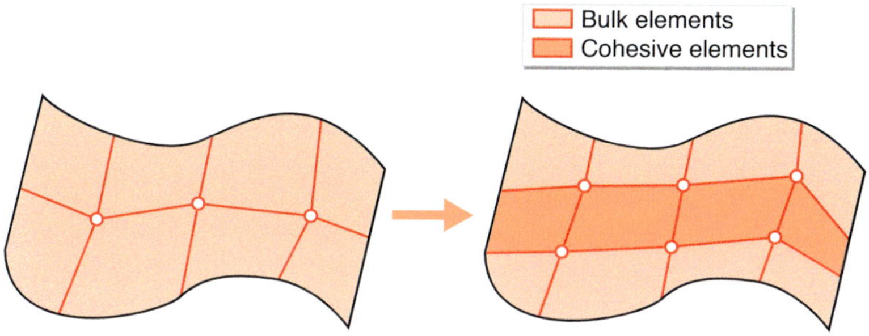

Figure 10.11 Cohesive element concept; cohesive element can have zero thickness

The concept of the cohesive zone in modeling was first introduced by Dugdale (1960) and Barenblatt (1962). The cohesive zone represents the cohesive forces which occur when two joining parts are pulled apart. The cohesive zone is not a material, so constitutive material properties are not needed. It can also have zero thickness. Essentially, it is an empirical model to represent the force transfer between the two joining surfaces. The cohesive zone can be defined by the traction force (T) and corresponding displacement (Δ) (Figure 10.12). This traction force displacement curve defines the force transfer through the adhesive interface. The area under the curve defines the energy needed to separate the two surfaces. The cohesive zone is often used to define the zone in which the crack propagates and the two joined materials separate. There are many variations and methods to define the traction-displacement relation for different materials [26], but the basic concept, involving a force-displacement curve and finite energy, stays the same.

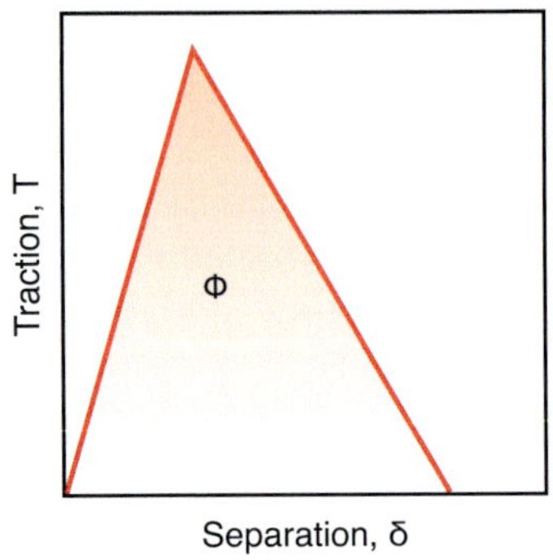

Figure 10.12 Traction–separation law; Δ, which is the area under the curve, represents energy

When finite element modeling is used to model the structure, the cohesive elements, which are a special type of elements, can be used to model joining behavior of the adhesive joint between two surfaces such as steel and composite. A method to define such cohesive zones in finite element models using cohesive element is well developed and available in most commercial codes. The parameters needed to define such cohesive zones can be established through experiments.

10.2.1 Background of Adhesive Modeling in Finite Element Models

The method to model the cohesive elements as adhesives has been evolving over the past few decades. Fredrik Fors [20] reported the FEA analysis of the adhesive joining of metal and composite in the aerospace industry using a cohesive zone model. Yu-Ping Yang [21] from the Edison Welding Institute (EWI) reported using a cohesive layer in Abaqus for the modeling of adhesive joining of composite and steel substrates for lap shear testing in the marine industry. Cohesive zone modeling has also been used to examine a wide spectrum of interface problems, such as glass/epoxy interfacial fracture and adhesion, delamination in stitched composites, plastic dissipation in thin de-bonding films, crack nucleation at bi-material corners, and peeling [22–24]. In cohesive zone modeling, the traction–separation rule and failure criterion is defined for the joining material such as adhesives. Often the cohesive elements completely degrade in shearing or peeling because of the deformation. Subsequently, it is important to obtain the damage initiation and evolution parameters in the traction–separation rule to characterize the interface properties of these adhesives.

The main purpose of this section is to demonstrate and validate the concept of cohesive zone modeling using cohesive elements to represent the adhesively bonded joints in finite element models. The interface properties of the adhesive joint for different substrates can be directly estimated using standard lap shear and peeling testing based on traction-separation laws. The adhesive properties used in the cohesive elements used in the finite element model can be estimated using reverse engineering from the experimental test results. The interface (adhesive) properties are then applied to the FEA model of a 3D part with a complex shape. In the next section this approach will be demonstrated with an example.

10.2.2 Sample Preparation and Testing

Three different types of adhesive joints, i.e., lap shear test, T-peeling test, and a complex shape combination of shear and peel are considered. The different substrates and three adhesives that are selected to study each joint type are presented in Table 10.1, which shows the various combinations of different substrates and adhesives. For lap shear and T-peeling conditions, three types of the substrate combination are considered: steel/steel, glass fiber-reinforced polypropylene (GFPP)/steel, and carbon fiber-reinforced ultramid polyamide (CFPA)/steel. Three types of adhesive materials (epoxy based, urethane based, and methyl methacrylate-MMA based) are used in the study to prepare the various combination of joints. All the adhesives used are commercial products available from Henkel^{TM} Corporation.

Table 10.1 Design of Experiments and the Test Matrix with Combinations of Three Different Substrates and Three Different Adhesives

Test case	Adherent A	Adherent B	Adhesives (Henkel Corporation) All the three adhesives are used
Lap shear test	Steel	Steel	MMA (Methacrylate) 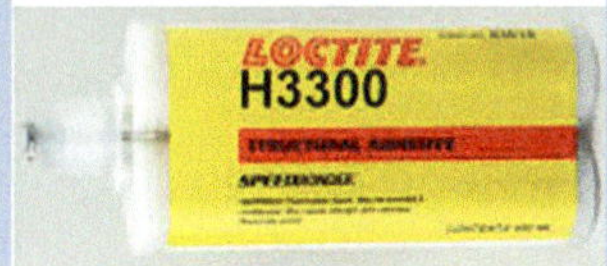
	GFPP	Steel	
	CFPA	Steel	
T-peeling test	Steel	Steel	
	GFPP	Steel	
	CFPA	Steel	Epoxy 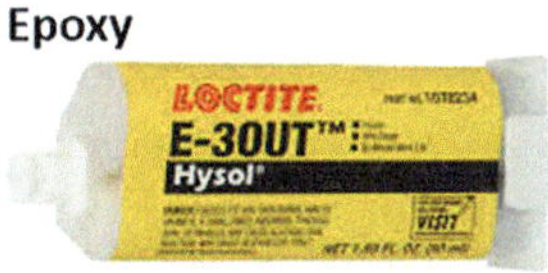
Complex shape part test	GFPP	Steel	Urethane 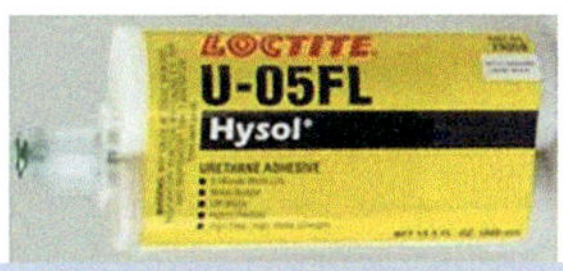

The thickness and material properties used for the substrate material used in this study are as follows:

- Steel: advanced high strength steel (HSLA 350/450), with a thickness of 1 mm.
- Glass fiber-reinforced polypropylene (GFPP): cut from injection molded panel, with 30% glass fiber weight percentage, with a thickness of 3 mm.
- Carbon fiber-reinforced polyamide 6 (CFPA): cut from injection molded panel with 30% carbon fiber weight percentage, with a thickness of 3 mm.

Before applying the adhesive, the samples used in this study were given surface treatment. The surface is cleaned using iso-propanol and dried at room temperature overnight. To obtain the constant thickness of the adhesives, glass beads with 0.5 mm diameter are also applied between the substrates where the adhesives are applied. Also, a spacer with thickness of 0.5 mm is applied on two sides of the lap shear joining to control the width of the adhesive joining. The adhesives are applied to the clean surface by the dispenser gun available from Henkel™.

10.2.3 Lap Shear Test

Following the ASTM D1002 standard, the lap shear testing sample is prepared as shown in Figure 10.13. The picture shows the test setup in an Instron machine for lap shear testing. The lap shear strength is calculated based on the ASTM D1002 standard, and the results for steel to steel for these three adhesives are compared with the reference data available from Henkel's product specifications. It can be seen that the experimental results agree closely with the specifications, as shown in Figure 10.14. At the same time, the lap shear strengths for different substrate combinations were compared. We can see that the adhesives applied to the combination of CFPA and steel substrate have higher lap shear strength than those of the GFPP and steel combination. In all cases, the failure location is at the joint (adhesive failure), and most of the adhesives fail on the side of the composite part. Here, the surface energy plays important role in the failure mode. There is a difference in the surface energy [25] between CFPA and GFPP. Since the surface energy of the CFPA is higher than that of GFPP, one gets better adhesion in the former case.

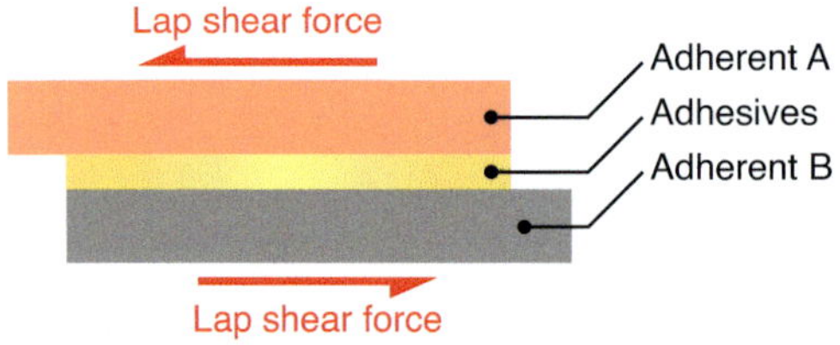

Figure 10.13 Lap shear test configuration and the sample in the Instron machine

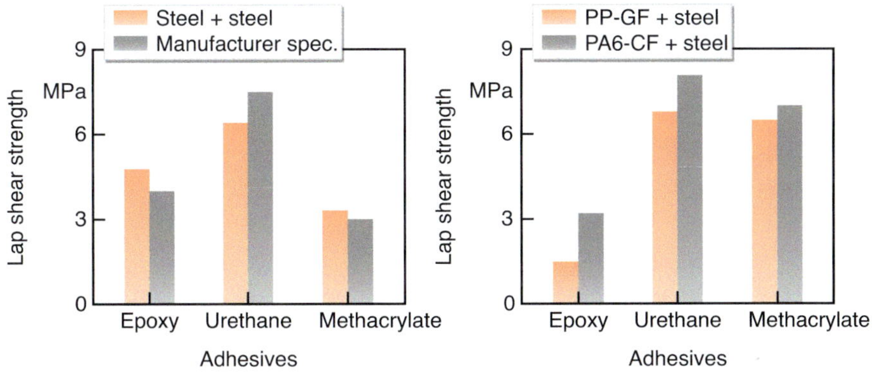

Figure 10.14 Lap shear test results. Left hand data shows test results are close to manufacturer specifications. Right hand data shows effect of different materials on joint strength

10.2.4 Peel Test

The peeling test is conducted according to ASTM 5109, using the same set of the substrates and adhesives. The test configuration and the actual sample layout are shown in Figure 10.15. The results are in Figure 10.16, which shows a close match between the test results and the reference from Henkel™'s specifications. Regarding the failure pattern, in each case the failure location of the adhesive is at the joint (adhesive failure), and most of the adhesives fail on the side of the composite part. The peeling strength is compared for the CFPA/steel and GFPP/steel combinations in the same figure. The difference in the peeling strengths between these combinations is explained by the surface energy difference between CFPA and GFPP materials.

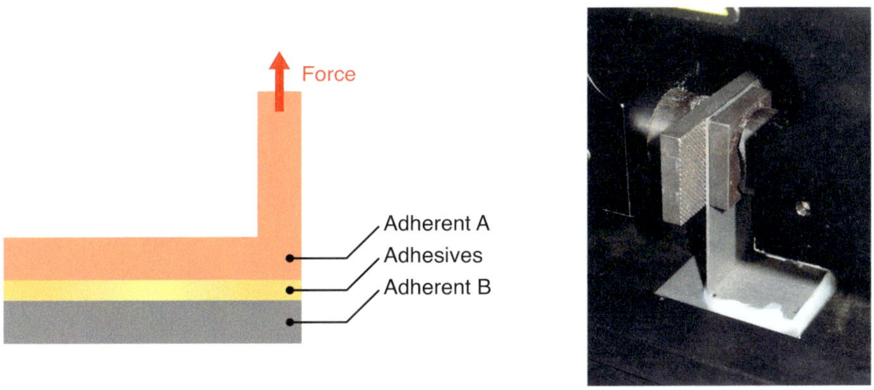

Figure 10.15 Peeling test configuration and the sample on the Instron test machine

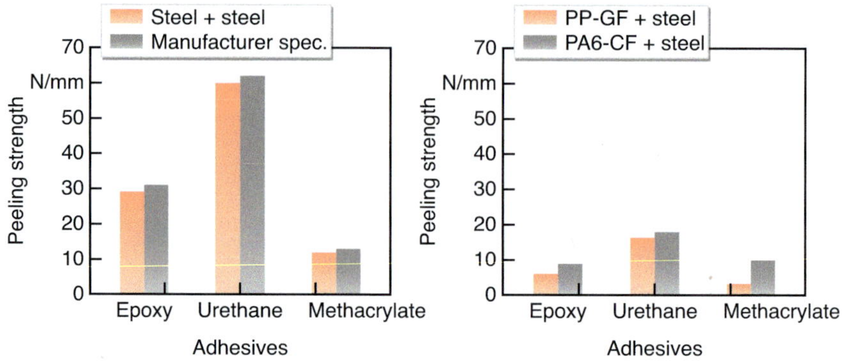

Figure 10.16 Peeling test results. Left hand data shows the test results are close to manufacturer specifications. Right hand data shows effect of different materials on joint strength

10.2.5 Complex Shaped Part Test

A U-shaped channel was chosen as base structure for the demonstration of the complex part. This part was made of glass fiber-reinforced composite, the same as the GFPP specimen used in the previous section. The U-shaped part is attached to the high strength steel panel, from the same material used in previous experiments. The shape and experimental configuration is shown in Figure 10.17. The comparison of the test results will be presented with the simulation in the next section.

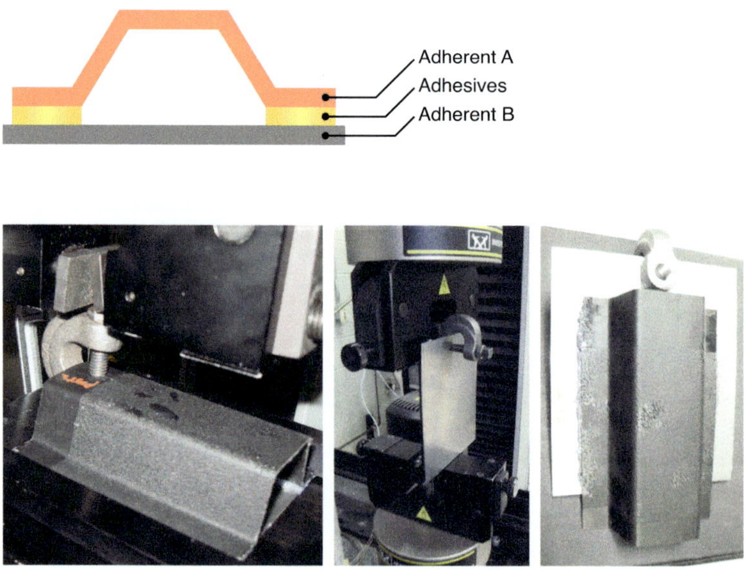

Figure 10.17 Complex shape part configuration and the corresponding adhesive joining test on the Instron machine

10.2.6 Modeling and Simulation

Before beginning with the FEA modeling of the adhesive for the test, it is necessary to introduce a basic concept of the cohesive elements used in the Abaqus model. A cohesive element concept in Abaqus is usually used for the interface properties modeling for adhesive joining/bonding. The concept level definition of cohesive element and its explanation is shown in Figure 10.18.

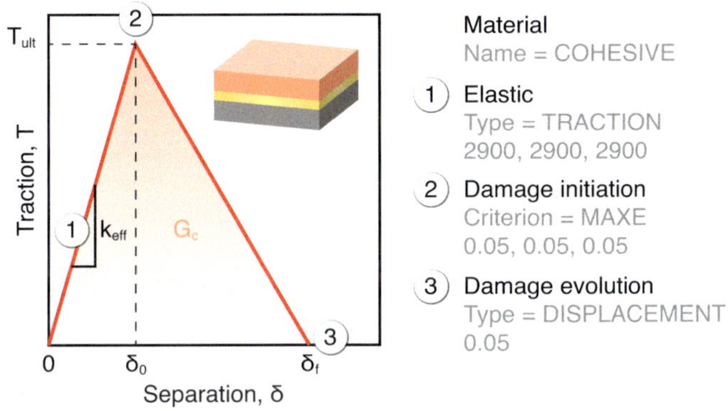

Figure 10.18 Cohesive element concept defined in Abaqus for this study

According to the Abaqus manual, a cohesive element is a special element used for the modeling of interface properties. The detailed parameters for the traction-separation model are shown in Figure 10.19. In Abaqus the model is assumed to have initial linear elastic behavior followed by the initiation and evolution of damage. The elastic behavior is written in terms of an elastic constitutive matrix that relates the nominal stresses to the nominal strains across the interface. The nominal traction stress vector, t, consists of three components: t_n, t_s, and t_t, which represent the normal and the two shear tractions (along the local 1- and 2-directions in three dimensions), respectively. The corresponding separations are denoted by δ_n, δ_s, and δ_t. Denoting by T_0 the original thickness of the cohesive element, the nominal strains can be defined as

$$\varepsilon_n = \frac{\delta_n}{T_0}, \varepsilon_s = \frac{\delta_s}{T_0}, \varepsilon_t = \frac{\delta_t}{T_0} \quad (10.1)$$

The elastic behavior can then be written as

$$t = \begin{Bmatrix} t_n \\ t_s \\ t_t \end{Bmatrix} = \begin{bmatrix} k_{nn} & k_{ns} & k_{nt} \\ k_{ns} & k_{ss} & k_{st} \\ k_{nt} & k_{st} & k_{tt} \end{bmatrix} \begin{Bmatrix} \varepsilon_n \\ \varepsilon_s \\ \varepsilon_t \end{Bmatrix} = K\epsilon \quad (10.2)$$

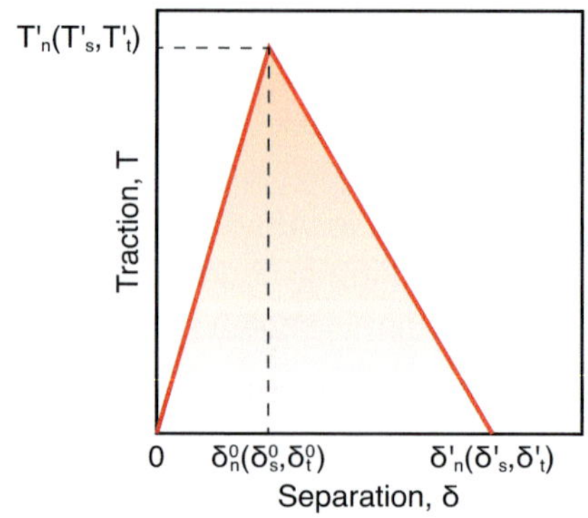

Figure 10.19 Traction-separation law in cohesive element concept

Figure 10.20 shows key steps used in the FEA model development for the adhesives. In the first step, a steel/steel substrate combination was used, and three adhesives were applied to it. The result is used to estimate the Young's modulus of the adhesives used in this study (K or E in the model). The Young's modulus of the adhesives obtained here will be used for all the other steps. In the second step, the lap shear test results were used (see Figure 10.21) for a GFPP/steel substrate to reverse engineer the shear direction damage initiation parameter and damage evolution parameter for this case $(\varepsilon_t, \varepsilon_s, \delta_t)$. This is repeated for all the three adhesives. In the third step, peeling test results are used for the GFPP/steel substrate to reverse engineer the damage initiation parameter for the peeling direction and its corresponding damage evolution parameter (ε_n). In the final step, i.e., step 4, the five key parameters (E or K, $\varepsilon_n \varepsilon_t, \varepsilon_s, \delta_t$), estimated from lap shear and peel test, are used to represent the adhesive in the complex shape part as cohesive elements. Comparisons between test and simulation for complex shape parts are presented in Figure 10.28 and Figure 10.29. In the Abaqus models, the maximum nominal strain criterion is used for the traction-separation law.

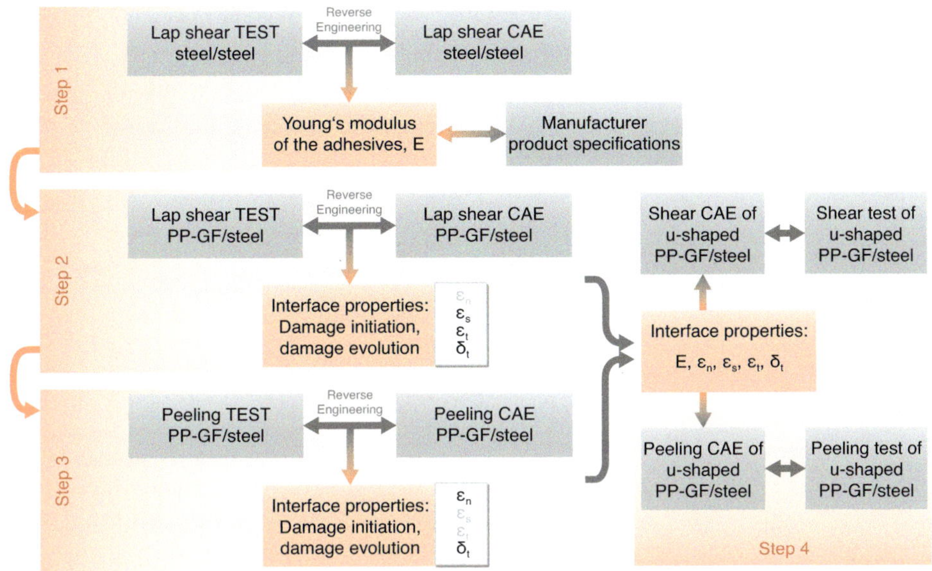

Figure 10.20 FEA analysis strategy and four steps roadmap for the adhesive joining in this section

10.2.7 Lap Shear FE Model

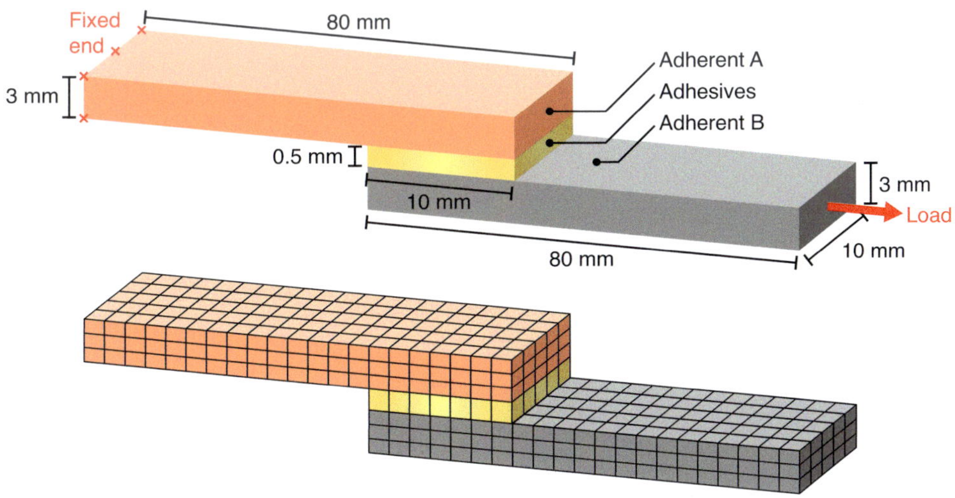

Figure 10.21 Detailed lap shear test layout and its Abaqus model

Based on the test sample geometry, the lap shear model is built using Abaqus/Standard. In the model, the cohesive element is used following the Abaqus manual. In this FEA model, as shown in Figure 10.21, the total number of nodes is

14678, the total number of elements is 8435, the adhesive element type is COH3D8, while the adherent element type for the substrate is C3D8I. Figure 10.22 shows the comparison of the test and FEA modeling curve after reverse engineering for lap shear testing for three adhesives for steel to steel substrate joining.

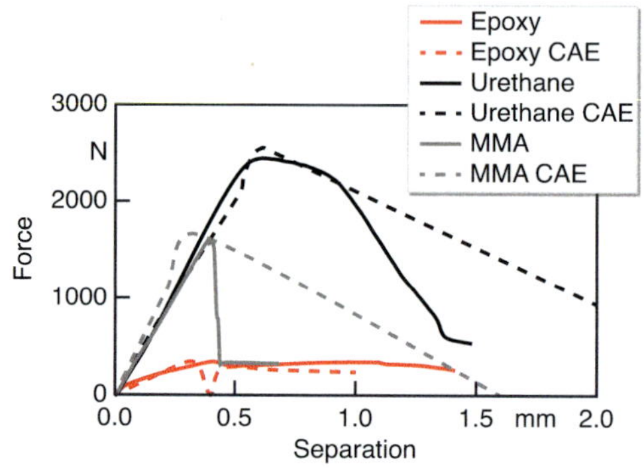

Figure 10.22 Lap shear test reverse engineering results for steel to steel

Similarly, for the peeling FEA modeling, the cohesive element is also used in the middle as for adhesives. In the model as shown in Figure 10.23, the total number of nodes is 7308, the total number of elements is 3444, the adhesive element type is COH3D8, and the standard element type C3D8I is used for the substrates. Figure 10.24 shows the comparison of the test and modeling for peeling test on these three adhesives on steel to steel joining. Table 10.2 shows the parameters from the reverse engineering analysis.

10.2.8 Peel Test FEM Modeling

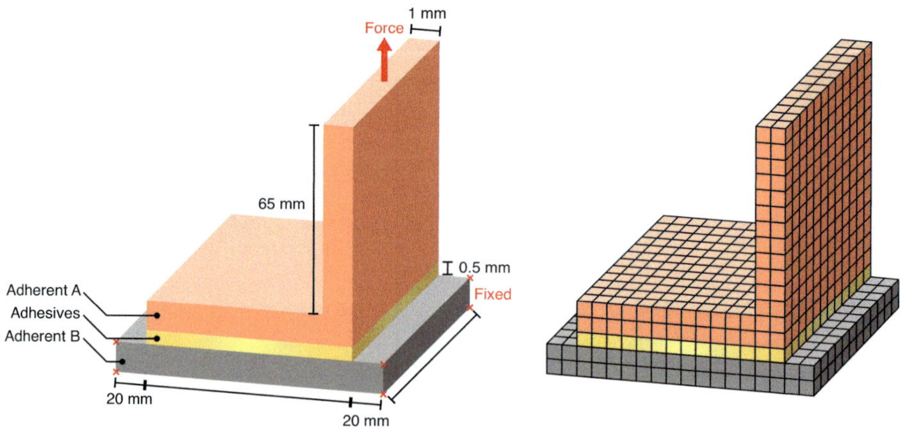

Figure 10.23 Detailed peeling test layout and its Abaqus model

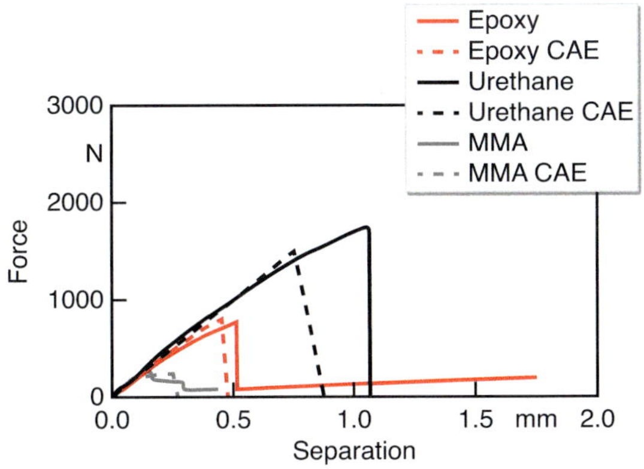

Figure 10.24 Peeling test reverse engineering results for steel to steel

Table 10.2 Reverse Engineering Based Properties of the Adhesives for Steel to Steel

	Adhesive	Young's modulus (GPa)		Damage initiation parameter	Damage evolution parameter (displacement)
		Reference	Reverse engineering	$\varepsilon_n, \varepsilon_s, \varepsilon_t$	
Case 1	Epoxy	2.788	2.8	0.0021, 0.008, 0.008	1.8
Case 2	MMA	2.6	2.4	0.001, 0.02, 0.02	0.2
Case 3	Urethane	0.199	0.3	0.05, 0.015, 0.015	2.6

Using the reverse engineering method, as was described in the last two sections, one can see that Young's modulus of the epoxy, MMA, and urethane closely match with the production specification in Table 10.2. At the same time, one can also obtain reverse engineered damage initiation and evaluation parameters.

Table 10.3 Reversed Engineering Adhesive Properties for GFPP and Steel

	Adhesive	Young's modulus (GPa)		Damage initiation parameter	Damage evolution parameter (displacement)
		Reference	Reverse engineering	$\varepsilon_n, \varepsilon_s, \varepsilon_t$	
Case 1	Epoxy	2.788	2.8	0.04, 0.005, 0.005	0.4
Case 2	MMA	2.6	2.4	0.02, 0.006, 0.006	0.5
Case 3	Urethane	0.199	0.3	0.05, 0.1, 0.1	2.0

Figure 10.25 and Figure 10.26 show the results using the similar reverse engineering method for the adhesive joining of GFPP and steel substrate. Based on the previous results, in the FEA model, the Young's modulus is used as it is from the steel to steel substrate cases. By adjusting the damage initiation parameters, we can

make the predicted lap shear simulation curves closely match with the test curves, i.e., reverse engineering; it is seen that the MMA has the highest shear force compared to the others, while urethane shows the highest damage evolution among these three. For the peeling test, the behavior of the adhesive appears different compared to the shear test, mainly because in this case the loading direction is totally different from the shearing direction. The urethane shows highest peeling force and the biggest damage evolution as well. A comprehensive list of the adhesive properties from the cohesive zone modeling is shown in Table 10.3. These properties will be input to the complex part FEA model for the validation of the prediction.

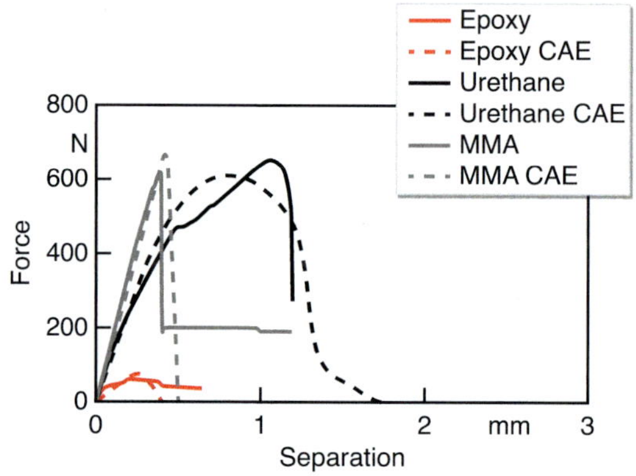

Figure 10.25 Lap shear test reverse engineering results for GFPP and steel

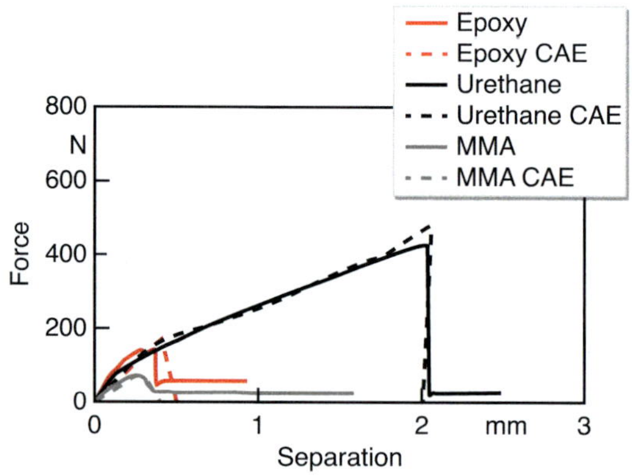

Figure 10.26 Peeling test reverse engineering results for GFPP and steel

10.2.9 Complex Shaped Part FEA Modeling

For the complex shape part as shown in Figure 10.27, the FEA model is similar to the lap shear and peeling model, except that the geometry of the part has different contours. The total number of nodes in the model is 100113, the total number of elements is 66725, the adhesive element type used is COH3D8, and the adherent element type is C3D8I. In this case, a similar cohesive element modeling method was used, but all the cohesive zone parameters used in the model are from previous reverse engineered properties from Table 10.3. Figure 10.28 shows the predicted results using the cohesive element parameters for the peeling test of the complex shape part. It shows the close agreement for the force-displacement curves for three different adhesives on the GFPP and steel substrate.

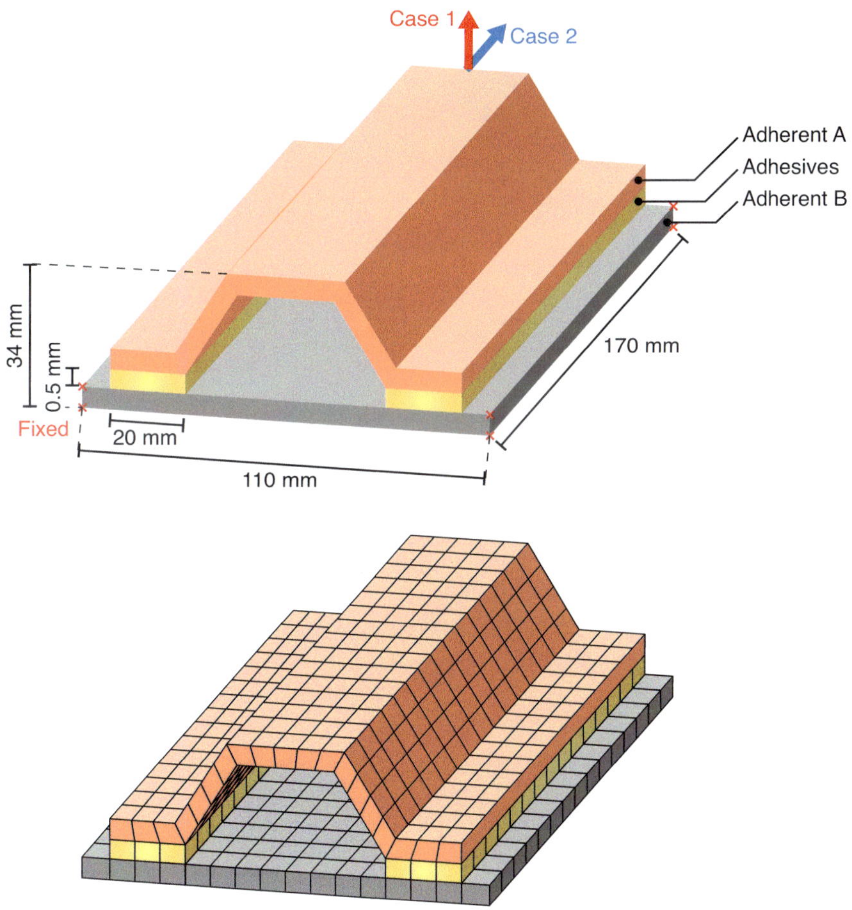

Figure 10.27 Detailed complex shape part geometry and the meshed Abaqus model

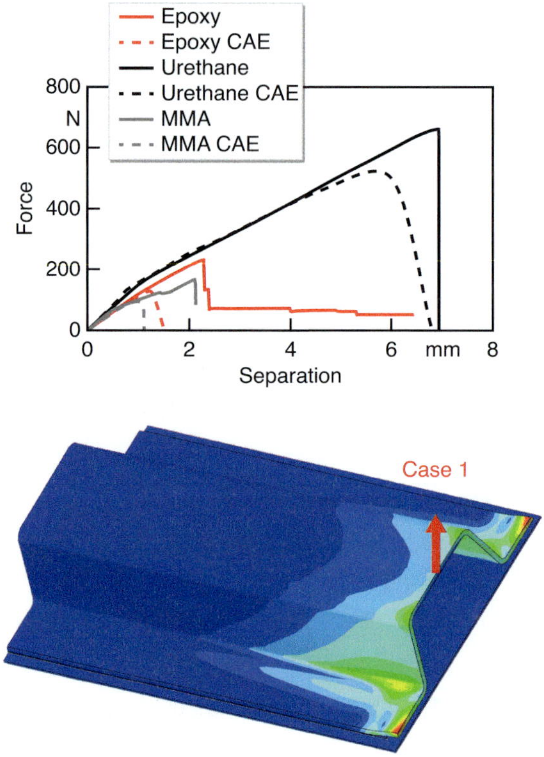

Figure 10.28 Comparison of the experimental results with FEA simulation using adhesive properties for peeling test

Similarly, the CAE and test results for the pull test for the complex part are compared in Figure 10.29. The results show close agreement for the force-displacement curve for three different adhesives on the GFPP and steel substrate. The comparison for urethane adhesives shows less traction behavior than the test results. This could be because of the discrepancy from the initial reverse engineering of the lap shear testing from Figure 10.22. Further study is needed for more accurate simulation in this case. The goal here was to demonstrate what approach can be used.

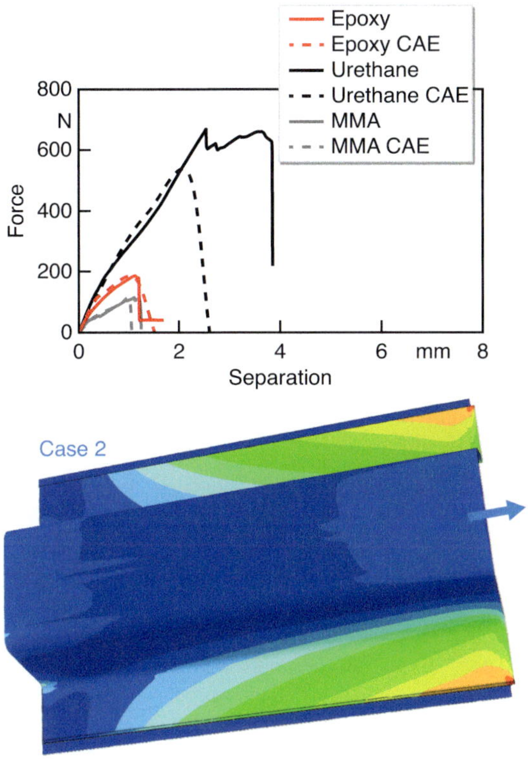

Figure 10.29 Comparison of the experimental results with FEA simulation using adhesive properties for pulling test

10.2.10 Conclusions for Adhesive Joining

This study provided a systematic method to use cohesive element method in Abaqus for adhesive joining of dissimilar materials. By combining with a reverse engineering approach on the simple peel and lap shear test results, the interface properties of the cohesive element model for the adhesives can be predicted. The finite element model based on predicted properties is test validated for a complex shaped 3D part.

The interface properties depend on the substrates and the adhesive joining as well. The adhesives used in this study from Henkel™ Corporation provided a good example for structural bonding for composite materials.

The approach presented here can be used to model joining with finite element analysis using the cohesive element of similar empirical models. The key is using simple tests to estimate the model parameters and then apply the parameters for complex shaped designs.

References

[1] Cole, G. S. and Sherman, A. M., Lightweight materials for automotive applications, *Materials Characterization*, 35(1), pp. 3–9 (1995)

[2] Das, S., *The cost of automotive polymer composites: a review and assessment of DOE's lightweight materials composites research* (p. 47), Oak Ridge National Laboratory, Oak Ridge, TN (2001)

[3] Bücheler, D. and Henning, F., Hybrid Resin Improves Position and Alignment of Continuously Reinforced Prepreg During Compression Co-Molding with Sheet Molding Compound. In ECCM17 17th European Conference on Composite Materials, Munich (2016)

[4] Bay, R. S. and Tucker III, C. L., Fiber orientation in simple injection moldings. Part I: Theory and numerical methods, *Polymer Composites*, 13(4), pp. 317–331 (1992)

[5] Gupta, M. and Wang, K. K., Fiber orientation and mechanical properties of short-fiber-reinforced injection-molded composites: Simulated and experimental results, *Polymer Composites*, 14(5), pp. 367–382 (1993)

[6] Adam, L. and Assaker, R., Integrated nonlinear multi-scale material modelling of fiber reinforced plastics with Digimat: application to short and continuous fiber composites. In Proceedings of the 11th World Congress on Computational Mechanics, pp. 20–25 (2014, July)

[7] Doghri, I. and Tinel, L., Micromechanical modeling and computation of elasto-plastic materials reinforced with distributed-orientation fibers, *International Journal of Plasticity*, 21(10), pp. 1919–1940 (2005)

[8] Tseng, H. C., Chang, R. Y., and Hsu, C. H., Numerical prediction of fiber orientation and mechanical performance for short/long glass and carbon fiber-reinforced composites, *Composites Science and Technology*, 144, pp. 51–56 (2017)

[9] Moldex3D user manual, CoreTech Company (2014)

[10] DIGIMAT user manual (2015)

[11] Abaqus user manual (2014)

[12] https://www.twi-global.com/technical-knowledge/published-papers/welding-technologies-for-polymers-and-composites/

[13] Lee, L. H. (Ed.), *Adhesive Bonding*, Springer Science and Business Media (2013)

[14] He, X., A review of finite element analysis of adhesively bonded joints, *International Journal of Adhesion and Adhesives*, 31(4), pp. 248–264 (2011)

[15] Banea, M. D. and da Silva, L. F., Adhesively bonded joints in composite materials: an overview, *Proceedings of the Institution of Mechanical Engineers, Part L: Journal of Materials Design and Applications*, 223(1), pp. 1–18 (2009)

[16] Blackman, B. R. K., Hadavinia, H., Kinloch, A. J., and Williams, J. G., The use of a cohesive zone model to study the fracture of fiber composites and adhesively-bonded joints, *International Journal of Fracture*, 119(1), pp. 25–46 (2003)

[17] Valoroso, N. and Fedele, R., Characterization of a cohesive-zone model describing damage and de-cohesion at bonded interfaces. Sensitivity analysis and mode-I parameter identification, *International Journal of Solids and Structures*, 47(13), pp. 1666–1677 (2010)

[18] Lee, M. J., Cho, T. M., Kim, W. S., Lee, B. C., and Lee, J. J., Determination of cohesive parameters for a mixed-mode cohesive zone model, *International Journal of Adhesion and Adhesives*, 30(5), pp. 322–328 (2010)

[19] Ghosh, D., Pancholi, L., and Sathaye, A., Comparative Studies of Adhesive Joints in Automotive (No. 2014-01-0788), SAE Technical Paper (2014)

[20] Fors, R., Analysis of Metal to Composite Adhesive Joints in Space Applications, LIU-IEI-TEK-A–10/00812–SE

[21] Yang, Y.-P., Ritter, G. W., and Speth, D. R., Finite Element Analyses of Composite-to-Steel Adhesive Joints, *Advanced Materials and Processes* (June 2011)

[22] Swadener, J. G. and Liechti, K. M., Asymmetric shielding mechanisms in the mixed-mode fracture of a glass/epoxy interface, *Journal of Applied Mechanics*, 65 (1), pp. 25–29 (1998)

[23] Wang, S. S., Fracture-mechanics for delamination problems in composite-materials, *Journal of Composite Materials*, 17 (3), pp. 210–223 (1983)

[24] Yang, Q. D. and Thouless, M. D., Mixed-mode fracture analyses of plastically-deforming adhesive joints, *International Journal of Fracture*, 110 (2), pp. 175–187 (2001)

[25] Kaelble, D. H., *Physical Chemistry of Adhesion*, Wiley-Interscience (1971)

[26] Chandra, N., Li, H., Shet, C., and Ghonem, H., Some Issues in the Application of Cohesive Zone Models for Metal-Ceramic Interfaces. *Int. J. Solids Structures*, vol. 39, pp. 2827–2855 (2002)

Index